Organoselenium Chemistry

Also of interest

Selenium and Tellurium Reagents.
In Chemistry and Materials Science
Risto Laitinen, Raija Oilunkaniemi (Eds.), 2019
ISBN 978-3-11-052794-0, e-ISBN 978-3-11-052934-0

Grignard Reagents and Transition Metal Catalysts.
Formation of C-C Bonds by Cross-Coupling
Cossy (Ed.), 2016
ISBN 978-3-11-035266-5, e-ISBN 978-3-11-035272-6

Bioorganometallic Chemistry
Weigand, Apfel (Eds.), 2019
ISBN 978-3-11-049650-5, e-ISBN 978-3-11-049657-4

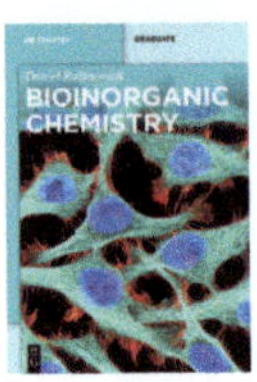

Bioinorganic Chemistry
Rabinovich, 2021
ISBN 978-3-11-049204-0, e-ISBN 978-3-11-049443-3

Organophosphorus Chemistry.
Novel Developments
Keglevich (Ed.), 2018
ISBN 978-3-11-053453-5, e-ISBN 978-3-11-053583-9

Organoselenium Chemistry

Edited by
Brindaban C. Ranu, Bubun Banerjee

DE GRUYTER

Editors
Prof. Brindaban C. Ranu
School of Chemical Sciences
Indian Association for the
Cultivation of Science
Kolkata 700032
India
bcranu@gmail.com

Dr. Bubun Banerjee
Department of Chemistry
Indus International University
Himachal Pradesh 174301
India
banerjeebubun@gmail.com

ISBN 978-3-11-062224-9
e-ISBN (PDF) 978-3-11-062511-0
e-ISBN (EPUB) 978-3-11-062229-4

Library of Congress Control Number: 2019955783

Bibliographic information published by the Deutsche Nationalbibliothek
The Deutsche Nationalbibliothek lists this publication in the Deutsche Nationalbibliografie; detailed bibliographic data are available on the Internet at http://dnb.dnb.de.

Cover image: theasis/iStock/Getty Images Plus
Typesetting: Integra Software Services Pvt. Ltd.
Printing and binding: CPI books GmbH, Leck

www.degruyter.com

Contents

List of Contributors

Laura Abenante
Laboratório de Síntese Orgânica
Limpa – LASOL – CCQFA, Universidade
Federal de Pelotas
RS 96160-000
Brazil

Diego Alves
Laboratório de Síntese Orgânica
Limpa – LASOL – CCQFA, Universidade
Federal de Pelotas
RS 96160-000
Brazil

Bubun Banerjee
Department of Chemistry
Indus International University
V.P.O. Bathu
Distt. Una
Himachal Pradesh 174301
India

Angelita M. Barcellos
Laboratório de Síntese Orgânica
Limpa – LASOL – CCQFA, Universidade
Federal de Pelotas
RS 96160-000
Brazil

Baitan Chakraborty
Department of Chemistry
Jadavpur University
Kolkata 700032
India

Aniruddha Das
Department of Chemistry
Jadavpur University
Kolkata 700032
India

Tubai Ghosh
School of Chemical Sciences
Indian Association for the Cultivation of science
Jadavpur
Kolkata 700032
India

Daniela Hartwig
Laboratório de Síntese Orgânica
Limpa – LASOL – CCQFA, Universidade
Federal de Pelotas
RS 96160-000
Brazil

Raquel G. Jacob
Laboratório de Síntese Orgânica
Limpa – LASOL – CCQFA, Universidade
Federal de Pelotas
RS 96160-000
Brazil

Swapnadeep Jalal
School of Chemical Sciences
Indian Association for the Cultivation of science
Jadavpur
Kolkata 700032
India

Umasish Jana
Department of Chemistry
Jadavpur University
Kolkata 700032
India

Mamoru Koketsu
Department of Chemistry and Biomolecular
Science
Faculty of Engineering
Gifu University
1-1 Yanagido
Gifu 501-1193
Japan

Eder J. Lenardão
Laboratório de Síntese Orgânica
Limpa – LASOL – CCQFA, Universidade
Federal de Pelotas
RS 96160-000
Brazil

https://doi.org/10.1515/9783110625110-203

Guilherme M. Martins
Departamento de Química
Universidade Federal de Santa Maria
97105–900, Santa Maria
RS
Brazil

Samuel R. Mendes
Laboratório de Síntese e Catálise – SINCA –
Departamento de Química
Universidade do Estado de Santa Catarina
89219-719 Joinville
SC
Brazil

Govindasamy Mugesh
Department of Inorganic and Physical
Chemistry
Indian Institute of Science
Bangalore 560012
India

Masayuki Ninomiya
Department of Chemistry and Biomolecular
Science
Faculty of Engineering
Gifu University
1-1 Yanagido
Gifu 501-1193
Japan

Pablo A. Nogara
Departamento de Bioquímica e Biologia
Molecular
Centro de Ciências Naturais e Exastas
Universidade Federal de Santa Maria
Santa Maria
RS
Brazil

Cláudia S. Oliveira
Instituto de Pesquisa Pelé Pequeno Príncipe
Curitiba
PR
Brazil
and
Faculdades Pequeno Príncipe
Curitiba
PR
Brazil

Subir Panja
School of Chemical Sciences
Indian Association for the Cultivation of science
Jadavpur
Kolkata 700032
India

Gelson Perin
Laboratório de Síntese Orgânica
Limpa – LASOL – CCQFA, Universidade
Federal de Pelotas
RS 96160-000
Brazil

Brindaban C. Ranu
School of Chemical Sciences
Indian Association for the Cultivation of Science
Jadavpur
Kolkata 700032
India

Kishorkumar M. Reddy
Department of Inorganic and Physical Chemistry
Indian Institute of Science
Bangalore 560012
India

João Batista T. Rocha
Departamento de Bioquímica e Biologia
Molecular
Centro de Ciências Naturais e Exastas
Universidade Federal de Santa Maria
Santa Maria
RS
Brazil

Marcio S. Silva
Laboratório de Síntese Orgânica
Limpa – LASOL – CCQFA, Universidade
Federal de Pelotas
RS 96160-000
Brazil

Amol D. Sonawane
Department of Chemistry and Biomolecular
Science
Faculty of Engineering
Gifu University
1-1 Yanagido
Gifu 501-1193
Japan

Brindaban C. Ranu, Tubai Ghosh, Subir Panja and Swapnadeep Jalal

1 Synthesis of Organoselenides by Coupling Reaction and C–H Activation – Recent Advances

1.1 Introduction

The area dealing with the synthesis of organoselenides has experienced a tremendous growth in last two decades. A number of methods have been reported. This chapter highlights the recent reports on various methods for the synthesis of selenides by cross-coupling and C–H activation. The metal-catalyzed as well as metal-free procedures have been discussed. Besides Pd, the catalysis by other metals such as Cu, Ni, Ru and Ca have also been addressed. Under nonmetallic catalysis I_2, (TBAI) and hypophosphorus acid (H_3PO_2) have been discussed.

Among the several organochalcogenides, organoselenium compounds are of considerable interest as selenides are found useful in the areas of agrochemicals, insecticides, and drugs (Figure 1.1) [1–6]. Apart from biological applications, they are also used as useful intermediates and catalysts in several organic transforma-

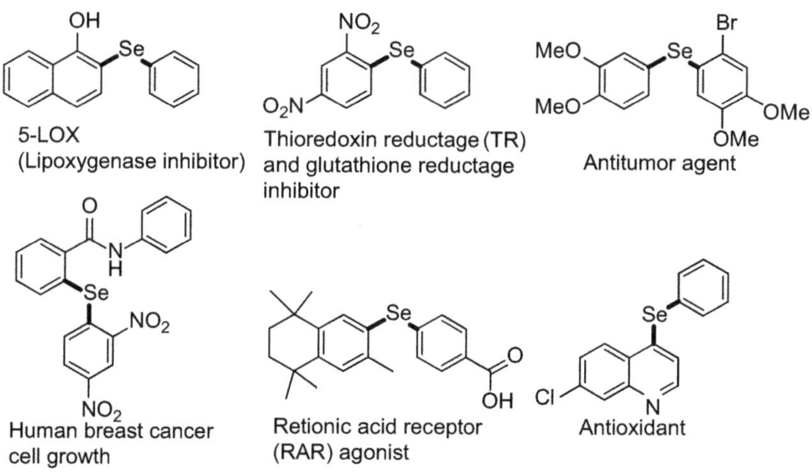

Figure 1.1: Examples of biologically active organoselenides.

Brindaban C. Ranu, Tubai Ghosh, Subir Panja and Swapnadeep Jalal, School of Chemical Sciences, Indian Association for the Cultivation of Science, Kolkata, India

https://doi.org/10.1515/9783110625110-001

tions [7–9]. They also show wide variety of applications as functional materials [10]. A large number of methods have been developed for the construction of C–Se bond both under metal free as well as in the presence of transition metal.

This chapter covers the synthesis of organoselenium compounds through C–H activation and cross-coupling reactions.

1.2 Reactions involving C–H activation

In recent years, transition metal-catalyzed/mediated C–H bond functionlization became an important tool for the synthesis of organic molecules [11–15]. The use of C–H bonds toward transformable functional group is advantageous because C–H bonds are the most abundant moieties in organic molecules. Thus, one-step conversion of these C–H bonds to the desired functionality minimizes the synthetic pathways, saves reagents and reduces waste of solvents and time.

In this context, Nishihara and coworkers reported the synthesis of benzoisoselenazolone and its derivatives by nickel-catalyzed dehydrogenative direct selenation of $C(sp^2)$–H bonds with elemental selenium powder under aerobic condition (Scheme 1.1) [16]. After optimization it was observed that a combination of 10 mol % of $Ni(OAc)_2{\cdot}4H_2O$, 20 mol % of PPh_3, 2 equiv. of Na_2CO_3 and 3 equiv. of n-Bu_4NCl along with 2 equiv. of selenium powder in DMF at 120 °C under air provided the best results among others. A range of functional groups including both electron-donating and electron-withdrawing substituents were compatible under the optimized oxidative conditions and the corresponding benzoisoselenazolone derivatives were formed in good-to-excellent yields. Apart from benzamides, acrylamides were also found to be suitable for the desired transformations. The newly formed benzoisoselenazolone derivatives were readily converted into a variety of useful organoselenium compounds. Based on the control experiments, the author proposed a mechanistic cycle that the reaction proceeds by the formation of nickelacycle(II) and single-electron oxidation of the stable nickelacycle(II) species under aerobic condition.

Synthesis of unsymmetrical ferrocene aryl chalcogenides by C–H activation of ferrocene amide using 8-aminoquinoline as a directing group has been developed by Sattar et al (Scheme 1.2) [17]. The reaction occurs in the presence of copper(II) salt as the active catalyst, silver acetate as an oxidant at 80 °C in DMSO. A range of diaryl diselenides containing electron-deficient and electron-withdrawing substituents underwent successful coupling with ferroceneamide under the optimized reaction conditions and the bis-arylselenylated ferrocenes were obtained. Although the reactions successfully produced the desired arylselenylated ferrocene using diaryl diselenides, dialkyl diselenides failed to initiate the reaction. The reaction went smoothly with other diaryl dichalcogenides too.

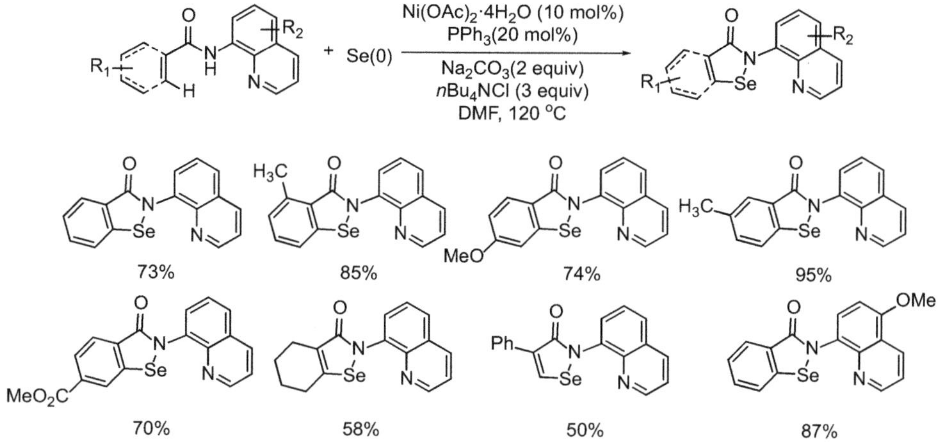

Scheme 1.1: Nickel-catalyzed dehydrogenative direct selenation of C(sp^2)–H bonds.

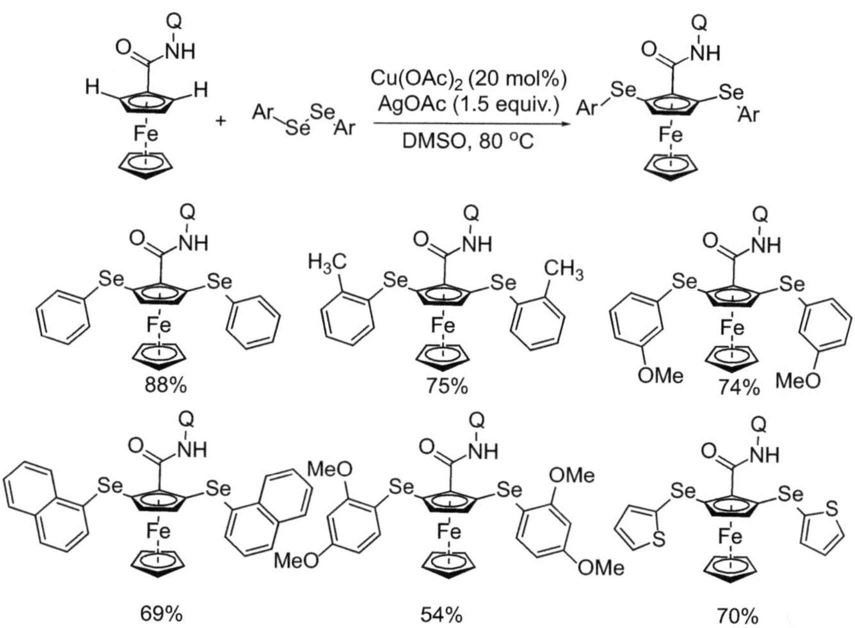

Scheme 1.2: Copper-catalyzed 8-aminoquinoline-assisted aryl selenylation of ferrocene amide.

C-4 Selenylated isoquinolin1(2H)-one was synthesized using a new and facile $AgSbF_6$-mediated procedure through a radical pathway (Scheme 1.3) [18]. The reaction occurs in the presence of equimolar amount of isoquinolin-1(2H)-ones and diaryl diselenide using $AgSbF_6$ as an oxidant in dichloroethane under reflux for 8 h. Among several other commonly used oxidizing agents, $AgSbF_6$ was found to be the most effective one. The reaction shows excellent regioselectivity and broad substrate scope. The corresponding C-4 selenylated derivatives were obtained in excellent yield. Apart from isoquinoline moiety, the reaction occurs without any difficulty when pyridin-2(1H)-one was used.

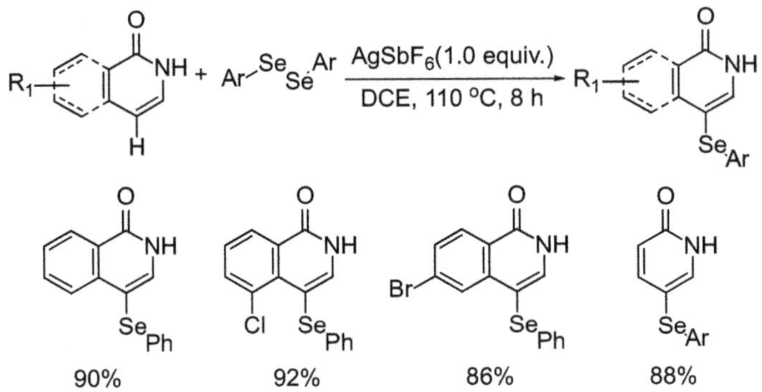

Scheme 1.3: $AgSbF_6$-mediated C-4 selenylation of isoquinolin-1(2H)-ones.

As recently reported by Ranu and his coworkers, when 2-naphthol and its derivatives were allowed to react with styrenyl selenocyanate/diaryl diselenide in the presence of a base at room temperature, selenylation occurs selectively at the 1-position of 2-naphthol unit (Scheme 1.4) [19]. The reaction occurs in the presence of Cs_2CO_3 as a base in DMSO solvent at room temperature in the absence of any transition metal or oxidants. Both electron-donating and electron-withdrawing styrenyl selenocyanate reacted with 2-naphthol and its derivatives at room temperature and the corresponding 1-styrenyl-selenylated naphthol derivatives were obtained in good yields (Scheme 1.4a). The electronically diverse diaryl diselenides also provided the desired 1-arylated derivatives without any difficulty (Scheme 1.4b). The reactions are relatively fast (2–4 h) and high yielding. The reaction is also feasible in gram scale. More importantly, the unaffected hydroxyl group (–OH) can be further functionalized. Initially, the base Cs_2CO_3 reacts with 2-naphthol to form naphtholate anion intermediate (**A**) (Scheme 1.4c), which reacts with styrenyl selenocyanate/diryl diselenide at the 1-position to form the species (**B**). The intermediate (**B**) then undergoes aromatization via proton elimination to form the desired 2-selenylated product.

Braga and his group developed a microwave (MW)-assisted C(Sp^2-H) bond functionalization protocol to allow access to several selenide bicyclic arenes. In this methodology I_2/DMSO was used as the catalyst [20]. The use of MW irradiation for C–Se bond formation produces a higher yield in shorter reaction time than conventional heating in the presence of I_2 (20 mol %) and 2 equivalent of DMSO. The reaction was screened for different levels of irradiation power and it was observed that 100 W was the ideal for this transformation. Different oxidizing agents, such as H_2O_2 and TBHP, were tested in place of DMSO but the result was not good as in DMSO. 2-Naphthol

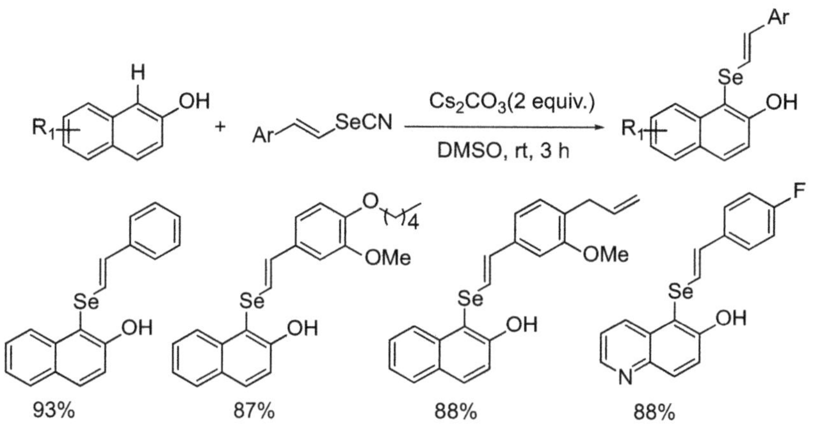

Scheme 1.4a: Transition metal-free base-mediated selenation of bicyclic arenes with styrenyl selenocyanate.

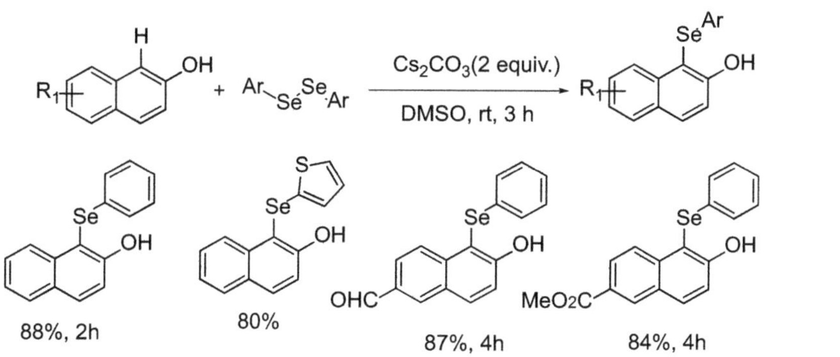

Scheme 1.4b: Transition metal-free base-mediated selenation of bicyclic arenes with diaryl diselenide.

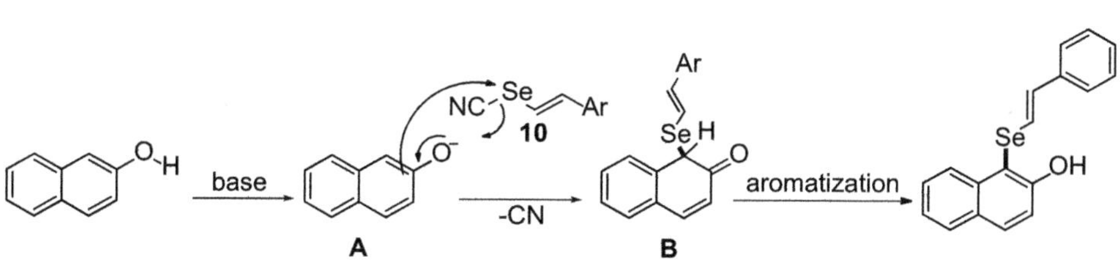

Scheme 1.4c: Possible reaction pathway.

and naphthalene-2-amines containing different functional groups afforded good-to-excellent yield (Scheme 1.5) in MW heating. 3-Hydroxyquinoline derivatives required longer times when compared to that needed by the 2-napthol. Electron-donating and -withdrawing group are compatible in the reaction condition. In the presence of a radical inhibitor 2, 2, 6, 6-tetra-methylpiperidine-1-oxyl (TEMPO), no remarkable change was observed. This result implies that the reaction does not proceed through radical intermediate.

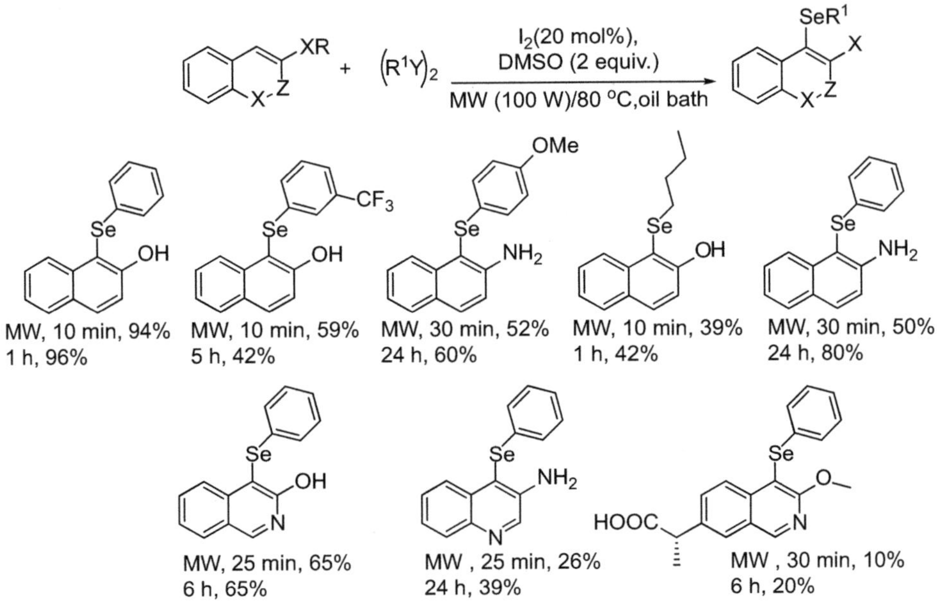

Scheme 1.5: Chalcogenation of bicyclic arenes using I_2/DMSO as nonmetallic catalytic system.

Alkynyl alkyl selenides are of particular importance, due to their applications in organic synthesis, presence as drug candidate and agrochemicals as well as being used in material science. Very recently, the preparation of alkynyl selenides from the reaction of alkynylseleno imidazolium salts with Grignard reagents has been reported. However, the limitations of the use of prepared selenating reagents and the expensive catalytic system, poor functional group tolerance, and harsh reaction conditions are not suitable for the late-stage functionalization of complex molecules. In this context, Ge and his coworkers developed an efficient method for the synthesis of alkynyl alkyl selenides via three-component coupling of terminal alkynes, Se, and epoxides [21]. Subsequently, a wide range of epoxides were examined under metal-free selenation in water medium. It was observed that the reaction of a mixture of alkynes (0.3 mmol), Se (0.6 mmol), cyclohex-ene oxide (0.9 mmol), tetrabutylammonium iodide (TBAI, 0.3 mmol), and KOH (0.6 mmol) in H_2O (2 mL) at 45 °C for 12 h under inert atmosphere provided the best yield. Functional groups including halogen, methoxyl, trifluoromethyl, and nitro group tolerate the reaction conditions (Scheme 1.6). This reaction is also applicable for bulk scale synthesis. To identify the role of elemental Se in this transformation and to gain more insight into the reaction mechanism, some control experiments were performed. Based on the experimental results and previous literature reports, a plausible reaction mechanism for double C–Se bond formation was proposed, as shown in Scheme 1.7. Initially, elemental selenium undergoes disproportionation [22] to generate a selenide anion under basic conditions, which reacts with an epoxide to form the ring-opened alkyl selenide anion species. The oxidative homo-coupling afforded the corresponding diselenide intermediate [23], and finally, the terminal alkyne underwent alkylselenation [24] in the presence of a base to provide the products.

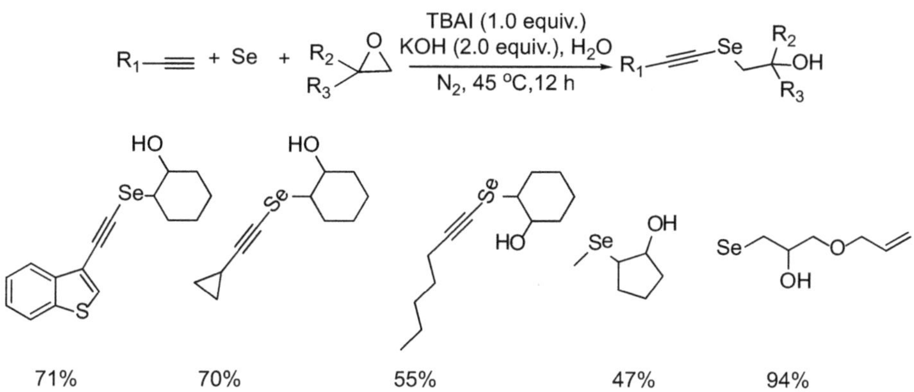

Scheme 1.6: Metal-free synthesis of alkynyl alkyl selenides.

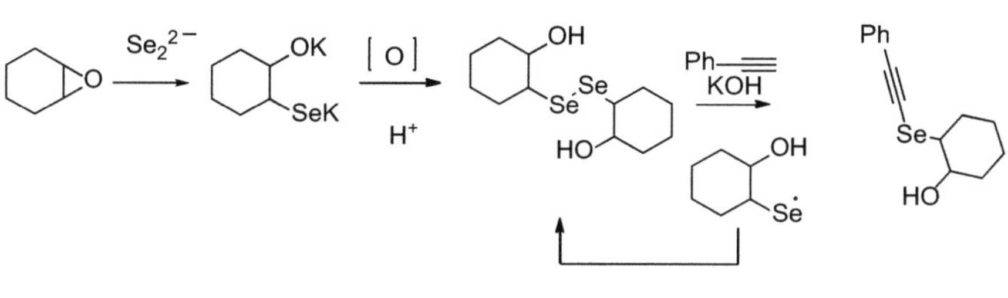

Scheme 1.7: Mechanism for double C–Se bond formation.

Lewis acid mediated ring-opening reactions of epoxides with the selenolate anions for the synthesis of selenides are of considerable interest. Liu and coworkers developed a convenient silver-catalyzed one-pot three-component selective synthesis of β-hydroxy selenides using organoboronic acids, selenium powder, and epoxide through regioselective and stereoselective ring opening of epoxides [25]. In general, aryl boronic acids containing both electron-rich and electron-deficient substituent were smoothly converted into the respective *trans*-β-hydroxy selenides in moderate-to-good yields (Scheme 1.8). Remarkably, *ortho*-functionalized substrates did not interfere with the reaction and yielded the corresponding products in good yield. A wide range of oxiranes were examined in silver-catalyzed double C–Se bond-forming reaction. Overall, both linear and branched aliphatic epoxides perform well, giving the desired product with excellent regioselectivities. Remarkably, functionalized styrene oxides were also acceptable under the reaction conditions. These compounds are usually challenging substrates in the SN^2 ring-opening reaction because of their propensity to yield inseparable regioisomers. The reaction was found to have a good tolerance for a variety of functional groups in oxiranes, including alkyl, alkoxy, alkenyl, and halogens. Application of the reaction was further demonstrated in the late-stage selenation of several bioactive compounds, such as vitamin E, sesamol, and thymol derivatives. Preliminary mechanistic studies suggest that the reaction proceeds through the silver-catalyzed radical selenation of the aryl boronic acids to generate aryl diselenide, and subsequent selenium-mediated selective ring-opening and arylselenation of epoxides.

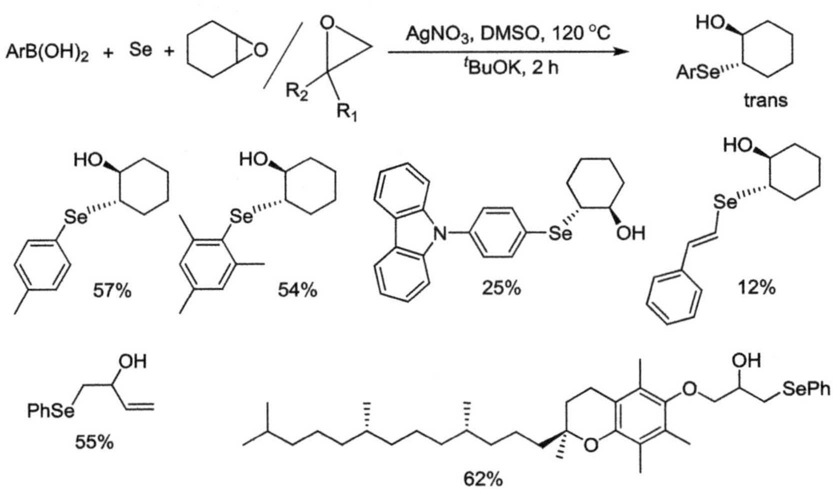

Scheme 1.8: Silver-catalyzed one-pot three-component synthesis of β-hydroxy selenides.

Another site-selective copper-catalyzed three-component coupling reaction of electron-deficient heterocycles with Se powder and aryl iodides was reported by Hu et al. [26]. The established protocol provides an efficient and practical pathway to access 2-arylselenated heterocycles via copper-catalyzed double C–Se bond formation (Scheme 1.9). The aforementioned reaction is highly regioselective. A variety of arylselenated 2-Ar-1,3,4-oxadiazole compounds were generated by this protocol. Generally, aryl iodides bearing electron-donating groups like methyl, methoxy, and trifluoromethoxy resulted in higher yields than those with electron-withdrawing groups like trifluoromethyl, ester, and nitro. Remarkably, hindered substrates showed higher reactivity and gave the desired products in shorter reaction times. Halogen-containing substrates were well tolerated, and these reactive groups provide an opportunity for the further functionalization. Furthermore, heterocyclic substrates such as pyrazole, thiophene, carbazole, and quinoline were accommodated under the current reaction conditions, further highlighting the generality of this coupling process. This transformation involves the use of elemental Se as the selenating reagent, a cost-effective catalytic system, and the late-stage selenation of bioactive compounds and natural products.

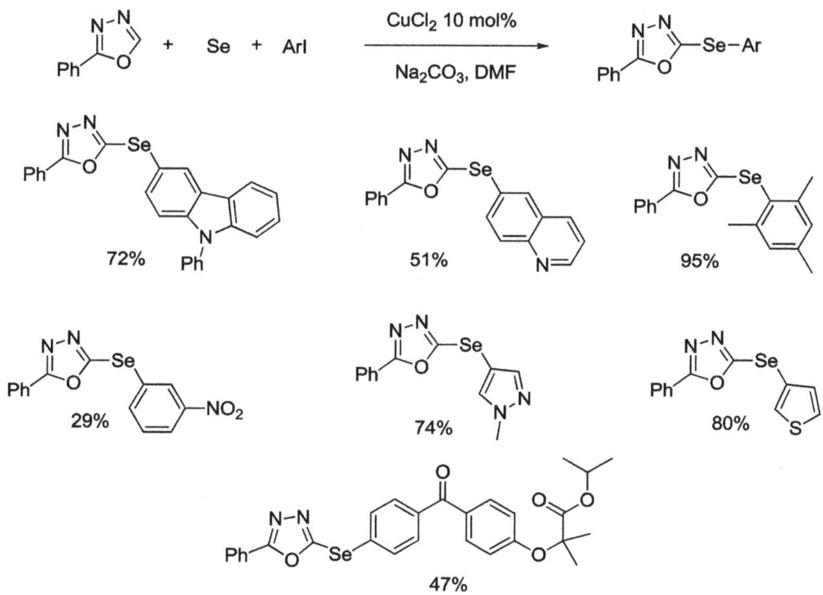

Scheme 1.9: Copper-catalyzed arylation of Se with aryl iodides and heterocycles.

An *ortho*-selective ammonium chloride-catalyzed selenylation of phenols was developed by Yeung and coworkers (Scheme 1.10) [27]. The advantages of the protocol are low catalyst loading and the reaction under mild conditions. The reaction shows very high regioselectivity.

Malemides are found as important structural motifs in natural products and drug molecules [28]. This scaffold also acts as a linker in protein, peptide, polymer, and so on [29], and as useful intermediate for the synthesis of heterocyclic frameworks such as pyrrolidines, succinimides, γ-lactams, lactims, and so on [30]. Baidya and coworkers demonstrated an unprecedented Ru(II)-catalyzed formation of C–Se bond in maleimides via weak coordinating carboxylate-assisted C–H functionalization keeping intact the olefin moiety (Scheme 1.11) [31]. For the first time they have reported Ru-catalyzed direct C–H selenation of alkene via an umpolung method. The selenylation

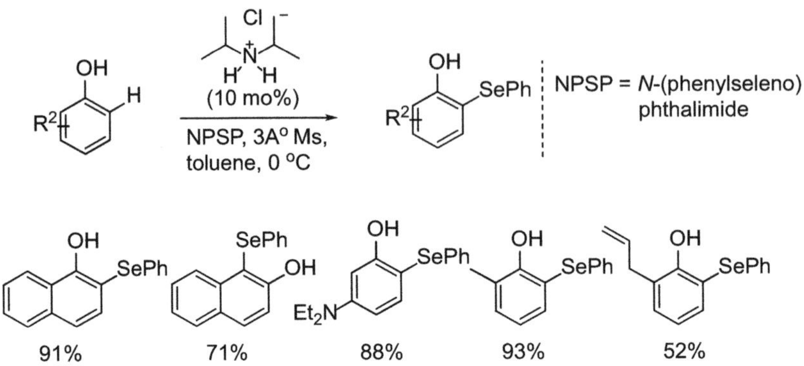

Scheme 1.10: Ammonium salt-catalyzed *ortho*-selenation of phenols.

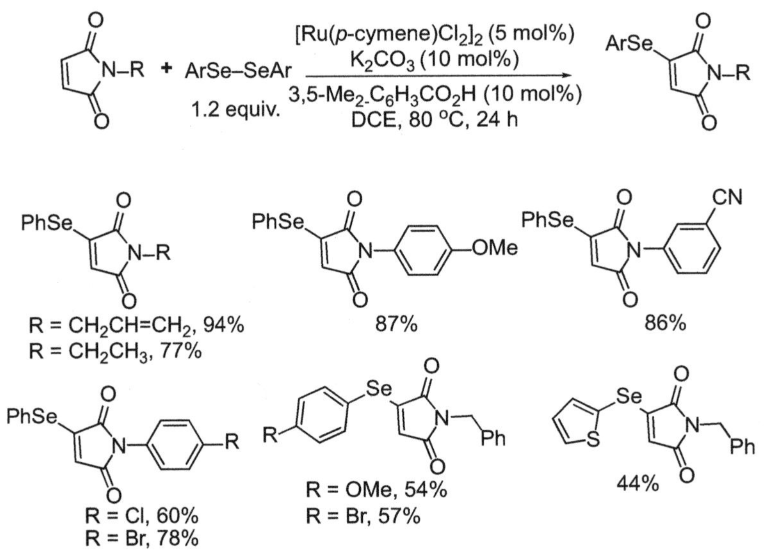

Scheme 1.11: Ru-catalyzed C–H selenylation of maleimides.

product was obtained following the optimized reaction conditions using diaryl diselenide (1.2 equiv.), catalyst [Ru(*p*-cymene)Cl_2]$_2$ (5 mol %), K_2CO_3 (10 mol %), and an acid additive 3,5-$Me_2C_6H_3CO_2H$ (10 mol %) in dichloroethane at 80 °C for 24 h. They explored the scope of the reaction. *N*-Substituted allyl and alkyl malemides provide the corresponding products with excellent yields. The electron-donating and electron-deficient functional group containing aromatic ring attached with N-atom underwent reaction efficiently to give the respective product exclusively. The halogen groups can be further functionalized. Various aryl and heteroaryl diselenides were also investigated. It was observed that the reaction did not go through a radical pathway and proceeded through the involvement of a direct C–H functionalization.

Baidya's group also reported the ruthenium(II)-catalyzed chalcogenation of aryl and heteroaryl acids via direct *ortho* C–H activation (Scheme 1.12) [32]. Without any metallic oxidant the reaction proceeds via weak *O*-coordination, avoiding the installation of an external directing group. The selenylation product, selenoxanthones, is important for biological activity [33]. In the general procedure, aromatic carboxylic acid was allowed to react with diaryl diselenide (2.0 equiv.) in the presence of catalyst [Ru(*p*-cymene)Cl_2]$_2$ (4 mol %), PCy_3 (8 mol %), $NaHCO_3$ (1 equiv.) in dimethylformamide at 100 °C for 48 h under aerobic condition. Benzoic acid containing electron-donating as well as electron-withdrawing groups at *para*-position smoothly reacted to give the diselenylated products in good yields. The *meta*-substituted benzoic acid gives monoselenylated product via preferential less hindered side due to the steric constraints. Only monoselenyted product was obtained when sterically hindered *ortho*-substituted benzoic acid and bicyclic 1- or 2-naphthoic acid were reacted. Heteroaryl carboxylic acid also provided excellent yields by this method. Various functionalized diaryl diselenides produced the corresponding products by this procedure.

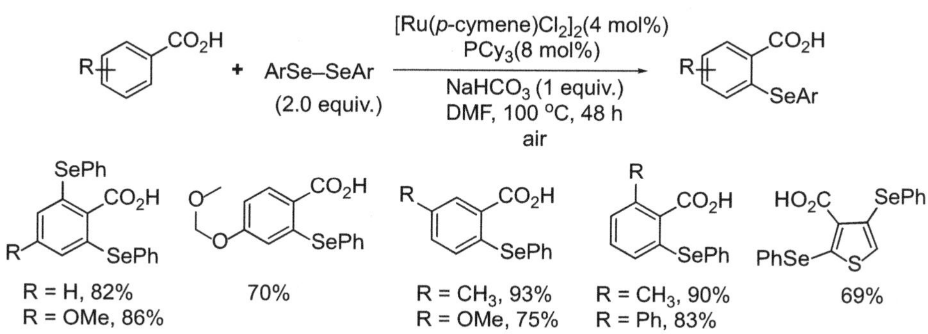

Scheme 1.12: Ru-catalyzed *ortho* C–H selenylation of aromatic carboxylic acid.

Aryl alkynoates are useful coupling partners in radical cascade atom transfer reaction. Based on the radical acceptor property of the activated alkyne functional group, synthesis of several molecular skeletons has been designed [34–36]. Recently Baidya et al. developed a route for the synthesis of substituted α,β-unsaturated acids containing chalcogen functionality at room temperature via a radical-based cascade reaction (Scheme 1.13) [37]. The oxidative difunctionalization of aryl alkynoate under metal-free condition produced the unsymmetrically tetrasubstituted acyclic olefin with CO_2 exclusion. In the presence of oxidant TBHP (2.0 equiv.), aryl alkynoates reacted with diaryl diselenides (2.0 equiv.) in acetonitrile at room temperature under argon atmosphere to give the corresponding product. A wide range of substituted aryl alkynoates were subjected to the reaction with diaryl diselenides by this procedure to produce the corresponding tetrasubstituted alkenes. O-Aryl ring of aryl alkynoate containing electron-releasing groups and electron-withdrawing substituents

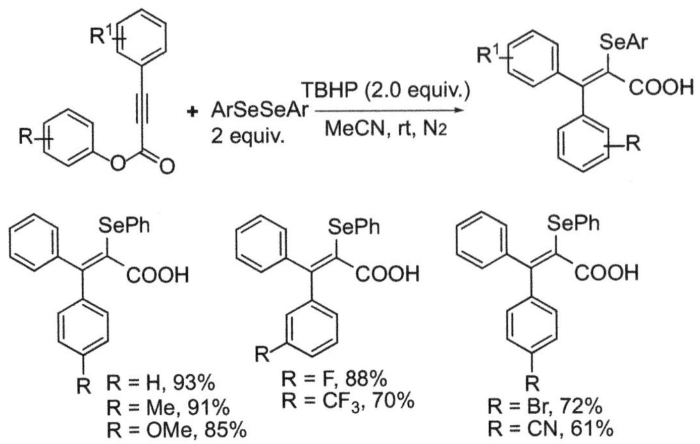

Scheme 1.13: Radical cascade selenylation of aryl alkynoates.

provided the corresponding products in high yields. The *para*-substituted alkyne aryl ring also gives the tetrasubstituted *α,β*-unsaturaterd acids in excellent yields. A gram-scale reaction was performed using this method. According to the results of the control experiments, a radical mechanism was proposed. The selenium radical induced cascade rearrangement with 1,4-aryl migration from oxygen center to carbon center. Through decarboxylative radical coupling, these products serve as building blocks to the synthesis of vinyl selenides, vinyl halides, geminal diselenoethers, 3,3- diaryl indanones. The carboxylic group can be further functionalized. A plausible reaction pathway is given below (Scheme 1.14). In the presence of TBHP, an aryl selenide radical is generated that reacts with the olefenic bond of aryl alkynoate to give the vinyl radical. This intermediate produces a cyclized product through an intramolecular pathway, which on aryl migration gives carboxyl radical. Finally, the carboxyl radical provided the product accepting hydrogen radical from TBHP.

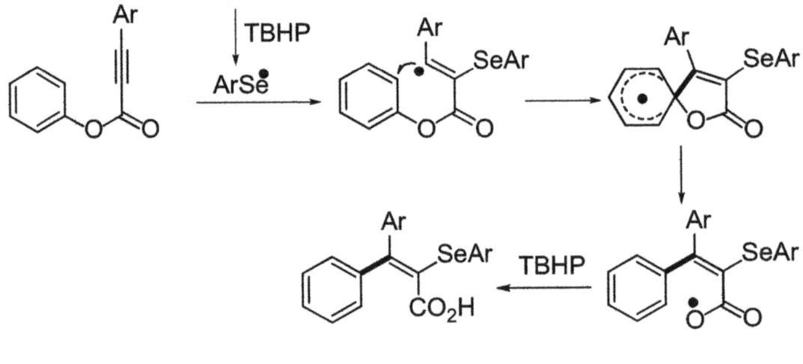

Scheme 1.14: Plausible reaction mechanism.

Very recently Cu-catalyzed four-component cross-coupling reaction of aryl iodides, Se powder, maleimides, and secondary amines was developed by Wu and coworkers to provide amino-arylselenated maleimides (Scheme 1.15) [38]. Vinyl selenides are useful synthetic intermediates for drug molecules and other biologically active compounds [1]. In this one-pot reaction, the best yield was obtained when maleimide was reacted with aryl iodide (3 equiv.), Se powder (Se_8) (3 equiv.), and an secondary amine (1.5 equiv.) in the presence of $Cu(OAc)_2$ (10 mol%), Na_2CO_3 (4.0 equiv.) in DMF at 120 °C under O_2 atmosphere for 18 h. Only secondary amine is effective where as primary alkyl amine, aniline, electron-deficient amide failed to initiate the reaction. Both cyclic and acyclic amines react. In case of substituted aryl iodide, electron-donating substituent provides better yields than electron-withdrawing one. Halogen-containing aryl iodides are well accepted and the halogen moieties can be further functionalized. Heteroaryl iodide also participates as a coupling partner in this method. A mechanistic pathway was proposed in Scheme 1.16.

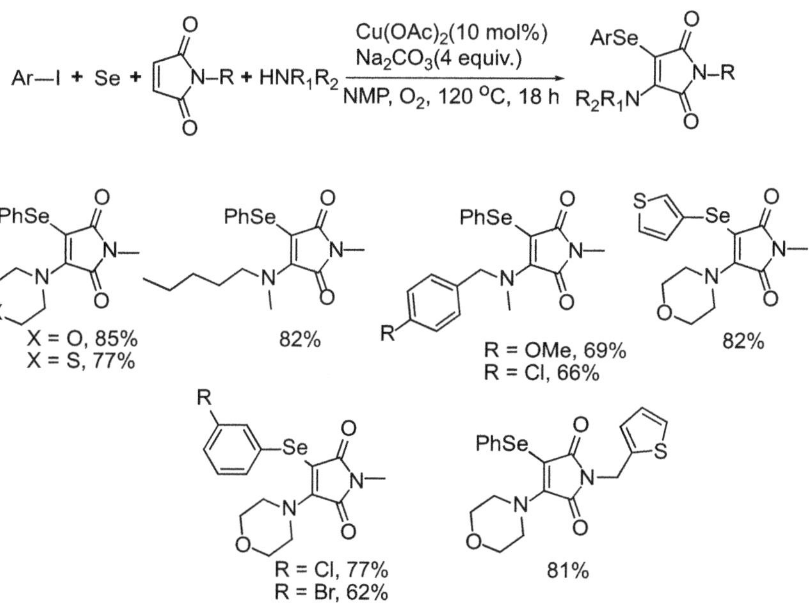

Scheme 1.15: Cu-catalyzed aminoarylselenation.

1.3 Reactions involving cross-coupling

Aromatic amines are used as precursor for the construction of C–Se bond on reaction with selenium sources. Easily available and cheap aromatic amines react with *tert*-butyl nitrite in neutral medium and subsequent reaction with diaryl/diheteroaryl/dialkyl diselenides under visible light irradiation at an ambient temperature, leading to

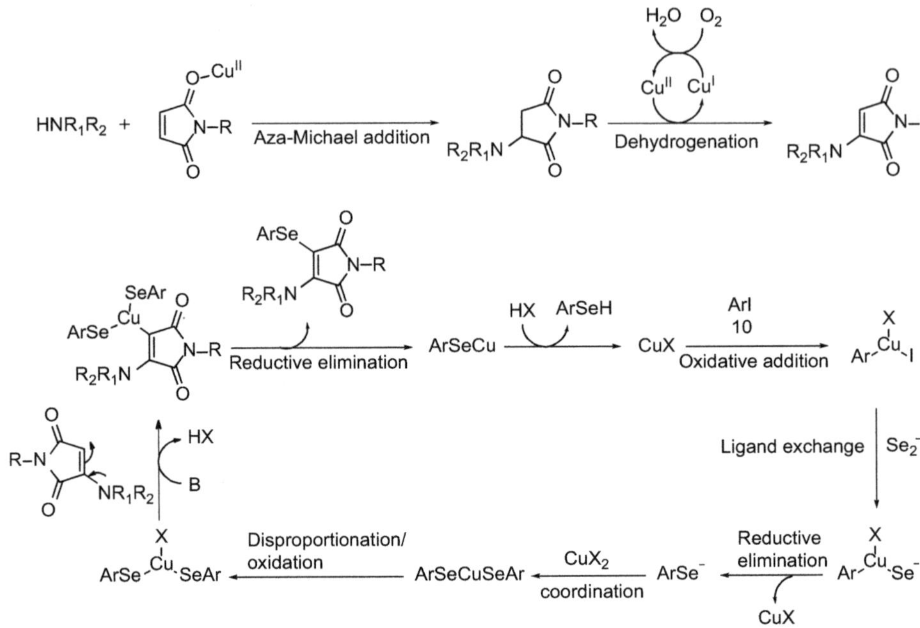

Scheme 1.16: Proposed mechanism.

selenylation of the corresponding amine (Scheme 1.17a) as reported by Ranu and his coworkers [39]. Among various photosensitizers, less expensive eosin Y is found to be the best choice. A variety of aryl/heteroary amines-bearing electron-donating and electron-withdrawing substituents reacted efficiently with diaryl/diheteroaryl/dialkyl diselenides to provide the corresponding unsymmetrical diaryl/diheteroaryl/dialkyl selenides in excellent yields. These organoselenides are of much interest in pharmaceutical industries. This reaction of C–Se bond formation was also applied for the synthesis of unsymmetrical *bis*-selenides and selenosulfides. The aryl amine undergoes in situ diazotization in the presence of *tert*-butyl nitrite and the corresponding diazonium ion formed immediately undergoes reduction by photo-activated eosin Y. The resulting aryl radical (**A**) interacts with diaryl diselenide to produce an intermediate (**B**), which is stabilized by aryl and selenium moiety (Scheme 1.17b). The radical (**B**) then leads to the formation of intermediate (**C**) via one-electron oxidation by eosin Y radical cation, which undergoes cleavage to provide the desired product. The involvement of radical pathway was confirmed by radical quenching experiment using TEMPO (radical quencher).

Alves and his coworkers reported a simple and catalyst-free method for the synthesis of diaryl selenides by the reaction of aryl selenols and diazonium fluoroborates (Scheme 1.18) [40]. Aryl selenols were generated in situ by the reaction of diaryl diselenides with hypophosphorous acid (H_3PO_2) at room temperature. The reaction successfully produces a large array of unsymmetrical diaryl selenides by the

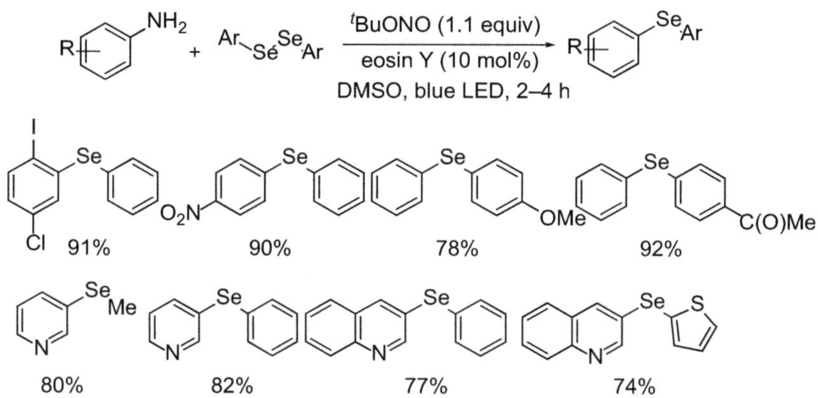

Scheme 1.17a: Visible light-photocatalyzed conversion of aryl-/heteroarylamines to selenides at room temperature.

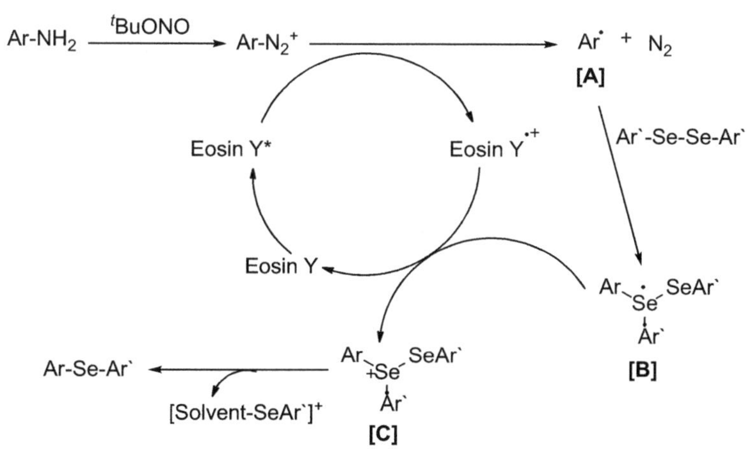

Scheme 1.17b: Possible reaction pathway.

reaction of arenediazonium tetrafluoroborates and diryl diselenides in the presence of hypophosphorous acid (H_3PO_2) at room temperature in THF. The reaction occurs via two steps. Initially diaryl diselenide reacts with hypophosphorous acid (H_3PO_2) at room temperature in THF to provide the corresponding aryl selenol moiety. Then, arenediazonium tetrafluoroborates were added to the reaction mixture at room temperature under inert atmosphere to produce the product. This reaction provides a new route for the preparation of diaryl selenides containing electron-withdrawing and electron-donating groups in moderate-to-good yields.

MW-assisted general, efficient, and green procedure for the synthesis of unsymmetrical diaryl selenides by the reaction of aryl diazonium fluoroborate and diaryl

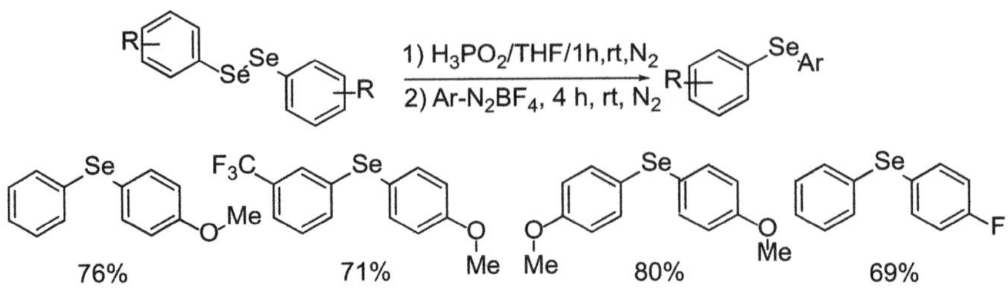

Scheme 1.18: Synthesis of diaryl selenides by the reaction of aryselenols and arenediazonium salts.

diselenides in dimethyl carbonate was developed by Ranu and his coworkers (Scheme 1.19) [41]. Zinc metal has been used as a reducing agent and the reaction is completed within a short time.

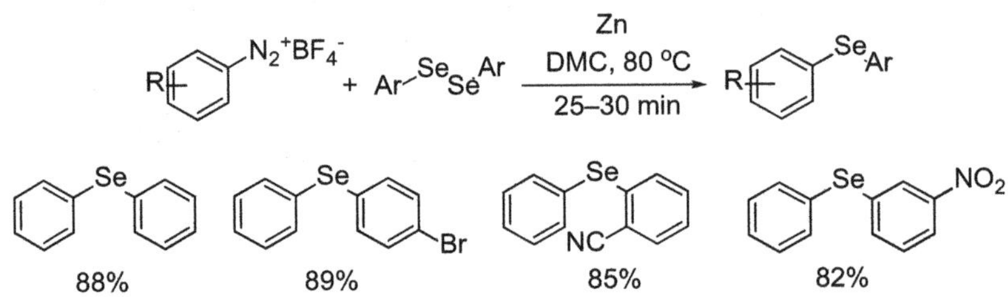

Scheme 1.19: Microwave-assisted reaction of aryl diazoniumfluoroborate and diaryl diselenides.

The reaction of arene diazonium tetrafluoroborates and diaryl diselenides on the surface of alumina under ball-milling without any external solvent or metal to produce unsymmetrical diaryl selenides was reported by Ranu et al. (Scheme 1.20) [42]. Usually, diaryl diselenides are more stable and are easily prepared and preferred over less stable and more toxic selenols as selenating agents [2]. Among several grinding auxiliaries such as neutral alumina, basic alumina, or silica, neutral alumina was the preferred choice. A mixture of diazonium tetrafluoroborate (1 mmol), diaryl diselenide (0.5 mmol), and KOH (0.75 mmol) on neutral alumina (3 g) was ball-milled at 600 rpm using six balls for 15–20 min. In conventional heating at 80 °C, the

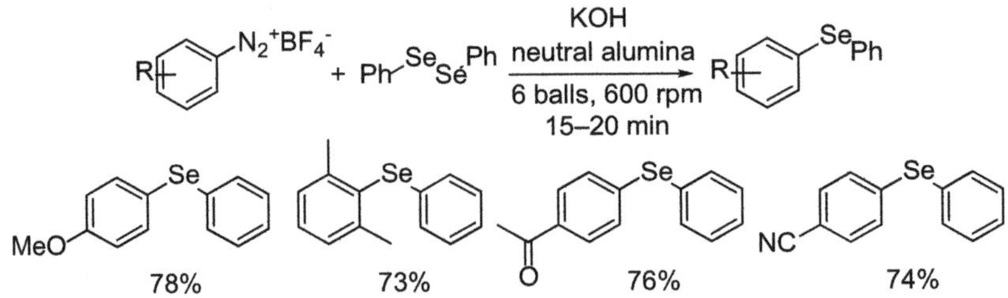

Scheme 1.20: Solvent-, ligand-, and metal-free synthesis of unsymmetrical diaryl selenides.

reaction required 8–10 h. Extraction of crude product by elution with ethanol or ethyl acetate followed by short column chromatography provided the desired product in a pure form. A range of electron-donating and electron-withdrawing diazonium salts underwent successful coupling with diphenyl diselenide under this condition to produce the products in moderate-to-good yields. This was the first report of synthesis using diaryl diselenides under ball milling in the absence of any metal or solvent.

In addition to the metal-free C–Se cross-coupling procedures, transition metal-catalyzed C–Se bond formations are well explored. Palladium-catalyzed addition reactions of diaryl diselenides to terminal alkynes proceed smoothly in room temperature ionic liquid, 1-butyl-3-methylimidazolium hexafluorophosphate ([bmim][PF_6]) as reported by Cai and coworkers (Scheme 1.21) [43]. Both aliphatic and aromatic terminal alkynes underwent successful coupling with diaryl diselenides and the corresponding (*Z*)-1,2-bis(arylthio)-1-alkenes or (*Z*)-1,2-bis(arylseleno)-1-alkenes were obtained in good-to-excellent yields. The ionic liquid, [bmim][PF_6], provides the advantages of rate acceleration, increase of yield, and a lower reaction temperature. For example, the addition reaction of Ph_2Se_2 to 1-hexyne in [bmim][PF_6] at 60 °C gave (*Z*)-1,2-bis(phenylseleno)-1-hexene in 96% yield after 2 h of reaction, but the same reaction when performed in benzene at 80 °C gave the desired product in 81% yield after 12 h.

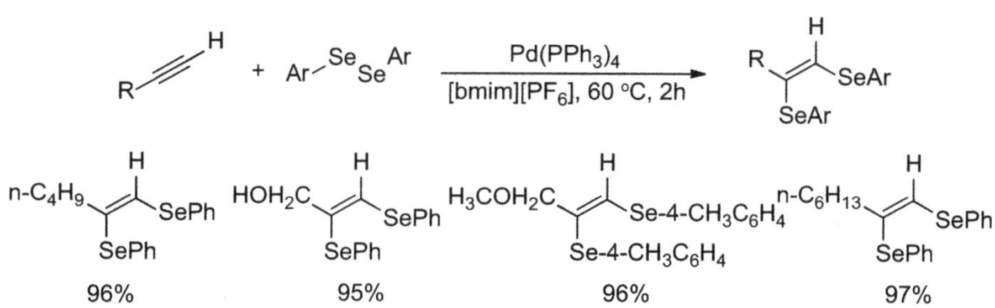

Scheme 1.21: Palladium-catalyzed addition of diaryl disulfides and diselenides to terminal alkynes in ionic liquids at room temperature.

The same group further developed palladium-catalyzed cross-coupling reaction of phenyl tributylstannyl selenide with aryl/alkyl halides at room temperature in ionic liquid, 1-butyl-3-methylimidazolium hexafluorophosphate ([bmim][PF_6]), for the synthesis of diorganyl selenides (Scheme 1.22) [44]. The coupling reaction was performed in [bmim] [PF_6] ionic liquid. Among several other palladium catalysts, $Pd(PPh_3)_4$ was proved to be the most effcient one. Best results were observed when the reaction was performed with 5 mol% of $Pd(PPh_3)_4$ in [bmim][PF_6] at 80 °C for 1 h.

Besides palladium catalyst, other metals such as nickel, copper, and calcium were also employed for the C–Se bond formation via cross-coupling reactions. For example, synthesis of selenium-containing diaryl retinoids has been reported by Millois

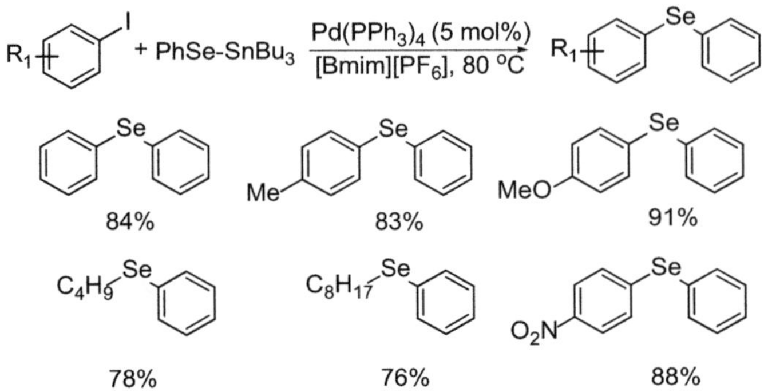

Scheme 1.22: Palladium-catalyzed cross-coupling of $PhSeSnBu_3$ with aryl and alkyl iodides in ionic liquid.

et al. (Scheme 1.23) [45] using nickel(II)-catalysts for the coupling of diaryl diselenide and an iodoarene in the presence of polymer-supported borohydride. Retinoids are synthetic [46] and natural analogues of all *trans*- or 9-*cis*-retinoic acid, which shows profound effects on cell differentiation and proliferation [47]. These biological properties are indicative of high potential for the treatment of hyperproliferative disorders such as psoriasis or cancer. Many of their biological effects are mediated by activation of nuclear receptors. A range of diaryl/aryl-heteroayl selenides were synthesized by this procedure using $(bpy)_2NiBr_2$ in a mixture of ethanol and THF (4:1) to improve the solubility at 65 °C for a period of 16 h.

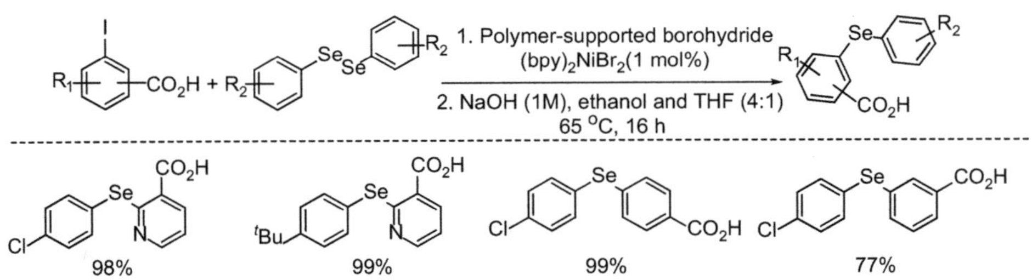

Scheme 1.23: Solution-phase synthesis of diaryl selenides using polymer-supported Ni catalyst.

The 1,3,4-oxadiazole moiety is a ubiquitous heterocycle, found in pharmaceuticals [47]. 1,3,4-Oxadiazole skeleton shows wide biological activity including, anti-inflammatory, analgesic, anticancer, anti-HIV, immune-stimulatory, anticonvulsant, and angiogenesis inhibition. An efficient copper-catalyzed three-component selenation of oxadiazoles with elemental selenium and aryl iodide for the synthesis of seleno-oxadiazoles was reported by Braga and his coworkers (Scheme 1.24) [48]. The reaction was performed under open atmosphere in the presence of cheap and easily available copper salt with minimum catalyst loading, which makes this one-pot $C(sp^2)$–H bond chalcogenation approach more attractive and practical. The reaction was effective for several iodo arenes containing both electron-donating and electron-withdrawing substituents. In general, the reaction tolerated the electronic effect (EDG, EWG) of the substituents at the *para-* and *meta*-positions, leading to the respective products in good-to-moderate yields. However, there was a negative influence on the yield of the *ortho*-substituted iodo arene due to steric hindrance. Substituted 1,3,4-oxadiazole (both aromatic and aliphatic substituents) were also investigated and the respective seleno-oxadiazoles were formed in good yields.

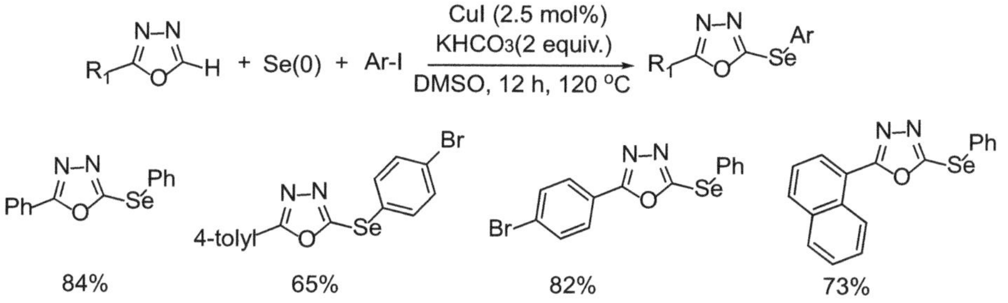

Scheme 1.24: Copper-catalyzed three-component reaction of oxadiazoles, elemental Se/S, and aryl iodides.

Along with the transition metals, group 2 alkaline earth metals, such as calcium, strontium, and barium, have also been employed for several important organic transformations. These metals show excellent Lewis acid property because of their large radii and electropositive character. In addition, their low toxicity and cost effciency have also made them more attractive to be used as catalysts for reaction. Ranu and his group has reported a calcium(II) chloride mediated C–F bond cleavage of electron deficient fluoroarenes followed by selenation in the absence of any additive, ligand, or organometallic reagents (Scheme 1.25) [49]. The reactions were performed in the presence of a mixture of fluoroarene (1.0 mmol), diaryl diselenide (0.6 mmol), calcium chloride (3.0 mmol), and Zn dust (1.1 mmol) in DMSO (3 mL) under an argon atmosphere at 110 °C for 12 h. After evaporation of DMSO, the crude product was extracted with ethyl acetate, which was then subjected to column chromatography over silica gel for pure product isolation. A wide range of diversely substituted diphenyl selenides have been employed in this procedure and the corresponding products were formed in good yields. The fluoroarenes containing electron-withdrawing substituents underwent successful coupling with diary diselenide, but unsubstituted and electron-donating group substituted fluoroarenes were not effective for this transformation. The role of calcium salt in the cleavage of the C–F bond, the probable mechanism as depicted in Scheme 1.26, has been suggested using DFT analysis.

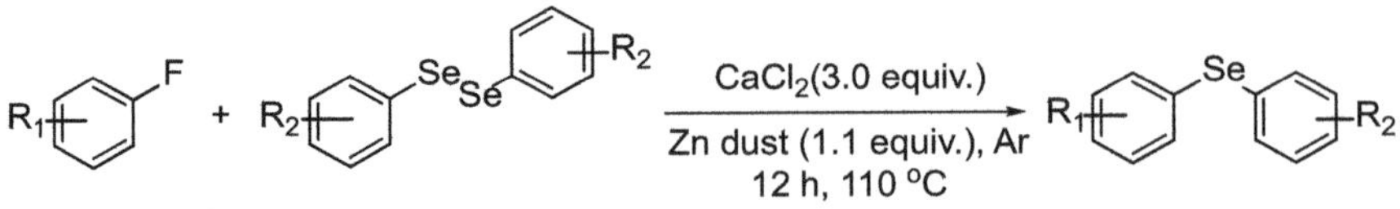

Scheme 1.25: Calcium-mediated C–F bond substitution in fluoroarenes and C–Se bond formation.

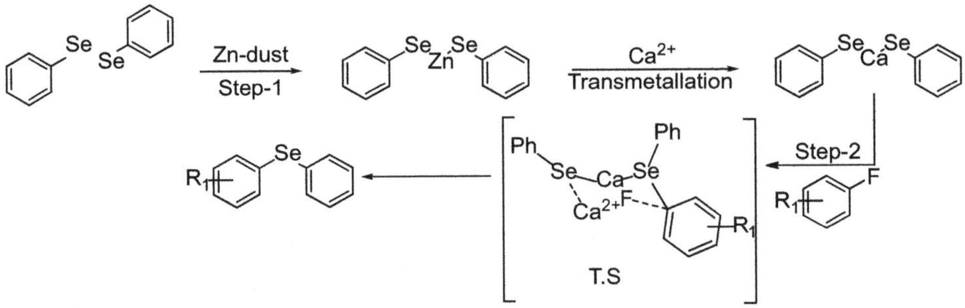

Scheme 1.26: Possible reaction pathway.

One-pot metal-free pathway for the synthesis of organoselenides and selenoglycosides involves alkylation, arylation, or alkynylation of selenium anions as developed by Townsend et al. [50]. The procedure is unique and it represents an umpolung approach to the synthesis of aryl-selenides through two-step strategy. The first step involves the arylation of potassium selenocyanate (KSeCN) with an iodonium reagent in the absence of a metal catalyst to produce an arylselenocyanate. In general, it was found that the yield of the reaction depends on the composition of iodonium salt (0.1 mmol-0.20 mmol) and KSeCN (0.11–0.40 mmol) in EtOAc (1–2 mL) at 80 °C for 24 h. In the second step, treatment with sodium borohydride unmasks a second selenium nucleophile that interacts with an aliphatic electrophile, iodonium reagent, or glycosyl halide. The *ortho-*, *para-*, *meta-*substituted electron-donating groups like –Me, –OMe as well as electron-withdrawing groups $-NO_2$, CO_2Et on aryl part led the reaction without any difficulty. The epoxide, ester, alkyne moieties also remained unaffected under the reaction conditions (Scheme 1.27).

A convenient protocol for the synthesis of styrenyl selenocyanates from readily available styrenyl bromides by the reaction with potassium selenocyanate in the presence of iodine under specified conditions has been developed by Ranu and his coworkers [51]. This method is of much significance as synthesis of styrenyl selenocyanates is reported for the first time. In addition, this strategy constitutes a one-pot reaction using commercially available chemicals. It was reported that a mixture of (*E*)-1-(2-bromovinyl)-4-methylbenzene (1.0 mmol), KSeCN (1.2 mmol), and catalyst (20 mol %) in dry DMSO (2.5 mL) at 90–100 °C under argon provided the best result (Scheme 1.28). This reaction is chemoselective. The iodo- and bromo-substituents

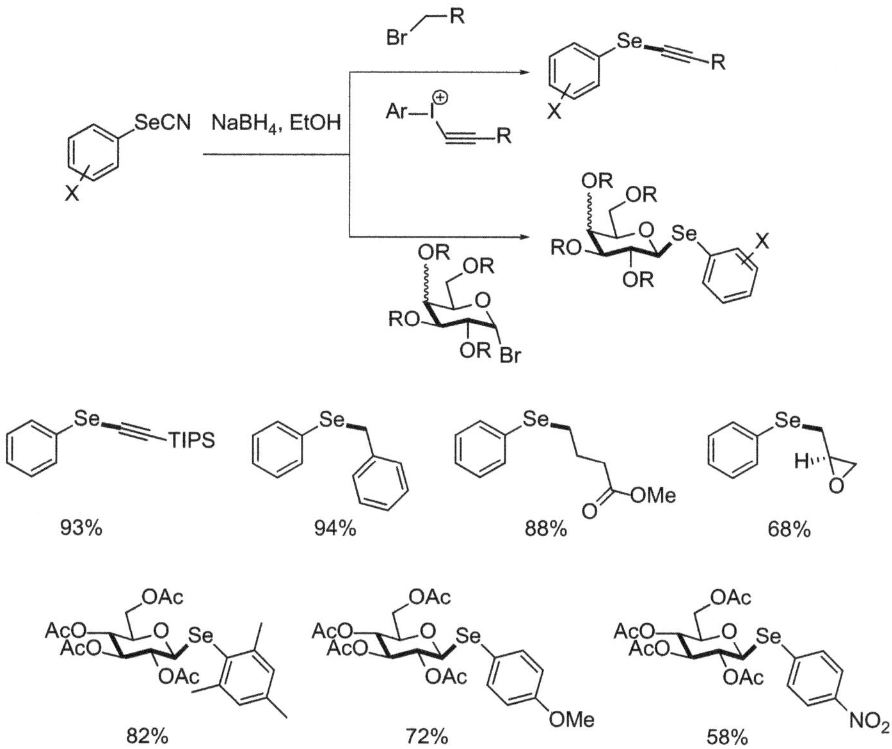

Scheme 1.27: Synthesis of unsymmetrical organoselenides and selenoglycosides.

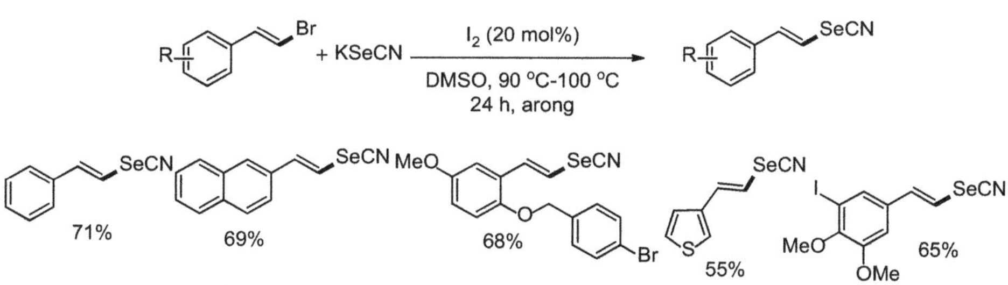

Scheme 1.28: Transition-metal-free iodine catalyzed selenocayanation of styrenyl bromides.

attached to the aromatic ring are compatible with the reaction conditions. Styrenyl bromides containing electron-rich groups or electron-poor groups undergo smooth reaction. This reaction is also applicable to naphthalene derivatives and *ortho*-substituted styrenyl bromides.

To find out whether the reaction proceeds via radical/ionic pathway, a representative reaction with 4-methyl styrenyl bromide and KSeCN was performed using

TEMPO ((2,2,6,6-tetramethylpiperidin-1-yl)oxyl) as a radical quencher. It was revealed that TEMPO does not have any effect on the reaction. This indicates that selenocyanate formation is likely to go through an ionic process rather than radical pathway. A possible reaction pathway has been outlined in Scheme 1.29.

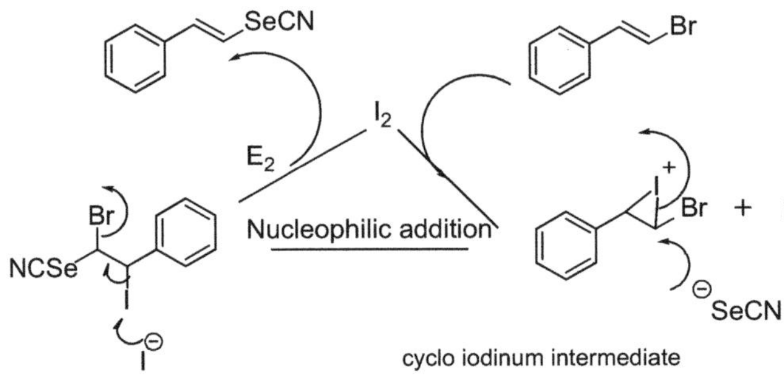

Scheme 1.29: Plausible reaction mechanism.

A simple and convenient approach for the synthesis of unsymmetrical diaryl selenides has been reported by Kumar and his group using copper-catalyzed cross-coupling reaction of boronic acid with diaryl diselenides in ethanol using $NaBH_4$ [52]. It was found that the reaction of boronic acid and diaryl diselenide in the presence of $CuSO_4.5H_2O$ along with 1,10-phenanthroline, $NaBH_4$ in EtOH provides the best yield. The methodology is very efficient for the synthesis of unsymmetrical diaryl selenides bearing various functionalities such as $-CF_3$, $-NO_2$, –F, –Br and the respective products were obtained in good-to-excellent yields (Scheme 1.30). Moreover, the symmetrical diaryl selenides have also been obtained from aryl boronic acids using selenium powder under optimized reaction conditions. The use of $NaBH_4$ is vital for the reaction, which enabled the formation of unsymmetrical diaryl selenides from boronic acids in ethanol at room temperature.

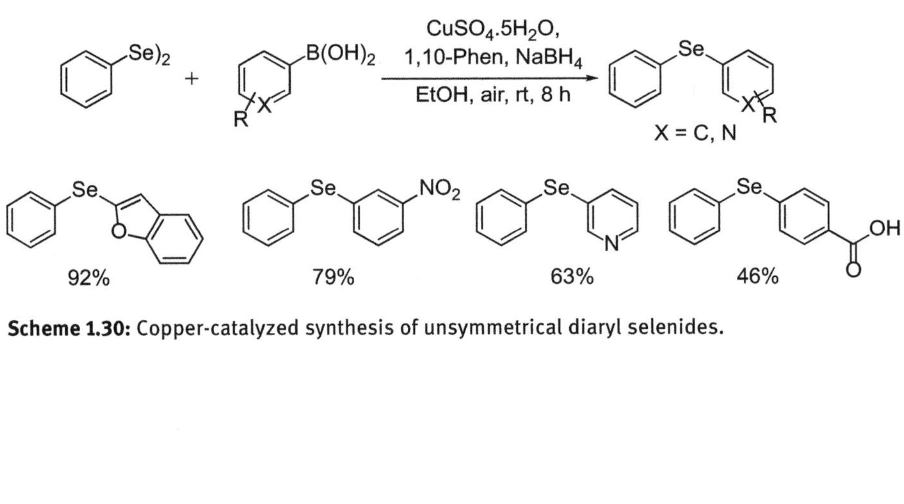

Scheme 1.30: Copper-catalyzed synthesis of unsymmetrical diaryl selenides.

Aryl bromides are used for the synthesis of diaryl selenides through the conversion to the corresponding aryl lithium or magnesium and subsequent reaction with diaryl diselenides [53]. However, this synthetic approach has several limitations, such as the availability and stability of the corresponding organometallic compounds, longer reaction time, lower yields, and so on. A new route to prepare unsymmetrical diaryl selenides has been reported by Beletskaya et al. [54]. It was found that Cu catalyst was better than Ni catalyst for the transformation. It was also observed that the reaction of aryl bromides and iodides with $Bu_3SnSeAr$ catalyzed by Cu(I) complexes like CuI·phen and $(Ph_3P)CuI$·phen in DMF under inert atmosphere produced the best yield. The di- and trisubstituted tributyltin aryl selenide and various aryl and heteroaryl bromides containing both electron-withdrawing and electron-donating substituents (Scheme 1.31) were employed in this protocol for C–Se bond formation. The yields of unsymmetrical diaryl selenides were high in all the cases. However, the arylselenylation of aryl bromides with electron-donating substituents required a longer reaction time.

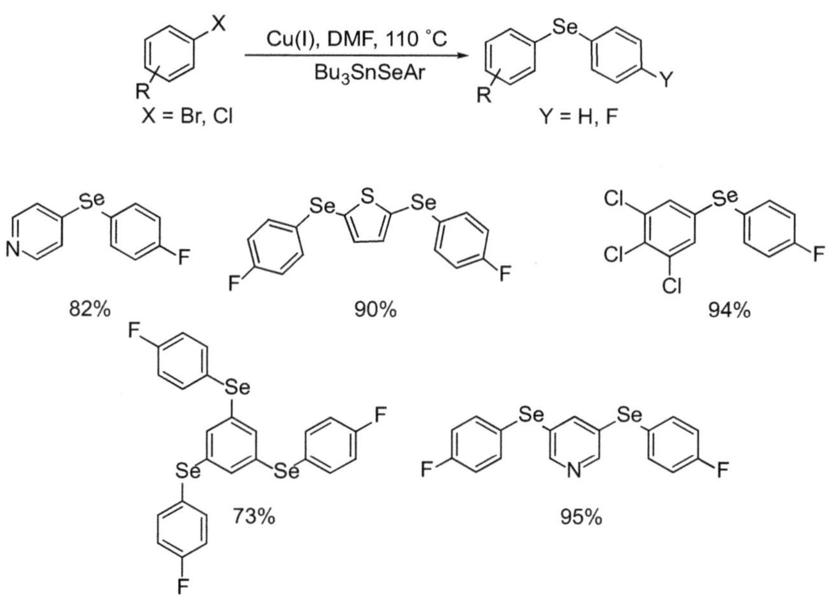

Scheme 1.31: Copper(I)-catalyzed arylselenylation of aryl bromides and iodides.

1.4 Conclusions

In this chapter we have highlighted the various methods for the synthesis of organoselenides reported during the recent period. The selenides are of much importance because of their useful applications as therapeutic agents, agrochemicals, materials, catalysts, and intermediates in organic synthesis. Thus, interest in the synthesis of these compounds is growing constantly. This chapter will cater the need of practicing chemists working in this field in academia as well as in industry.

Acknowledgment: Indian National Science Academy, New Delhi, is gratefully acknowledged for the support to one of us, B C Ranu.

References

[1] Mugesh, G., duMont, W W., and Sies, H. Chemistry of biologically important synthetic organoselenium compounds. Chem Rev 2001, 101, 2125–2180.

[2] Nogueira, C W., Zeni, G., and Rocha, J B T. Organoselenium and organotellurium compounds: toxicology and pharmacology. Chem Rev 2004, 104, 6255–6286.

[3] Woods, J A., Hadfield, J A., Mcgown, A T., and Fox, B W. Bioactivity and molecular modelling of diphenylsulfides and diphenylselenides. Bioorg Med Chem 1993, 1, 333–340.

[4] Engman, L., Stern, D., Frisell, H., Vessman, K., Berglund, M., Ek, B., and Anderson, C M. Synthesis, antioxidant properties, biological activity and molecular modelling of a series of chalcogen analogues of the 5-lipoxygenase inhibitor DuP 654. Bioorg Med Chem 1995, 3, 1255–1262.

[5] Bernardon, J.-M., and Diaz, P. Diarylselenide compounds and their use in human or veterinary medicine and in cosmetics. PCT Int Appl 1999, WO 1999065872, A1.

[6] Millois, C., and Diaz, P. Solution-phase synthesis of diaryl selenides using polymer-supported borohydride. Org Lett 2000, 2, 1705–1708.

[7] Ando, T., Kwon, T S., Kitagawa, A., Tanemura, T., Kondo, S., Kunisada, H., and Yuki, Y. Synthesis and free radical polymerization of p-methylseleno- and p-phenylselenostyrenes. Macromol Chem Phys 1996, 197, 2803–2810.

[8] Khanna, P K., and Das, B K. Novel synthesis of silver selenide nano-powder from silver nitrate and organo-selenium compound. Mater Lett 2004, 58, 1030–1034.

[9] Frizon, T E., Rampon, D S., Gallardo, H., Merlo, A A., Schneider, P H., Rodrigues, O E D., and Braga, A L. Selenides and diselenides containing oxadiazoles: A new class of functionalised materials. Liq Cryst 2012, 39, 769–777.

[10] Okamoto, Y. In The Chemistry of Organic Selenium and Tellurium Compounds, Patai, S., Rappoport, Z, Eds., Wiley, Chichester, U.K, 1986.

[11] Labinger, J A., and Bercaw, J E. Understanding and exploiting C-H bond activation. Nature 2002, 41, 7507–7514.

[12] Godula, K., and Sames, D. C-H bond functionalization in complex organic synthesis. Science 2006, 312, 67–72.

[13] McDonald, R I., Liu, G S., and Stahl, S S. Palladium(II)-catalyzed alkene functionalization via nucleopalladation: stereochemical pathways and enantioselective catalytic applications. Chem Rev 2011, 111, 2981–3019.

[14] Arockiam, P B., Bruneau, C., and Dixneuf, P H. Ruthenium(II)-catalyzed C–H bond activation and functionalization. Chem Rev 2012, 112, 5879–5918.

[15] Rouquet, G., and Chatani, N. Catalytic functionalization of C(sp^2)-H and C(sp^3)-H bonds by using bidentate directing groups. Angew Chem, Int Ed 2013, 52, 11726–11743.

[16] Iwasaki, M., Miki, N., Tsuchiya, Y., Nakajima, K., and Nishihara, Y. Synthesis of benzoisoselenazolone derivatives by nickel-catalyzed dehydrogenative direct selenation of $C(sp^2)$–H bonds with elemental selenium in air. Org Lett 2017, 19, 1092–1095.

[17] Sattar, M., Shareef, M., Patidar, K., and Kumar, S. Copper-catalyzed 8-aminoquinoline assisted aryl chalcogenation of ferroceneamide with aryl disulfides, diselenides, and ditellurides. J Org Chem 2018, 83, 8241–8249.

[18] Zhu, Y.-Q., He, J.-L., Niu, Y.-X., Kang, H.-Y., Han, T.F., and Li, H.-Y. $AgSbF_6$-mediated selective thiolation and selenylation at C-4 position of isoquinolin-1(2H)-ones. J Org Chem 2018, 83, 9958–9967.

[19] Ghosh, T., Mukherjee, N., and Ranu, B C. Transition metal- and oxidant-free base-mediated selenation of bicyclic arenes at room temperature. ACS Omega 2018, 3, 17540–17546.

[20] Silva, L T., Azeredo, J B., Saba, S., Rafique, J., Bortoluzzi, A J., and Braga, A L. Solvent- and metal-free chalcogenation of bicyclic arenes using I_2/DMSO as non-metallic catalytic system. Eur J Org Chem 2017, 4740–4748.

[21] Ge, W., Lin, M., Hongchen, L., Wenxia, G., Jinchang, D., Xiaobo, H., Miaochang, L., and Huayue, W. Metal-free synthesis of alkynyl alkyl selenides via three-component coupling of terminal alkynes, Se, and epoxides. Green Chem 2018, 20, 1560–1563.

[22] Jiang, Y., Qin, Y., Xie, S., Zhang, X., Dong, J., and Ma, D. A general and efficient approach to aryl thiols: CuI-catalyzed coupling of aryl iodides with sulfur and subsequent reduction. Org Lett 2009, 11, 5250–5253.

[23] Chen, C., Chu, L., and Qing, F L. Metal-free oxidative trifluoromethylthiolation of terminal alkynes with CF_3SiMe_3 and elemental sulfur. J Am Chem Soc 2012, 134, 12454–12457.

[24] Prasad, C D., Satter, M., and Kumar, S. Transition-metal-free selective oxidative $C(sp^3)$–S/Se coupling of oxindoles, tetralone, and arylacetamides: synthesis of unsymmetrical organochalcogenides. Org Lett 2017, 19, 774–777.

[25] Leng, T., Wu, G., Zhou, Y B., Gao, W., Ding, J., Huang, X., Liu, M., and Wu, H. Silver-catalyzed one-pot three-component selective synthesis of β-hydroxy selenides. Adv Synth Catal 2018, 360, 4336–4340.

[26] Dehua, H., Miaochang, L., and Huayue, W., Wenxia, G., and Wu, G. Copper-catalyzed diarylation of Se with aryl iodides and heterocycles. Org Chem Front 2018, 5, 1352–1355.

[27] Xiaodong, X., and Ying-Yeung, Y. Ammonium salt-catalyzed highly practical ortho-selective monohalogenation and phenylselenation of phenols: scope and applications. ACS Catal 2018, 8, 4033–4043.

[28] Bin, J W., Wong, I L K., Hu, X., Yu, Z X., Xing, L F., Jiang, T., Chow, L M C., and Biao, W S. Structure–activity relationship study of permethyl ningalin B analogues as P-glycoprotein chemosensitizers. J Med Chem 2013, 56, 9057–9070.

[29] Youziel, J., Akhbar, A R., Aziz, Q., Smith, M E B., Caddick, S., Tinker, A., and Baker, J R. Bromo- and thiomaleimides as a new class of thiol-mediated fluorescence 'turn-on' reagents. Org Biomol Chem 2014, 12, 557–560.

[30] Molleti, N., Bjornberg, C., and Kang, J Y. Phospha-Michael addition reaction of maleimides employing N-heterocyclic phosphine-thiourea as a phosphonylation reagent: Synthesis of 1-aryl-2,5-dioxopyrrolidine-3-yl-phosphonate derivatives. Org Biomol Chem 2016, 14, 10695–10704.

[31] Dana, S., Mandal, A., Sahoo, H., and Baidya, M. Ru(II)-catalyzed C–H functionalization on maleimides with electrophiles: a demonstration of umpolung strategy. Org Lett 2017, 19, 1902–1905.

[32] Mandal, A., Dana, S., Sahoo, H., Grandhi, G., and Baidya, M. Ruthenium(II)-catalyzed ortho-C–H chalcogenation of benzoic acids via weak O-coordination: synthesis of chalcogenoxanthones. Org Lett 2017, 19, 2430–2433.

[33] Engman, L., Cotgreave, I., Angulo, M., Taylor, C. W., Paine-Murrieta, G. D., and Powis, G. Diaryl chalcogenides as selective inhibitors of thioredoxin reductase and potential antitumor agents. Anticancer Res 1997, 17, 4599–4605.
[34] Ni, S., Zhou, J., Mei, H., and Han, J. Radical reactions of aryl alkynoates in organic synthesis: recent advances. Tetrahedron Lett 2018, 59, 309–316.
[35] Mi, X., Wang, C., Huang, M., Zhang, J., and Wu, Y. Silver-catalyzed synthesis of 3-phosphorated coumarins via radical cyclization of alkynoates and dialkyl H-phosphonates. Org Lett 2014, 16, 3356–3359.
[36] Hua, H.-L., He, Y.-T., Qiu, Y.-F., Li, Y.-X., Song, B., Gao, P., Song, X.-R., Guo, D.-H., Liu, X.-Y., and Liang, Y.-M. Copper-catalyzed difunctionalization of activated alkynes by radical oxidation–tandem cyclization/dearomatization to synthesize 3-trifluoromethyl spiro[4.5] trienones. Chem Eur J 2015, 21, 1468–1473.
[37] Sahoo, H., Singh, S., and Baidya, B. Radical cascade reaction of aryl alkynoates at room temperature: synthesis of fully substituted α,β-unsaturated acids with chalcogen functionality. Org Lett 2018, 20, 3678–3681.
[38] Gao, X., Tang, L., Huang, L., Huang, Z. -S., Ma, Y., and Wu, G. Oxidative aminoarylselenation of maleimides via copper-catalyzed four-component cross-coupling. Org Lett 2019, 21, 745–748.
[39] Kundu, D., Ahammed, S., and Ranu, B C. Visible light photocatalyzed direct conversion of aryl-/ heteroarylamines to selenides at room temperature. Org Lett 2014, 16, 1814–1817.
[40] Balaguez, R A., Ricordi, V G., Freitas, C S., Perin, G., Schumacher, R F., and Alves, D. Simple and catalyst-free method for the synthesis of diaryl selenides by reactions of arylselenols and arenediazonium salts. Tetrahedron Lett 2014, 55, 1057–1061.
[41] Kundu, D., Ahammed, S., and Ranu, B C. Microwave-assisted reaction of aryl diazonium fluoroborate and diaryl dichalcogenides in dimethyl carbonate: a general procedure for the synthesis of unsymmetrical diaryl chalcogenides. Green Chem 2012, 14, 2024–2030.
[42] Mukherjee, N., Chatterjee, T., and Ranu, B C. Reaction under ball-milling: solvent-, ligand-, and metal-free synthesis of unsymmetrical diaryl chalcogenides. J Org Chem 2013, 78, 11110–11114.
[43] Cai, M., Wang, Y., and Hao, W. Palladium-catalyzed addition of diaryl disulfides and diselenides to terminal alkynes in room temperature ionic liquids. Green Chem 2007, 9, 1180–1184.
[44] Zhao, H., Hao, W., Xia, Z., and Cai, M. Palladium-catalyzed cross-coupling of $PhSeSnBu_3$ with aryl and alkyl halides in ionic liquids: A practical synthetic method of diorganyl selenides. New J Chem 2011, 35, 2661–2665.
[45] Millois, C., and Diaz, P. Solution-phase synthesis of diaryl selenides using polymer-supported borohydride. Org Lett 2000, 2, 1705–1708.
[46] Charpentier, B., Bernardon, J M., Eustache, J., Millois, C., Martin, C., Michel, S., and Shroot, B. Synthesis, Structure-Affinity Relationships, and Biological Activities of Ligands Binding to Retinoic Acid Receptor Subtypes, J Med Chem 1995, 38, 4993–5006.
[47] Sporn, M., Roberts, A B., and Goodman, D S. In The Retinoids: Biology, Chemistry, and Medicine, Raven Press, New York, 1994.
[48] Peterle, M M., Scheide, M R., Silva, L T., Saba, S., Rafique, J., and Braga, A L. Copper-catalyzed three-component reaction of oxadiazoles, elemental Se/S and aryl iodides: synthesis of chalcogenyl (Se/S)-oxadiazoles. Chemistry Select 2018, 3, 13191–13196.
[49] Maity, P., Ahammed, S., Manna, R N., and Ranu, B C. Calcium mediated C–F bond substitution in fluoroarenes towards C–Chalcogen bond formation. Org Chem Front 2017, 4, 69–76.
[50] Yong, G., and Steven, D T. Metal-free synthesis of unsymmetrical organoselenides and selenoglycosides. Org Lett 2017, 19, 5252–5255.

[51] Maity, P., Paroi, B., and Ranu, B C. Transition-metal-free iodine catalyzed selenocayanation of styrenyl bromides and an easy access to benzoselenophenes via intermediacy of styrenyl selenocyanate. Org Lett 2017, 19, 5748–5751.
[52] Kumar, A., and Kumar, S. A convenient and efficient copper-catalyzed synthesis of unsymmetrical and symmetrical diaryl chalcogenides from arylboronic acids in ethanol at room temperature. Tetrahedron 2014, 70, 1763–1772.
[53] Nakanishi, W., Haiyashi, S., Sakaue, A., Ono, G., and Kawada, Y. Attractive interaction caused by the linear F ··· Se-C alignment in naphthalene peri positions. J Am Chem Soc 1998, 120, 3635–3640.
[54] Beletskaya, I P., Alexander, S S., Alexander, P S., and Petrovskii, P V. Copper(I)-catalyzed arylselenylation of aryl bromides and iodides. Tetrahedron Lett 2003, 44, 7039–7041.

Amol D. Sonawane and Mamoru Koketsu

2 Synthesis of Organoselenium Scaffolds through Selenium Radical Formation

2.1 Introduction

Selenium chemistry has proved to be a powerful tool for the synthesis of organic molecules over several decades. The radical species are known as one of the most important reactive species in organic synthesis. The selenium-containing molecules are used in a variety of synthetic radical reactions such as radical precursors, for the insertion of selenium into target product, and also as other important applications such as radical trapping. The main focus in the radical chemistry has been on reactions of carbon radicals, because carbon radicals readily react with various radical acceptors. Most of the heterocyclic compound syntheses were achieved by intramolecular radical formation. There are several radical formations like *O*-centered or *N*-centered radicals, and all of them have different stability and also have different reactivity. Many complex natural products were synthesized by reactions of *C*-centered radicals. The bond formation also occurs by radical coupling reactions between heteroatom-centered radicals and other radicals. In addition, many heteroatom-centered radicals are capable for directly abstracting hydrogen atoms from sp^3-carbons to form *C*-centered radicals.

The chapter is divided into several sections; we basically focused on the synthesis of organoselenium scaffolds through selenium radical formation. In the literature, selenium-containing heterocycles and the selenium radical chemistry have been well reviewed [1–5].

2.2 Synthesis of selenide ether by decarboxylation

The decarboxylation reaction makes practical and effective use of cheap and stable carboxylic acids as starting materials. It becomes an attractive methodology to form carbon–carbon as well as carbon–heteroatom bonds in modern organic synthesis. Generally, the decarboxylative coupling reactions required transition metal catalysts [6, 7]. Phenyl acetic acids can be used as starting materials in the construction of new bonds not only through decarboxylation but also by sp^3 C-H functionalization [8]. The development of new decarboxylation methods under metal-free conditions is also important. Herein, the method for the formation of C–Se bond through

Amol D. Sonawane and Mamoru Koketsu, Department of Chemistry and Biomolecular Science, Faculty of Engineering, Gifu University, Gifu, Japan

https://doi.org/10.1515/9783110625110-002

base-promoted decarboxylative coupling reaction was carried out by using diselenide and carboxylic acid derivative as staring materials [9].

The synthesis of selenide ether **3** by decarboxylation with mechanistic pathway is shown in Figure 2.1 [9]. Initially, phenyl acetic acid was converted to anion intermediate species **4** via a decarboxylation reaction in the presence of Cs_2CO_3. Further the species **4** reacted with diphenyl diselenide to afford the product **5** with by-product PhSeH. The intermediate **5** was converted to the secondary radical **6** when reacted with Fe(III) in oxygen atmosphere [8]; finally by a radical process, the secondary radical **6** reacted with diselenide to obtain selenide ether **3** in moderate-to-good yields.

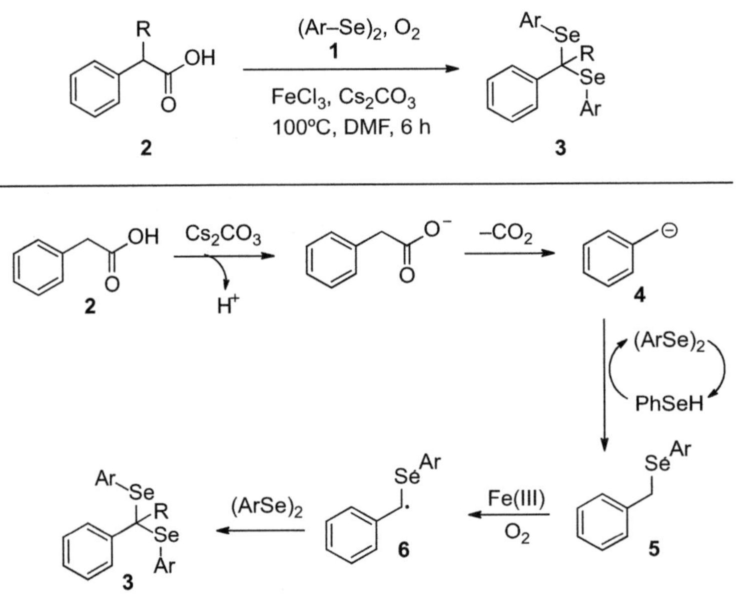

Figure 2.1: Plausible mechanism for the synthesis of selenide ether **3** by decarboxylation.

The mechanistic study involves the effect of electron donating as well as electron-withdrawing substituents on phenyl acetic acids, and the substituents on the aromatic diselenides. The reactions of aromatic diselenides with phenyl acetic acids bearing both strong electron-donating groups such as malonic acid and electron-withdrawing groups on the aryl ring afforded the good yields. Unfortunately, cyanoacetic acid and acetoacetic acid substrates failed to afford the desired products. Aliphatic carboxylic acids including acetic acid, propionic acid, hydrocinnamic acid and 1, 2-diphenyldiselane also did not react under these reaction conditions. It is observed that aromatic diselenides with strong electron-withdrawing groups on the phenyl ring provided the products in higher yields than those bearing electron-donating groups. Similarly, the halogen (F, Cl, and Br)-substituted phenylacetic acids reacted readily to afford the desired products in good yields. Finally, it concludes that the substituent on aromatic phenylacetic acids does not affect the product yield **3**; on the contrary, the substituents on aromatic selenides affect the product yields **3** [9].

2.3 Imidoyl selenide

The C=C centered radicals having α-N-atom are called *imidoyl radicals*. The synthetic use of imidoyl radicals is known for the formation of nitrogen heterocycles. Imidoyl selenides are known to act as precursors for imidoyl radicals, and these precursors are suitable for developing synthetic protocols [10–13]. They have been efficiently used in cyclizations, annulations, and cascade reactions, leading to the construction of various nitrogen-containing heterocyclic compounds [14]. They have also been used as key intermediates in the synthesis of carbonyl compounds, amides, and nitriles and as precursors of alkyl radicals in Fe-free reactions [15]. In addition, the imidoyl selenides would require reductive radical procedures, which were potentially more efficient for preventing termination reactions [16]. Radical reagents such as tributyltin hydride are well known for the abstraction of the phenylselenide group in S_N2 reactions and these precursors would yield the required imidoyl radicals.

The synthesis of imidoyl selenide precursors **8** by imidoyl chloride **7** is shown in Figure 2.2 [15]. Generally $PhSe^-$ anion is prepared in methanol in the presence of sodium borohydride, but the imidoyl chloride **7** readily reacts with methanol. Therefore, phenyl selenide anion was prepared in nonreactive solvent like THF in the presence of K-selectride (solution of 1.0 M potassium tri-*sec*-butylborohydride in THF).

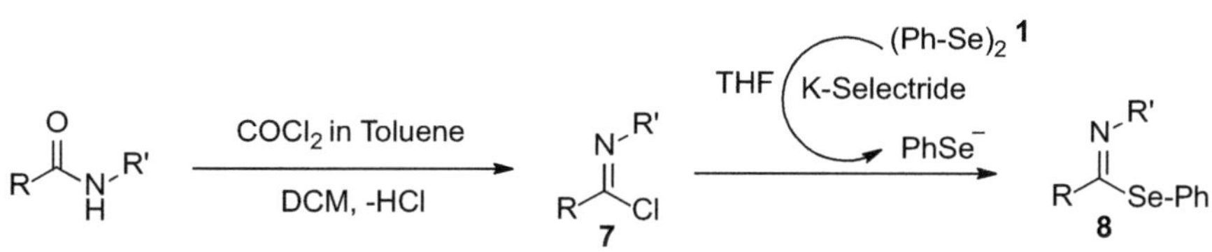

Figure 2.2: Synthesis of imidoyl selenide precursors **8**.

2.3.1 Synthesis of carbapenem framework

Diphenyl diselenide is an encouraging candidate for heteroatom compounds for selective three-component coupling reaction with an electron-poor alkyne and an electron-rich alkene [17, 18]. Diphenyl diselenide showed highly selective sequential addition to ethyl propiolate **9** and isocyanides **10**, providing the corresponding three-component coupling products bearing both vinyl and imidoyl selenide functional group **15** (Figure 2.3) [19].

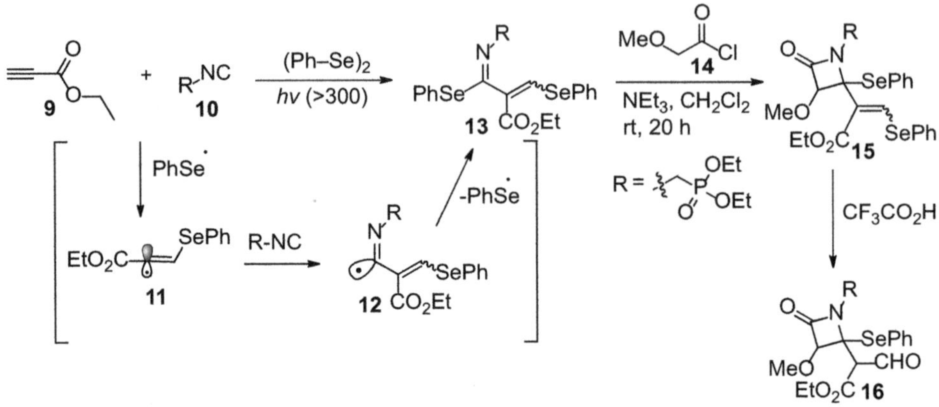

Figure 2.3: Plausible mechanism for construction of carbapenem **16**.

A possible mechanistic pathway involves irradiation with near-UV light. The diphenyl diselenide undergoes homolytic dissociation to generate PhSe˙, which selectively adds to ethyl propiolate **9**, forming β-seleno-substituted vinylic radical **11**. The vinylic radical **11** reacts with isocyanides **10** to produce imidoyl radical intermediate **12**, which is trapped with $(PhSe)_2$ yielding the three-component coupling product **13** with regeneration of PhSe. Further, synthesized imidoyl selenide **13** was treated with α-methoxyacetyl chloride **14** in the presence of triethylamine, successfully providing the corresponding β-lactam **15**. Further, the reaction of the β-lactam **15** with trifluoroacetic acid successfully removed the vinylic selenium group from **15**, giving the corresponding aldehyde **16** as a precursor for the construction of carbapenem framework [19].

2.3.2 Double chalcogenation of isocyanide

A selective method for introducing the seleno group into a variety of isocyanide is shown in Figure 2.4 [20]. Selectivity of the final product **20** is based on the relative reactivities of organic dichalcogenides and chalcogen-centered free radicals. When the reactions of aromatic isocyanide (Ar-NC) **17** with organic disulfides (R′S-SR′) **18** and diselenides (R″Se-SeR″) **1** are conducted upon irradiation with a tungsten lamp through Pyrex (*hv* > 300 nm), simultaneous introduction of both thio- and seleno-groups into the isocyanides takes place to provide the corresponding thioselenation products (R′S-C(=NAr)-SeR″) **20** by thio-imidoyl radical intermediate **19** in good yields with excellent selectivity.

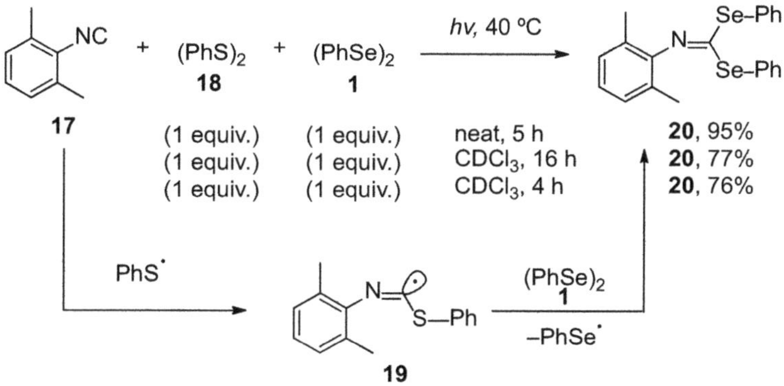

Figure 2.4: Possible pathways for thioselenation of isocyanide **17**.

When the [2 + 2] cyclization reaction was attempted with **20** by using excess amount (10 equiv.) of methoxyacetyl chloride **21**, the desired β-lactum **22** was obtained in 81% yield (Figure 2.5) [20].

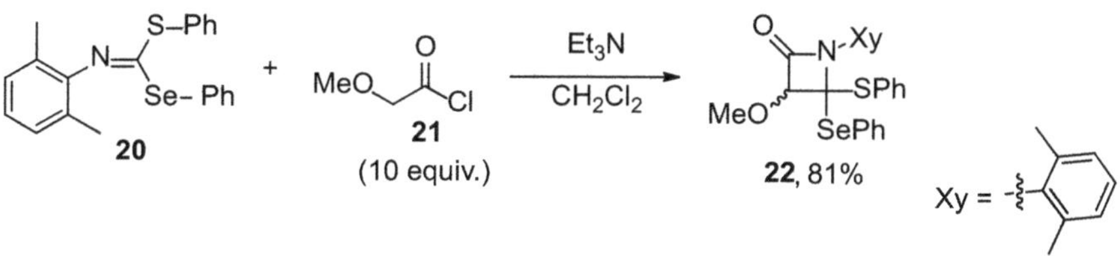

Figure 2.5: Synthesis of β-lactam **22**.

2.4 Selenourea and carbamimidoselenoate

The indole skeletons are known as important building blocks in organic synthesis and are prevalent in a series of natural products and pharmaceuticals [21–23]. The reactions catalyzed by 2,2,6,6-tetramethylpiperidin-1-yl)oxyl (TEMPO) give O_2 as the green oxidant and this protocol provides a practical route for the synthesis of 3-selenylindole derivatives **34**. An electron spin-resonance (ESR) study indicates that the formation of selenoates involves nitrogen-centered radicals and selenium radicals, leading to in situ oxidation of selenoates. The selenium functionalization of indole was carried out by using selenium powder as selenium source (Figure 2.6) [24].

Initially, the reaction of isocyanide **23** with elemental selenium **24** generates isoselenocyanate **25** under the basic conditions. Next, **25** reacts with an amine to produce selenoate **26**, resulting in equilibrium between selenium anion intermediate **26**

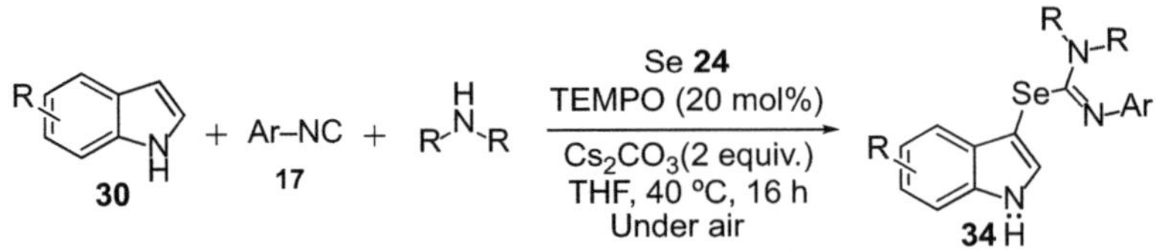

Figure 2.6: Synthesis of 3-substituted Indole carbamimidoselenoates **34**.

and nitrogen anion intermediate **27**. Subsequently, the oxidation of intermediate **27** by TEMPO in oxygen atmosphere generates nitrogen-centered radical **28**, which resonates to the more active selenium intermediate **29**. Meanwhile, deprotonation of indole **30** in the presence of Cs_2CO_3 gives nitrogen anion intermediate **31** and carbon anion intermediate **32**. Next, the cross-coupling of radical **29** with **32** affords nitrogen radical cation **33**. Finally, the single electron transfer of intermediate **33** furnishes the desired product **34** (Figure 2.7) [24].

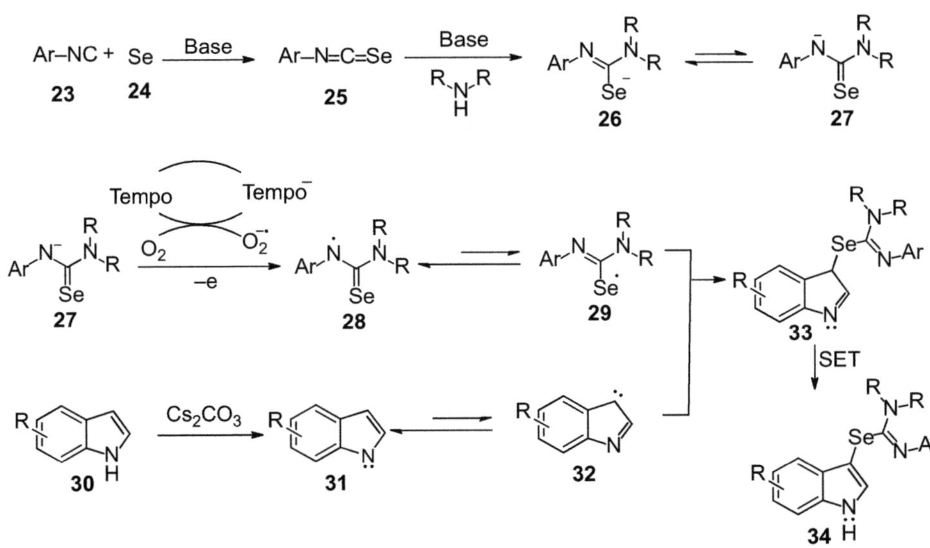

Figure 2.7: Plausible mechanism for synthesis of 3-substituted indole carbamimidoselenoate **34**.

The reactions of aryl isocyanides **35** having electron-withdrawing groups proceeded smoothly to afford the desired products **36** in moderate-to-good yields. Substrates bearing strong electron-withdrawing groups, such as nitro, cyano, and amido, gave better results, affording the corresponding products in 67–82% yields. The copper catalysis also accommodated electron-withdrawing acetyl, ester, and bromo substituents, which showed slightly lower efficiency of the transformations. The use of chloro substituent produced the product in 71% yield. However, aryl isocyanides bearing electron-donating groups, such as 4-methoxybenzene, 2,6-dimethyl benzene **35**, failed to lead the desired products **36** (Figure 2.8) [25].

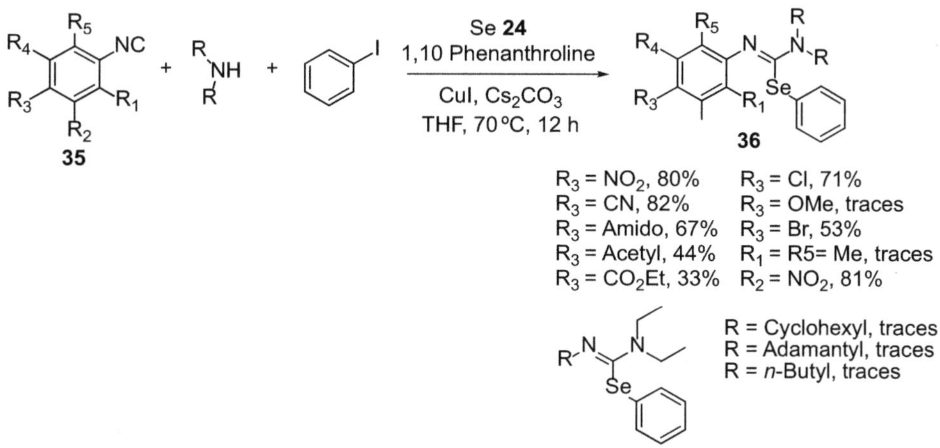

Figure 2.8: Synthesis of substituted carbamimidoselenoate **36**.

2.5 Synthesis of unsymmetrical diaryl selenide

The synthesis of unsymmetrical diaryl chalcogenide, acyl chalcogenide [26], and oxyacyl chalcogenide [27] by selenium radical pathways is highly important in organic synthesis [28–34]. Photoinduced synthesis of unsymmetrical diaryl selenide **38** from triarylbismuthine **37** and diaryl diselenide **1** is depicted in Figure 2.9 [35]. The arylation reactions proceed with triarylbismuthines upon photoirradiation in the absence of transition-metal catalysts. A variety of unsymmetrical diaryl selenides can be conveniently prepared by using this arylation method [35].

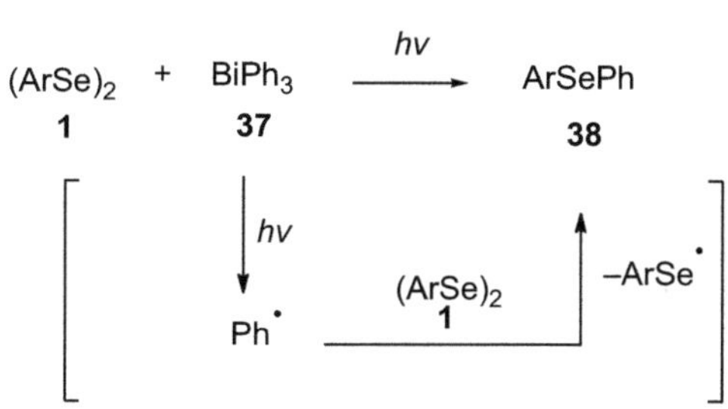

Figure 2.9: Photoinduced radical reaction.

2.5.1 Unsymmetrical diaryl selenide from arylhydrazine

The synthesis of unsymmetrical diaryl selenides from aryl hydrazine and stoichiometric amounts of diaryl diselenides under mild reaction conditions overcomes the disadvantage of the HAS (the use of excess amounts of radical accepters) reaction for practical synthesis [36, 37]. A method for the atom-economical synthesis of unsymmetrical diaryl selenides from aryl hydrazine hydrochlorides and stoichiometric amounts of diselenides was successfully developed by using air as the oxidant. This process avoids the use of transition meal and makes the procedure more practical to access unsymmetrical selenides (Figure 2.10) [38].

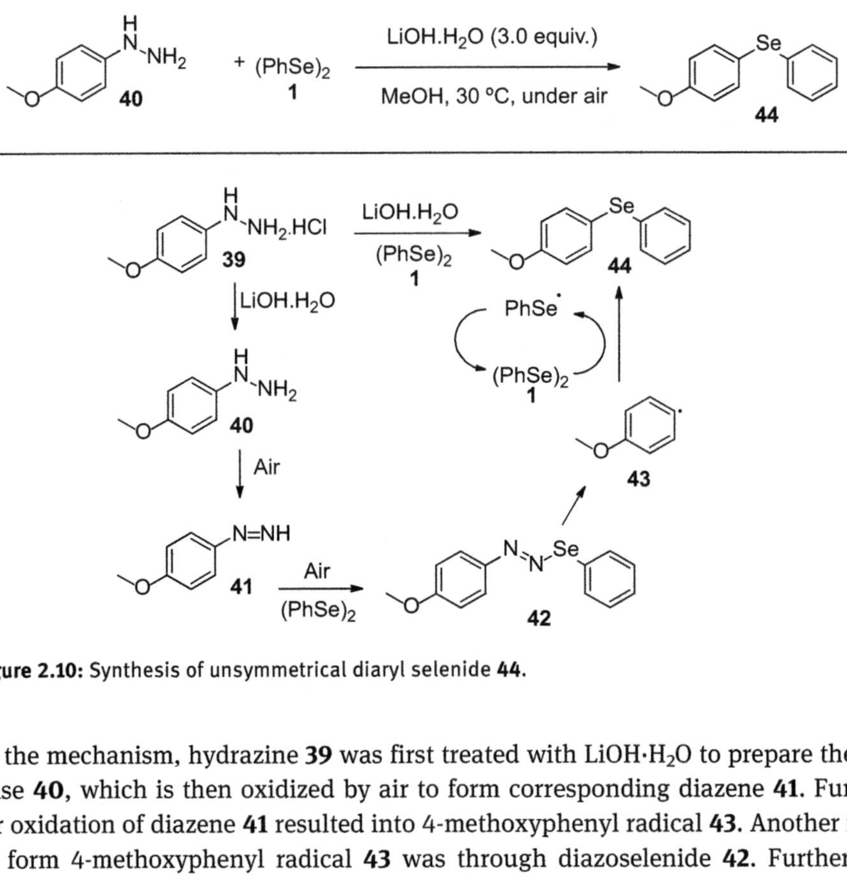

Figure 2.10: Synthesis of unsymmetrical diaryl selenide **44**.

In the mechanism, hydrazine **39** was first treated with $LiOH{\cdot}H_2O$ to prepare the free base **40**, which is then oxidized by air to form corresponding diazene **41**. Further, air oxidation of diazene **41** resulted into 4-methoxyphenyl radical **43**. Another route to form 4-methoxyphenyl radical **43** was through diazoselenide **42**. Further, the radical intermediate **43** was trapped by diphenyl diselenide **1** to afford desired product **44** in good yield [35, 39].

2.5.2 Aryl di-selenation of ferroceneamide

The synthesis of unsymmetrical ferrocene aryl chalcogenide **46** by C–H activation of ferrocene amide 8-aminoquinoline **45** as a directing group is shown in Figure 2.11 [40]. The reaction was carried out in the presence of silver acetate oxidant, aryl dichalcogenides, and copper (II) catalyst at 80°C in DMSO.

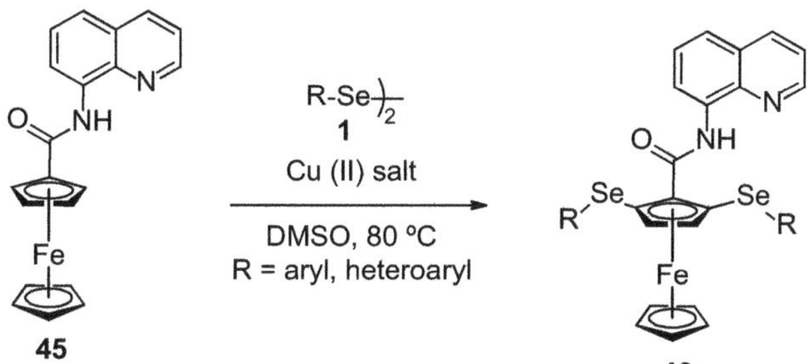

Figure 2.11: Directed ferrocene C–H bond functionalization.

In the mechanistic consideration, the Cu (II) species is important for the catalytic cycle. $Cu(OAc)_2$ would undergo substitution reaction with ferrocene carboxamide **45** to provide copper-ferroceneamidate **I** (Figure 2.12) [40]. Intramolecular interaction of copper with the C–H bond would activate it, followed by proton abstraction by the AcO^- ligand enable cyclometalated intermediate **II.** Ligand substitution by PhSe–SePh followed by homolytic cleavage of PhSe–SePh bond would form copper–chalcogenolate **III** and $PhSe^•$, which subsequently dimerizes to PhSe–SePh. Alternatively, the intermediate **III** could be generated by the disproportionation of **II** into Cu (III) and Cu (I) species. Oxidation of Cu (I) by PhSe–SePh led to **III** and $PhSe^+$. The substitution in Cu (III) by $PhSe^-$ would furnish intermediate **III.** The reductive elimination in copper (III) led to the desired C–SeR bond and release of Cu (I), which upon oxidation by AgOAc would regenerate to active Cu (II) species. Consequently, monochalcogenated product **IV** would furnish dichalcogenated product **46.**

2.6 Preparation of benzoselenophene

Benzothiophene and its derivatives represent a highly important and valuable class of heterocyclic compounds widely present in many medicinally relevant molecules [41]. A variety of synthetic methods for the preparation of substituted benzothiophene and benzoselenophene have been developed [42]. The preparation of benzothiophenes by radical cascade reactions has attracted considerable attention since

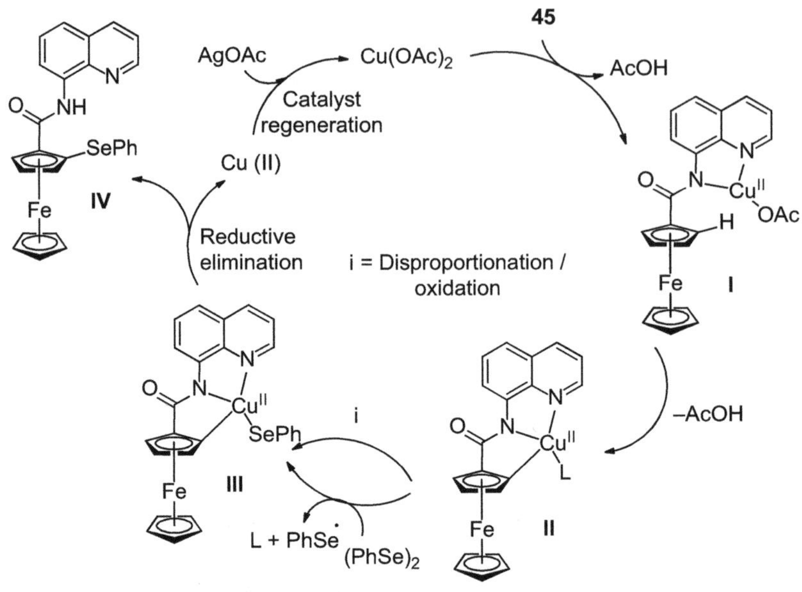

Figure 2.12: Plausible reaction pathway.

it offers a simple and efficient approach to construct the benzothiophene ring [43, 44]. The synthesis of benzoselenophenes was successfully achieved by readily prepared *O*-methylselanyl-arylamines and alkynes under metal free conditions (Figure 2.13). Reactions with *O*-methylselanyl-arylamines carrying electron-donating and electron-withdrawing substituents proceed well, and the corresponding products **54** were obtained in good yields. Moreover, a variety of alkynes-bearing different substituents were tested. The electron-donating (methyl, methoxy) and also electron-withdrawing substituents (halides) were well tolerated, affording the corresponding products **54** in good yields. With heteroaryl, alkyl, and TMS substituents on the alkynes, moderate yields were obtained [45].

A plausible mechanism is proposed in Figure 2.13 [43, 46, 47]. First, arylamine **47** was treated with *t*-BuONO to the corresponding nitrosamine **48**, which undergoes self-condensation to generate diazo anhydride **49**. N–O homolysis of **49** provides aryl radical **50** along with azoxy radical **51** and nitrogen. The addition of **51** to alkyne **52** leads to the vinyl radical **53**, which reacts by intramolecular homolytic substitution at the sulfur atom or selenium atom to form the final product **54** along with R radical [48]. The R radical can further react by H-abstraction from the solvent. Notably, these transformations occur efficiently without the help of any transition metal or additive. The applications of the method are demonstrated by the synthesis of the key intermediates of the drug raloxifene and AT1 receptor antagonist [45].

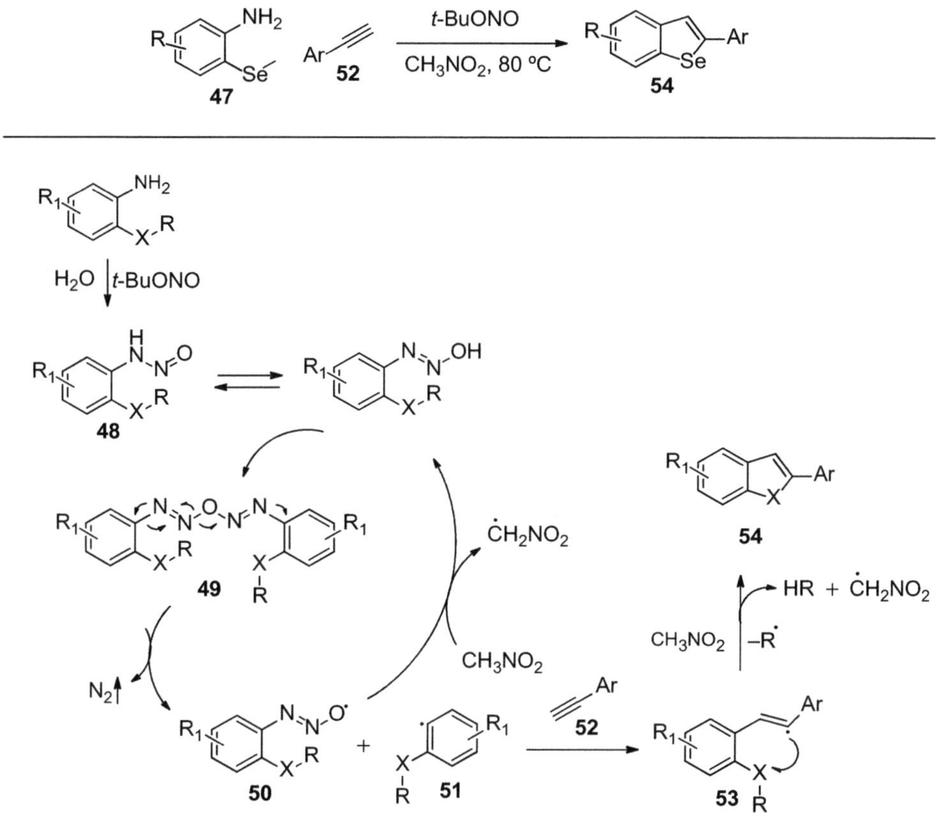

Figure 2.13: Synthesis of 2-substituted benzoselenophene **54** by radical cascade reaction.

2.7 Selenylation of imidazo[1,2-*a*]pyridine

The development of convenient and novel approaches for imidazo [1,2-*a*]pyridine synthesis and functionalization has received much attention in organic chemistry [49]. The controllable radical chemistry to construct diverse organic molecules has been of growing interest [50, 51].

The copper-catalyzed convenient and efficient approach for the construction of nitrogen heterocycle-fused imidazo[1,2-*a*]pyridine and benzo[*b*]selenophenes has been developed by direct selenation of readily available 2-(2-bromophenyl)imidazo [1,2-*a*]pyridines by regioselective cleavage of C(sp^2)-Br and C(sp^2)-H bonds using readily available selenium powder as the selenating reagents under ligand- and base-free conditions in air (Figure 2.14) [52]. To study this selenation process, a series of mechanistic studies by ESR spectra were performed [51]. The plausible mechanism, showed in Figure 2.14 [52], indicates that the Cu (I) is oxidized to Cu (II) by oxygen in air. The

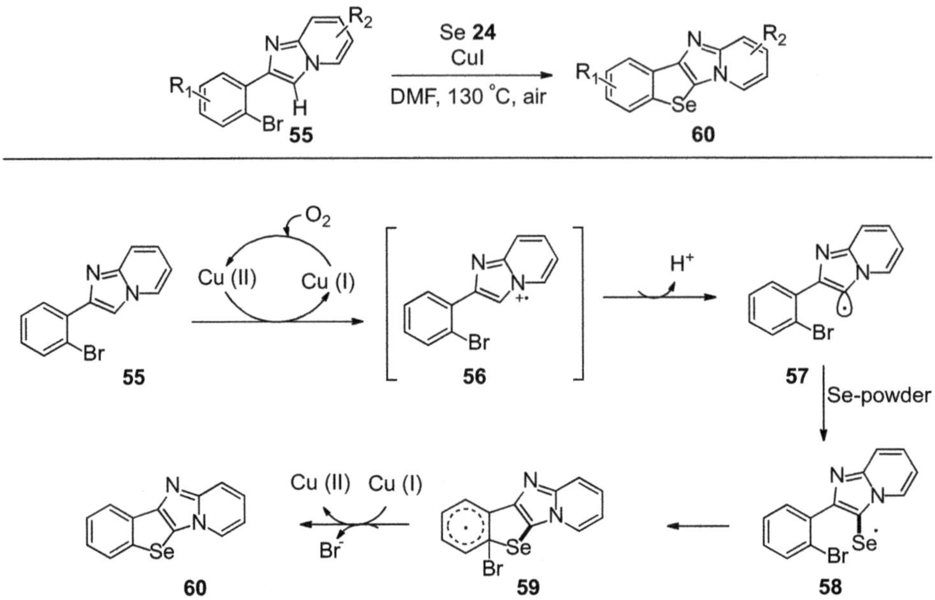

Figure 2.14: Mechanistic pathway for selenylation of 2-(2-bromophenyl) imidazo [1,2-*a*] pyridine **60**.

single electron transfer takes place between **55** and Cu (II) that generates radical cation **56**. Further, the intermediate **56** loses a proton leading to vinyl radical intermediate **57**, which reacts with elemental Se-power to give a selenium-free radical **58**. Subsequently, the radical **58** undergoes an intramolecular cyclization to generate radical intermediate **59**. Finally, Cu (I)-mediated bromine abstraction of radical intermediate **59** takes place, releasing the product **60** with Br^-, and Cu (II).

2.8 Selenosulfonation

2.8.1 Synthesis of (*E*)-β-selenovinyl sulfone

Selenide group transfer from sulfonyl selenide is useful due to the weak S–Se bond and is widely used for organic synthesis [53–55]. A copper-catalyzed high regio- and stereo-specific selenosulfonation of alkynes with arylsulfonohydrazides **61** and diphenyl diselenide are shown in Figure 2.15 [56]. The three-component reaction proceeded under mild reaction conditions, providing a wide range of (*E*)-β-selenovinyl sulfones **64** in good-to-excellent yields.

To study the reaction mechanism, some control experiments were carried out [56]. This result showed that the selenosulfonation might not involve a selenosulfonate intermediate. The results suggested that the transformation might proceed via a radical pathway. Initially, the sulfonyl radical **62** was generated in the presence of copper salt and $K_2S_2O_8$ by single electron transfer and deprotonation process, along with the release of N_2 [57, 58]. Furthermore, the sulfonyl radical **62** was added to alkynes, forming a relatively stable β-sulfonyl vinyl radical **63**. Subsequently, the phenyl selenol group transferred from diphenyl diselenide to **63** affords thermodynamically stable (*E*)-β-selenovinyl sulfone **64**. Alternatively, the phenyl selenol group transfer might be much more rapid than the inversion of **63**, thus resulting in a single isomer **64** (Figure 2.15) [56, 59].

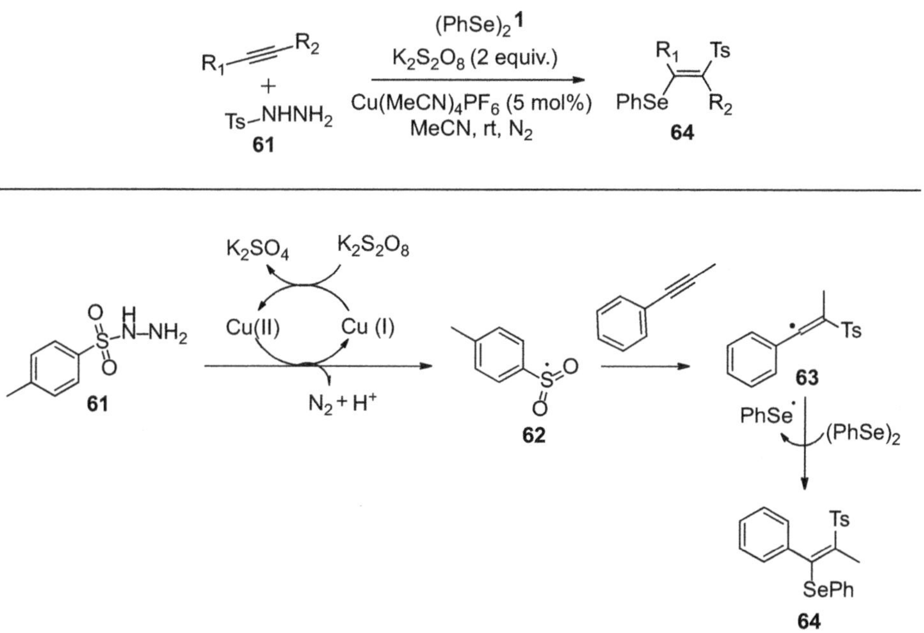

Figure 2.15: Mechanism for selenosulfonation.

The four component selenosulfonation of alkyne is shown in Figure 2.16 [60]. Various aryl diazonium tetrafluoroborates with different functional groups were used including electron-donating (OMe), electron-neutral (H), and electron-withdrawing (F and Cl) delivering the corresponding products in high yield.

The radical mechanism for this multicomponent reaction is proposed in Figure 2.16 [60]. The combination of a aryldiazonium cation with DABSO (1,4-diazabicyclo[2.2.2]octane bis-sulfur dioxide adduct) [61, 62] generates the complex **A**, which provides sulfur dioxide, nitrogen, an aryl radical, and tertiary amine radical cation **B**. Then, the addition of the aryl radical to sulfur dioxide produces arylsulfonyl radical **C**, which subsequently, regiospecifically adds to alkyne **1** to form a relatively stable β-sulfonyl vinyl radical **D**. Finally, with the help of $K_2S_2O_8$,

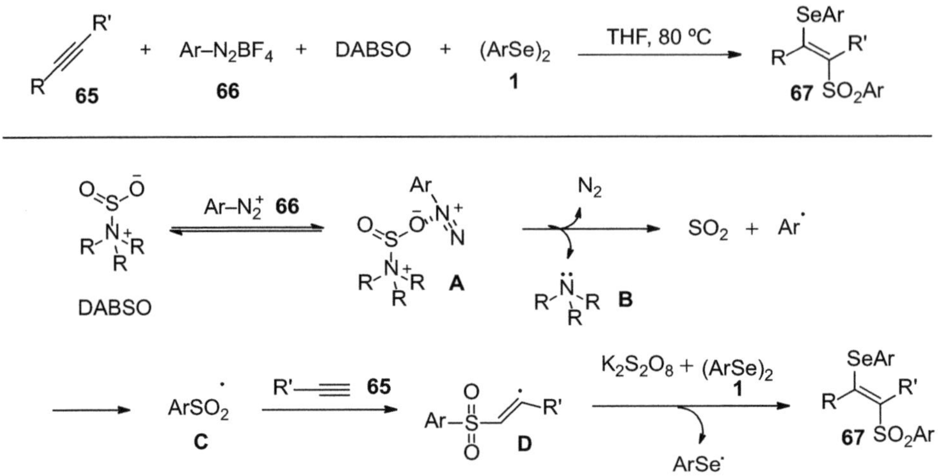

Figure 2.16: Plausible catalytic cycle for (*E*)-β-selenovinyl sulfone **67**.

a phenyl selenol radical is generated and couples with radical **C** to afford the thermodynamically stable (*E*)-β-selenovinyl sulfone **67**.

2.8.2 Selenosulfonation of 1,7-enyne

The three-component selenosulfonation of 1,7-enynes **68** with sulfinic acids **69** and diphenyl diselenides for the formation of multifunctional 3,4-dihydroquinolin-2 (1H)-ones **70** was developed in a batch flow process. The reaction carried out at the room temperature provides a highly efficient diversified selenosulfones in moderate-to-excellent yields with a broad scope of substrates (Figure 2.17) [63]. The reaction proceeds via radical-induced 6-*exo-dig* cyclization.

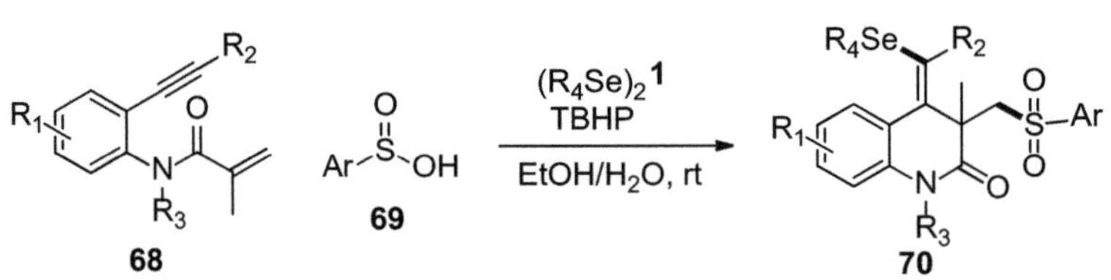

Figure 2.17: Selenosulfonation of 1,7-enyne **70**.

The mechanism involved the formation of sulfonyl radical **71** through oxidation of the arylsulfinic acid in the presence of TBHP via a single electron transfer process. Similar to this procedure, diphenyl diselenide generates a phenylselenyl radical

[64–66]. Subsequently, the sulfonyl radical **71** attacks the terminal olefin of 1,7-enyne **68** to give radical **72**, followed by a 6-*exo-dig* cyclization to form vinyl radical intermediate **73**. In the presence of phenylselenyl radicals, intermediate **73** is transformed to the final 3,4-dihydroquinolin-2(1H)-ones **72** by radical coupling (Figure 2.18) [63].

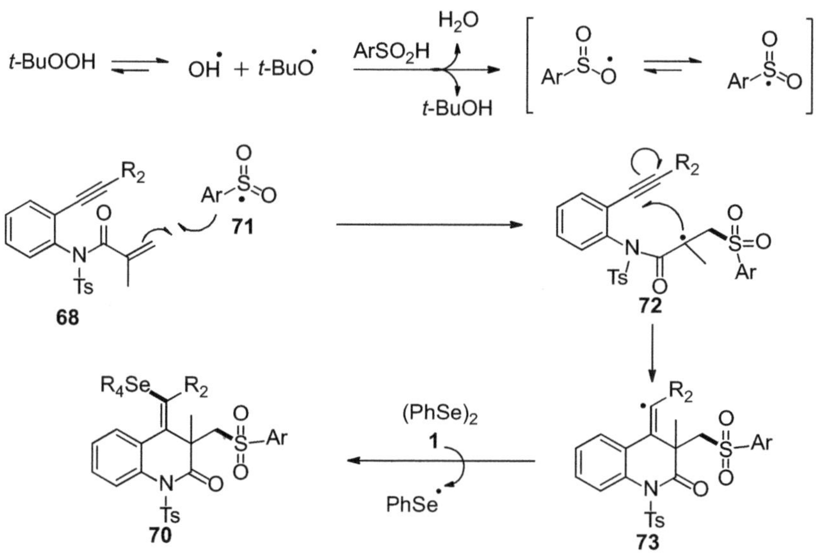

Figure 2.18: Proposed mechanism.

2.8.3 Synthesis of selenocarbamate

The synthesis of secondary selenocarbamate **75** through metal-free multicomponent reactions of isocyanide **17**, selenosulfonate **74**, and water is shown in Figure 2.19 [67].

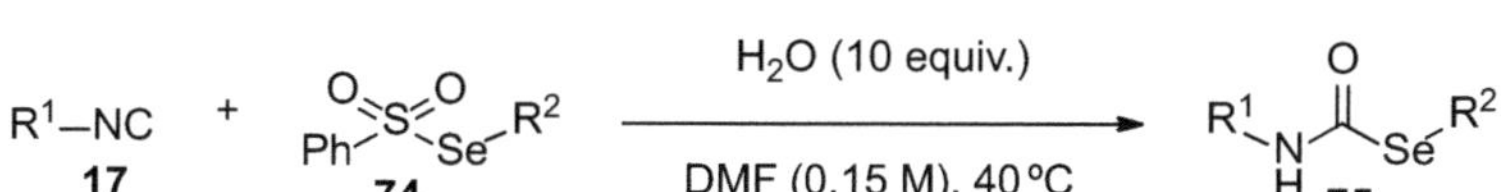

Figure 2.19: Synthesis of selenocarbamate.

The homolysis of selenosulfonate **74a** will occur in the mixture to deliver benzenesulfonyl radical **A** and selenium radical **B**. Next, **A** reacts with isocyanide **1a** to give the intermediate **D**, which can be trapped by TEMPO. Subsequently, **D** reacts with **2a** to give intermediate **E** and regenerates **A** (major path). Alternatively, the diselenide **C** reacts with **D** to produce **E** and regenerate **B** (minor path). Finally the hydrolysis of intermediate **E** provided product **75** Figure 2.20 [67].

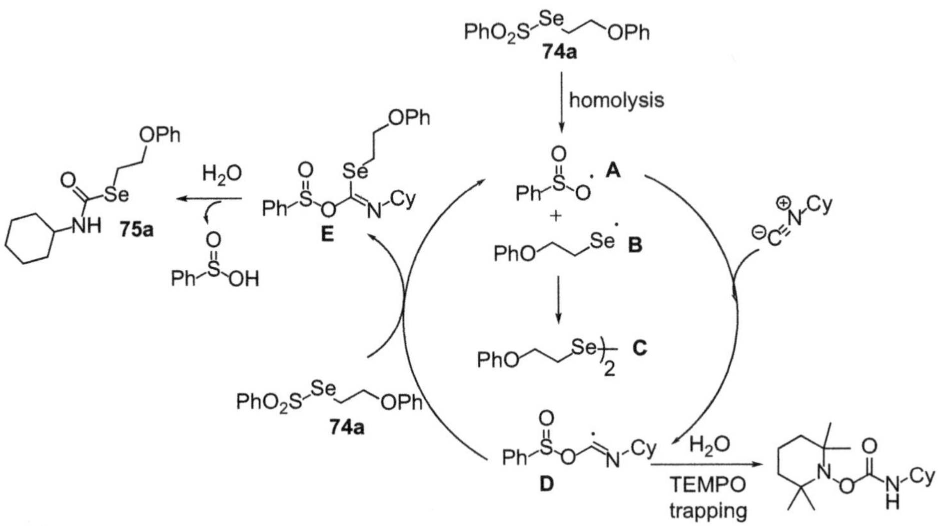

Figure 2.20: Plausible reaction mechanism.

2.8.4 Synthesis of seleno-sulfonated 1-indenone

The TBHP-catalyzed direct selenosulfonylation of β-alkynyl propenone **76** by combining sulfinic acid **69** and diphenyl diselenide **1** is shown in Figure 2.21 [68]. The protocol features a broad substrate scope, high functional group tolerance, and mild reaction conditions.

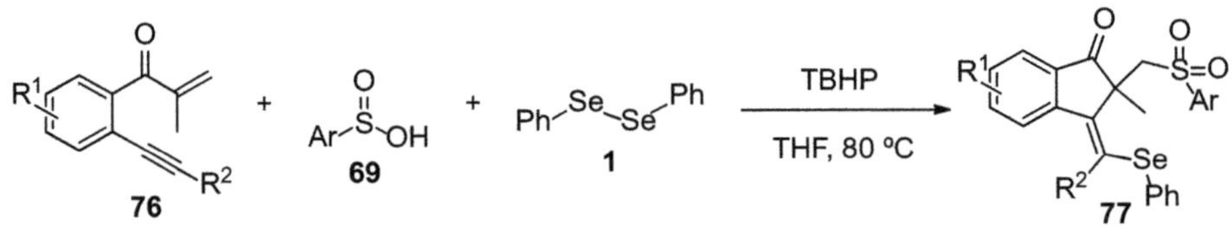

Figure 2.21: Synthesis of sulfonated 1-indenone **77**.

The radical addition of aryl sulfonyl radical **71** into β-alkynyl propenone **76** gives intermediate **78**, followed by *5-exo-dig cyclization* to vinyl radicals **79**, which are trapped by diphenyl diselenide radical to afford the product **77** (Figure 2.22) [69].

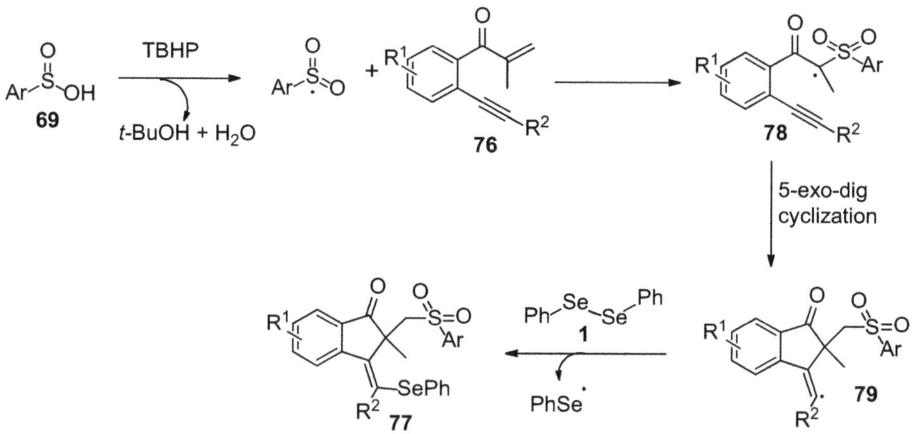

Figure 2.22: Plausible reaction pathway.

2.8.5 Radical cyclization of 1,6-diene

The polystyrene-supported selenosulfone **81** was prepared from 1% cross-linked polystyrene resin **80** and further applied to free radical cyclization reactions of 1,6-diene **82**, followed by the subsequent "traceless" resin release by oxidation–elimination reaction, which gives a convenient method for the synthesis of methylenecyclopentane **83** (Figure 2.23) [70].

The alcohol product can be formed in the polymer-supported oxidation–elimination reaction, which was not observed in solution phase synthesis under the same conditions. By applying a different oxidation–elimination procedure, methylenecyclopentanes can be formed as the only products, while cyclopentanyl methyl alcohols were obtained as the only product employing an oxidation–hydroboration–oxidation sequence [70].

2.9 Synthesis of 2-aryl-1,3-benzoselenazole

The reaction of 2-iodoaniline **86** with *n*-BuLi, subsequent trapping of the lithium anion with elemental selenium **24** and potassium ferrocyanide oxidation, afforded the corresponding diselenide **87** (Figure 2.24) [71].

The reaction between bis(2-aminophenyl) diselenide **87** and substituted benzaldehyde **88** was carried out by using $Na_2S_2O_5$ as the reducing agent and DMSO as the solvent. The reaction was carried out at 120 °C for 24 h. A possible reaction mechanism for the synthesis of 2-aryl-1,3-benzoselenazole **93** is shown in Figure 2.25 [71]. The amino group of bis(2-aminophenyl) diselenide **87** initially reacts with the aryl

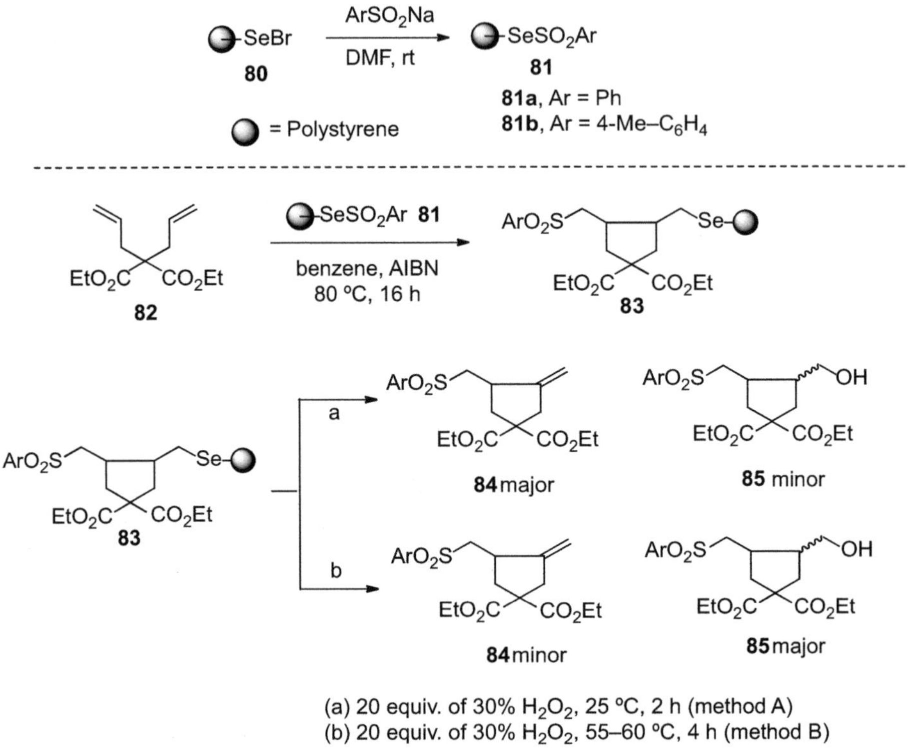

Figure 2.23: Radical addition of polystyrene-supported selenosulfone **81** with 1,6-diene **82**.

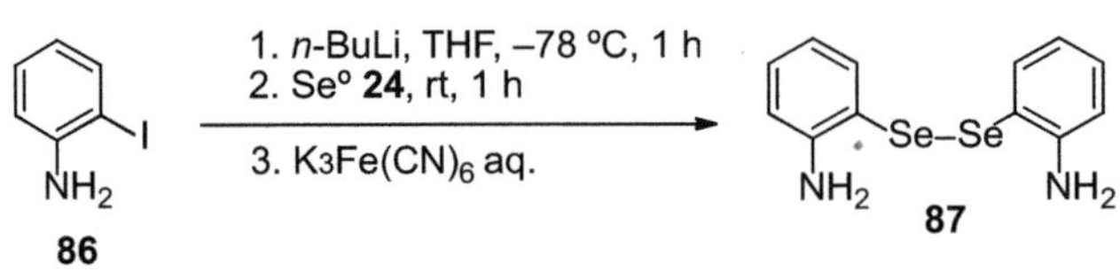

Figure 2.24: Synthesis of bis(2-aminophenyl) diselenide **87**.

aldehyde **88** to form the imine diselenide compound **89**. Next, the Se–Se bond was cleaved by the radical anion SO_2 generated form $S_2O_5^{2-}$ by heating [72] to afford the intermediates **90** and **91**. The intermediate **90** can be reoxidized to starting imine diselenide **89**. Further, the radical **91** undergoes to the intramolecular cyclocondensation leading to the aminyl radical **92**. Finally, the oxidation of intermediate **92** provides the target product **93**.

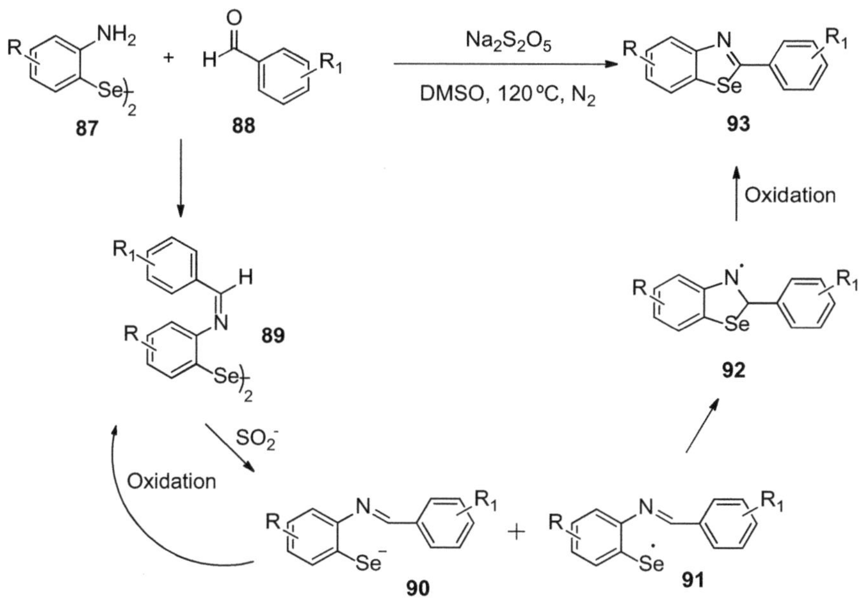

Figure 2.25: Synthesis of 2-aryl-1,3-benzoselenazole **93** prompted by $Na_2S_2O_5$.

2.10 Aerobic radical-cascade cycloaddition

The synthesis of 1,2,4-selenadiazole-5-amine derivatives **101** was successfully achieved by multicomponent reaction of isocyanide **17**, selenium powder **24**, and imidamide compounds **94** under metal-free conditions. The reaction proceeded under mild conditions with O_2 as green oxidant and no extra catalysts or oxidants were required (Figure 2.26) [73].

In the plausible mechanism, the reaction of isocyanide **17** with selenium powder **24** gives isoselenocyanate **25** under the action of DIPEA (Figure 2.26, eq. **5**) [73]. Next, **25** reacts with imidamide **94** to produce the intermediate **95**. Subsequently, **95** is oxidized by O_2 or superoxide anion radical to give selenium radical intermediate **96** (Figure 2.26, eq. **6**) [73]. Further, **96** may undergo two possible pathways to generate the desired product **101**. In the first pathway, diselenide intermediate **97** is generated by homocoupling of **96** [74], followed by deprotonation under the action of DIPEA or peroxide anion, to afford product **101** (Figure 2.26, *route a*) [73]. Intramolecular cyclization of **96** is involved in the second pathway for the formation of intermediate **98**, and **99** undergoes deprotonation under the influence of DIPEA or peroxide anion to yield **101** (Figure 2.26, *route b*) [73]. The *TEMPO* can trap intermediate **98** to generate **100**, which is unstable under the reaction conditions. As a result, a molecule of *TEMPOH* will be eliminated by DIPEA or peroxide anion to produce **101**. The peroxide anion and proton, which is formed in the reaction,

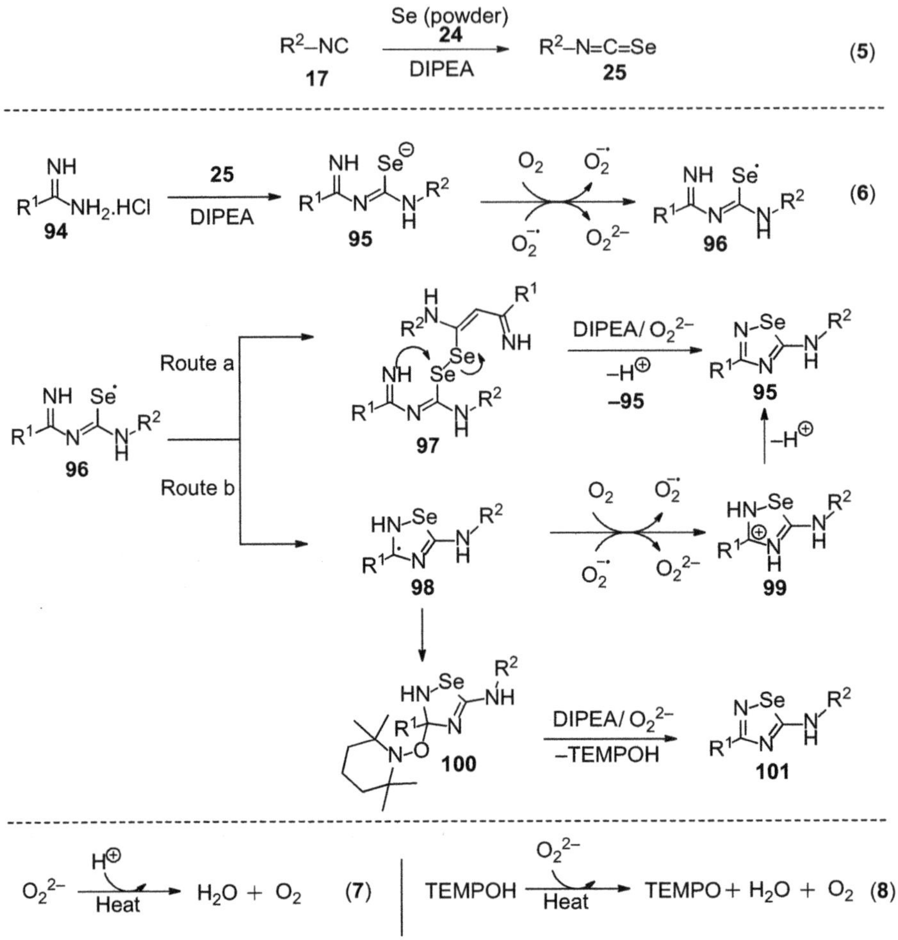

Figure 2.26: The plausible reaction mechanism.

will combine and thermally decompose to H_2O and O_2 (Figure 2.26, eq. **7**) [73]. In addition, *TEMPO* is regenerated from *TEMPOH* under the action of peroxide anion (Figure 2.26, eq. **8**) [73, 75].

2.11 Conclusions

This chapter offers an updated overview on the synthesis of organoselenium scaffolds through selenium radical pathway. Here we have reported the radical pathway for the synthesis of carbapenem and double chalcogenation of isocyanide. In addition, this chapter involved the synthesis of selenide ether, selenourea, and carbamimidoselenoate. The cascade reactions including 2-aryl-1,3-benzoselenazole and selenosulfonation of 1,7-enyne, unsymmetrical diaryl selenide, synthesis of sulfonated 1-indenone, selenocarbamate, and cyclization of 1,6-diene were reported by selenium radical pathway.

References

[1] Thomas, W. Organoselenium chemistry; synthesis and reactions, Wiley-VCH Verlag and Co, KGaA, ed., ISBN: 978-3-527-32944-1, 2011.
[2] Back, TG., and Press, DJ. Advances in free radical reactions of organoselenium and organotellurium compounds. In: Rappoport Z., ed. *PATAI's chemistry of functional groups*. John Wiley & Sons, Ltd., 2011. Doi: 10.1002/9780470682531.pat0581.
[3] Thomas, T., Bonifac, M., Naumov, S., and Asmus, KD. Time-resolved study on the reactions of organic selenides with hydroxyl and oxide radicals, hydrated electrons, and H-atoms in aqueous solution, and DFT calculations of transients in comparison with sulfur analogues. Phys Chem Chem Phys 2010, 12, 6750–6758.
[4] Belyaev, AV., Sozin, AY., Aleksandrov, YA., and Churbanov, MF. Homogeneous oxidation of hydrogen selenide with ozone in an argon-oxygen medium. Russian J Applied Chem 2012, 85, 1919–1922.
[5] Mardyukov, A., Tsegaw, YA., Sander, W., and Schreiner, PR. The phenylselenyl radical and its reaction with molecular oxygen. Phys Chem Chem Phys 2017, 19, 27384–27388.
[6] Dawes, GJS., Scott, EL., Nktre, JL., Sanders, JPM., and Bitter, JH. Deoxygenation of biobased molecules by decarboxylation and decarbonylation – a review on the role of heterogeneous, homogeneous and bio-catalysis. Green Chem 2015, 17, 3231–3250.
[7] Guntreddi, T., Vanjari, R., and Singh, KN. Decarboxylative thioamidation of arylacetic and cinnamic acids: a new approach to thioamides. Org Lett 2014, 16, 3624–3627.
[8] Chen, XL., Chen, TQ., Ji, FY., Zhou, YB., and Yin, SF. Iron-catalyzed aerobic oxidative functionalization of sp^3 C-H bonds: a versatile strategy for the construction of *N*-heterocycles. Catal Sci Technol 2015, 5, 2197–02.
[9] Cui, FH., Chen, J., Su, SX., Xu, YL., Wang, HS., and Pana, YM. Regioselective synthesis of selenide ethers through a decarboxylative coupling reaction. Adv Synth Catal 2017, 359, 3950–3961.
[10] Fujiwara, SI., Matsuda, T., Maeda, H., Shinike, T., Kambe, N., and Sonoda, N. Imidoyl radicals as synthons of unstable acyl radicals. J Org Chem 2001, 66, 2183–2185.
[11] Bachi, MD., and Denenmark, D. Cyclizations of ene radicals. Imidoyl radicals as intermediates in the synthesis of heterocyclic compounds. J Am Chem Soc 1989, 111, 1886–1888.
[12] Bachi, MD., Balanov, A., Bar-Ner, N., Bosch, E., Denenmark, D., and Mizhiritskii, M. Synthesis of non-aromatic heterocyclic compounds through free-radical reactions. Pure Appl Chem 1993, 65, 595–01.
[13] Lei, J., Huang, J., and Zhu, Q. Recent progress in imidoyl radical-involved reactions. Org Biomol Chem 2016, 14, 2593–2602.
[14] Minozzi, M., Nanni, D., and Spagnolo, P. Imidoyl radicals in organic synthesis. Current Org Chem 2007, 11, 1366–1384.

[15] Bowman, WR., Fletcher, AJ., Pedersen, JM., Lovell, PJ., Elsegood, MRJ., Hern, E., Opez, AL., McKee, V., and Potts, GBS. Amides as precursors of imidoyl radicals in cyclisation reactions. Tetrahedron 2007, 63, 191–193.
[16] Review Zard, SZ. Radicals in organic synthesis, Renaud P. Sibi MP, Eds, Wiley-VCH, Weinheim, 2001, Vol. 1, 90–98.
[17] Ogawa, A., Doi, M., Ogawa, I., and Hirao, T. Highly selective three-component coupling of ethyl propiolate, alkenes, and diphenyl diselenide under visible-light irradiation. Angew Chem Int Ed 1999, 38, 2027–2029.
[18] Ogawa, A., Ogawa, I., and Sonoda, N. A novel three-component coupling of alkynes, vinylcyclopropanes, and diphenyl diselenide under visible-light irradiation. J Org Chem 2000, 65, 7682–7685.
[19] Ogawa, A., Doi, M., Tsuchii, K., and Hirao, T. Selective sequential addition of diphenyl diselenide to ethyl propiolate and isocyanides upon irradiation with near-UV light. Tetrahedron Lett 2001, 42, 2317–2319.
[20] Tsuchii, K., Tsuboi, Y., Kawaguchi, SI., Takahashi, J., Sonoda, N., Nomoto, A., and Ogawa, A. Highly selective double chalcogenation of isocyanides with disulfide-diselenide mixed systems. J Org Chem 2007, 72, 415–423.
[21] Humphrey, GR., and Kuethe, JT. Practical methodologies for the synthesis of indoles. Chem Rev 2006, 106, 2875–2911.
[22] Hassam, M., Basson, AE., Liotta, DC., Morris, L., Van Otterlo, WAL., and Pelly, SC. Novel cyclopropyl-indole derivatives as HIV non-nucleoside reverse transcriptase inhibitors. ACS Med Chem Lett 2012, 3, 470–475.
[23] Tanaka, H., Noguchi, H., and Abe, I. Enzymatic formation of indole-containing unnatural cyclic polyprenoids by bacterial squalene: Hopene cyclase. Org Lett 2005, 7, 5873–5876.
[24] Liu, H., Fang, Y., Wang, SY., and Ji, SJ. TEMPO-catalyzed aerobic oxidative selenium insertion reaction: Synthesis of 3-selenylindole derivatives by multicomponent reaction of isocyanides, selenium powder, amines, and indoles under transition-metal-free conditions. Org Lett 2018, 20, 930–933.
[25] Liu, H., Fang, Y., Yin, L., Wang, SY., and Ji, SJ. Copper (I)-catalyzed ligand-promoted multicomponent reactions of isocyanides, selenium, amines, and iodoarenes: access to highly functionalized carbamimidoselenoates. J Org Chem 2017, 82, 10866–10874.
[26] Horvat, SM., and Schiesser, CH. An *ab initio* and DFT study of homolytic substitution reactions of acyl radicals at sulfur, selenium, and tellurium. New J Chem 2010, 34, 1692–1699.
[27] Horvat, SM., and Schiesser, CH. An *ab initio* and DFT study of homolytic substitution reactions by oxyacyl radicals at sulfur, selenium, and tellurium. Tetrahedron 2012, 68, 10482–10488.
[28] Perkins, MJ., and Turner, ES. S_N2 reactions of diphenyl diselenide; preparation and reactions of bridgehead selenides. J Chem Soc Chem Commun 1981, 3, 139–140.
[29] Russell, GA., and Tashtoush, H. Free-radical chain-substitution reactions of alkylmercury halides. J Am Chem Soc 1983, 105, 1398–1399.
[30] Kawaguchi, SI., and Ogawa, A. The development of highly selective addition reactions of tetraphenyldiphosphine to carbon-carbon unsaturated bonds. J Synth Org Chem Jpn 2010, 68, 705–717.
[31] Kawaguchi, SI., Ohe, T., Shirai, T., Nomoto, A., Sonoda, M., and Ogawa, A. Highly selective phosphinotelluration of terminal alkynes using a $(Ph_2P)_2$-$(PhTe)_2$ mixed system upon visible light irradiation: straightforward access to 1-phosphino-2-telluro-alkenes. Organometallics 2010, 29, 312–316.

[32] Arnauld, T., Barton, DHR., and Normant, JF. New and facile synthesis of symmetrical and unsymmetrical diaryl chalcogenides using trivalent organobismuth derivatives. J Org Chem 1999, 64, 3722–3725.
[33] Buriak, JM., and Sikder, Md DH. From molecules to surfaces: radical-based mechanisms of Si-S and Si-Se bond formation on silicon. J Am Chem Soc 2015, 137, 9730–9738.
[34] Ouchi, A., Liu, S., Li, Z., Kumar, S. A., Suzuki, T., Hyugano, T., and Kitahara, H. Factors controlling photochemical cleavage of the energetically unfavorable Ph-Se bond of alkyl phenyl selenides. J Org Chem 2007, 72, 8700–8706.
[35] Kobiki, Y., Kawaguchi, SI., Ohe, T., and Ogawa, A. Photoinduced synthesis of unsymmetrical diaryl selenides from triarylbismuthines and diaryl diselenides. Beilstein J Org Chem 2013, 9, 1141–1147.
[36] Wetzel, A., Ehrhardt, V., and Heinrich, MR. Synthesis of amino- and hydroxybiphenyls by radical chain reaction of arenediazonium salts. Angew Chem Int Ed 2008, 47, 9130–9133.
[37] Jasch, H., Scheumann, J., and Heinrich, MR. Regioselective radical arylation of anilines with arylhydrazines. J Org Chem 2012, 77, 10699–10706.
[38] Taniguchi, T., Akinori, M., Motonori, T., Takumi, M., Akihiro, N., and Akiya, O. Atom-economical synthesis of unsymmetrical diaryl selenides from arylhydrazines and diaryl diselenides. Eur J Org Chem 2017, 2017, 4928–4934.
[39] Ogawa, A., Ogawa, I., Obayashi, R., Umezu, K., Doi, M., and Hirao, T. Highly selective thioselenation of vinylcyclopropanes with a $(PhS)_2$-$(PhSe)_2$ binary system and its application to thiotelluration. J Org Chem 1999, 64, 86–92.
[40] Sattar, M., Shareef, M., Patidar, K., and Kumar, S. Copper-catalyzed 8-aminoquinoline assisted aryl chalcogenation of ferroceneamide with aryl disulfides, diselenides, and ditellurides. J Org Chem 2018, 83, 8241–8249.
[41] Qin, Z., Kastrati, I., Chandrasena, REP., Liu, H., Yao, P., Petukhov, PA., Bolton, JL., and Thatcher, GRJ. Benzothiophene selective estrogen receptor modulators with modulated oxidative activity and receptor affinity. J Med Chem 2007, 50, 2682–2692.
[42] Godoi, B., Schumacher, RF., and Zeni, G. Synthesis of heterocycles via electrophilic cyclization of alkynes containing heteroatom. Chem Rev 2011, 111, 2937–2980.
[43] Hari, DP., Hering, T., and Kccnig, B. Visible light photocatalytic synthesis of benzothiophenes. Org Lett 2012, 14, 5334–5337.
[44] Liu, K., Jia, F., Xi, H., Li, Y., Zheng, X., Guo, Q., Shen, B., and Li, Z. Direct benzothiophene formation *via* oxygen-triggered intermolecular cyclization of thiophenols and alkynes assisted by manganese/PhCOOH. Org Lett 2013, 15, 2026–2029.
[45] Zang, H., Sun, JG., Dong, X., Li, P., and Zhang, B. Preparation of benzothiophenes and benzoselenophenes from arylamines and alkynes *via* radical cascade reactions. Adv Synth Catal 2016, 358, 1746–1752.
[46] Leardini, R., Pedulli, GF., Tundo, A., and Zanardi, G. Reaction pathways for the cyclization of ortho-thioalkyl and ortho-thioaryl substituted phenyl radicals with alkynes. Reaction of o-methylthioarenediazonium tetrafluoroborates with alkynes to give 2-substituted benzo[*b*] thiophenes. J Chem Soc Chem Commun 1985, 20, 1390–1391.
[47] Cantillo, D., Mateos, C., Rincon, JA., Frutos de, O., and Kappe, CO. Light-induced C-H arylation of (hetero)arenes by *in situ* generated diazo anhydrides. Chem Eur J 2015, 21, 12894–12896.
[48] Kampmeier, JA., and Evans, TR. Radical reactions of organic sulfides. II.[1] Behavior of the 2-(2′-Methylthio)biphenylyl radical[2]. J Am Chem Soc 1966, 88, 4096–4097.
[49] Wang, H., Wang, Y., Liang, D., Liu, L., Zhang, J., and Zhu, Q. Copper-catalyzed intramolecular dehydrogenative aminooxygenation: direct access to formyl-substituted aromatic *N*-heterocycles. Angew Chem Int Ed 2011, 50, 5678–1.

[50] Zhang, C., Tang, C., and Jiao, N. Recent advances in copper-catalyzed dehydrogenative functionalization *via* a single electron transfer (SET) process. Chem Soc Rev 2012, 41, 3464–3484.

[51] Narayanam, JMR., and Stephenson, CR. Visible light photoredox catalysis: applications in organic synthesis. Chem Soc Rev 2011, 40, 102–113.

[52] Sun, P., Jiang, M., Wei, W., Min, Y., Zhang, W., Li, W., Yang, D., and Wang, H. Copper-catalyzed selenylation of imidazo[1,2-*a*]pyridines with selenium powder *via* a radical pathway. J Org Chem 2017, 82, 2906–2913.

[53] Huang, S., Chen, Zhili., Mao, H., Hu, F., Li, D., Tan, Y., Yang, F., and Qin, W. Metal-free difunctionalization of alkynes to access tetrasubstituted olefins through spontaneous selenosulfonylation of vinylidene ortho-quinone methide (VQM). Org Biomol Chem 2019, 17, 1121–1129.

[54] Luo, S., Zhang, N., Wang, Z., and Yan, H. Enantioselective addition of selenosulfonates to α, β-unsaturated ketones. Org Biomol Chem 2018, 16, 2893–01.

[55] Kawaguchi, SI., Shirai, T., Ohe, T., Nomoto, A., Sonoda, M., and Ogawa, A. Highly regioselective simultaneous introduction of phosphino and seleno groups into unsaturated bonds by the novel combination of $(Ph_2P)_2$ and $(PhSe)_2$ upon photoirradiation. J Org Chem 2009, 74, 1751–1754.

[56] Yang, L., Guangfan, Z., Qian, Z., Yan, L., and Qian, Z. Copper-catalyzed three component regio- and stereospecific selenosulfonation of alkynes: synthesis of (*E*)-β-selenovinyl sulfones. J Org Chem 2017, 82, 2269–2275.

[57] Taniguchi, T., Sugiura, Y., Zaimoku, H., and Ishibashi, H. Iron-catalyzed oxidative addition of alkoxycarbonyl radicals to alkenes with carbazates and air. Angew Chem Int Ed 2010, 49, 10154–10157.

[58] Wei, W., Liu, C., Yang, D., Wen, J., You, J., Suoc, Y., and Wang, H. Copper-catalyzed direct oxysulfonylation of alkenes with dioxygen and sulfonylhydrazides leading to β-ketosulfones. Chem Commun 2013, 49, 10239–10241.

[59] Miura, T., and Kobayashi, M. Addition of phenyl areneselenosulphonates to acetylenes: a route to acetylenic sulphones. J Chem Soc Chem Commun 1982, 8, 438–439.

[60] Sun, K., Shi, Z., Liu, Z., Luan, B., Zhu, J., and Xue, Y. Synthesis of (*E*)-β-selenovinyl sulfones through a multicomponent regio- and stereospecific selenosulfonation of alkynes with insertion of sulfur dioxide. Org Lett 2018, 20, 6687–6690.

[61] Zheng, D., An, Y., Li, Z., and Wu, J. Metal-free aminosulfonylation of aryldiazonium tetrafluoroborates with DABCO·$(SO_2)_2$ and hydrazines. Angew Chem Int Ed 2014, 53, 2451–2454.

[62] Zheng, D., Yu, J., and Wu, J. Generation of sulfonyl radicals from aryldiazonium tetrafluoroborates and sulfur dioxide: the synthesis of 3-sulfonated coumarins. Angew Chem Int Ed 2016, 55, 11925–11929.

[63] Qiu, JK., Shan, C., Wang, DC., Wei, P., Jiang, B., Tu, SJ., Li, G., and Guo, K. Metal-free radical-triggered selenosulfonation of 1,7-enynes for the rapid synthesis of 3,4-dihydroquinolin-2 (1H)-ones in batch and flow. Adv Synth Catal 2017, 359, 4332–4339.

[64] Prasad, CD., Sattar, M., and Kumar, S. Transition-metal-free selective oxidative C(sp^3)-S/Se coupling of oxindoles, tetralone, and arylacetamides: synthesis of unsymmetrical organochalcogenides. Org Lett 2017, 19, 774–777.

[65] Sun, K., Wang, X., Fu, F., Zhang, C., Chen, Y., and Liu, L. Metal-free selenosulfonylation of alkynes: rapid access to β-(seleno)vinyl sulfones *via* a cationic-species-induced pathway. Green Chem 2017, 19, 1490–1493.

[66] Sun, K., Wang, X., Lv, YH., Li, G., Jiao, HZ., Dai, CW., Li, YY., Zhang, C., and Liu, L. Peroxodisulfate-mediated selenoamination of alkenes yielding amidoselenide-containing sulfamides and azoles. Chem Commun 2016, 52, 8471–8474.

[67] Fang, Y., Liu, C., Wang, F., Ding, R., Wang, SY., and Ji, SJ. Radical-chain reaction of isocyanides with selenosulfonates and water: facile synthesis of selenocarbamates under metal-free conditions. Org Chem Front 2019, 6, 660–663.
[68] Shen, ZJ., Wu, YN., He, CL., He, L., Hao, WJ., Wang, AF., Tu, SJ., and Jiang, B. Stereoselective synthesis of sulfonated 1-indenones *via* radical-triggered multi-component cyclization of β-alkynyl propenones. Chem Commun 2018, 54, 445–448.
[69] Liu, F., Wang, JY., Zhou, P., Li, G., Hao, WJ., Tu, SJ., and Jiang, B. Merging [2+2] cycloaddition with radical 1,4-addition: metal-free access to functionalized cyclobuta[*a*]naphthalene-4-ols. Angew Chem Int Ed 2017, 56, 15570–15574.
[70] Qian, H., and Huang, X. Radical cyclization of 1,6-diene using polystyrene-supported selenosulfones. J Comb Chem 2003, 5, 569–576.
[71] Radatz, CS., Alves, D., and Schneider, PH. Direct synthesis of 2-aryl-1,3-benzoselenazoles by reaction of bis(2-aminophenyl) diselenides with aryl aldehydes using sodium metabisulfite. Tetrahedron 2013, 69, 1316–1321.
[72] Janzen, EG. ESR studies of thermal decomposition mechanisms. II. Electron spin resonance study of the $SO_2^{\cdot}$ formation in the thermal decomposition of sodium dithionite, sodium and potassium metabisulfite, and sodium hydrogen sulfite. J. Phys. Chem 1972, 76, 157–162.
[73] Fang, Y., Zhu, ZL., Xu, P., Wang, SY., and Ji, SJ. Aerobic radical-cascade cycloaddition of isocyanides, selenium and imidamides: facile access to 1,2,4-selenadiazoles under metal-free conditions. Green Chem 2017, 19, 1613–1618.
[74] Kobayashi, K., and Yokoi, Y. A convenient synthesis of 2-sulfanylbenzoselenazole derivatives *via* the reaction of 2-lithiophenyl isothiocyanates with selenium. Helvetica Chimica Acta 2012, 95, 761–765.
[75] Sen, VD., Golubev, VA., and Efremova, NN. Kinetics and mechanism of the WO_4^{2-}-catalyzed oxidation of di-tert-alkylamines and di-tertalkylhydroxylamines to nitroxyl radicals by hydrogen peroxide. Bull Acad ScinUSSR Div Chem Sci 1982, 31, 53–63.

Amol D. Sonawane and Mamoru Koketsu

3 Role of Isoselenocyanates for the Synthesis of Selenium-Containing Heterocycles

3.1 Introduction

Over the past decade, the chemistry of organoselenium compounds has attracted much attention because of their importance as synthetic tools [1, 2]. In particular, selenium-containing heterocycles have contributed to the remarkable growth of interest in the organoselenium chemistry. Preparations of selenium-containing heterocycles often involve the use of toxic Se-reagents, which are difficult to handle and to store. Isoselenocyanates are widely employed in the synthesis of selenium-containing heterocycles as a result of their convenient preparation, low toxicity, relative stability, and excellent reactivity [3–5].

In this chapter, we focused on the synthesis of organoselenium-containing organic heterocycles via isoselenocyanate intermediate. Here isoselenocyanate was used as versatile synthons in the preparation of Se-containing organic heterocycles. The chapter includes preparation of isoselenocyanate, further reactions with different nucleophiles, and synthesis of various important Se-heterocycles.

3.2 Preparation of organic isoselenocyanate

3.2.1 By formation of a C–Se bond

The general procedure for the synthesis of organic isoselenocyanate **3** was achieved by reaction of isonitrile **1** with selenium powder **2** (Figure 3.1) [6–10].

The reaction of formamide **4** with selenium powder **2** in the presence of triphosgene or phosgene or $(Cl_3CO)_2CO$ and triethylamine is shown in Figure 3.2 [9, 11, 12]. Phosgene is colorless gas, it is very poisonous, and it was used as a chemical weapon during World War I, where it was responsible for 85,000 deaths. Triphosgene is used as a safer substitute for phosgene because of solid material.

The synthesis of aryl isoselenocyanate **3** can be achieved by the reaction of selenoamide **7** with arylhydroximoyl chloride **5** in the presence of triethylamine (Figure 3.3) [13]. This reaction was proposed to proceed through cycloaddition of

Amol D. Sonawane and Mamoru Koketsu, Department of Chemistry and Biomolecular Science, Faculty of Engineering, Gifu University, Gifu, Japan

https://doi.org/10.1515/9783110625110-003

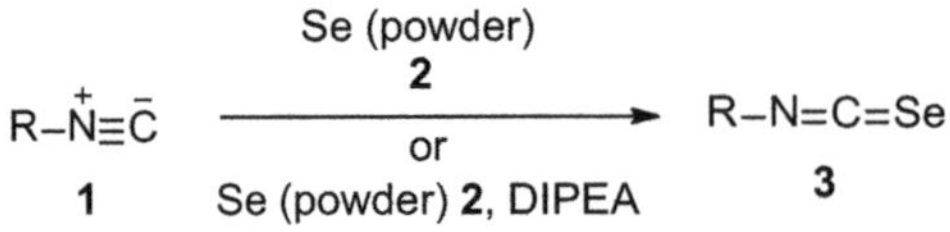

Figure 3.1: Synthesis of isoselenocyanate **3** by isonitrile **1** and Se (powder) **2**.

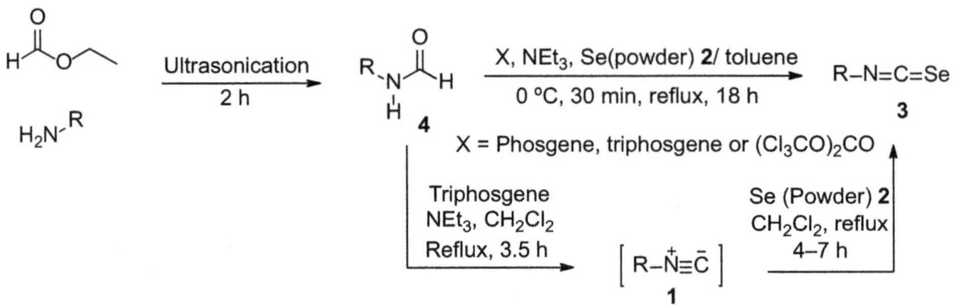

Figure 3.2: Preparation of organic isoselenocyanate **3**.

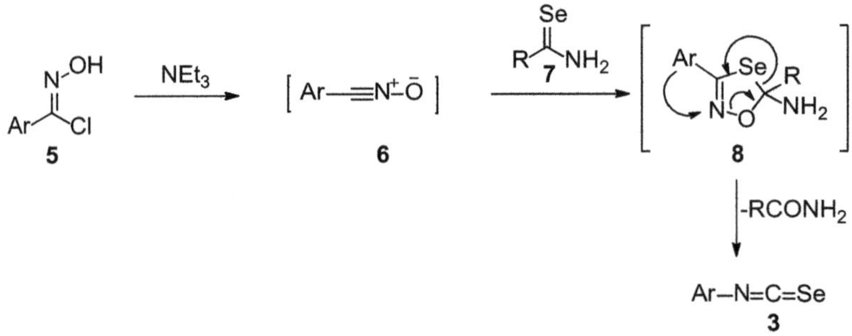

Figure 3.3: The synthesis of aryl isoselenocyanate **3**.

the selenocarbonyl group of selenoamide with an intermediate nitrile oxide **6**, formed in situ from aryl hydroximoyl chloride **5** and amine, followed by the elimination of amide accompanied by the migration of the aryl group **8** from carbon to nitrogen [14].

3.2.2 By substitution of halogen atom

The reaction of 5-bromo-1,2,3,4,5-pentaphenylcyclopentadiene **9** with KSeCN in acetonitrile solvent afforded the 1,2,3,4,5-pentaphenylcyclopentadien-5-yl isoselenocyanate **10** (Figure 3.4) [15]. Similar reactions proceeded with three- and seven-membered

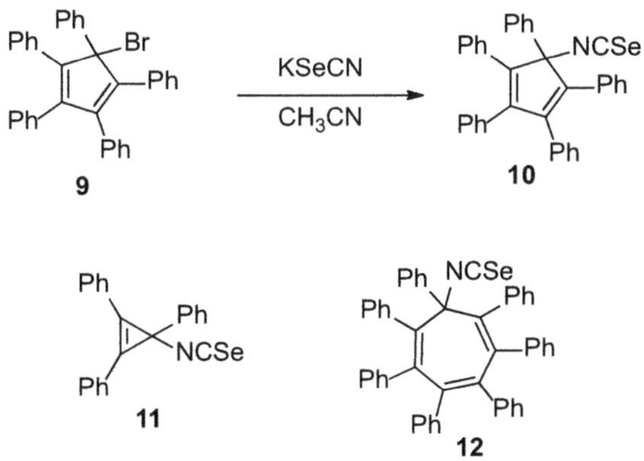

Figure 3.4: Synthesis of isoselenocyanate **10** by substitution of halogen atom.

analogues to afford 1,2,3-triphenylcyclopropen-3-yl isoselenocyanate **11** [16], and 1,2,3,4,5,6,7-heptaphenylcycloheptatrien-7-yl isoselenocyanate **12** [17].

The reaction of *N*-phenyl or *N*-benzyl-imidoyl chloride **13** with KSeCN proceeded to afford the corresponding imidoyl isoselenocyanates **14** (eq. (3.1)) [18, 19]. Similarly, the reaction of acid chloride **15** and thiocarbamoyl chloride **17** (eqs. (3.2) and (3.3)) [20] was reported to produce the corresponding isoselenocyanates **16** and **18**, respectively. Although the isoselenocyanates formed in these reactions were not isolated, their formations were clearly indicated by the spectral data and the products of the subsequent reactions [21].

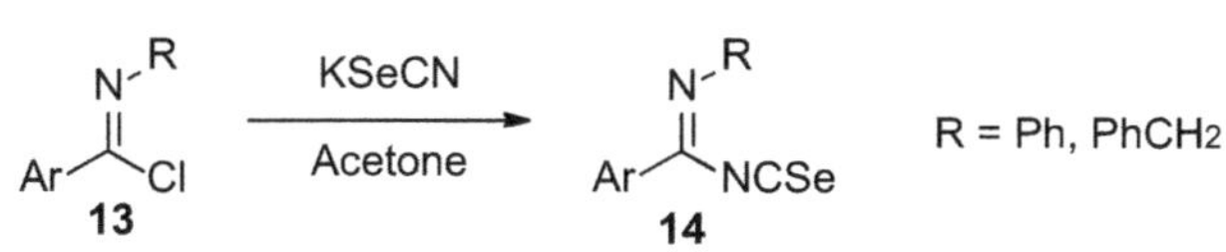

(3.1)

(3.2)

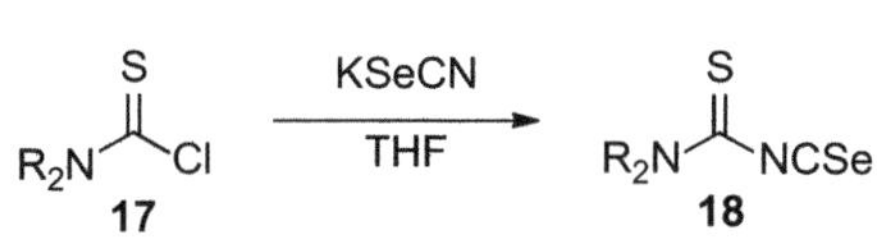

(3.3)

3.2.3 By using phase-transfer catalysis

The synthesis of isoselenocyanates was achieved by two methods under phase-transfer conditions (50% aq NaOH, CH_2Cl_2, Aliquat 336). The first started from isocyanides **19** and selenium gave isoselenocyanates, while the second started from

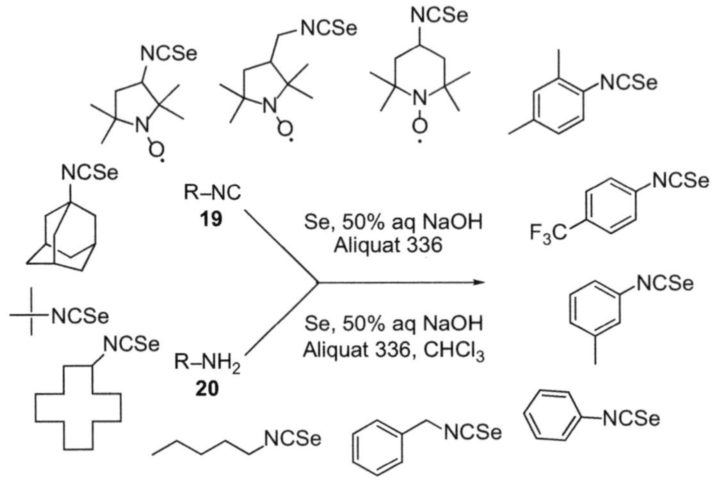

Figure 3.5: Synthesis of isoselenocyanates by using phase-transfer catalysis.

amines **20** with chloroform and selenium by sequentially applying the Hofmann isonitrile synthesis and the addition of selenium (Figure 3.5) [22].

The isoselenocyanate synthesis involves the use of an aqueous/organic biphasic system operating under strong alkaline conditions for the addition of selenium to aromatic, aliphatic, and cycloaliphatic (including a cage adamantyl moiety as well as five- and six-membered nitroxides) isocyanides **19** or directly to the corresponding amines **20** (Figure 3.5) [22]. Isocyanides **19** were synthesized by the Hofmann isonitrile synthesis using amines as starting materials together with chloroform and 50% aqueous sodium hydroxide solution in the presence of the ammonium salt Aliquat 336. The selenium reacted with aqueous sodium hydroxide solution and underwent disproportionation according to eq. (3.4) [23]:

$$3\text{Se} + 6\text{HO}^- \rightarrow \text{SeO}_3{}^{2-} + 2\text{Se}^{2-} + 3\text{H}_2\text{O} \tag{3.4}$$

According to eq. (3.4), the Se^{2-} anion may be considered as a direct agent, which is the effective nucleophilic agent attacking the isocyanide carbon atom A. A characteristic, strong, broad absorption at 2,100–2,160 cm^{-1} is visible in the IR spectra that confirms the presence of the NCSe group (asymmetrical NCSe stretching vibration) [24]. Recording of NMR spectra in the presence of phenylhydrazine is impossible due to the reaction of isoselenocyanate group with phenylhydrazine. It should be noted that in the ^{13}C-NMR spectra, no signals were attributed to an isoselenocyanate group [24, 25]. The signal of isoselenocyanate group in ^{77}Se NMR was observed at $\delta = -278$ to -298 for Ar-NCSe and at $\delta = -345$ to -358 for aliphatic and cycloaliphatic isoselenocyanates [26].

3.3 Imidoyl isoselenocyanate

The imidoyl isoselenocyanate derivatives **21** were prepared from *N*-phenylimidoyl chlorides and potassium selenocyanate [27]. The reaction of **21** with primary or secondary amines leads to selenourea derivatives, which on further reaction with activated bromomethylene compounds and treatment with a strong base yields 2-amino-1,3-selenazoles **22** (Figure 3.6) [28]. Imidoyl isoselenocyanates **21**, Ar^2 = 2-(chloromethyl)phenyl, react with amines to give 6*H*-5,1,3-benzoselenadiazocines **23** [29]. In both cases, the initial reaction is a nucleophilic attack of the amine onto the isoselenocyanate **21**, followed by the addition of the intermediate thiourea derivative with an electrophile in an inter- or intramolecular manner. With amidines **24** and **26**, 1,3,5-triazineselones **25** and **27**, respectively, are formed smoothly at room temperature (Figure 3.6) [27, 30]. Similarly, 2-amino-4,5-dihydro-1,3-thiazole **28** reacts with **21** to give the fused triazine-selones **29**. The formation of the triazine-selones **25**, **27**, and **29** has been rationalized by the initial formation of a selenourea derivative, followed by cyclization and the elimination of aniline (Ar^2 = Ph).

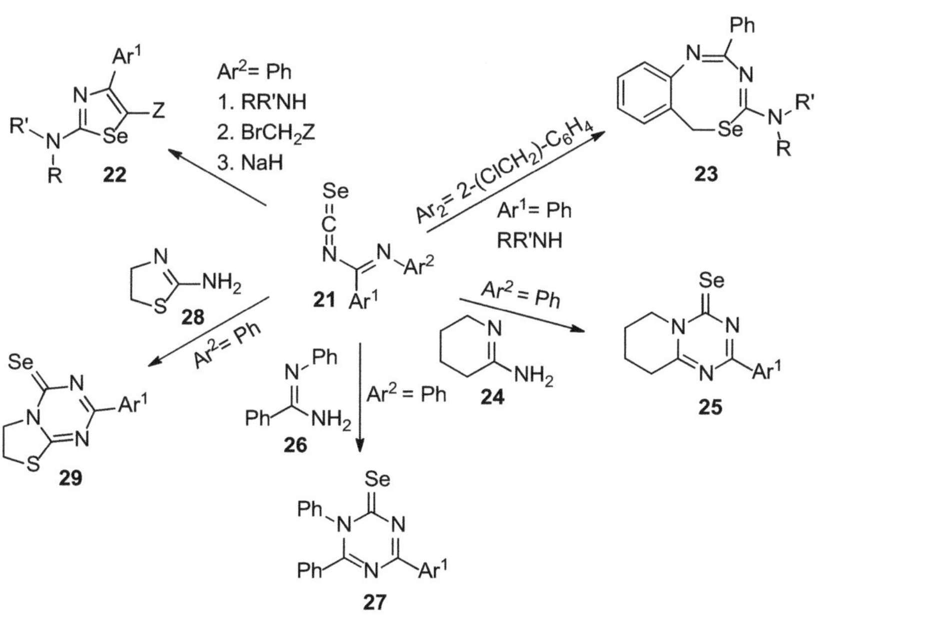

Figure 3.6: Reaction of imidoyl isoselenocyanates **21** with aromatic 2-amino-*N*-heterocycles.

The next experiment was carried out with imidoyl isoselenocyanate **30** and 2-amino-3-methylpyridine **31**. The reaction worked out in acetone at room temperature; after stirring for 3 h, the mixture was poured into H_2O, leading to a yellowish, crystalline product **34** in ca. 65% yield. The NMR spectra clearly indicated that no aniline was eliminated, and also the elemental analyses were in accordance with a 1:1 adduct [28].

The unexpected formation of **34** may be due to the initial addition of the amino *N*-atom of **31** onto the isoselenocyanate **30** to yield the intermediates selenoureas **32** and **33**. A spontaneous oxidation then leads to the isolated product **34**. In contrast to the reaction of imidoyl isoselenocyanate **30** with **35**, in which the ring *N*-atom was the most nucleophilic center, the NH_2 group of **35** acted as nucleophile, which proceeds through the reaction intermediates **36**, **37**, and **38**, respectively, to afford the product **39**. This observation may be rationalized by the higher aromaticity of the pyridine ring **31** compared with that of the 1,3-thiazole ring **35** (Figure 3.7) [28].

The reaction mechanism for the unexpected formation of **45** is proposed in Figure 3.8 [28]. The nucleophilic addition of imidazole **40** to the isoselenocyanate **30** gives a zwitterionic intermediate **41**. Further, the reaction undergoes cyclization to give the fused triazine-selone derivative **42**, which then forms a new zwitterion **43** via ring opening. Subsequent ring closure may form the fused 1,3-diazetidine **44**, which, by elimination of $NCSe^-$, leads to the isolated imidazolium salt **45**.

3.4 1,3-Selenazole

1,3-Selenazole is a five-membered ring system containing selenium and nitrogen atoms in the 1- and 3-positions, respectively. A synthesis of 1,3-selenazoles using isoselenocyanate has been developed. Reactions of allenyl isoselenocyanate **46** with carbon-, nitrogen-, oxygen-, or seleno-containing nucleophiles afford the corresponding 1,3-selenazoles [31]. The allenyl isoselenocyanate **46** have different reactivity with different nucleophile as shown in Figure 3.9 [32–34].

3.4.1 Synthesis of 2-amino-1,3-selenazole

The reaction of amidinoselenourea **48**, prepared by the treatment of isoselenocyanate **3** with *N*,*N*-diethyl-amidine **47** [35] under basic conditions, with halomethylene/phenacyl bromide **49**, formed 2-amino-1,3-selenazoles **50** (Figure 3.10) [36].

A plausible reaction mechanism for the formation of these trisubstituted selenazole derivatives is depicted in Figure 3.10 [36]. The amidinoselenourea **48** formed by the nucleophilic addition of the *N*,*N*-diethyl-amidine **47** to the isoselenocyanate **3** has one donor (C=Se) site and one acceptor (C-NEt_2) site. The selenium atom present in **48** displaces the bromine atom from phenacyl bromide to form intermediate **49**. At this stage, triethylamine abstracts a proton from the methylene group present in **49**, which attacks the carbon atom to which the electron-pulling quaternary nitrogen atom is attached to furnish the intermediate **51**. This intermediate is converted into the desired 2-amino-1,3-selenazole **50** by elimination of diethylamine.

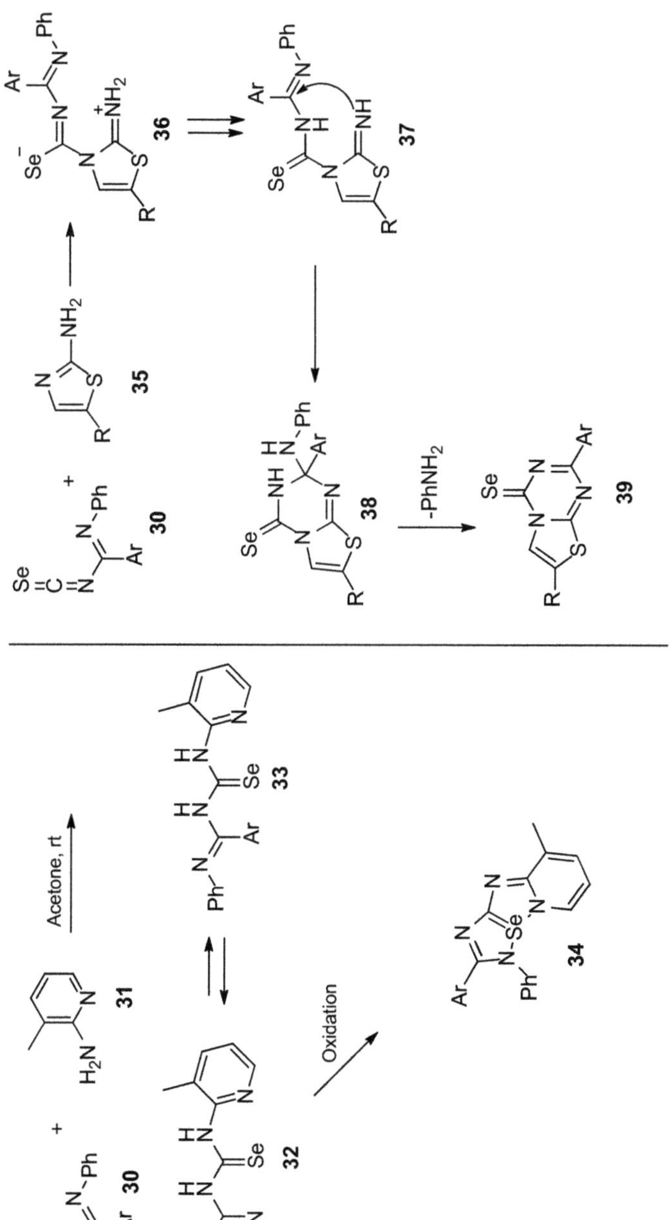

Figure 3.7: Synthesis of 2-aryl-4H-1,3-thiazolo[3,2-a][1,3,5]triazole-4-selones 39 and unexpected formation of **34**.

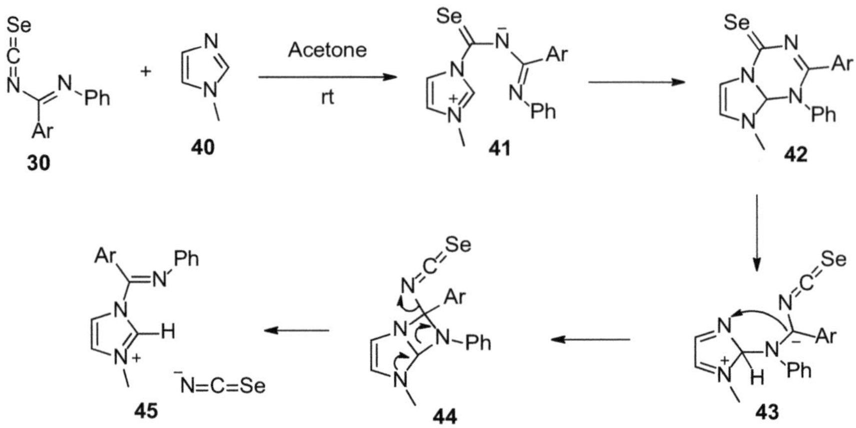

Figure 3.8: Synthesis of unexpected imidazolium salt **45**.

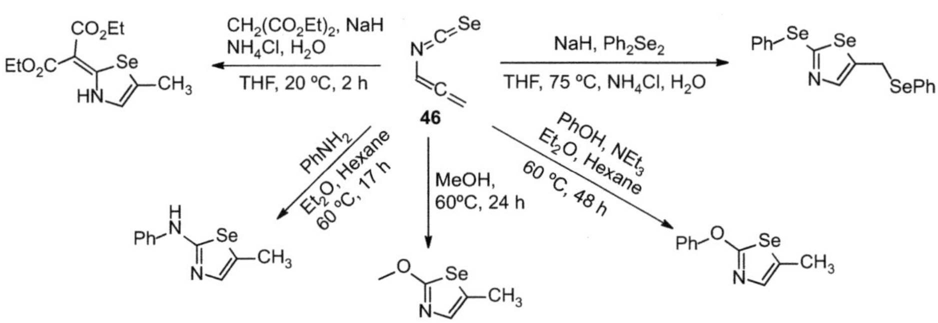

Figure 3.9: The reactivity of allenyl isoselenocyanate **46** with the different nucleophiles.

3.4.2 Synthesis of 5-arylamino-1,3,4-selenadiazol-2(3*H*)-one

The reaction of aryl isoselenocyanate **3** and hydrazine hydrate **52** was stirred in CH_2Cl_2 at room temperature at 7.5:1 ratio of $NaHCO_3$ to BTC {Bis (trichloromethyl) carbonate} for 2h. Under these optimized conditions, 5-phenylamino-1,3,4-selenadiazol-2(3*H*)-one **54** was obtained (Figure 3.11) [37].

Nucleophilic addition of the amino group of the hydrazine **52** to aryl isoselenocyanates **9** leads to the adduct intermediate **53**, which then reacts with the bis-electrophilic reagent BTC to form intermediate **55**. In the presence of sodium bicarbonate, the intermediate **56** undergoes a cyclization to afford the final heterocycle **54** (Figure 3.11) [37].

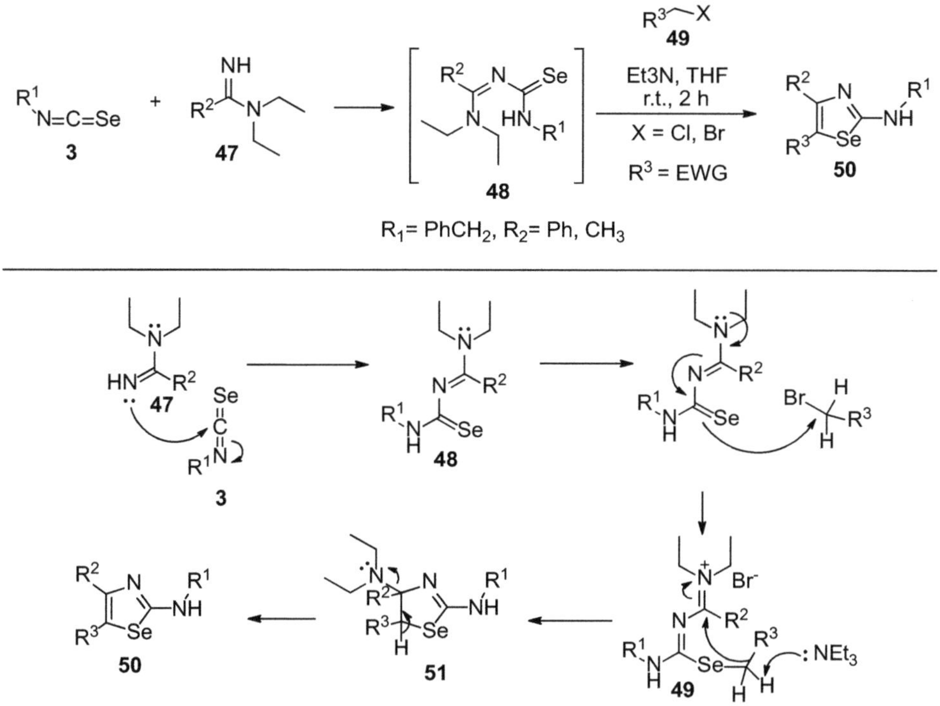

Figure 3.10: Synthesis of 2-amino-1,3-selenazole **50**.

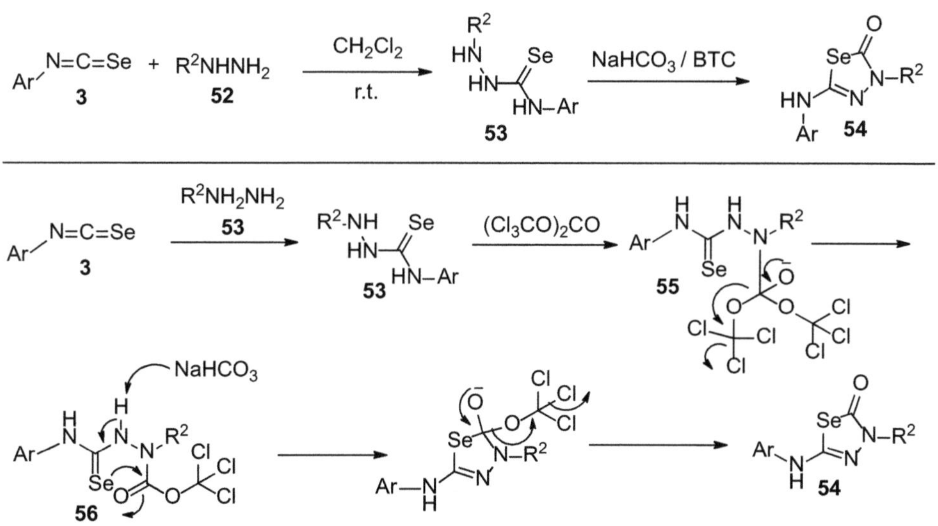

Figure 3.11: One-pot synthesis of 5-arylamino-1,3,4-selenadiazol-2(3*H*)-one **54**.

3.4.3 Synthesis of 2-imino-(1,3-selenazolidin-4-one and 1,3,4-selenadiazin-5-one)

The efficient and regioselective synthesis of 2-imino-1,3-selenazolidin-4-one **59** and 2-amino-1,3,4-selenadiazin-5-one **60** was achieved by one-pot reaction of isoselenocyanate **3**, hydrazine **52**, and ethyl chloroacetate **57** or chloroacetyl chloride **58** (Figure 3.12) [38].

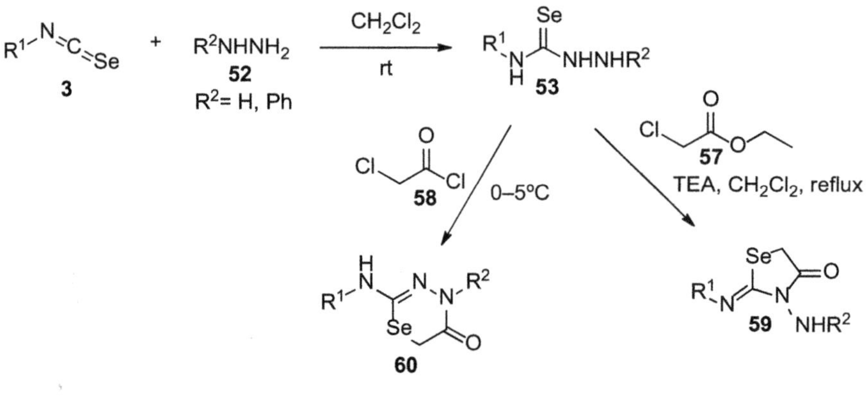

Figure 3.12: Preparation of 2-imino-1,3-selenazolidin-4-one **59** and 2-amino-1,3,4-selenadiazin-5-one **60**.

The plausible mechanisms were proposed for the formation of **59** and **60**, respectively (Figure 3.13) [38]; selenosemicarbazide **53** obtained from isoselenocyanate **3** reacted with ethyl chloroacetate **57** to form intermediate **61**, which then cyclized

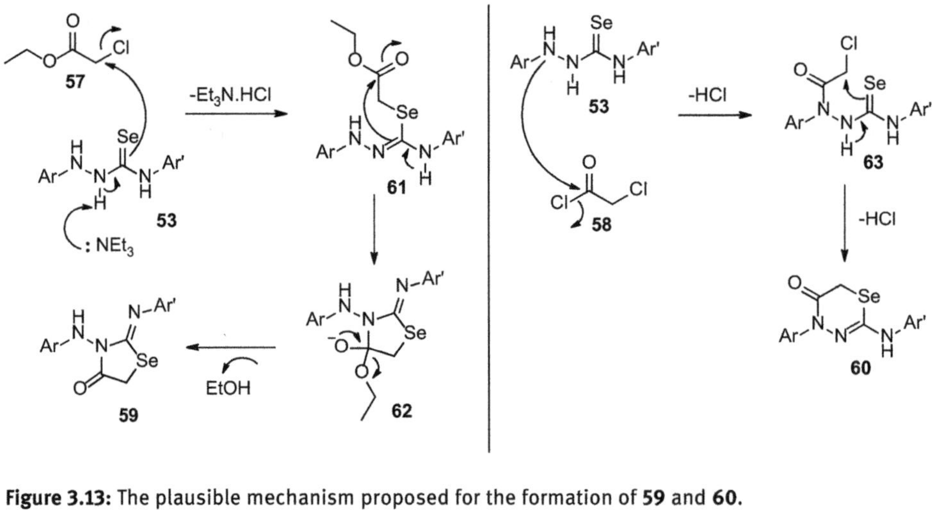

Figure 3.13: The plausible mechanism proposed for the formation of **59** and **60**.

to **62**. Successively, **62** was converted to the final product **59** by intramolecular elimination. Compared with ethyl chloroacetate **57**, chloroacetyl chloride **58** is much more reactive. Thus, selenosemicarbazide **53** reacted with chloroacetyl chloride **58** easily to afford intermediate **63**. Further, intermediate **63** was rapidly cyclized to the six-membered product **60**.

In conclusion, the one-pot condensation of isoselenocyanate **3**, hydrazine **52**, and ethyl chloroacetate **57** afforded 2-imino-1,3-selenazolidin-4-one **59**, while the reaction of isoselenocyanate **3**, hydrazine **52** and chloroacetyl chloride **58** provided 2-amino-1,3,4-selenadiazin-5-one **60**.

3.4.4 Synthesis of benzoselenazole and benzoselenazole-2(3*H*)-thione

The one-pot preparation of the 2-aminobenzoselenazole **67** by the phenylselenoureas **65** has been accomplished by the copper-catalyzed ligand-free reaction of the 2-iodoaniline **64** and isoselenocyanate **3** (Figure 3.14) [39, 40]

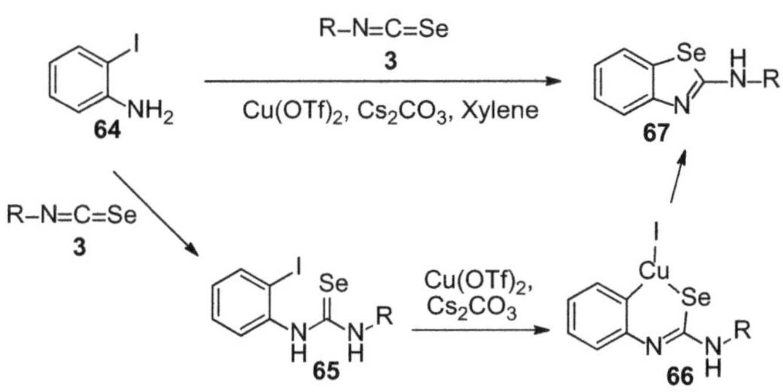

Figure 3.14: Synthesis of the 2-aminobenzoselenazole **67**.

A possible mechanism for the formation of 2-aminobenzoselenazole **67** from 2-iodoaniline **64** with isoselenacyanate **3** is shown in Figure 3.14 [39]. The successful copper-catalyzed intramolecular cyclization of the initial adduct, phenylselenourea **65**, proceeded by the intermediate **66** to give the benzoselenazole **67**. Evidence for the generation of the phenylselenourea **65** was confirmed by the isolation.

2-Bromophenyl isothiocyanates **68** were treated with BuLi in THF at –78 °C to generate 2-lithiophenyl isothiocyanates **69**, which were then allowed to react with Se. Attack of these carbanion on Se and the subsequent ring closure of the resulting 2-isothiocyanatobenzeneselenide by the intramolecular addition of selenide to the isocyanate C-atom proceeded rapidly at this temperature to lead to the formation of lithium benzoselenazole-2-thiolate intermediates **70**, which were protonated by acidic aqueous workup to give **71** (Figure 3.15) [41, 42].

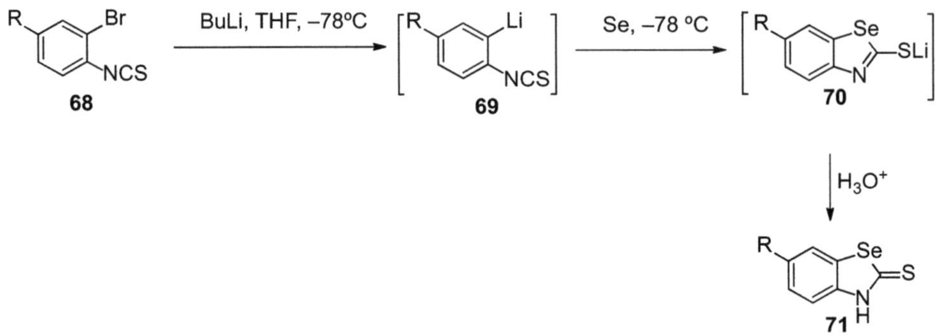

Figure 3.15: Synthesis of 2-sulfanyl-benzoselenazole **71**.

3.5 Multicomponent reactions (MCRs) using isoselenocyanate

3.5.1 Synthesis of chiral 2-iminoselenazoline

Multicomponent reactions (MCRs) are a powerful synthetic tool for the rapid and efficient construction of complicated molecular frameworks. MCRs are strategically amenable with modern synthetic tools such as microwave irradiation, ultrasonication, polymer-, and ionic liquid-supported synthesis. The reaction of *L*-amino ester **72**, isoselenocyanate **3**, and α-substituted bromoketone **74** was carried out in one pot for the regioselective synthesis of enantiopure 2-iminoselenazole **75** under ultrasonication. (Figure 3.16) [12]. The isolated yields were much higher than those previously reported on the application of sonication [43, 44].

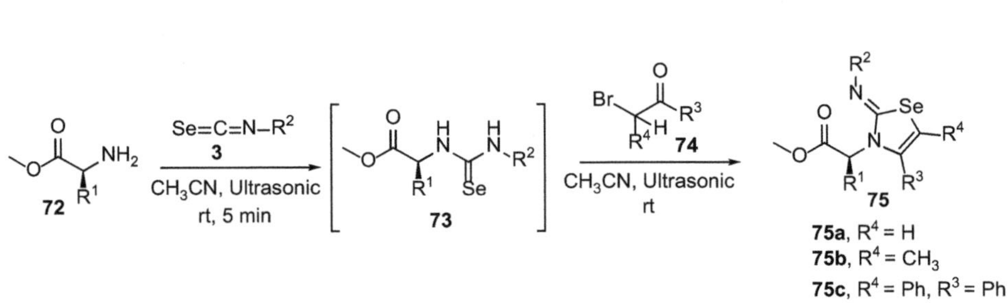

Figure 3.16: One-pot synthesis of 2-iminoselenazole **75**.

The plausible mechanism involves the loss of a proton from N_1 atom of **73** and the intramolecular nucleophilic attack of N_2 (hard nucleophile) on the carbonyl group of the ketone (hard electrophile) to deliver **76**, which, upon subsequent dehydration of **78**, releases the observed 2-iminothiozole **80** (Figure 3.17) [12]. The N_2 was more reactive than N_1 because N_1 was near the electron withdrawing carbonyl group of the amino ester, causing regioselectivity. Therefore, the formation of the 2-iminoselenazole **80** rather than **81** is due to the preferential attack of selenium because of its enhanced nucleophilicity, and the driving force for the selective N_1 attack on the alkyl carbon is to eliminate a stable hydrobromide salt. Furthermore, the influence of steric factors on the reaction time has been effective to characterize the intermediate 2-imino-5-selenol **76**, which undoubtedly confirmed that selenourea reacts via a soft nucleophile group (i.e., the selenium atom).

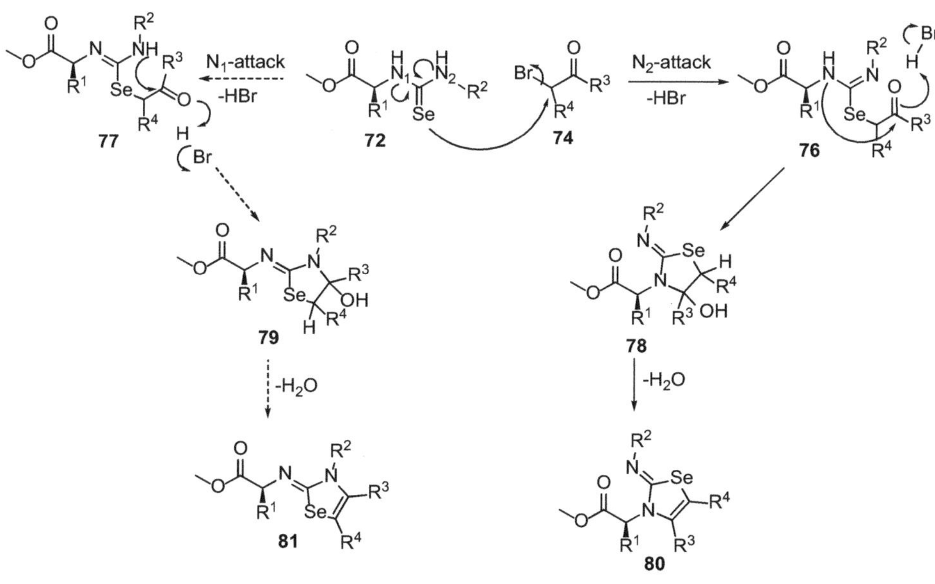

Figure 3.17: A plausible mechanism for the formation of 2-iminoselenazole **80**.

3.5.2 Synthesis of 2-aminobenzo[*d*][1, 3]selenazine

The MCR of *O*-functionalized aryl isocyanide **82**, elemental selenium, and amine by isoselenocyanate formation **83** and subsequently intramolecular Michael addition reaction under the metal-free condition led to 2-aminobenzo[*d*][1,3]selenazine **84** (Figure 3.18) [45].

The reaction of methyl (*E*)-3-(2-isocyanophenyl)acrylate **82** with elemental selenium and piperidine, in the presence of NEt_3, was performed in dichloroethane solvent for 12 h at room temperature. This approach provides a direct construction of selenazine derivatives with potential biological and medicinal activities under mild conditions. Based on the experimental results and literature reports [46–48], the plausible mechanism was shown in Figure 3.18 [45]. Initially, an isoselenocyanate **83** is generated in situ by the reaction of isocyanide **82** with elemental selenium in the presence of Et_3N. Furthermore, amine **85** reacts with isoselenocyanate **83** to give carbamimidoselenoate **86** through intermolecular nucleophilic attack. Following a subsequent Michael addition **87** and further protonation, **84** was formed.

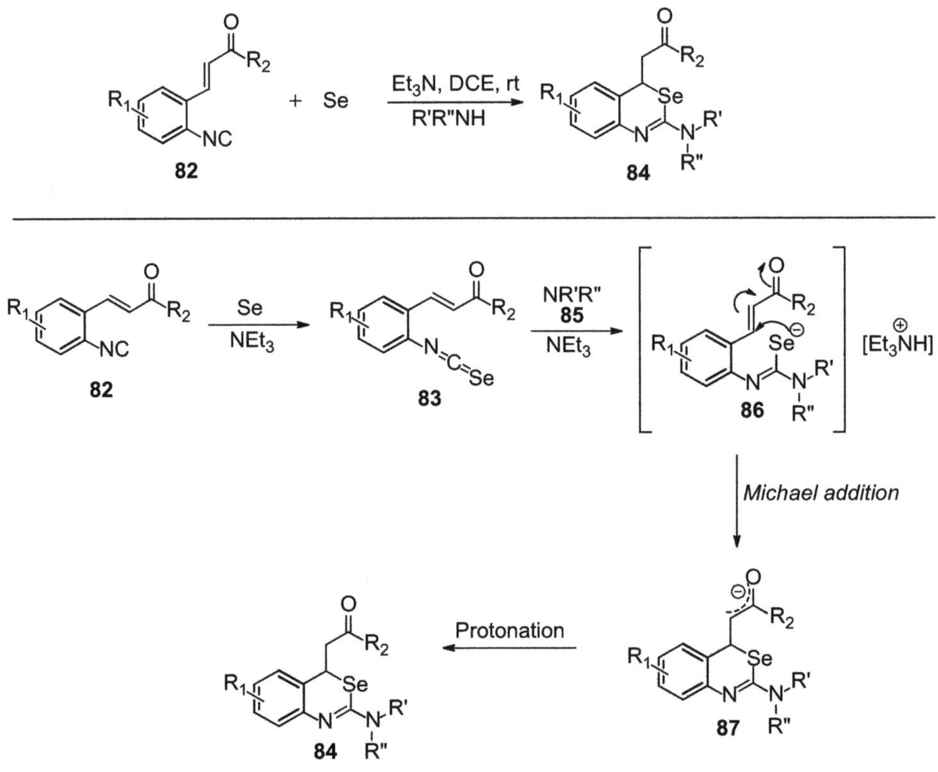

Figure 3.18: Synthesis of 2-aminobenzo[*d*][1,3]selenazine **84**.

3.6 Selenoimidoylation of alcohol

The compounds containing a selenoimidoyl skeleton (Se–C=N) have been synthesized and used as precursors of imidoyl radicals. The selenoimidates **89** were prepared by (i) alkylation of selenoamides with alkyl halides [49, 50], (ii) reaction of imidoyl chlorides with selenols [51], (iii) reaction of imidoyl radicals with diaryl diselenides [52], (iv) three-component radical coupling reactions of diselenides with isocyanides and alkynes [53], (v) transition-metal-catalyzed addition of diaryl diselenide to isocyanide [54], and (vi) reaction of oxime sulfonates with organoaluminum selenolate. For the selenoimidoylation of an alcohol **88**, the reaction was carried out in methanol (2 mmol), cyclohexyl isocyanide (1 mmol), and selenium (1 mmol) in THF (1 mL) in the presence of DBU (1 mmol) at room temperature. The black suspension became a yellow homogeneous solution within 3 h. After 20 h the reaction mixture was quenched with Bu–I and subsequent workup afforded the expected selenocarbonimidate **89** (Figure 3.19) [47, 55].

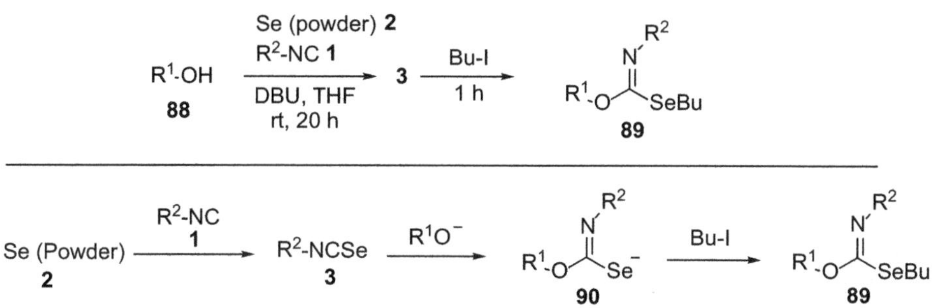

Figure 3.19: Synthesis of selenoimidate **89**.

The plausible mechanism shows that the reaction of alcohol with selenium and isocyanide in the presence of DBU gives oxyimidoylselenoate **90**. Trapping of **90** with Bu–I resulted in the high-yield formation of selenocarbonimidate **89**. The 1-amino-2-alkyne **91** reacted with selenium and carbon monoxide in the presence of DBU to yield 5-alkylideneselenazolidin-2-one **92** (eq. (3.5)). In addition, alk-2-yn-1-ol **93** was allowed to react with selenium and isocyanide under similar conditions and new selenium-containing heterocycle, 2-imino-4-alkylidene-1,3-oxaselenolane **94** (eq. (3.6)), was obtained by cycloaddition of oxyimidoylselenoate **96** (Figure 3.20) [47] generated in situ by intramolecular addition of selenolate to carbon–carbon triple bonds.

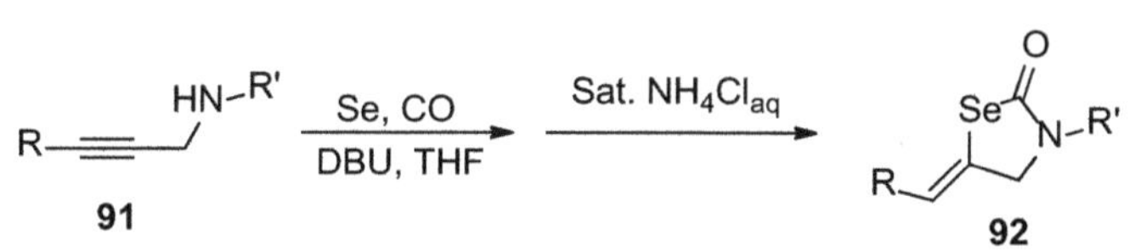

(3.5)

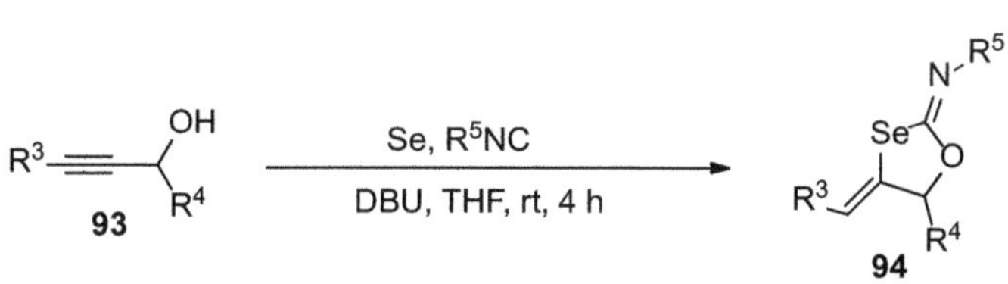

(3.6)

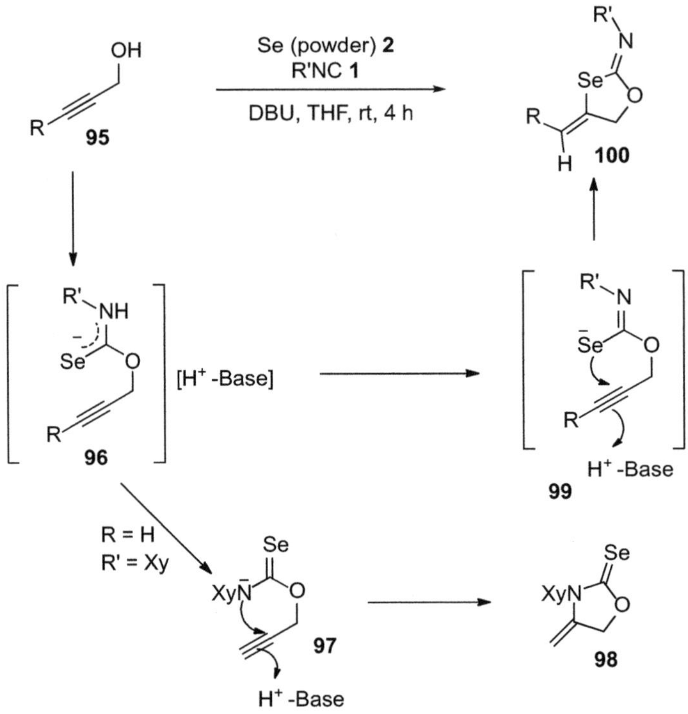

Figure 3.20: Reaction pathway for the formation of *Z*-imino-4-alkylidene-1,3-dioxaselenolane **100**.

A reaction mechanism for this transformation of **100** is suggested in Figure 3.20 [47]. First, alk-2-yn-1-ol **95** undergoes selenoimidoylation by the reaction with selenium and isocyanide to yield oxyimidoylselenoate **96.** The stereoselectivity of the C=C double bonds of the products can be explained by a trans-addition mechanism (**96→99→100**) where proton coordination to the carbon–carbon triple bond facilitates nucleophilic addition of selenium to the triple bond from the opposite side. 2-Selenoxo-1,3-oxolidine **98** is formed by the nucleophilic addition of nitrogen to the carbon–carbon triple bond of **97.**

3.7 Synthesis of benzo[*c*]selenophene

The synthesis of the benzo[*c*]selenophene **107** has been of considerable interest by various pathways [56–58]. The *O*-bromoethynylbenzene **101** was lithiated with 1.2 equiv. of *t*-BuLi in anhydrous THF at 80 °C under an argon atmosphere, followed by treatment with 1.5 equiv. of aryl isoselenocyanate **3** at room temperature, and further quenched with ethanol to give the desired (*Z*)-3-methylidenebenzo[*c*]selenophene **107.**

The product **107** was obtained in good yield via one-pot *5-exo-dig* mode cyclization instead of *6-endo-dig* cyclization **106** [59]. A plausible mechanism for the formation of (*Z*)-3-methylidenebenzo[*c*]selenophene **107** from *O*-bromoethynylbenzene **101** with aryl isoselenocyanate **3** is shown in Figure 3.21 [59]. The initial adduct, **103** or **104**, is generated by the attack of the carbanion **102** on the sp carbon of the aryl isoselenocyanate **9** and the protonation of **103** or **104** was carried out by EtOH to give unstable selenol **105**. Further, the selenol **105** was cyclized to afford the *5-exo-dig* mode product **107** instead of *6-endo-dig* mode cyclization to give product **106**. In the absence of a proton source, no product formation occurs and starting isoselenocyanates were obtained. EtOH as a proton source gave the best results. Therefore, it is clear that the addition of carbanion **102** to the aryl isoselenocyanate **3** gave the adduct **103** or **104**, which gradually decomposed during the isolation operation or on standing.

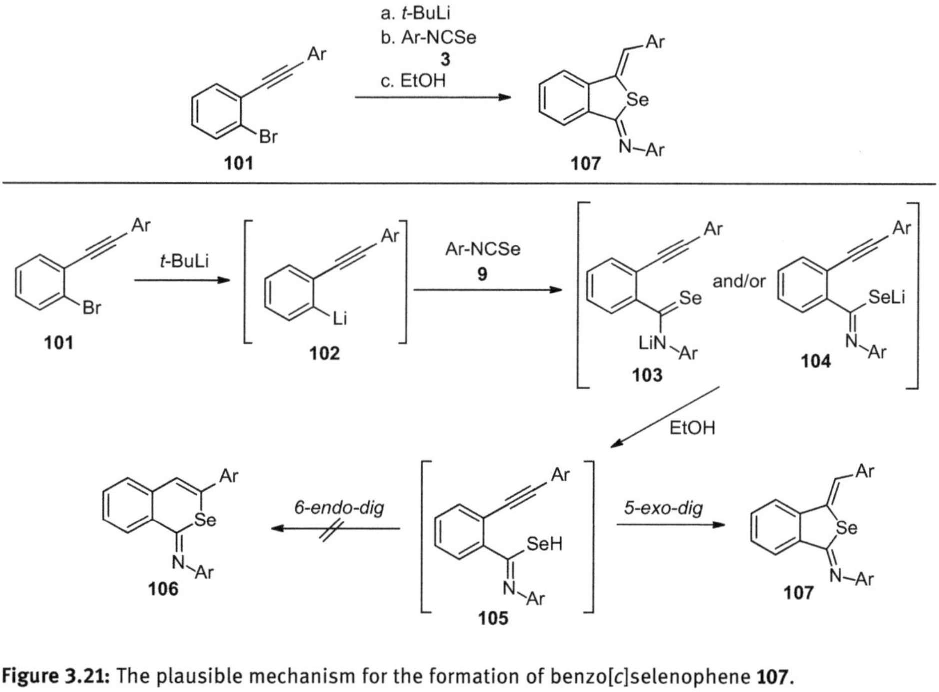

Figure 3.21: The plausible mechanism for the formation of benzo[c]selenophene **107**.

The iodocyclization reaction of *O*-ethynylphenyl lithium **109** with cyclohexyl isoselenocyanate **110** was proceeding stereoselectively, affording the (*E*)-10-iodo-3-methylidenebenzo[*c*]selenophene **111** as shown in Figure 3.22 [59]. *O*-ethynylphenyl lithium **109** generated from **108** was similarly treated with cyclohexyl isoselenocyanate **110**, and then protonated with *t*-BuOH, followed by iodination with I_2, giving the desired (*E*)-10-iodobenzo[*c*]selenophene **111**. The NIS gave lower yield compared

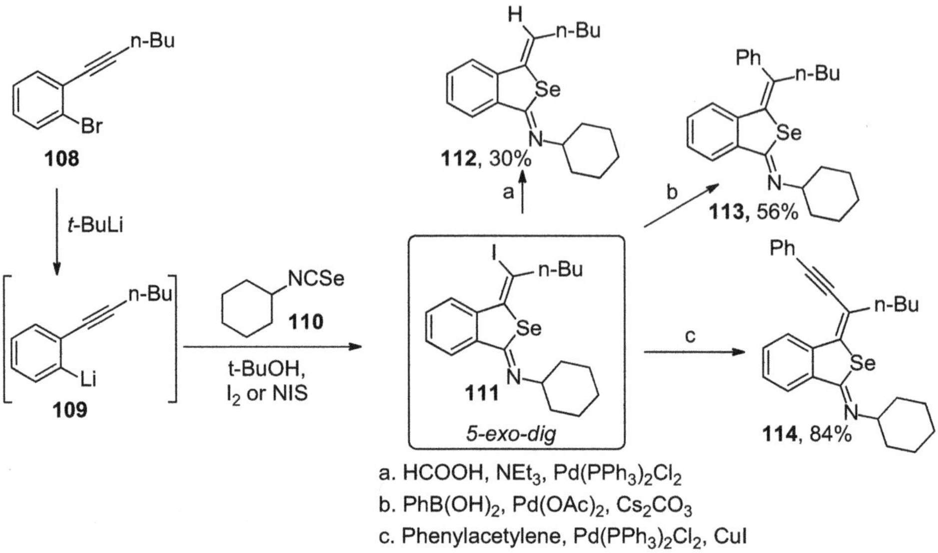

Figure 3.22: Functionalization of iodocyclization product **111**.

to that of iodine. Finally, iodobenzo[*c*]selenophene **111** can be reduced to **112** by treatment with HCOOH/Et_3N in the presence of a palladium catalyst. The (*E*)-10-phenyl-3-methylidenebenzo[*c*]selenophene **113** was produced in 56% yield by the Suzuki cross-coupling of **111** with phenylboronic acid. The Sonogashira reaction of **111** with phenylacetylene gave **114** (84%) [59].

3.8 Cycloaddition of carbodiimide to selenazetidine

The synthesis of four-membered Se-containing heterocycles was reported using isoselenocyanate as starting materials [60]. The easy preparation of 1,3-selenazetidine-2,4-diimines from isoselenocyanate **3** and carbodiimide **115** was shown in Figure 3.23 [61]. The formation of product **116** is formal [2 + 2] cycloaddition; it is favorable that a two-step mechanism is responsible for the formation of **116**. The nucleophilic attack of the Se-atom at carbodiimide C-atom leads to a zwitterion of type **A**, which by ring closure resulted in the product **116**.

The selenazetidine **116a** is planar. Both the imido group R^1 and R^2 are (*Z*)-configured and the relative configuration was confirmed by X-ray crystallography. The IR spectrum of **116a** shows strong absorption for the C=N group at calculated 1,690 and 1,674 cm^{-1}. The ^{13}C-NMR spectrum shows two signals for C=N at 133.1 and 137.1 ppm, and CI-MS (NH_3), *m/z* 390 $[M+1]^+$ and 207 ($[(C_6H_{11})_2CN_2]^{+\cdot}$). Further, **116b-116f** bearing an arylimino group also have the (*Z,Z*)-configuration. However, the symmetrically

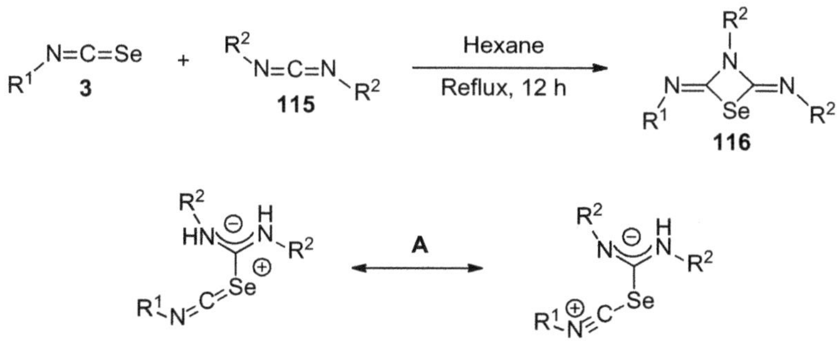

Compound	R^1	R^2	Yield (%) 116
116a	Ph	Cyclohexyl	98
116b	Ph	i-Pr	84
116c	4-ClC_6H_4	Cyclohexyl	88
116d	4-ClC_6H_4	i-Pr	97
116e	4-BrC_6H_4	Cyclohexyl	98
116f	4-BrC_6H_4	i-Pr	99
116g	Cyclohexyl	Cyclohexyl	99
116h	Cyclohexyl	i-Pr	88

Figure 3.23: Preparation of selenazetidine **116.**

substituted **116g** with two cyclohexylimido moieties shows three signals for CH and 12 signals for CH_2 groups of the cyclohexyl substituents. Therefore, the structure cannot be symmetric. A likely interpretation is that mixture of (*Z,Z*)- and (*E,Z*)-**116g** is present in varying ratios. The similar observation has been made in the case of **116h**, which, in the NMR spectra, also shows some doubling of signals.

Thiocarbamoyl isoselenocyanate **118** was prepared by reactions of thiocarbamoyl chloride **117** with KSeCN [17, 62, 63]. Reactions of the thiocarbamoyl isoselenocyanate **118** with imines **119** were carried out at reflux in THF for 5 h. The reactions gave formal [2 + 2] cycloadducts, 2-imino-1,3-selenazetidines **120** (Figure 3.24) [20].

The [4 + 2] cycloadduct product, 6-amino-2*H*-1,3,5-thiadiazine-4-selone **121,** was ruled out by the X-ray crystal analysis of the product. The signal in ^{77}Se NMR spectra of selenazetidines was observed around δ 750, while chemical shifts of selenoureas

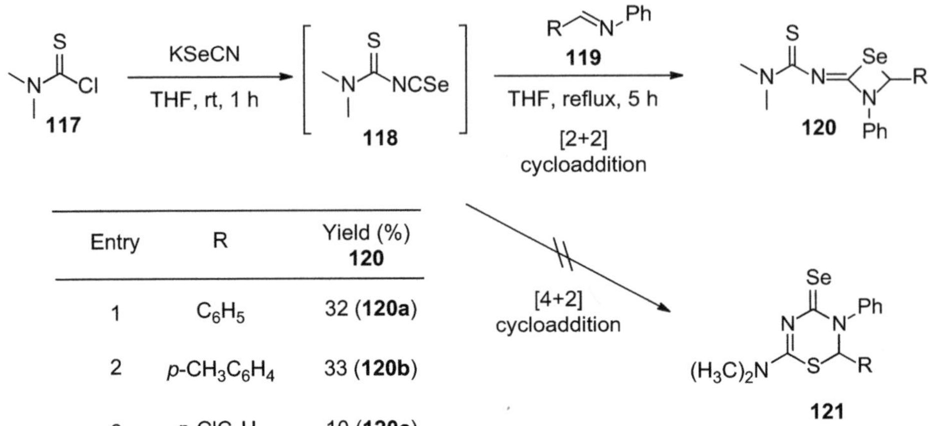

Entry	R	Yield (%) **120**
1	C_6H_5	32 (**120a**)
2	*p*-$CH_3C_6H_4$	33 (**120b**)
3	*p*-ClC_6H_4	10 (**120c**)
4	*p*-$(CH_3)_2CHC_6H_4$	20 (**120d**)

Figure 3.24: Preparation of 2-imino-1,3-selenazetidines **120**.

appear in the range *δ* 170-340 in ^{77}Se NMR spectra. In the case of compound **120**, the chemical shifts should be in a range of chemical shifts of selenoureas. Chemical shifts of ^{77}Se NMR spectra for product **120** appeared at *δ* 756.9 ± 2.24, which give evidence of 1,3-selenazetidine structure **120** (Table 3.1).

Table 3.1: Chemical shifts in ^{77}Se NMR of 1,3-selenazetidines **120** and selenoureas [20].

	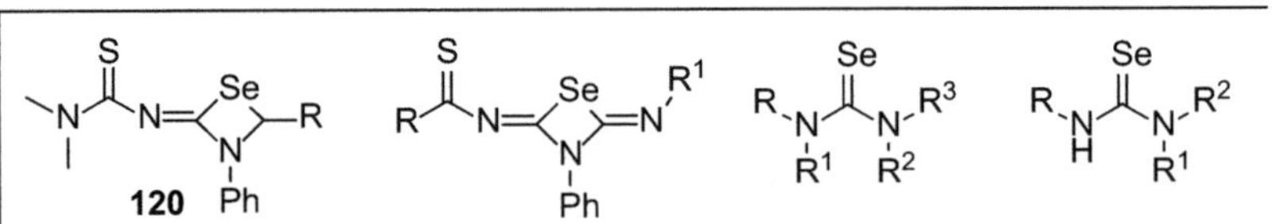			
^{77}Se NMR (δ)	756.9±2.24	753.2±1.97	336.3±20.3	177.7±7.71

	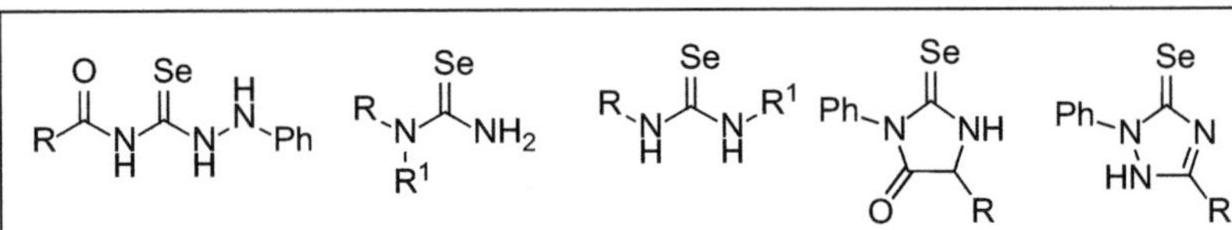				
^{77}Se NMR (δ)	298.3±8.19	218.4±21.6	189.2±32.5	267.8±16.1	88.6±1.07

3.9 Synthesis of 1,3-oxaselinane, 1,3-oxaselenepane and 1,3-oxaselenolane

The preparation of 1-oxa-3-selenaheterocycle was achieved by the reaction of 3-chloropropanol **122** with aryl isoselenocyanate **3**. The mixture of aryl isoselenocyanate **3** and **122** in dichloromethane at room temperature was treated with an equimolar amount of sodium hydride, and the mixture was stirred for 3–4 h. After chromatographic workup, an oily product, which contains aromatic as well as aliphatic H-atoms (^{1}H-NMR), was obtained in 36–60% yield. Mass spectrometry and elemental analysis confirmed that the two starting materials had reacted to yield the product by the elimination of HCl. The addition of the anion of **122** to aryl isoselenocyanate **3** gives the intermediate **123**. There are two possibilities for the cyclization reaction: nucleophilic substitution of the chloride by the selenide would lead to 2-imino-1,3-oxaselinane **124** (*path a*), whereas the analogous cyclization via the *N*-atom would yield 1,3-oxazinane-2-selone **125** (*path b*) [62, 64, 65]. Both pathways are based on 6-*exo-tet* cyclization (Figure 3.25) [66, 67]. On the basis of their spectroscopic data, the structure for the product was determined as **124**. For example, **124** shows a strong IR absorption at 1,662 cm^{-1} (C=N); the corresponding absorption of the *N*-analogue **125** appears at 1,630 cm^{-1}.

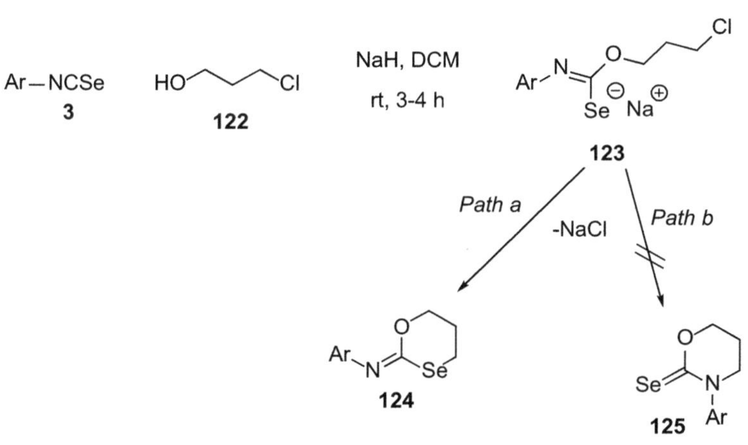

Figure 3.25: Synthesis of 1,3-oxaselinane **124**.

The synthesis of 1,3-oxaselenepane **127** was achieved by the reaction of aryl isoselenocyanate **3** with 4-bromobutanol **126**. The reaction was carried out with NaH in THF solvent (Figure 3.26) [68].

The synthesis of 1,3-oxaselenolane **130** was achieved by iodocyclization reaction. The reaction of (*Z*)- or (*E*)-O-allylselenocarbamate **129**, obtained by reactions

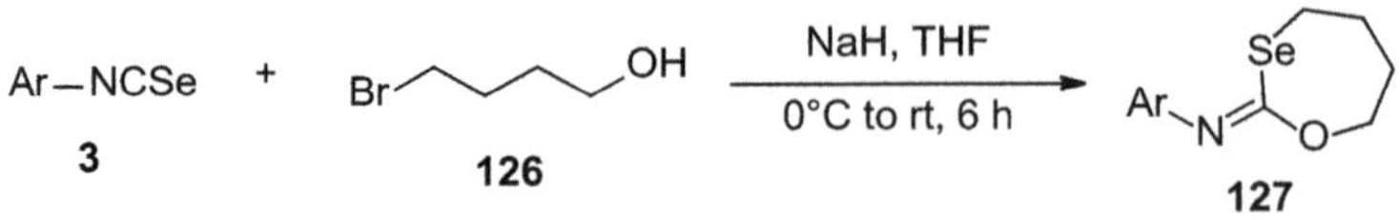

Figure 3.26: Synthesis of 1,3-oxaselenepane 127.

of allyl alcohol **128** and aryl isoselenocyanate **3,** with iodine or NIS gave five-membered ring 1,3-oxaselenolane **130**, which on treatment with DBU resulted in the formation (*Z*)- or (*E*)-4-alkylidene-2-imino-1,3- oxaselenolane **131** (Figure 3.27) [69].

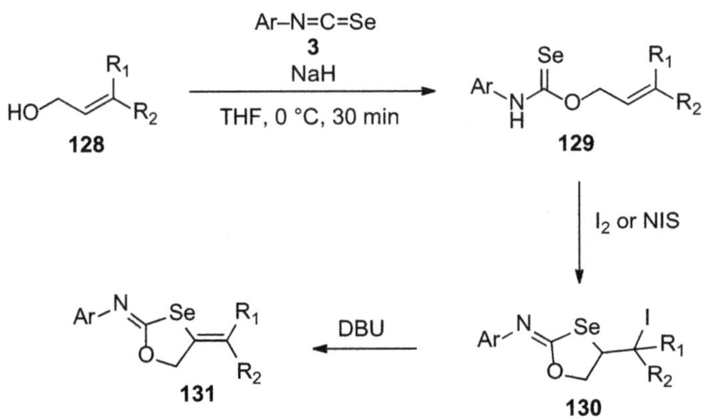

Figure 3.27: Synthesis of 1,3-oxaselenolane 130 via iodocyclization reaction.

3.10 Syntheses of 1,3,4-diazaselinan-2-imine and 2-amino-1,3,4-oxadiazole

The aryl isoselenocyanates **3** are the convenient precursor for the introduction of selenium into four- [61], five- [70], six- [66, 71], and seven-membered selenaheterocycles [72]. The reaction pathway for the syntheses of 1,3,4-diazaselinan-2-imine **137** was shown in Figure 3.28 [67]. The addition of a nucleophile **132**, which has a leaving group, leads to the intermediate **133.** Further, the nucleophilic Se-atom attacks to form the cyclized product **134.** Alternatively, the adduct **135** of a bis-nucleophile, for example, hydrazine, reacts with a bis-electrophile to give **136**, which undergoes a ring-closure to yield 1,3,4-diazaselinan-2-imine **137.**

The synthesis of nonselenium-containing compounds from isoselenocyanates has much attention via deselenization processes [11]. The cyclodeselenization of selenosemicarbazide derivatives, which were obtained by the reaction of various isoselenocyanates with hydrazide, produced 2-amino-1,3,4-oxadiazoles. The one-pot

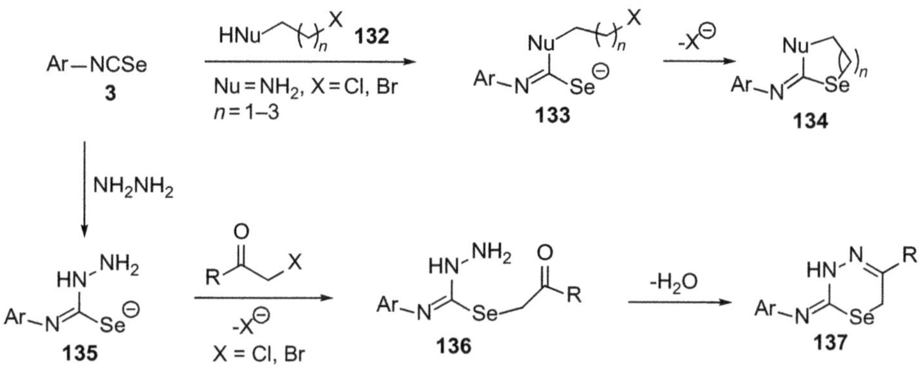

Figure 3.28: Syntheses of 1,3,4-diazaselinan-2-imine **137**.

reaction of isoselenocyanate **3** and hydrazide **138** was performed in DMF at 90 °C and 2-amino-1,3,4-oxadiazole **139** was obtained in good yield (Figure 3.29) [73]. The nature of R_1 group on the isoselenocyanates **3** affects the reaction rate. Electron-donating group substituted compounds showed slightly higher reactivity, giving better yields than those with electron-withdrawing groups. However, when the methyl or ethyl group was attached to the ortho position of the phenyl isoselenocyanate, the reaction yields were decreased due to the steric effect.

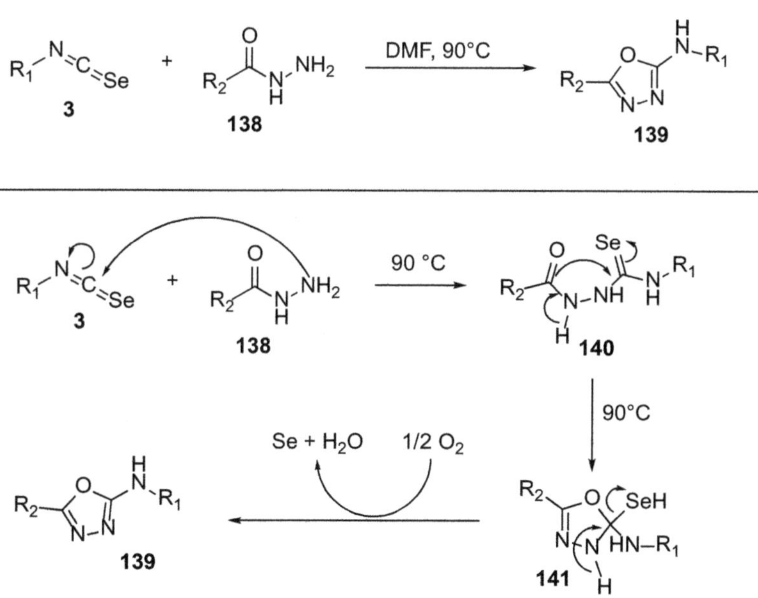

Figure 3.29: The plausible mechanism for the formation of 2-amino-1,3,4-oxadiazole **139**.

A plausible mechanism is proposed for the formation of 2-amino-1,3,4-oxadiazole **139** (Figure 3.29) [73]. First, selenosemicarbazide **140** was generated from isoselenocyanate **3** and hydrazide **146**. Subsequently, the intramolecular cyclization of **140** produced the intermediate **141** via attack of oxygen atom on carbonyl to selenocarbonyl. The intermediate **141** was then converted to the 2-amino-1,3,4-oxadiazole **139** via cyclodeselenization with the aid of oxygen. To confirm this process, the reaction was performed in a nitrogen-protected vessel and no desired product and selenium powder were detected. These results indicated that the oxygen was crucial to this reaction [74, 75].

The reaction of isoselenocyanate **3** with dihydrazide **142** was carried out. The desired product bis(1,3,4-oxadiazole) **143** was successfully obtained in excellent yield in DMF solvent at 90 °C (Figure 3.30) [73].

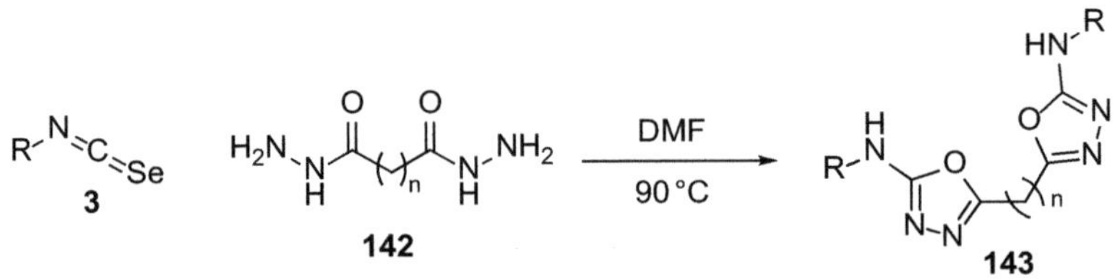

Figure 3.30: Synthesis of bis(1,3,4-oxadiazole) **143**.

3.11 Synthesis of 5-amino-2-selenoxo-1,3-imidazole-4-carboselenoamide

The one-pot synthesis of 5-amino-2-selenoxo-1,3-imidazole-4-carboselenoamide **145** was accomplished by the reaction of isoselenocyanate **3** with 2-aminoacetonitrile **144**. The reaction of *p*-tolylisoselenocyanate **146** with 3-aminopropionitrile **147** gave the corresponding acyclic selenoureas **148** in quantitative yield (Figure 3.31) [76]. The present reaction required pyridine and 2 equiv. of isoselenocyanate **3** to obtain 2-selenoxo-1,3-imidazole-5-carboselenoamide **145**.

The plausible mechanism for the formation of **145** is explained in Figure 3.32 [76]. The reaction of isoselenocyanate **3** with 2-aminoacetonitriles **144** is initiated by the nucleophilic addition of the nitrogen atom of **146** to the electrophilic carbon atom of isoselenocyanate **3** affording the selenourea **149**, which is subsequently cyclized to give 4-imino-1,3-imidazoline-2-selenone **150**. Next, the carbanion of 1,3-imidazolidine ring **150** attacks the electrophilic carbon of another molecule of isoselenocyanate **3** to form a carboselenoamide **151**, which subsequently tautomerizes to the more stable amine **145**.

Reported chemical shifts of selenocarbonyl groups in selenoureas and selenoamides are summarized in Table 3.2 [76].

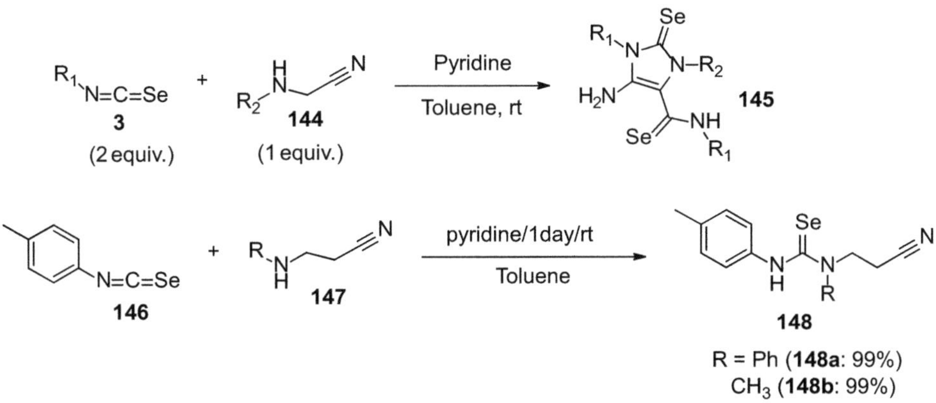

Figure 3.31: Synthesis of 5-amino-2-selenoxo-1,3-imidazole-4-carboselenoamide **145** and selenoureas **148**.

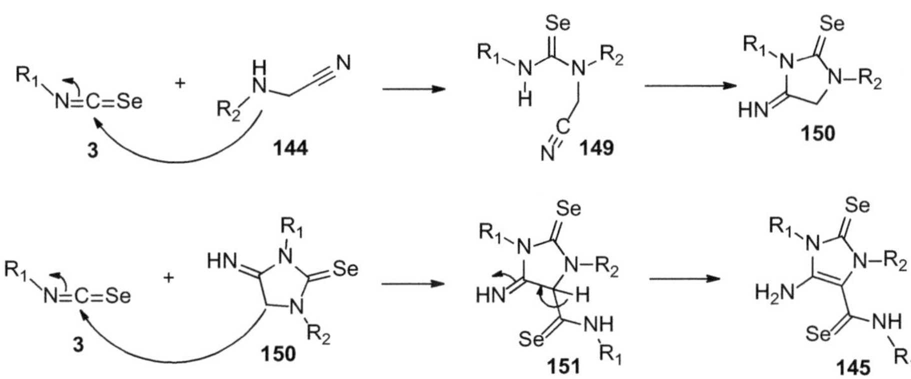

Figure 3.32: Plausible reaction mechanism for 5-amino-2-selenoxo-1,3-imidazole-4-carboselenoamide **145**.

It is easy to distinguish between selenoamide and selenourea by comparison of their chemical shift differences in the ^{77}Se-NMR spectra. In ^{13}C-NMR spectra, most chemical shifts for carbonyl carbons of selenocarbonyl groups in selenoureas are observed at higher fields than those of selenoamides; however, the difference is minimal. In contrast, the chemical shift in ^{13}C NMR of the C-atom of the selenocarbonyl group in selenoureas **A** (193.4 ppm ± 3.1), **B** (180.2 ppm ± 0.6), **C** (177.3 ppm ± 1.6), and **E** (183.6 ppm ± 0.8) is observed at lower fields than the chemical shift (174.9 ppm ± 0.7 or 179.4 ppm ± 0.1) of Se(2) of the selenoamide group in compound **145.** In the case of selenoureas including compounds **145,** ^{77}Se-NMR signals for Se-atoms are observed in the range of 88–336 ppm. The chemical shift values for selenium in selenoamides fall in the range of 491–702 ppm, which is obviously lower than those for

Table 3.2: Chemical shifts of selenocarbonyl group in selenoureas and selenoamides [76].

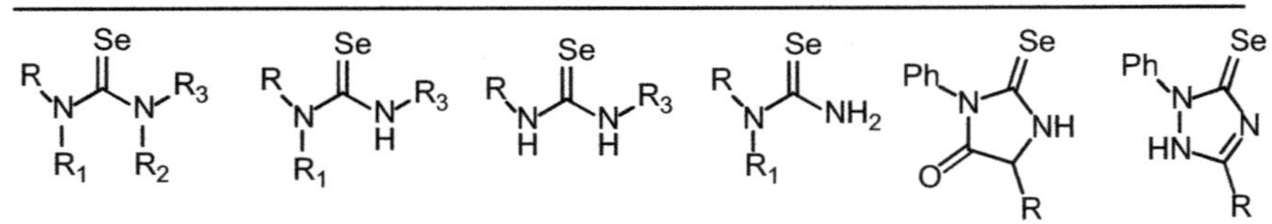

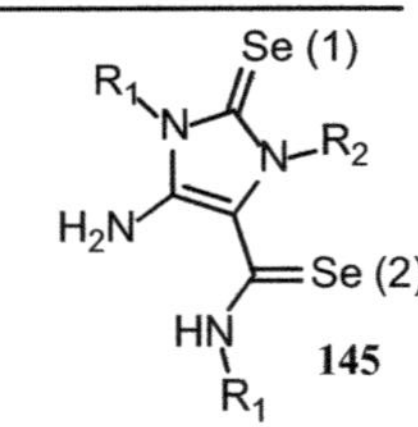

	Selenoureas						Se (1)	
	A	B	C	D	E	F	145, R_2 = Ph	145, R_2 = CH_3
Se[a]	336.2±20.3	177.7±7.7	194.7±53.0	225.7±13.3	267.8±16.1	88.7±1.1	137.9±24.3	94.5±19.3
Se=C[b]	193.4±3.1	180.2±0.6	177.3±1.6	176.4±1.7	183.6±0.8	158.5±1.4	161.8±0.3	158.6±0.9

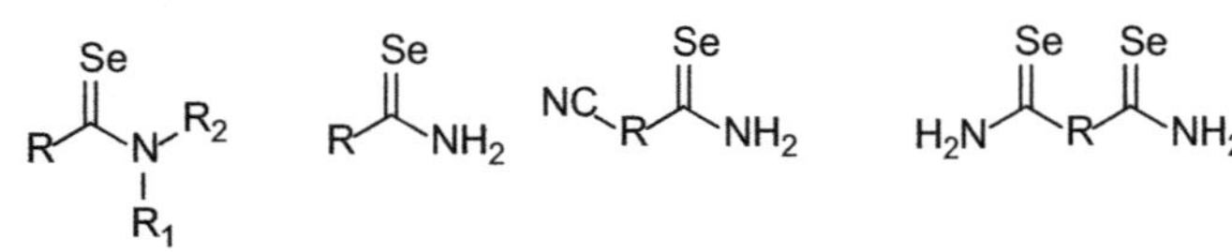

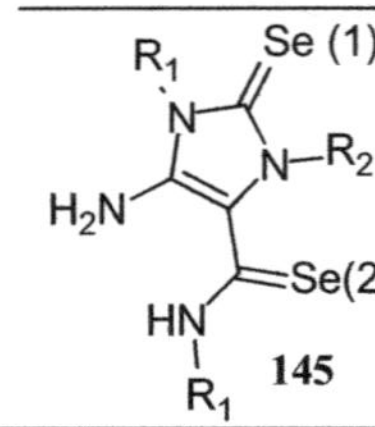

	Selenoamides				Se (2)	
	G	H	I	J	145, R_2=Ph	145, R_2=CH_3
Se[a]	701.5 ± 41.7	629.3 ± 53.7	614.9 ± 83.7	543.6 ± 62.7	491.3±72.6	544.3±19.2
Se = C[b]	203.3 ± 1.8	212.4 ± 6.7	210.6 ± 4.4	208.4 ± 3.7	174.9±0.7	179.4±0.1

[a] Average chemical shift of selenium in ^{77}Se NMR spectra.

[b] Average chemical shift of selenocarbonyl carbon in ^{13}C-NMR spectra.

selenoureas. The average chemical shift (94.5 ppm ± 19.3) of Se (1) (urea) for compound 3 (R_2 = methyl) is at a higher field than that of **145** (R_2 = phenyl, 137.9 ppm ± 24.3). The average chemical shift (544.3 ppm ± 19.2) of Se (2) (amide) for compound **145** (R_2 = methyl) is also at lower field than that of **145** (R_2 = phenyl, 491.3 ppm ± 72.6). This clearly reflects the variable pattern of electron density provided by the alkyl and aryl substituents around the selenium atom in **145**. Furthermore, the selenoamide and selenourea fragments can be distinguished by their significant differences in their chemical shifts in the ^{77}Se-NMR spectra [13, 77–83].

3.12 Conclusions

This chapter offers an updated overview on the synthesis of isoselenocyanates, and its applications for the synthesis of selenium-containing heterocycles. This chapter includes preparation of isoselenocyanates by various methods, imidoyl isoselenocyanates, and their reactions. Isoselenocyanates were used as versatile synthons in the preparation of Se-containing organic heterocycles. The synthesis of 1, 3-selenazole via MCRs using isoselenocyanate was also described. The selenoimidoylation of alcohol was successfully achieved via isoselenocyanate intermediates. Also this chapter showed the synthesis of benzo[*c*]selenophene, 1,3-oxaselenan-2-imine, 1,3-oxaselenepane, 2-amino-1,3,4-oxadiazoles, 5-amino-2-selenoxo-1,3-imidazole-4-carboselenoamide and cycloaddition of carbodiimide.

References

[1] Banerjee, B., and Koketsu, M. Recent developments in the synthesis of biologically relevant selenium-containing scaffolds. Coord Chem Rev 2017, 339, 104–127.

[2] Ninomiya, M., Garud, DR., and Koketsu, M. Biologically significant selenium-containing heterocycles. Coord Chem Rev 2011, 255, 2968–2990.

[3] Garud, DR., Koketsu, M., and Ishihara, H. Isoselenocyanates: A powerful tool for the synthesis of selenium-containing heterocycles. Molecules 2007, 12, 504–535.

[4] Sommen, GL., Linden, A., and Heimgartner, H. Selenium-containing heterocycles from isoselenocyanates: Synthesis of 1,3-selenazolidine and perhydro-1,3-selenazine derivatives. Eur J Org Chem 2005, 2005, 3128–3137.

[5] Garud, DR., and Koketsu, M. Synthesis of 3-selena-1-dethiacephems and selenazepines *via* iodocyclization. Org Lett 2008, 10, 3319–3322.

[6] Bulka, E. In the chemistry of cyanates and their thio derivatives, part 2, Ed., S. Patai, 1977, chap. 19, John Wiley & Sons, Ltd, Chichester.

[7] Toshimitsu, A., and Uemura, S. In the chemistry of organic selenium and tellurium compounds, Vol. 2, Ed., S. Patai, 1987, chap. 14, John Wiley & Sons, Ltd, Chichester.

[8] Ninomiya, M., Garud, DR., and Koketsu, M. Selenium-containing heterocycles using selenoamides, selenoureas, selenazadienes, and isoselenocyanates. Heterocycles 2010, 81, 2027–2055.

[9] Barton, DHR., Parekh, SI., Tajbakhsh, M., Theodorakis, EA., and Chi-Lam, T. A convenient and high yielding procedure for the preparation of isoselenocyanates. Synthesis and reactivity of *O*-alkylselenocarbamates. Tetrahedron 1994, 50, 639–654.
[10] Fang, Y., Zhu, ZL., Xu, P., Wang, SY., and Ji, SJ. Aerobic radical-cascade cycloaddition of isocyanides, selenium and imidamides: Facile access to 1,2,4-selenadiazoles under metal-free conditions. Green Chem 2017, 19, 1613–1618.
[11] Fernandez-Bolanos, JG., Lopez, O., Ulgar, V., Maya, I., and Fuentes, J. Synthesis of *O*-unprotected glycosyl selenoureas. A new access to bicyclic sugar isoureas. Tetrahedron Lett 2004, 45, 4081–4084.
[12] Chang, WJ., Kulkarni, MV., and Sun, CM. Regioselective one-pot three component synthesis of chiral 2-iminoselenazolines under sonication. RSC Adv 2015, 5, 97113–97120.
[13] Koketsu, M., Takahashi, A., and Ishihara, H. A facile preparation of selenohydantoins using isoselenocyanate. J Heterocycl Chem 2007, 44, 79–81.
[14] Koketsu, M., Suzuki, N., and Ishihara, H. Preparation of isoselenocyanate and synthesis of carbodiimide by oxidation of selenourea. J Org Chem 1999, 64, 6473–6475.
[15] Dushenko, GA., Mikhailova, OI., Mikhailov, IE., Minyaev, RM., and Minkin, VI. Rearrangements of cyclopentadienyl cyanates, isocyanates and their thio-, seleno-, and telluro-analogs. Russ Chem Bull 2009, 58, 1713–1723.
[16] Mikhailov, IE., Dushenko, GA., Reck, G., Schulz, B., Mikhailova, OI., and Minkin, VI. Structure and rearrangements of 3-iso(thio,seleno)cyanato-1,2,3-triarylcyclopropenes. Russ J Org Chem 2006, 42, 1630–1635.
[17] Duschenko, GA., Mikhailov, IE., Zschunke, A., Reck, G., Schulz, B., and Minkin, VI. Structure and rearrangements of 7-(1,2,3,4,5,6,7-heptaphenylcycloheptatrienyl) isocyanate, isothiocyanate and isoselenocyanate. Mendeleev Commun 1999, 9, 222–225.
[18] Atanassov, PK., Zhou, Y., Linden, A., and Heimgartner, H. Synthesis of bis(2,4-diarylimidazol-5-yl) diselenides from *N*-benzylbenzimidoyl isoselenocyanates. Helv Chim Acta 2002, 85, 1102–1117.
[19] Zhou, Y., and Heimgartner, H. Selenium-containing heterocycles from isoselenocyanates: Synthesis of 1,2,3-selenadiazole derivatives. Helv Chim Acta 2000, 83, 539–553.
[20] Koketsu, M., Otsuka, T., and Ishihara, H. Synthesis of 1,3-selenazetidine derivatives from imines and thiocarbamoyl isoselenocyanate. Heterocycles 2006, 68, 2107–2112.
[21] Toshimitsu, A. Organic selenocyanates, tellurocyanates and related compounds. In: Rappoport Z., ed. *PATAI's Chemistry of Functional Groups*, John Wiley & Sons, Ltd, 2013. Doi: 10.1002/9780470682531.pat0723
[22] Zakrzewski, J., Huras, B., and Kiełczewska, A. Synthesis of isoselenocyanates. Synthesis 2016, 48, 85–96.
[23] Fan, H., Wang, Z., Liu, X., Zheng, W., Guo, F., and Qian, Y. Controlled synthesis of trigonal selenium crystals with different morphologies. Solid State Commun 2005, 135, 319–2.
[24] Braverman, S., Cherkinsky, M., and Birsa, ML. Science of Synthesis, Knight, J. G, Ed., Geog Thieme Verlag, Stuttgart, 2005, Vol. 18, 65–320. see p. 248 in particular.
[25] Zakrzewski, J., and Krawczyk, M. Synthesis and pesticidal properties of thio and seleno analogs of some common urea herbicides. Phosphorus Sulfur Silicon Relat Elem 2009, 184, 1880–1903.
[26] Duddeck, H. ^{77}Se NMR spectroscopy and its applications, in chemistry annual reports on NMR spectroscopy, Vol. 52, Elsevier, Amsterdam, 2004, 105–166.
[27] Zhou, Y., Linden, A., and Heimgartner, H. Selenium-containing heterocycles from isoselenocyanates: Synthesis of 1,3-selenazoles from *N*-phenylimidoyl isoselenocyanates. Helv Chim Acta 2000, 83, 1576–1598.

[28] Zhou, Y., Linden, A., and Heimgartner, H. Reactions of imidoyl isoselenocyanates with aromatic 2-amino *N*-heterocycles and 1-methyl-1H-imidazole. Helv Chim Acta 2011, 94, 1575–1585.

[29] Atanassov, PK., Linden, A., and Heimgartner, H. Derivatives from isoselenocyanates: Synthesis of 2-phenyl-6H-[5,1,3]benzoselenadiazocine. Helv Chim Acta 2004, 87, 1452–1466.

[30] Zhou, Y., Linden, A., and Heimgartner, H. Synthesis of 1,3,5-triazineselones from imidoyl isoselenocyanates and amidines. Heterocycles 2011, 84, 493–504.

[31] Banert, K., and Toth, C. Synthesis and reactions of vinyl isoselenocyanates. Angew Chem Int Ed Engl 1995, 34, 1627–1629.

[32] Banert, K., Huckstadt, H., and Vrobel, K. Synthesis and reactions of isothiocyanate-substituted allenes and 1,3-butadienes. Angew Chem Int Ed Engl 1992, 31, 90–92.

[33] Koketsu, M., and Ishihara, H. 1,3-Selenazoles. In: Katritzky AR, Ramsden CA, Scriven EFV, Taylor RJK. ed. *Comprehensive Heterocyclic Chemistry III*, Vol. 4, Elsevier Ltd., Oxford, UK, 2008, Chap. 4.08, 791–821.

[34] Purseigle, F., Dubreuil, D., Marchand, A., Pradere, JP., Goli, M., and Toupet, L. Synthesis and reactivity of *N*-selenoacylamidines precursors of selenoheterocycles. Tetrahedron 1998, 54, 2545–2562.

[35] Hullin, RP., Miller, J., and Short, WF. Amidines. Part V. Preparation of amidines from cyanides and substituted aminomagnesium halides. J Chem Soc 1947, 394–396.

[36] Ranjan, A., Yerande, R., Jadhav, M., Yerande, SG., and Dethe, DH. One-pot synthesis of 2-amino -1,3-selenazole *via* an intermediary amidinoselenourea. Eur J Org Chem 2015, 2015, 3230–3234.

[37] Xie, Y., Yang, P., and Chen, X. One-pot synthesis of 5-arylamino-1,3,4-selenadiazol-2(3*H*)-ones from arylisoselenocyanates. Journal of Chem Res 2012, 36, 421–424.

[38] Xie, Y., Liu, J., and Li, J. Selective synthesis of novel 2-imino-1,3-selenazolidin-4-ones and 2-amino-1,3,4-selenadiazin-5-ones from isoselenocyanates. Tetrahedron Lett 2011, 52, 932–935.

[39] Kaname, M., and Minoura Mao, Sashida H. One-pot copper-catalyzed tandem addition-cyclization of 2-iodoanilines with isoselenocyanates for the practical preparation of 2-aminobenzoselenazoles. Tetrahedron Lett 2011, 52, 505–508.

[40] Guo, YJ., Tang, RY., Zhong, P., and Li, JH. Copper-catalyzed tandem reactions of 2-halobenzenamines with isothiocyanates under ligand- and base-free conditions. Tetrahedron Lett 2010, 51, 649–652.

[41] Kobayashi, K., and Yokoi, Y. A convenient synthesis of 2-sulfanylbenzoselenazole derivatives *via* the reaction of 2-lithiophenyl isothiocyanates with selenium. Helv Chim Acta 2012, 95, 761–765.

[42] Benz, S., Mareda, J., Besnard, C., Sakai, N., and Matile, S. Catalysis with chalcogen bonds: Neutral benzodiselenazole scaffolds with high-precision selenium donors of variable strength. Chem Sci 2017, 8, 8164–8169.

[43] Lopez, O., Maza, S., Ulgar, V., Maya, L., and Fernandez-Bolanos, J. Synthesis of sugar-derived isoselenocyanates, selenoureas, and selenazoles. Tetrahedron 2009, 65, 2556–2566.

[44] Galina, K., Jaroslav, R., Zdenk, P., Zdenek, N., Katerina, V., and Alexandr, H. One-pot synthesis of 1-substituted-5-alkylselanyl-1H-tetrazoles from isoselenocyanates: Unexpected formation of *N*-alkyl-*N*-arylcyanamides and (*Z*)-Se-alkyl-*N*-cyano-*N*, *N′*-diarylisoselenoureas. Tetrahedron 2013, 69, 8798–8808.

[45] Fang, Y., Wang, SY., Shen, XB., and Ji, SJ. Base-promoted cascade reaction of isocyanides, selenium and amines: A practical approach to 2-aminobenzo[*d*][1,3]selenazines under metal-free conditions. Org Chem Front 2015, 2, 1338–1341.

[46] Fujiwara, SI., Shikano, Y., Shin-ike, T., Kambe, N., and Sonoda, N. Stereoselective synthesis of new selenium-containing heterocycles by cyclocarbonylation of aminoalkynes with carbon monoxide and selenium. J Org Chem 2002, 67, 6275–6278.

[47] Asanuma, Y., Fujiwara, SI., Shin-ike, T., and Kambe, N. Selenoimidoylation of alcohols with selenium and isocyanides and its application to the synthesis of selenium-containing heterocycles. J Org Chem 2004, 69, 4845–4848.
[48] Dong, B., Lu, S., Jiang, J., Zhang, WJ., Zhu, WM., Tang, MS., and Yang, GY. Aerobic oxidative radical coupling catalyzed by TEMPO for the preparation of 2,2′-dibenzothiazole disulfide. J Chem Tech Biotech 2012, 87, 341–345.
[49] Cava, MP., and Saris, LE. Synthesis of anhydro-2,3,5-triphenyl-4-hydroxyselenazolium hydroxide, a mesoionic selenium heterocycle. J Chem Soc Chem Commun 1975, 15, 617–618.
[50] Meese, CO., Walter, W., Mrotzek, H., and Mirzai, H. Selenoimidsaureester. Chem Ber 1976, 109, 956–964.
[51] Detty, MR., and Wood, GP. Substitution reactions of thallous thiophenoxide and thallous phenyl selenide with halogen-bearing substrates. J Org Chem 1980, 45, 80–89.
[52] Yamago, S., Miyazoe, H., Goto, R., and Yoshida, J. Radical-mediated imidoylation of telluroglycosides. Insertion of isonitriles into the glycosidic carbon-tellurium bond. Tetrahedron Lett 1999, 40, 2347–2350.
[53] Ogawa, A., Doi, M., Tsuchii, K., and Hirao, T. Selective sequential addition of diphenyl diselenide to ethyl propiolate and isocyanides upon irradiation with near-UV light. Tetrahedron Lett 2001, 42, 2317–2319.
[54] Kuniyasu, H., Maruyama, A., and Kurosawa, H. Palladium-catalyzed reaction of diselenide with isocyanide. Organometallics 1998, 17, 908–913.
[55] Fujiwara, SI., Maeda, H., Matsuya, T., Shin-ike, T., Kambe, N., and Sonoda, N. Imidoylation of acidic hydrocarbons with selenium and isocyanides: A new synthetic method for preparation of selenoimidates. J Org Chem 2000, 65, 5022–5025.
[56] Aqad, E., Lakshmikantham, MV., Cava, MP., Broker, GA., and Rogers, RD. Synthesis of benzo [*c*]selenophene and derivatives *via* new routes. Org Lett 2003, 5, 2519–2521.
[57] Mohanakrishnan, AK., and Amaladass, P. Synthesis of 1,3-diarylbenzo[*c*]selenophenes. Tetrahedron Lett 2005, 46, 7201–7204.
[58] Mohanakrishnan, AK., Kumar, NS., and Amaladass, P. Synthesis and characterization of 9,9-dialkylfluorene capped benzo[*c*]thiophene/benzo[*c*] selenophene analogs as potential OLEDs. Tetrahedron Lett 2008, 49, 4792–4795.
[59] Kaname, M., and Sashida, H. Tandem addition-cyclization of o-ethynylphenyllithiums and isoselenocyanates: A convenient preparation of functionalized benzo[*c*]selenophenes. Tetrahedron Lett 2011, 52, 3279–3282.
[60] Atanassov, PK., Linden, A., and Heimgartner, H. Synthesis of 3-acetyl-*N*-aryl-4-diethylaminoselenet-2(2H)-imines from 4-diethylamino-3-butyn-2-one and aryl selenocyanates. Heterocycles 2004, 62, 521–533.
[61] Sommen, GL., Linden, A., and Heimgartner, H. Selenium-containing heterocycles from isoselenocyanates: Cycloaddition of carbodiimides to selenazetidines. Helv Chim Acta 2005, 88, 766–773.
[62] Sommen, GL., Linden, A., and Heimgartner, H. Synthesis of 2-selenoxo-1,3-thiazolidin-4-ones and 2-selenoxo-1,3-thiazinan-4-ones from isoselenocyanates. Heterocycles 2005, 65, 1903–1915.
[63] Kanda, T., Aoki, H., Mizoguchi, K., Shiraishi, S., Murai, T., and Kato, S. Selenotellurocarbamic acid Te-alkyl esters: First isolation and characterization. Organometallics 1996, 15, 5753–5755.
[64] Favero, F., Sommen, GL., Linden, A., and Heimgartner, H. Synthesis of 5-selenoxo-1,2,4-triazole-1-carboxylates from isoselenocyanates and azodicarboxylates. Heterocycles 2006, 67, 749–762.
[65] Baldwin, JE. Rules for ring closure. J Chem Soc Chem Commun 1976, 18, 734–736.

[66] Sommen, GL., Linden, A., and Heimgartner, H. Selenium-containing heterocycles from isoselenocyanates: Use of hydrazine for the synthesis of 1,3,4-selenadiazine derivatives. Helv Chim Acta 2006, 89, 1322–1329.
[67] Sommen, GL., and Heimgartner, H. Synthesis of 1,3-oxaselenan-2-imines from isoselenocyanates. Pol J Chem 2007, 81, 1413–1418.
[68] Garud, DR., Toyoda, Y., and Koketsu, M. First synthesis of 1,3-oxaselenepanes. Tetrahedron Lett 2009, 50, 3035–3037.
[69] Garud, DR., Makimura, M., Ando, H., Ishihara, H., and Koketsu, M. First regioselective iodocyclization of *O*-allylselenocarbamates. Tetrahedron Lett 2007, 48, 7764–7768.
[70] Sommen, GL., Linden, A., and Heimgartner, H. Selenium-containing heterocycles from isoselenocyanates: Synthesis of 2-methylidene-1,3-selenazolidine derivatives. Tetrahedron 2006, 62, 3344–3354.
[71] Sommen, GL., Linden, A., and Heimgartner, H. Selenium-containing heterocycles from isoselenocyanates: Synthesis of 1,3-selenazinane and 1,3-selenazinanone derivatives. Helv Chim Acta 2007, 90, 472–487.
[72] Sommen, GL., Linden, A., and Heimgartner, H. First synthesis of a selenazepane. Tetrahedron Lett 2005, 46, 6723–6725.
[73] Xie, Y., Liu, J., Yang, P., Shi, X., and Li, J. Synthesis of 2-amino-1,3,4-oxadiazoles from isoselenocyanates *via* cyclodeselenization. Tetrahedron 2011, 67, 5369–4.
[74] Xie, Y., Zhang, F., Li, JJ., and Shi, XJ. Novel synthesis of 2-aminobenzimidazoles from isoselenocyanates. Synlett 2010, 6, 901–904.
[75] Xie, Y., Zhang, F., Cheng, XD., and Li, JJ. An efficient synthesis of 2-aminobenzoxazoles and 2-aminobenzothiazoles from 2-aminophenols or 2-aminothiophenols and isoselenocyanates. Heterocycles 2010, 81, 2087–2096.
[76] Tanahashi, N., and Koketsu, M. Synthesis of 5-amino-2-selenoxo-1,3-imidazole-4-carboselenoamides by the reaction of isoselenocyanates with aminoacetonitriles. Tetrahedron Lett 2011, 52, 4650–4653.
[77] Koketsu, M., Fukuta, Y., and Ishihara, H. Reaction of *N, N*-dimethylselenocarbamoyl chloride with nucleophiles. Preparation of diselenocarbamates, selenothiocarbamates, and selenoureas. J Org Chem 2002, 67, 1008–1011.
[78] Koketsu, M., Takakura, N., and Ishihara, H. Efficient synthesis of selenoureas from the corresponding carbodiimides. Synth Commun 2002, 32, 3075–3079.
[79] Koketsu, M., Fukuta, Y., and Ishihara, H. Preparation of *N, N*-unsubstituted selenoureas and thioureas from cyanamides. Tetrahedron Lett 2001, 42, 6333–6335.
[80] Koketsu, M., Yamamura, Y., and Ishihara, H. Synthesis of selenosemicarbazides and 1,2,4-Triazoles. Heterocycles 2006, 68, 1191–1200.
[81] Koketsu, M., Okayama, Y., and Ishihara, H. Facile synthesis of *N, N*-dialkylselenoamides from amides. Heteroat Chem 2002, 13, 195–198.
[82] Ishihara, H., Yosimura, K., and Kouketsu, M. A facile preparation of aliphatic and aromatic primary selenoamides using 4-methylselenobenzoate as a new selenating reagent. Chem Lett 1998, 27, 1287–1288.
[83] Koketsu, M., Takenaka, Y., and Ishihara, H. Syntheses of cyanoselenoamides and diselenoamides: Conversion into selenazoles and selenazines. Heteroat Chem 2003, 14, 106–110.

Masayuki Ninomiya and Mamoru Koketsu

4 Selenoureas and Their Applications

4.1 Introduction

Selenoureas are organoselenium compounds having two amine groups at both sides of a selenocarbonyl (C=Se) functional group. The simplest example is a selenium analogue of normal urea and its formula is $SeC(NH_2)_2$. In 1884, the synthesis of selenourea ($SeC(NH_2)_2$) from hydrogen selenide and cyanamide was reported for the first time by the French chemist, Auguste Victor Louis Verneuil [1]. Later, a number of the synthetic procedures have been developed by several research groups in the world [2, 3]. In addition, selenoureas have been used as precursors for the synthesis of selenium-containing heterocyclic compounds [4–6]. The synthetic studies using selenoureas are being actively investigated because of unique chemical behavior of Se atom. Recently, medicinal and pharmaceutical studies of selenium-containing compounds derived from selenoureas are becoming increasingly interesting due to their potent and diverse biological properties [7–9]. There are several excellent books and reviews concerned with selenoureas and their applications [2–9]. This chapter deals with methods for the preparation of selenoureas, the reactions using selenoureas, and their valuable applications in various fields of research.

4.2 Preparation of selenoureas

4.2.1 Using isoselenocyanate

The most common method for the preparation of selenoureas is based on the reaction of alkyl and aryl isoselenocyanates with amines (Figure 4.1). In 1937, Irwin B. Douglass for the first time reported the synthesis of acylselenoureas via in situ generation of isoselenocyanates using RCOCl and KSeCN [10]. However, intermediates were not detected. Isoselenocyanates have been extensively used as the starting materials in the synthesis of selenoureas [11, 12].

Selenoureido analogues (**1**) of 4-(4-fluorophenylureido)benzenesulfonamide (SLC-0111) were synthesized as carbonic anhydrase inhibitors (Figure 4.2) [13]. The selenoureido compounds (**1**) were obtained by standard coupling reactions of aromatic isoselenocyanates with benzenesulfoamides in MeCN. Selenoureidodipeptides (**2**) were prepared by the reaction of isoselenocyanates with amino acid esters. These reactions were clean and complete within 30 min at room temperature [14].

Masayuki Ninomiya and Mamoru Koketsu, Department of Chemistry and Biomolecular Science, Faculty of Engineering, Gifu University, Gifu, Japan

https://doi.org/10.1515/9783110625110-004

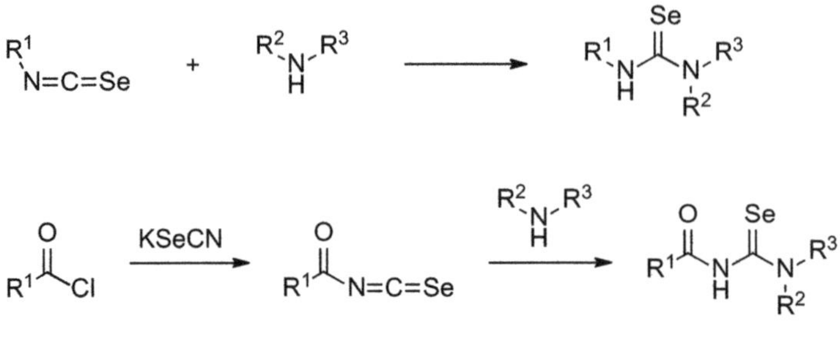

Figure 4.1: Preparation of selenoureas using isoselenocyanates.

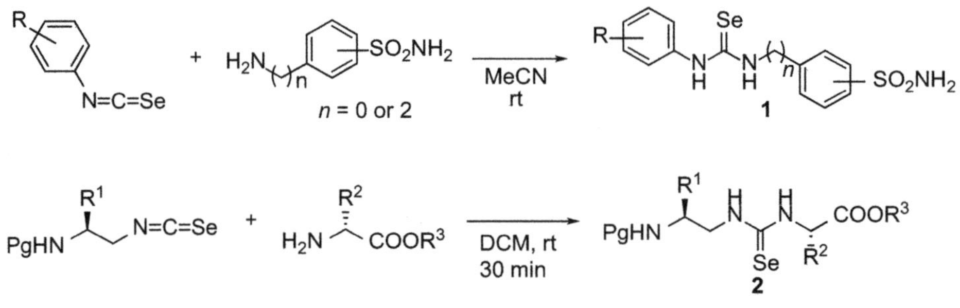

Figure 4.2: Synthesis of selenoureido analogues (1) of 4-(4-fluorophenylureido) benzenesulfonamide (SLC-0111) and selenoureidodipeptides (2).

β-D-Glycopyranosyl selenoureas with *gluco* and *manno* configurations (3) were prepared from the corresponding β-D-glycopyranosylamines with phenyl isoselenocyanates in aqueous pyridine (Figure 4.3) [15]. This procedure did not require the protection of hydroxy groups in sugar.

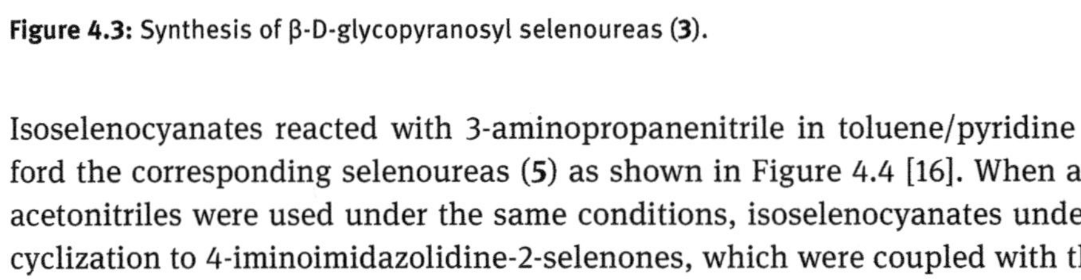

Figure 4.3: Synthesis of β-D-glycopyranosyl selenoureas (3).

Isoselenocyanates reacted with 3-aminopropanenitrile in toluene/pyridine to afford the corresponding selenoureas (5) as shown in Figure 4.4 [16]. When aminoacetonitriles were used under the same conditions, isoselenocyanates underwent cyclization to 4-iminoimidazolidine-2-selenones, which were coupled with the additional molecule of aminoacetonitriles, giving 5-amino-2-selenoxo-1,3-imidazole-4-carboselenoamides (4: cyclic selenoureas).

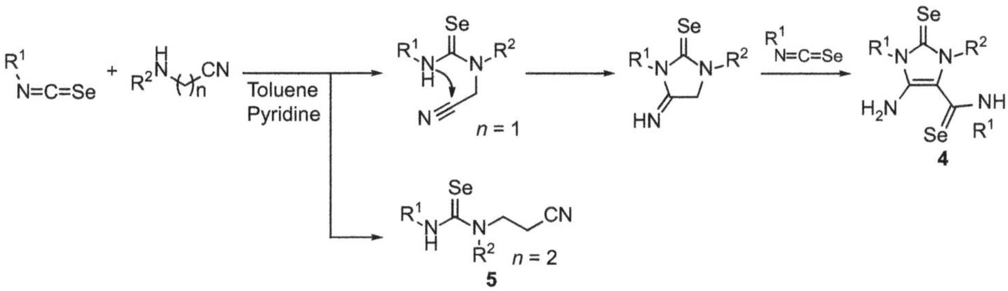

Figure 4.4: Reactions of isoselenocyanates with 3-aminopropanenitrile and aminoacetonitriles.

4.2.2 Using elemental selenium

The simplest way for the synthesis of selenoureas is by using elemental selenium, which is easy to handle and affordable. *N,N′*-Dimethyl-*N,N′*-diphenylselenourea (**6**) was prepared by fusion of elemental selenium with *N,N′,N″*-trimethyl-*N,N′,N″*-triphenylmethanetriamine (Figure 4.5) [17]. In the other way, refluxing orthoformic acid derivatives in toluene or xylene with elemental selenium furnished tetramethylselenourea (**7**) [18]. The condensation of 2-bromophenyl isocyanide with elemental selenium and piperidine in the presence of 1,8-diazabicyclo[5.4.0]undec-7-ene in THF afforded *N*-(2-bromophenyl)selenourea (**8**) [19]. In addition, the three-component condensation reactions of elemental selenium with triethyl orthoformate and primary or secondary diamines were reported [20]. These reactions were conducted under solvent-free conditions at high temperature, giving *N,N'*-disubstituted cyclic selenoureas (**9** and **10**).

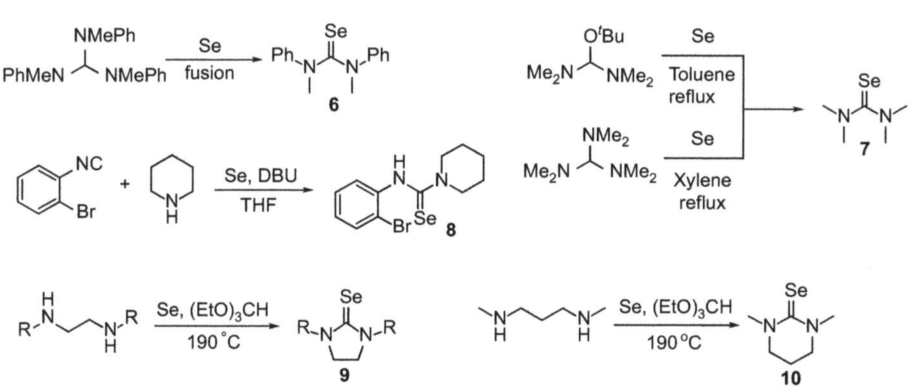

Figure 4.5: Synthesis of various selenoureas (6–10) using elemental selenium.

N-methylimidazole-based selenoureas (**11**) were synthesized by the treatment of *N*-methylimidazole with appropriate dibromoalkanes, followed by reaction with elemental selenium in the presence of K_2CO_3 by MeOH refluxing (Figure 4.6) [21]. The selenoureas could rapidly inhibit peroxynitrile- and peroxidase-mediated nitration of protein tyrosine residues. The reaction of selenourea (n = 3) with peroxynitrile or hydrogen peroxide produced the corresponding seleninic acid derivative, which upon elimination of selenous acid afforded the *N*-methylimidazole derivative (Figure 4.7).

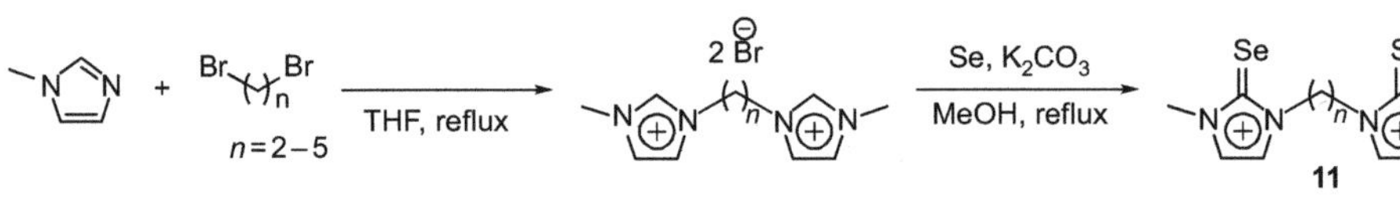

Figure 4.6: Synthesis of *N*-methylimidazole-based selenoureas (**11**).

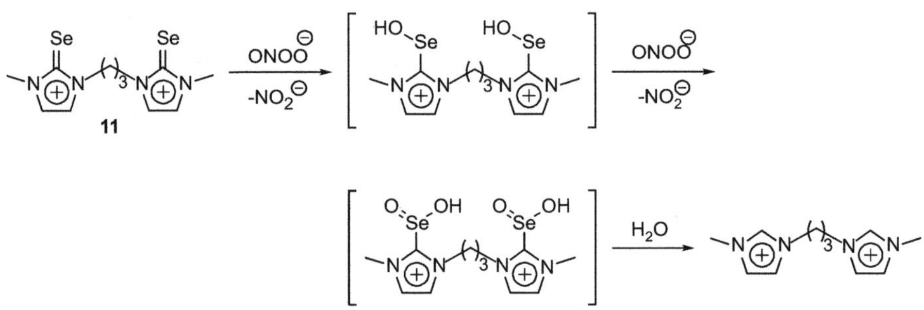

Figure 4.7: Reaction of *N*-methylimidazole-based selenourea (n = 3) **11** with peroxynitrile.

4.2.3 Using sodium hydroselenide

In 1956, selenopyrimidine containing a selenourea moiety (**12**) was prepared from 2,4-dichloropyrimidine with the nucleophilic selenium species sodium hydroselenide (NaSeH) (Figure 4.8) [22]. In the last 1960s, the preparation of tetramethylselenourea (**7**) by the replacement of S atom by HSe^- ion was reported by Daniel L. Klayman and Robert J. Shine [23]. Treatment of tetramethylthiourea with methyl iodide gave the pentamethyl analogue, which was immediately added to an aqueous ethanolic solution containing NaSeH [24]. With loss of methyl mercaptan, this reaction produced tetramethylselenourea (**7**).

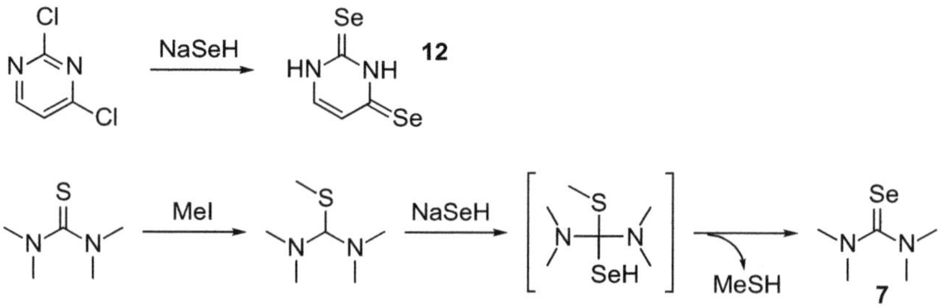

Figure 4.8: Synthesis of selenopyrimidine (**12**) and tetramethylselenourea (**7**) using NaSeH.

Similarly, a selenouridine derivative (**13**) was prepared from a 2-thioiridine derivative with maintenance of *tert*-butyldimethylsilyl protection groups via *S*-methylation and nucleophilic substitution using NaSeH generated in situ from Se and $NaBH_4$ in EtOH (Figure 4.9) [25].

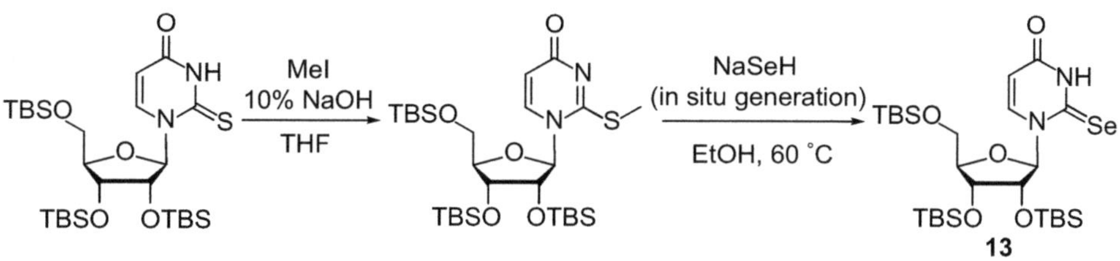

Figure 4.9: Synthesis of 2-selenouridine (**13**).

Very recently, pyrimidinic selenoureas (**14**) were prepared via *S*-methylation of thioureas, followed by nucleophilic substitution with NaSeH by EtOH refluxing. The NaSeH did not influence other functionalities (Figure 4.10) [26].

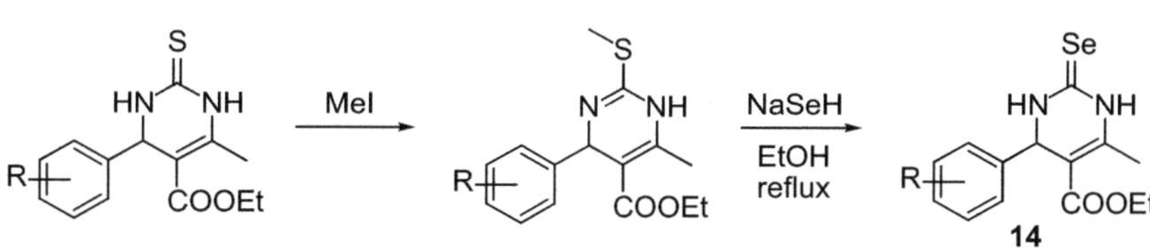

Figure 4.10: Synthesis of pyrimidinic selenoureas (**14**).

4.2.4 Using lithium aluminum hydride hydroselenide

Lithium aluminum hydride hydroselenide (LiAlHSeH) is a useful selenating reagent involved in the preparation of various organoselenium compounds. The LiAlHSeH was prepared by our group (Mamoru Koketsu and Hideharu Ishihara) from $LiAlH_4$ with elemental selenium in 2001 [27]. The LiAlHSeH has advantages because of its high efficiency and wide-ranging utility [28, 29]. In addition, this reagent can be applied for the synthesis of selenoureas. Following activation of cyanamides and carbodiimides using HCl in Et_2O, treatment with LiAlHSeH affords the corresponding selenoureas (Figure 4.11).

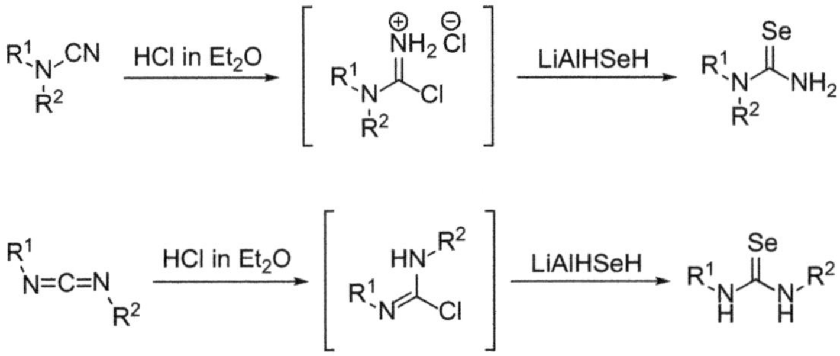

Figure 4.11: Preparation of selenoureas using LiAlHSeH.

Recently, the facile one-pot synthesis of selenoureidopeptides (**15**) employing LiAlHSeH through the Staudinger aza-Wittig-type reaction was reported (Figure 4.12) [30]. The general mechanism of the carbodiimide synthesis by aza-Wittig condensation

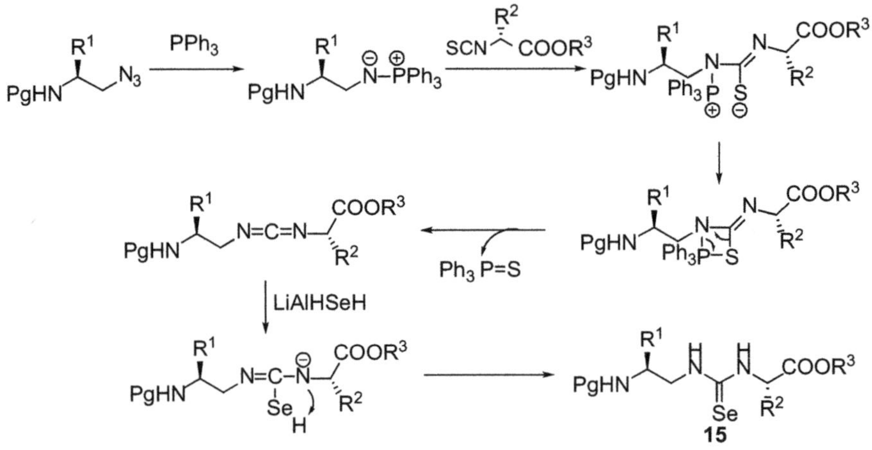

Figure 4.12: Synthesis of selenoureidopeptides (**15**) via the Staudinger aza-Wittig-type reaction.

proceeds through the reaction of alkyl azides with isothiocyanates to form betaine intermediates. The ring-opening of thiaphosphetanes with loss of triphenylphosphine sulfide gave carbodiimide intermediates, which are subjected to reaction with LiAlHSeH. The carbodiimidyl selenol species then rearranged to form the desired selenoureidopeptides (**15**).

4.2.5 Using Woollins' reagent

Woollins' reagent is a highly efficient selenating reagent [31–37]. As its name suggests, Woollins' reagent was developed by John Derek Woollins using $(PhP)_5$, or Li_2Se with $PhPCl_2$, and elemental selenium between the late 1980s and the beginning of the 1990s. In recent years, his research group has proposed the synthetic methods of selenoureas (**18**) from cyanamides with Woollins' reagent (Figure 4.13) [38].

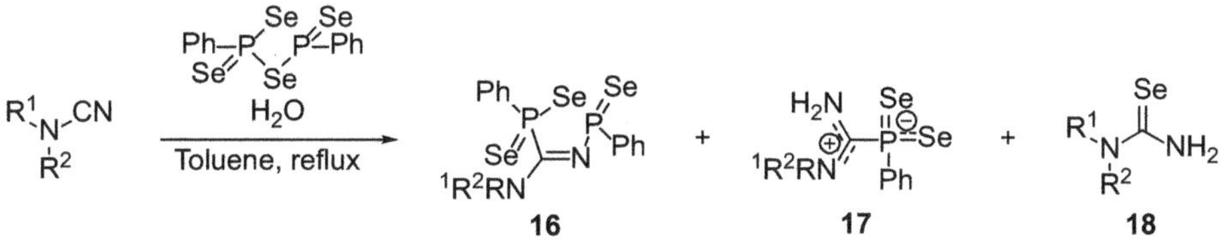

Figure 4.13: Preparation of selenoureas (**18**) using Woollins' reagent.

Reactions of cyanamides and Woollins' reagent in toluene at reflux with post-water-treatment furnished selenazadiphosphoaminediselenides (**16**), carbamidoyl (phenyl)phosphinodiselenoic acids (**17**), and selenoureas (**18**). His research group explained the possible mechanism of the formation of selenoureas as follows: Woollins' reagent at elevated temperature was in equilibrium with a diselenaphosphorane $PhP(Se)_2$, which was believed to be a true reactive species in refluxing toluene. The initial step was a typical [2 + 2] cycloaddition of a P=Se bond from diselenaphosphorane $PhP(Se)_2$ across the C≡N bond of cyanamide to give intermediates as three tautomeric forms. The hydrolysis of the intermediate **B** with one molecule of H_2O can afford selenoureas (**18**) by further loss of $(PhPO_2)_3$ and Se (Figure 4.14).

4.2.6 Using bis(dimethylaluminium) selenide

Bis(dimethylaluminium) selenide $(Me_2Al)_2Se$ is a useful selenating reagent, which can transform various carbonyl compounds [39–43]. This reagent was developed by

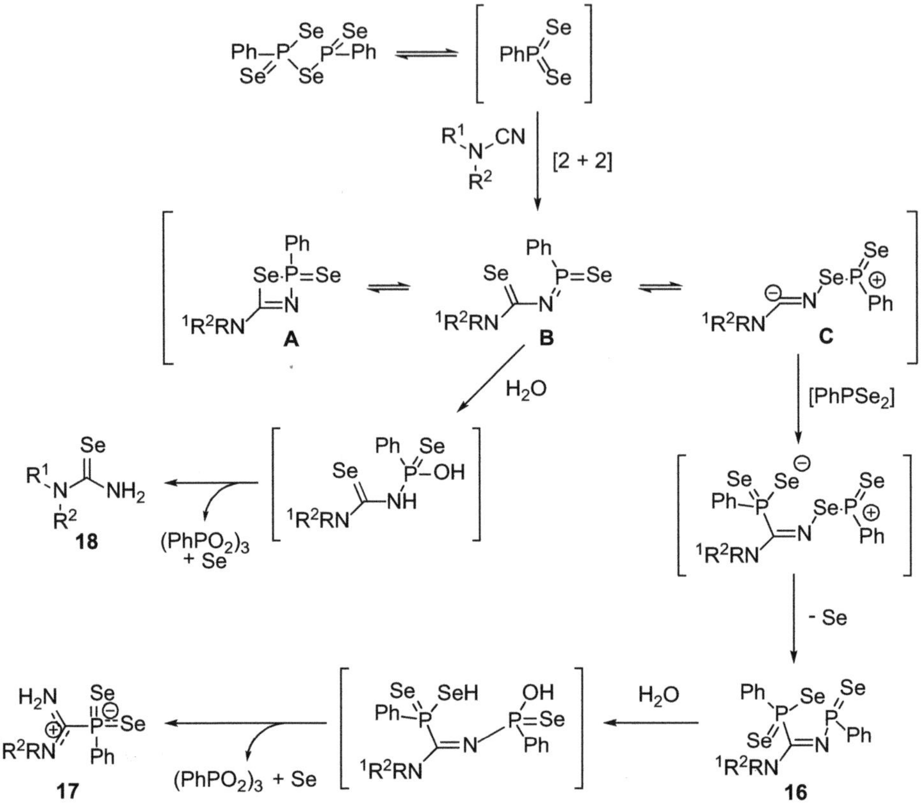

Figure 4.14: Proposed mechanism for the formation of selenoureas (18) using Woollins' reagent.

Masahito Segi's research group in 1990s. His group reported the one-pot synthesis of selenoureas via selenation of isocyanates with $(Me_2Al)_2Se$ (Figure 4.15) [44]. This reaction proceeded through in situ generation of isoselenocyanates, followed by the addition of amines, affording both of symmetrical and unsymmetrical selenoureas.

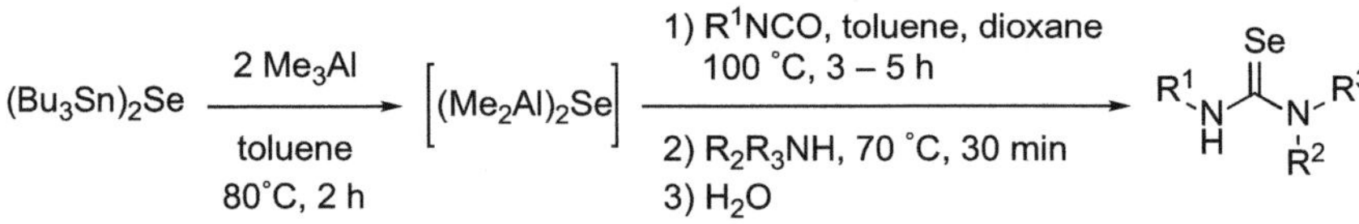

Figure 4.15: Preparation of selenoureas using $(Me_2Al)_2Se$.

4.2.7 Using tetraethylammonium tetraselenotungstate

Tetraethylammonium tetraselenotungstate $[Et_4N]_2WSe_4$ is a versatile selenium transfer reagent in organoselenium chemistry. Samuel O'Neal and Joseph W. Kolis for the first time prepared this reagent by treatment of K_2Se_3 with $W(CO)_6$ and Et_4NBr in 1981 [45], and it has been successfully utilized by Srinivasan Chandrasekaran's research group [46–51]. His group presented the one-pot protocol for the synthesis of *N,N*-dimethylselenoureas under mild conditions by the reaction of primary and secondary amines with Viehe's iminium salt (phosgene iminium chloride) and $[Et_4N]_2WSe_4$ (Figure 4.16). The proposed pathway involves a nucleophilic addition of amines to Viehe's iminium salt, followed by displacement of one of the chloride ions by the amines. The $[Et_4N]_2WSe_4$ attacks the intermediates, giving the corresponding selenoureas with elimination of WSe_3 (Figure 4.17).

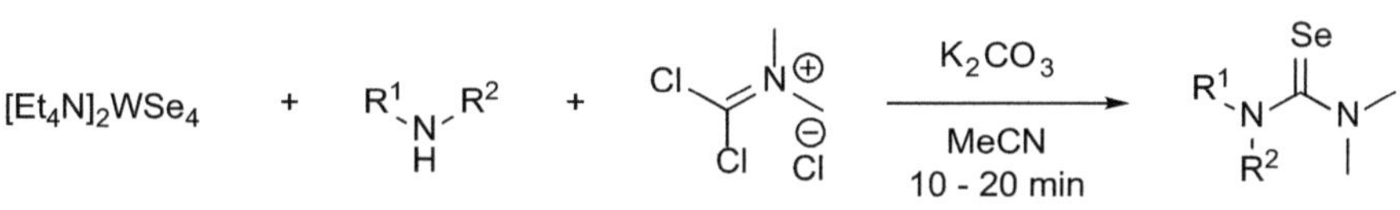

Figure 4.16: Preparation of selenoureas using $[Et_4N]_2WSe_4$.

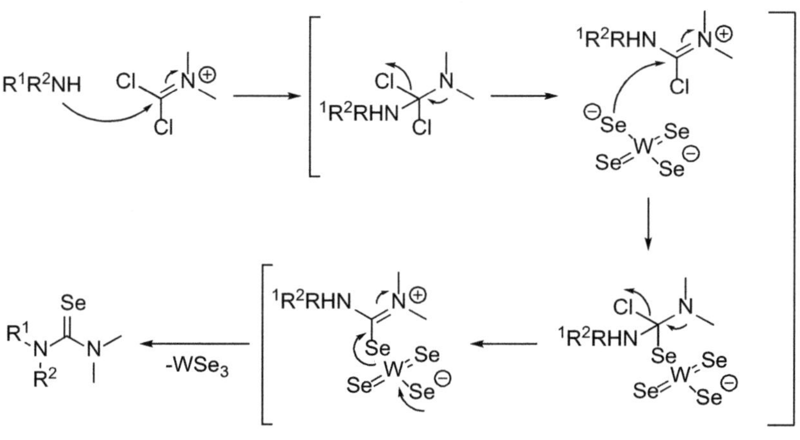

Figure 4.17: Proposed mechanism for the formation of selenoureas using $[Et_4N]_2WSe_4$.

4.3 Reactions using selenoureas

4.3.1 Synthesis of 1,3-selenazoles

In recent years, various Se-containing five-membered ring compounds have been extensively studied in organic synthesis and also medicinal chemistry. Selenazoles, first reported in 1889 by Hofmann G [52], contain one Se atom and one N atom in their rings. The selenazole moiety is present in many pharmacologically active substances. Considerable interest in the synthesis and biological activities of selenazoles was grown due to their potentials for practical applications. Especially, 1,3-selenazoles (N atom at the 3-position and two double bonds) are the most interesting heterocycles that are being reported for their potential pharmacological profile [8].

For the construction of the 1,3-selenazole core, selenoureas and α-halocarbonyl compounds are commonly employed [53, 54]. In the classical Hantzsch reaction, ring closure involves amine and carbonyl groups of alkylated isoselenoureas with the formation of dihydroselenazoles. Through dehydration, the desired 1,3-selenazoles are produced (Figure 4.18).

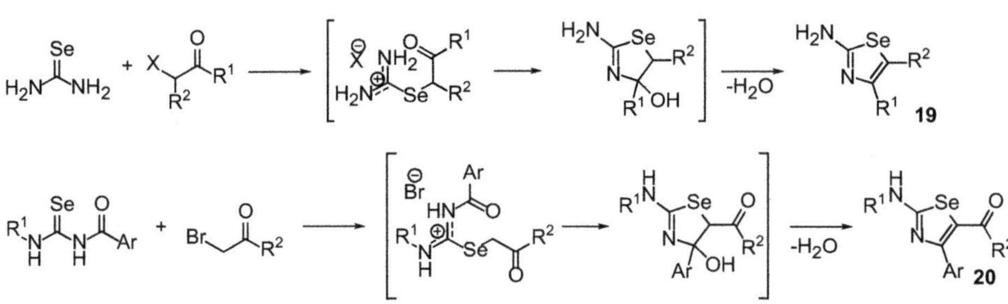

Figure 4.18: Preparation of 1,3-selenazoles (**19** and **20**) by the reaction of selenoureas with α-halocarbonyl compounds.

Numerous examples for the synthesis of 2-amino-1,3-selenazoles derived from *N*,*N′*-unsubstituted selenourea and α-halocarbonyl compounds were reported (Figure 4.19) [55–58]. The selenourea reacted with dichloroacetone, 2-bromo-2-arenesulfonyl-1-phenylethanones, hydrazonoyl bromides, 2-bromomalonaldehyde, ethyl 3-bromo-2-oxopropanate, and methyl 2-chloro-3-oxopropanoate, giving 4-chloromethyl-1,3-selenazol-2-amine (**21**), 5-arenesulfonyl-4-phenyl-1,3-selenazol-2-amines (**22**), 5-aryldiazenyl-4phenyl-1,3-selenazol-2-amines (**23**), 2-amino-1,3-selenazole-5-carbaldehyde (**24**), ethyl 2-amino-1,3-selenazole-4-carboxylate (**25**), and methyl 2-amino-1,3-selenazole-5-carboxylate (**26**), respectively.

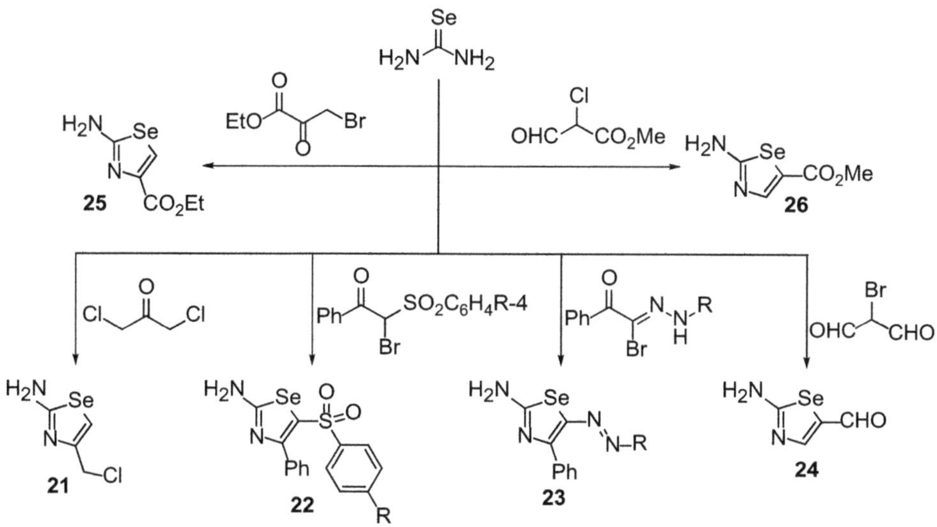

Figure 4.19: A series of 2-amino-1,3-selenazoles (**26–26**) from selenoureas.

There are several reports regarding eco-friendly preparation of 1,3-selenazoles starting from selenoureas [59–61]. A simple and environmentally benign synthesis of 2-amino-1,3-selenazoles (**27**) was achieved by microwave irradiation using 1-butyl-3-methylimidazolium tetrafluoroborate ([Bmim]BF_4) ionic liquid (Figure 4.20) [62].

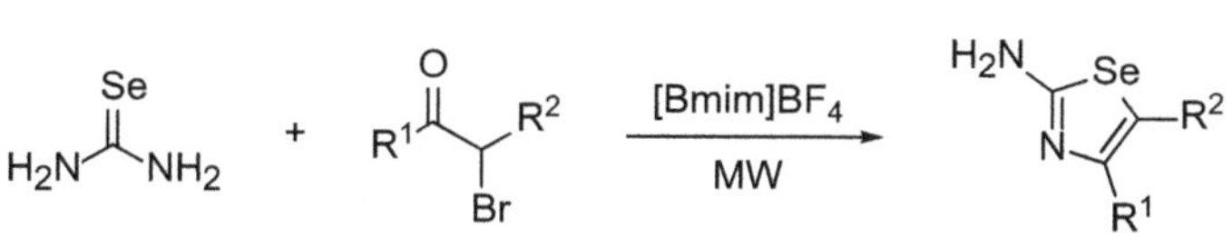

Figure 4.20: Eco-friendly preparation of 2-amino-1,3-selenazoles (**27**).

A novel and efficient biomimetic conversion of β-keto esters into 2-amino-1,3-selenazoles (**28**) using easily accessible *N*-bromosuccinimide and the appropriate selenourea was demonstrated (Figure 4.21) [63]. The similar tandem conversion using phenylacetylenes was successful [64]. In these reactions, β-cyclodextrin plays a significant role as a promoter in water. 2-Amino-4-aryl-1,3-selenazoles (**30**) were also synthesized from α-bromo ketones and *N*,*N′*-unsubstituted selenourea in the presence of β-cyclodextrin in water [65].

Under solvent-free conditions, 3-(2-amino-1,3-selenazol-4-yl)-2*H*-chromen-2-ones (**31**) were prepared by the reaction of *N*,*N′*-unsubstituted selenourea with

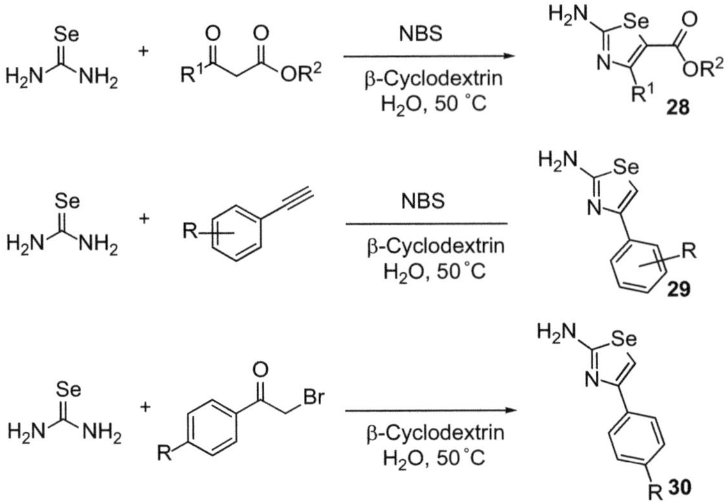

Figure 4.21: Synthesis of 2-amino-1,3-selenazoles (**28–30**) in the presence of β-cyclodextrin in water.

3-bromoacetylcoumarins in the presence of $CuPy_2Cl_2$ as an efficient Lewis acid catalyst (Figure 4.22) [66].

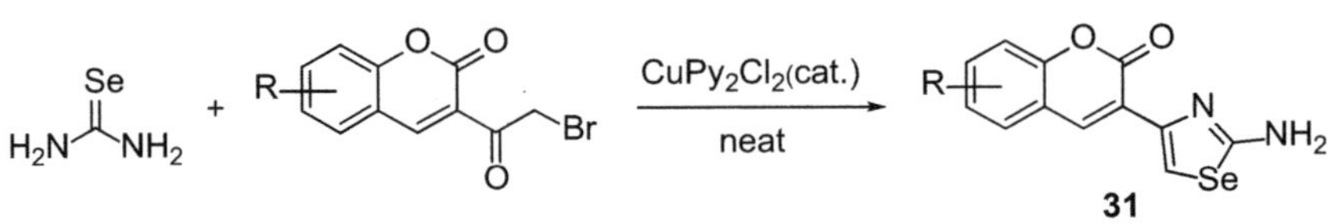

Figure 4.22: Synthesis of 3-(2-amino-1,3-selenazol-4-yl)-2*H*-chromen-2-ones (**31**) using $CuPy_2Cl_2$.

Ferric chloride ($FeCl_3$) is an efficient catalyst for the synthesis of 2-amino-1,3-selenazoles (**32–36**) by the reaction of *N,N*-disubstituted selenoureas with α,β-unsubstituted aldehydes, α,β-unsubstituted ketones, α-diketones, and ketones in boiling EtOH (Figure 4.23) [67–70]. The $FeCl_3$ interacts as a Lewis acid with the selenocarbonyl group, and EtOH as a solvent participates in these reactions.

1,3-Selenazol-4-ones are a subclass of 1,3-selenazoles and have a remaining carbonyl group in the C4 position. Reaction of *N,N′*-dialkylselenoureas with α-haloacyl halides in pyridine gave 2-imino-1,3-selenazol-4-ones **37** (Figure 4.24). This type of selenazoles can exist in a tautomeric equilibrium between 2-amino- and 2-iminoselenazol-4-ones [71].

2-Piperidinoselenourea reacted with chloroacetonitrile to give 2-piperidino-4,5-dihydro-1,3-selenazol-4-iminium chloride, which upon water refluxing afforded

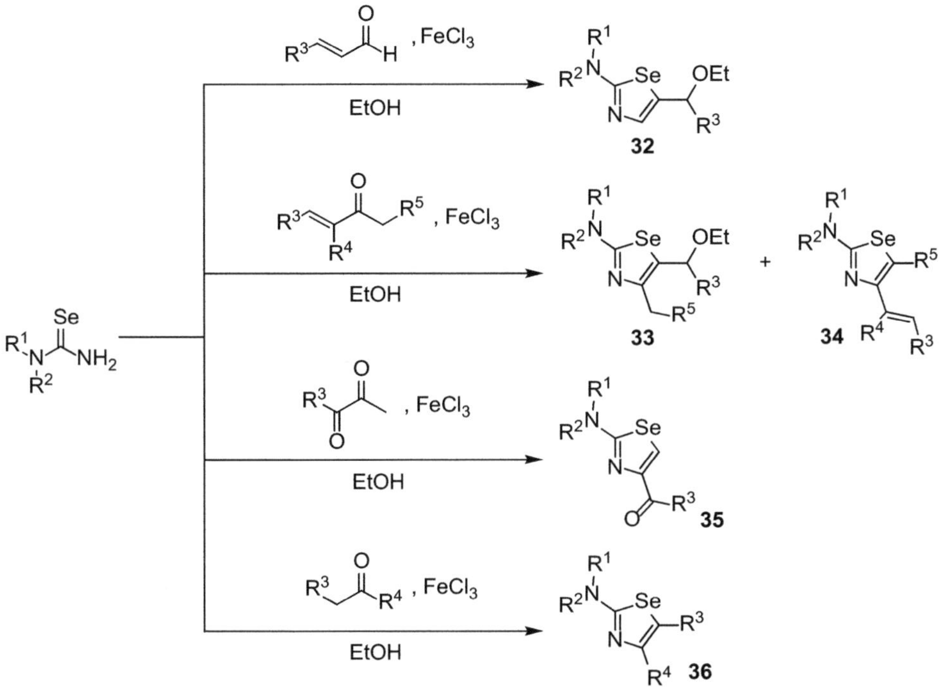

Figure 4.23: A series of 2-amino-1,3-selenazoles (**32–36**) from selenoureas using $FeCl_3$ in EtOH.

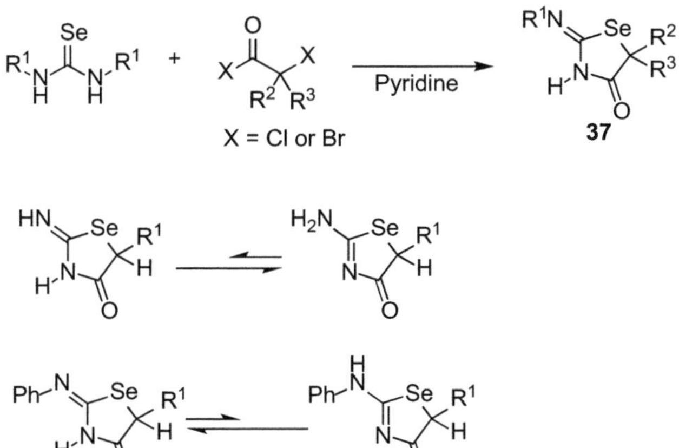

Figure 4.24: Preparation of 1,3-selenazol-4-ones (**37**) by the reaction of selenoureas with α-haloacyl halides.

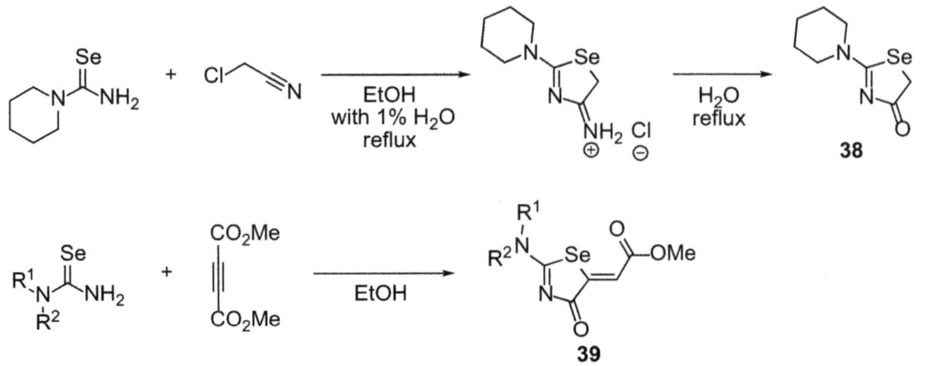

Figure 4.25: Synthesis of 1,3-selenazol-4-ones (**38** and **39**) using chloroacetonitrile and DMAD.

2-piperidinoselenazol-4-one (**38**) (Figure 4.25) [72]. Treatment of *N,N*-disubsituted selenoureas with dimethyl acetylenedicarboxylate (DMAD) without any catalyst yielded 2-aminoselenazol-4-ones (**39**) [73].

4.3.2 Synthesis of 1,3-selenazines

1,3-Selenazines are Se-containing six-membered ring compounds having one N atom at the 3-position. 1,3-Selenazines gain much attention from organic and medicinal chemists. In 1968, 2-chloro-1,3-benzoselenazin-4-one was for the first time prepared by Simchen G [74]. 1,3-Selenazinium salts (**40**) were synthesized from *N*, *N*-dialkylselenoureas with chloropropenylidene imminium salt in 1976 (Figure 4.26) [75].

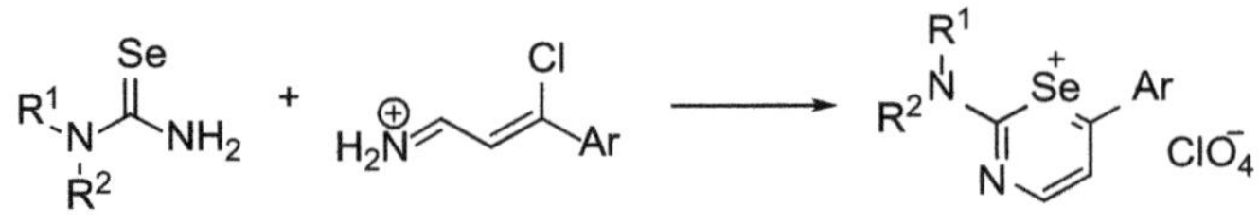

Figure 4.26: Synthesis of 1,3-selenazinium salts (**40**) from selenoureas.

The synthesis of 2-(*N,N*-diphenylamino)-5,6-diphenyl-5,6-dihydro-4*H*-1,3-selenazin-4-one (**41**) by the reaction of *N,N*-diphenylselenourea with diphenylcyclopropane at reflux was demonstrated (Figure 4.27) [76]. 2-Amino-4*H*-5,6-dihydro-1,3-selenazin-4-ones (**42**) were obtained by the reaction of *N,N'*-substituted selenoureas with α, β-unsaturated acid chlorides [77, 78].

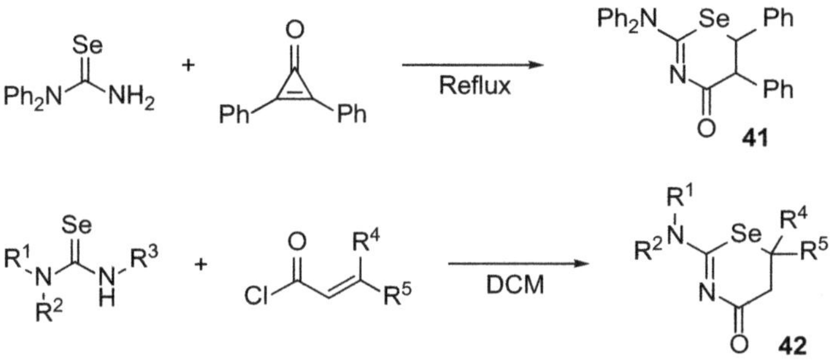

Figure 4.27: Synthesis of 1,3-selenazin-4-ones (41 and 42) from selenoureas.

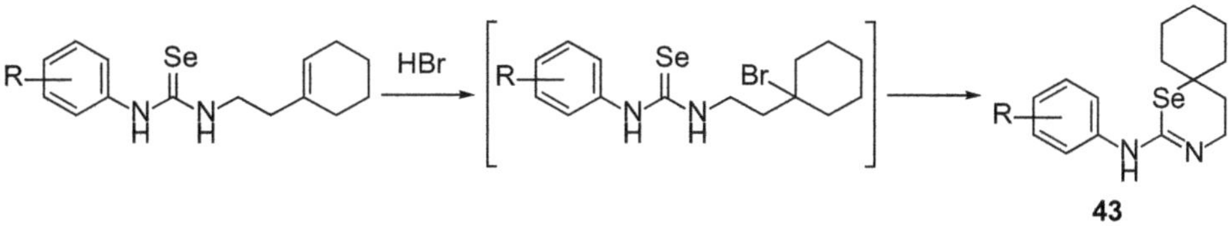

Figure 4.28: Synthesis of spiro derivatives of 1,3-selenazines (43).

A synthetic approach to spiro derivatives of 1,3-selenazines (**43**) based on intramolecular cyclization of selenoureas containing a γ,σ-unsaturated fragment was achieved (Figure 4.28) [79].

4.3.3 Synthesis of 1,3-selenazepines

Selenazepines are Se-containing seven-membered heterocycles having one N atom in the ring. The number of their reports is quite limited [80–83]. An iodocyclization of β-allene-selenoureas afforded the 3-selena-1-dethiacephems (**44**) and 1,3-selenazepines (**45**) in Figure 4.29 [84]. With the assistance of iodine anion, intramolecular nucleophilic attack of selenium in the sekenourea group on the terminal carbon of allene (when R^1 = Me, Et, or *c*-Hex) in the favored 7-*endo* mode affords the corresponding selenazepines (**44**), accompanied by the simultaneous elimination of hydrogen iodide.

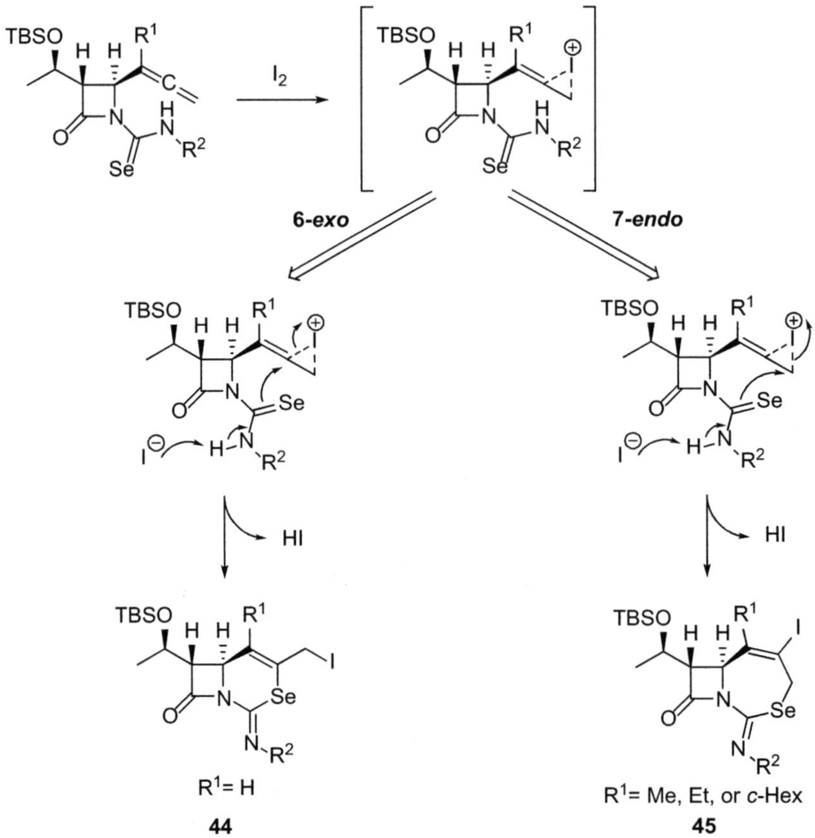

Figure 4.29: Synthesis of 3-selena-1-dethiacephems (**44**) and 1,3-selenazepines (**45**) via iodocyclization.

4.3.4 Synthesis of 1,3-diselenetanes and 1,4-diselenafulvenes

There are only a few reports for the synthesis of 1,3-diselenetane (**46**) and 1,4-diselenafulvenes (**47** and **48**) from selenoureas. A novel reaction of *N,N′*-unsubstituted selenourea with benzoylbromoacetylene in the presence of triethylamine was described (Figure 4.30) [85].

4.3.5 Synthesis of selenolactones

Selenoureas are unsuitable for the formation of selenolactones. A selenolactone-based fluorescent chemodosimeter (**49**) for monitoring mercury/methylmercury species was prepared from Rhodamine B using *N,N′*-unsubstituted selenourea

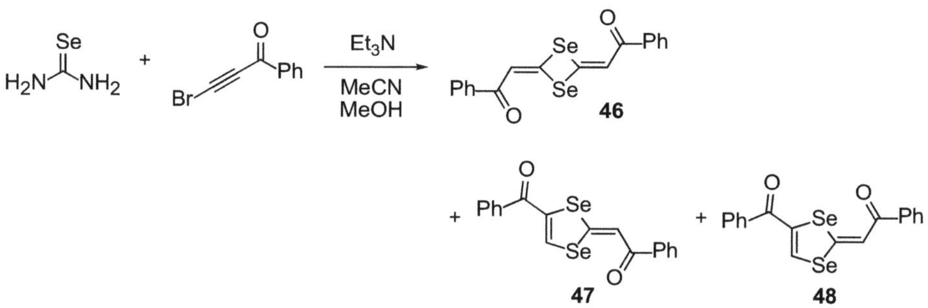

Figure 4.30: Synthesis of 1,3-diselenetane (**46**) and 1,4-diselenafulvenes (**47** and **48**) from selenoureas.

(Figure 4.31) [86]. On the detection mechanism, Hg^{2+}-induced spiro ring-opening followed by deselenation is likely to be responsible for the fluorescence enhancement (Figure 4.32).

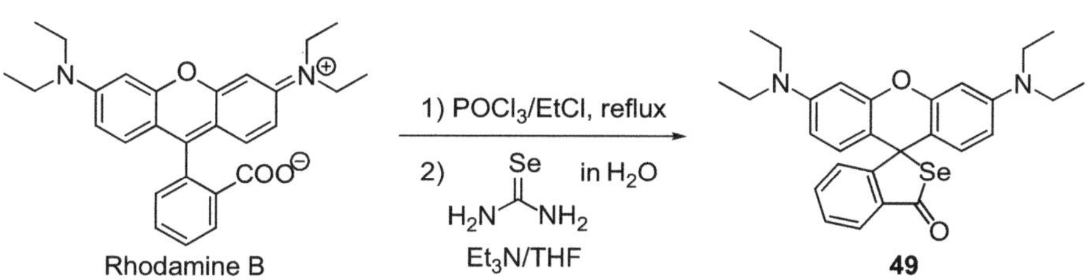

Figure 4.31: Synthesis of a selenolactone-based fluorescent chemodosimeter (**49**).

4.3.6 Deselenative reactions of selenoureas

There are several reports on the deselenative reactions of selenoureas. A brief oxidation of selenoureas with $NaIO_4$ afforded the corresponding carbodiimides (**50**) in Figure 4.33 [87]. Other oxidants such as $NaClO_4$, $KMnO_4$, and $NaCrO_4$ could not yield the carbodiimides. This synthesis of carbodiimides from selenoureas by iodine-mediated deselenation was reported for the first time [88].

N-Acetylureas were synthesized regioselectively from *N,N'*-disubstituted selenoureas and zinc acetate ($Zn(OAc)_2$) in Figure 4.34 [89]. The regioselectivity was dependent on the pK_a of the amine attached to the selenourea and occurred toward the amine with the lower pK_a. A plausible mechanism was proposed as follows: Triethyl amine interacts with *N*-phenyl-*N'*-cyclohexylselenourea followed by reaction with $Zn(OAc)_2$. Similarly, the second attack of Et_3N gives the corresponding carbodiimide with elimination of Se atom. Then the liberated AcO^- ion attacked the

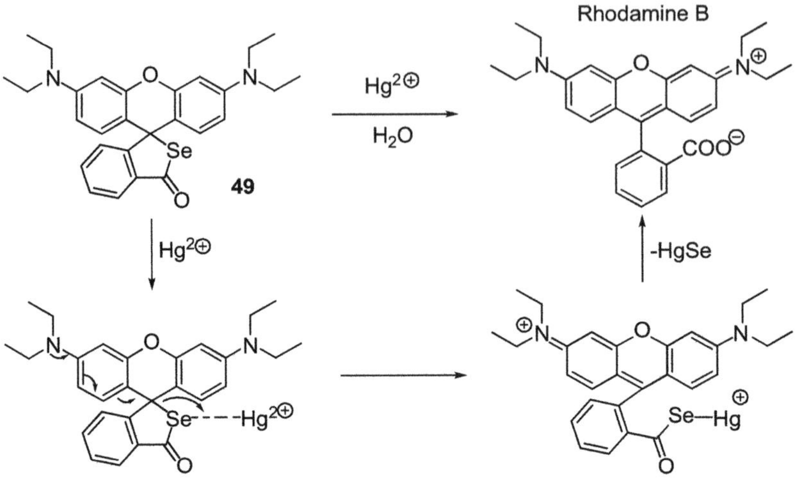

Figure 4.32: Proposed mechanism of deselenation of a selenolactone-based fluorescent chemodosimeter (**49**) with Hg^{2+} ion.

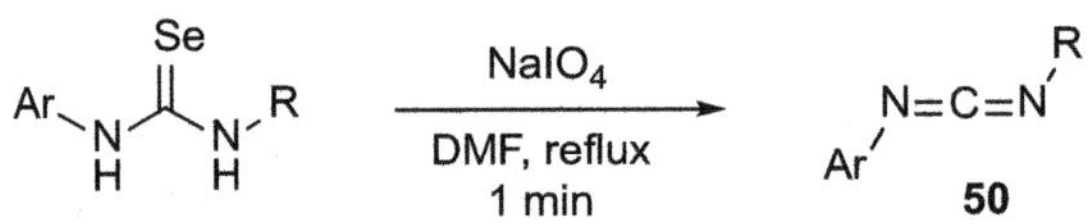

Figure 4.33: Deselenation of selenoureas using $NaIO_4$.

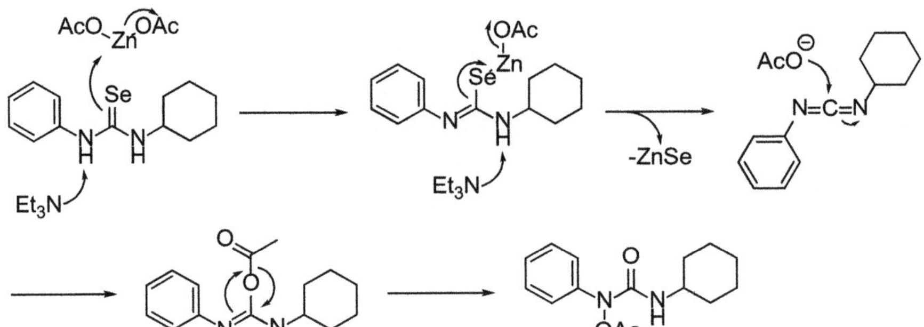

Figure 4.34: Deselenative synthesis of *N*-acetylureas (**51**) using $Zn(OAc)_2$.

carbodiimide to produce isourea, which has a higher pK_a after protonation towards the amine without affecting the imine group of the other side. Lastly, consequential isourea underwent intramolecular acetyl migration to give the regioselective *N*-acetylurea (**51**). The similar conversion employing ion-supported hypervalent iodine reagent $[dibmim]^+[BF_4]^-$ instead of $Zn(OAc)_2$ was successful [90].

4.4 Applications

4.4.1 As biologically active substances

Interest of selenium-containing therapeutics has grown over last 30 years. The most successful ones are ebselen (2-phenyl-1,2-benzisoselenazol-3(2*H*)-one) as an anti-inflammatory antioxidant agent and selenazofurin (2-β-D-ribofuranosylselenazol-4-carboxamide) as an antitumor agent. There are several reviews regarding biologically important selenoureas [91–96]. This section shows some examples.

By using a 1,1-diphenyl-2-picrylhydrazyl (DPPH) assay, sugar-derived selenoureas (**52**) in Figure 4.35 are excellent free radical scavengers compared to ascorbic acid, Trolox, and BHT [97]. This report suggested that the increasing order of antioxidant profile could be: selenohydantoins < selenocarbamates < selenoureas. Selenoureas can act as free radical scavengers by donating a H atom from an NH group to DPPH (Figure 4.36).

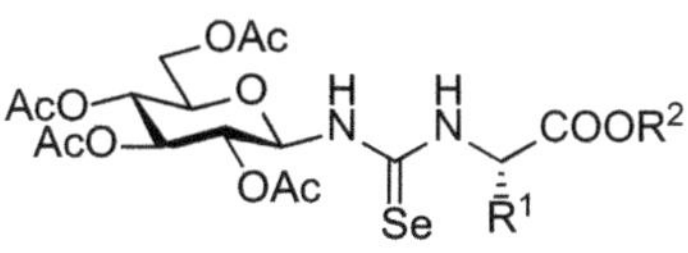

52

Figure 4.35: Sugar-derived selenoureas (**52**).

Superoxide anion-scavenging activity of *N,N*-unsubstituted selenoureas (**53**–**57**) in Figure 4.37 was studied [98]. Pyrrolidine, piperidine, and morpholine analogues (**55**–**57**) exerted high activity with IC_{50} values of 125 nM, 142 nM, and 121 nM, respectively. Their potency was greater than that of L-ascorbic acid (IC_{50}: 227 nM).

Inhibitory effects of *N,N*-unsubstituted selenourea derivatives (**53**–**57**) as in Figure 4.37 on mushroom tyrosinase and their depigmenting effect in melan-a cells were investigated [99]. A piperidine analogue (**56**) at a concentration of 200 μM exhibited 55.5% of inhibition on dopa oxidase activity of mushroom tyrosinase. This inhibitory effect was higher than that of kojic acid (39.4%), a well-known tyrosinase inhibitor. Moreover, the analogue (**56**) was identified as a noncompetitive inhibitor

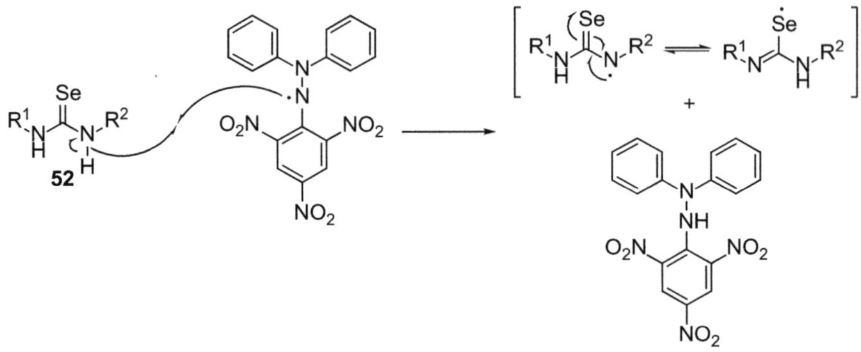

Figure 4.36: Mechanism for the scavenging of free radicals by sugar-derived selenoureas (52).

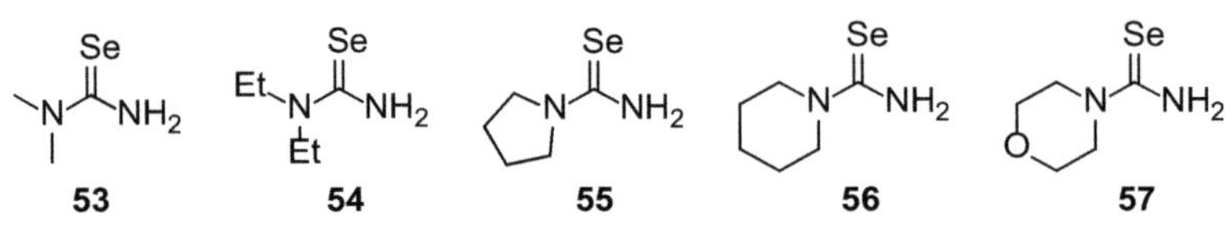

Figure 4.37: *N*,*N*-unsubstituted selenoureas (53–57).

by Lineweaver–Burk plot analysis. In addition, 1-selenocarbamoylpiperidine (**56**) also inhibited the melanin production in melan-a cells.

Acetylcholinesterase inhibitory effects of selenourea-containing tacrine derivatives (**58**) in Figure 4.38 were studied [100]. These derivatives showed a remarkable increase of activity compared to parent tacrine. Presumably, the planar selenoureido motif also undergoes favorable interactions with the aromatic residues of the enzyme peripheral domain.

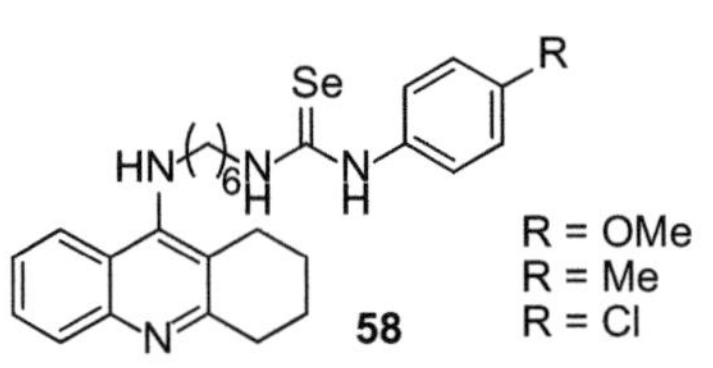

Figure 4.38: Selenourea-containing tacrine derivatives (58).

A series of acyl selenoureido benzensulfonamides (**59**) as in Figure 4.39 was evaluated as carbonic anhydrase (CA, EC 4.2.1.1) inhibitors against two *Vibrio cholerae* such as enzymes (VchCAα over VchCAβ) belonging to the α- and β-classes, potential

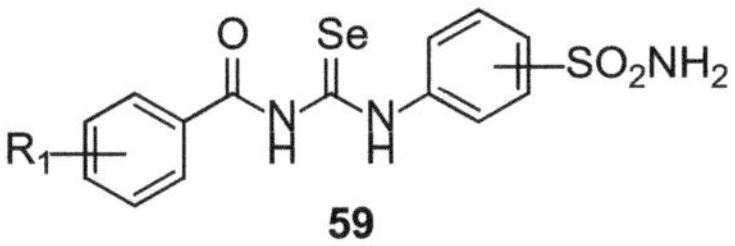

59

Figure 4.39: Acyl selenoureido benzensulfonamides (**59**).

novel targets for anti-infective drugs development [101]. These compounds showed strong inhibitory action against VchCAα over VchCAβ and excellent selectivity over the human off-target isoforms hCA I and II.

Selenomerocyanine 56 (**60**) in Figure 4.40 has been identified as a possible modulator of apoptosis in leukemia and solid tumor cells [102]. Selenobarbitutic acid analogues (**61**–**63**) in Figure 4.5 showed in vitro antiproliferative effects against melanoma CHL-1 and UACC 903 cells with μM IC_{50} values [103].

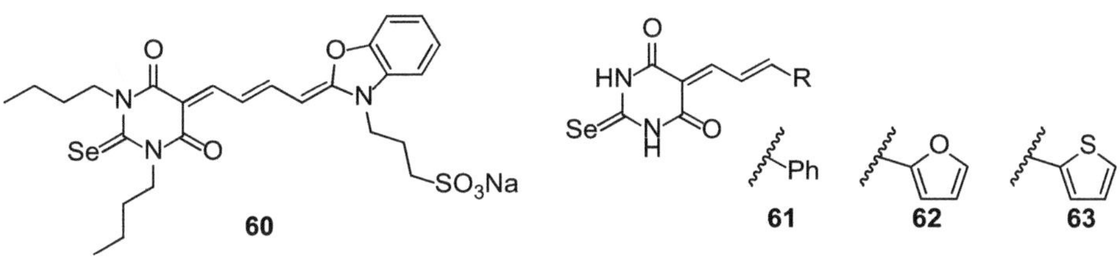

60 61 62 63

Figure 4.40: Selenomerocyanine 56 (**60**) and selenobarbitutic acid analogues (**61–63**).

Ferrocene incorporated selenoureas (**64**) in Figure 4.41 have been reported as anticancer agents by Raja Azadar Hussain and Amin Badshah research group [9, 104, 105]. Cancer termination of the ferrocene incorporated selenoureas (**64**) is proved by their cytotoxic activities against neuroblastoma, hepa 1c1c7 and MCF-7 cancerous cell lines. Mostly those ferrocene incorporated selenoureas (**64**) that had *ortho* substitution on the phenyl ring attached with the carbonyl carbon were found most active.

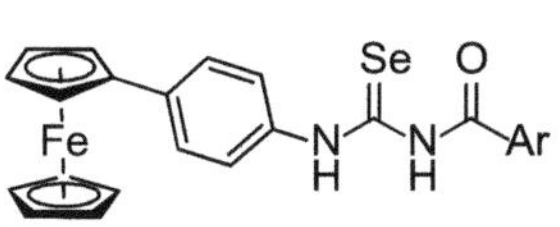

64

Figure 4.41: Ferrocene incorporated selenoureas (**64**).

4.4.2 Others

1-Selenocarbamoylpiperidine (**65**) chemoselectively cleaves the *O*-chloroacetyl group in the presence of other acyl groups such as acetyl, pivaloyl, and Fmoc without the assistance of a base (Figure 4.42) [106]. Makoto Kiso research group applies this selenourea (**65**) as a dechloroacetyl reagent to the total synthesis of complex gangliosides [107–109].

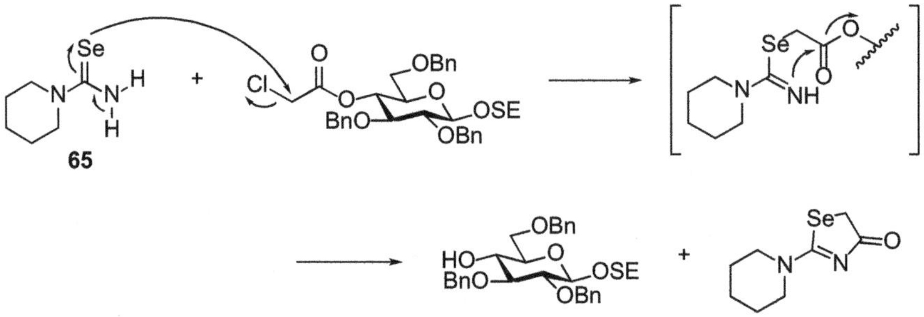

Figure 4.42: Proposed mechanism of dechloroacetylation with 1-selenocarbamoylpiperidine (**65**).

A chemosensor containing a selenourea moiety (**66**) was reported [110]. The chemosensor (**66**) is able to colorimetrically sense the presence of CN^- and S^{2-} in H_2O/MeCN (Figure 4.43). Detailed spectroscopic studies resulted to support a sensing mechanism based on the formation of a diselenide derivative (**67**). Moreover, when the chemosensor (**66**) was loaded into functionalized mesoporous silica nanoparticles an increase in the selectivity toward S^{2-} occurred via a selective fluorescence response.

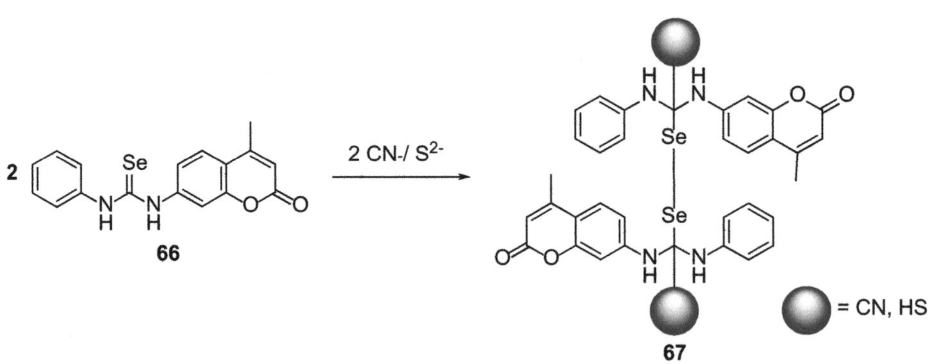

Figure 4.43: Proposed detection mechanism of a chemosensor containing a selenourea moiety (**66**) for CN^- and S^-.

Narrow-band gap IV–VI group semiconductors, such as PbS, PbSe, CdSe, and PbTe, provide unique properties for investigating the effects of strong confinement on electrons and phonons [111]. PbSe can be potentially employed in various areas such as laser materials, solar cells, infrared detectors, near-IR luminescence, photographic planes, photovoltaic absorbers, and thermoelectric devices [112]. CdSe is an n-type semiconductor with direct band gap (1.73 eV). Several studies on the electrochemical behavior of CdSe involve cyclic voltammetry, both in dispersed solution and on thin films, differential pulse voltammetry, spectroelectrochemistry, and electrochemiluminescence [113]. In the preparation processes of PbSe and CdSe, selenourea can be used as selenium source replacing SeO_2 and toxic selenium powder. In addition, Sfeir MY and Owen JS research group reported a tunable library of *N,N,N′*-trisubstituted selenoureas and their reaction with lead oleate to form carboxylate-terminated PbSe nanocrystals in quantitative yields (Figure 4.44) [114]. The conversion kinetics of PbSe nanocrystals can be finely controlled by adjusting the substitution pattern of selenourea precursors.

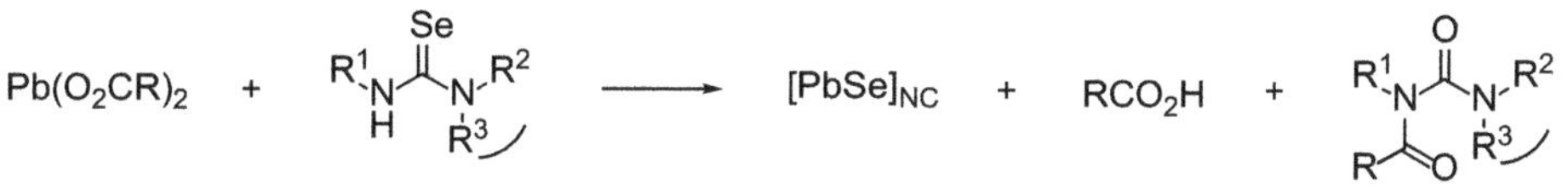

Figure 4.44: Synthesis of PbSe nanocrystals.

4.5 Conclusions

This chapter provides advances in the preparation of selenoureas and development of *Se*-containing molecules using them. Great potentials of selenoureas for wide applications continuously attract extensive attention of researchers.

References

[1] Hope, A. A note on the synthesis of selenourea. Acta Chem Scand 1964, 18, 1800.
[2] Koketsu, M., and Ishihara, H. Thiourea and selenourea and their applications. Curr Org Synth 2006, 3, 439–455.
[3] Thomas, W. Organoselenium chemistry; Synthesis and Reactions, Wiley-VCH Verlag and Co. KGaA, ed.,, ISBN: 978-3-527-32944-1, 2011.
[4] Mugesh, G. du Mont, WW., and Sies, H. Chemistry of biologically important synthetic organoselenium compounds. Chm Rev 2001, 101, 2125–2179.

[5] Ninomiya, M., Garud, DR., and Koketsu, M. Selenium-containing heterocycles using selenoamides, selenoureas, selenazadienes, and isoselenocyanates. Heterocycles 2010, 81, 2027–2055.
[6] Sekov, IV., Serova, TM., Proshin, AN., and Bachurin, SO. Synthesis of selenoureas and heterocycles based thereon. Russ J Org Chem 2015, 51, 471–490.
[7] Koketsu, M., and Ishihara, H. Synthesis of 1,3-selenazine and 1,3-selenazole and their biological activities. Curr Org Chem 2003, 7, 175–185.
[8] Ninomiya, M., Garud, DR., and Koketsu, M. Biologically significant selenium-containing heterocycles. Coord Chem Rev 2011, 255, 2968–2990.
[9] Hussain, RA., Badshah, A., and Shah, A. Synthesis and biological applications of selenoureas. Appl Organometal Chem 2014, 28, 61–73.
[10] Douglass, IB. Acylselenoureas. J Am Chem Soc 1937, 59, 740–742.
[11] Garud, DR., Koketsu, M., and Ishihara, H. Isoselenocyanates: A powerful tool for the synthesis of selenium-containing heterocycles. Molecules 2007, 12, 504–535.
[12] Heimgartner, H., Zhou, Y., Atanassov, PK., and Sommen, GL. Isoselenocyanates as building blocks for selenium-containing heterocycles. Phosphorus Sulfur Silicon Relat Elem 2008, 183, 840–855.
[13] Angeli, A., Tanini, D., Peat, TS., Di Cesare Mannelli, L., Bartolucci, G., Capperucci, A., Ghelardini, C., Supuran, CT., and Carta, F. Discovery of new selenoureido analogues of 4-(4-fluorophenylureido)benzenesulfonamide as carbonic anhydrase inhibitors. ACS Med Chem Lett 2017, 8, 963–968.
[14] Chennakrishnareddy, G., Nagendra, G., Hemantha, HP., Das, U., Guru Row, TN., and Sureshbabu, VV. Isoselenocyanates derived from Boc/Z-amino acids: Synthesis, isolation, characterization, and application to the efficient synthesis of unsymmetrical selenoureas and selenoureidopeptidomimetics. Tetrahedron 2010, 66, 6718–6724.
[15] López, Ó., Maza, S., Ulgar, V., Maya, I., and Fernández-Bolaños, JG. Synthesis of sugar-derived isoselenocyanates, selenoureas, and selenazoles. Tetrahedron 2009, 65, 2556–2566.
[16] Tanahashi, N., and Koketsu, M. Synthesis of 5-amino-2-selenoxo-1,3-imidazole-4-carboselenoamides by the reaction of isoselenocyanates with aminoacetonitriles. Tetrahedron Lett 2011, 52, 4650–4653.
[17] Schönberg, A., Singer, E., and Stephan, W. C=C-Doppelbindungen mit extremer reaktivität, VI tris(methylphenylamino)methan – Zur Chemie der Orthoamide. Chem Ber 1987, 120, 1581–1588.
[18] Kantlehner, W., Hauber, M., and Vettel, M. Orthoamide. IL. Umsetzungen von orthoamid-Dderivaten mit schwefel und selen, synthesen von 1,3-thiazol- und 1,3-selenazolderivaten. J Prakt Chem 1996, 338, 403–413.
[19] Fujiwara, S., Asanuma, Y., Shin-ike, T., and Kambe, N. Copper(I)-catalyzed highly efficient synthesis of benzoselenazoles and benzotellurazoles. J Org Chem 2007, 72, 8087–8090.
[20] Zhou, Y., and Denk, MK. Synthesis and reactivity of subvalent compounds. Part 13: Reaction of triethyl orthoformate with amines and selenium – A convenient one-step three-component synthesis for selenoureas. Tetrahedron Lett 2003, 44, 1295–1299.
[21] Bhabak, KP., Satheeshkumar, K., Jayavelu, S., and Mugesh, G. Inhibition of peroxynitrite- and peroxidase-mediated protein tyrosine nitration by imidazole-based thiourea and selenourea derivatives. Org Biomol Chem 2011, 9, 7343–7350.
[22] Mautner, HG. The synthesis and properties of some selenopurines and selenopyrimidines. J Am Chem Soc 1956, 78, 5292–5294.
[23] Klayman, DL., and Shine, RJ. A new synthesis of 1,1,3,3-tetramethyl-2-selenourea. Chem Commun (London) 1968, 372–373.

[24] Klayman, DL., and Shine, RJ. Synthesis of selenoureas and selenothiocarbamic esters from thioureas. J Org Chem 1969, 34, 3549–3551.
[25] Kogami, M., Davis, DR., and Koketsu, M. An efficient synthesis of 2-selenouridine and its phosphoramidite precursor. Heterocycles 2016, 92, 64–74.
[26] Barbosa, FAR., Siminski, T., Canto, RFS., Almeida, GM., Mota, NSRS., Ourique, F., Pedrosa, RC., and Braga, AL. Novel pyrimidinic selenourea induces DNA damage, cell cycle arrest, and apoptosis in human breast carcinoma. Eur J Med Chem 2018, 155, 503–515.
[27] Ishihara, H., Koketsu, M., Fukuta, Y., and Nada, F. Reaction of lithium aluminium hydride with elemental selenium. Its application as a selenating reagent into organic molecules. J Am Chem Soc 2001, 123, 8408–8409.
[28] Koketsu, M., Fukuta, Y., and Ishihara, H. Reaction of *N, N*-dimethylselenocarbamoyl chloride with nucleophiles. Preparation of diselenocarbamates, selenothiocarbamates, and selenoureas. J Org Chem 2002, 67, 1008–1011.
[29] Koketsu, M., Takakura, N., and Ishihara, H. Efficient synthesis of selenoureas from the corresponding carbodiimides. Synthetic Commun 2002, 32, 3075–3079.
[30] Basavaprabhu, Sharanabai KM., Prabhu, G., Panduranga, V., and Sureshbabu, VV. A facile one-pot synthesis of selenoureidopeptides employing LiAlHSeH through Staudinger aza-Wittig-type reaction. Synthesis 2015, 47, 801–806.
[31] Woollins, JD. How not to discover a new reagent. The evolution and chemistry of Woollins' reagent. Synlett 2012, 1154–1169.
[32] Fitzmaurice, JC., Williams, DJ., Paul, T., Wood, PT., and Woollins, JD. Organo-phosphorus-selenium heterocycles. J Chem Soc, Chem Commun 1988, 741–743.
[33] Baxter, I., Hill, AF., and Malget, JM. Selenoketenyl and selenoalkyne complexes *via* the reactions of ketenyl complexes with Woollins' reagent. Chem Commun 1997, 2049–2050.
[34] Gray, IP., Bhattacharyya, P., Slawin, AMZ., and Woollins, JD. A new synthesis of $(PhPSe_2)_2$ (Woollins reagent) and its use in the synthesis of novel P–Se heterocycles. Chem Eur J 2005, 11, 6221–6227.
[35] Hua, G., and Woollins, JD. Formation and reactivity of phosphorus-selenium rings. Angew Chem Int Ed Engl 2009, 48, 1368–1377.
[36] Hua, G., Du, J., Slawin, AMZ., and Woollins, JD. Synthesis and structural study of novel selenation derivatives of *N, N*-dialkylcyanamides. Chemistryselect 2016, 1, 6810–6817.
[37] Hua, G., Du, J., Slawin, AMZ., and Woollins, JD. Synthesis and single crystal structures of substituted-1,3-selenazol-2-amines. Molecules 2016, 22, 46. (13 pages).
[38] Hua, G., Du, J., Li, Y., Slawin, AMZ., and Woollins, JD. Novel heterocyclic selenazadiphospholaminediselenides, zwitterionic carbamidoyl(phenyl)-phosphinodiselenoic acids and selenoureas derived from cyanamides. Tetrahedron 2009, 65, 6074–6082.
[39] Segi, M., and Suga, S. Selenating reagent: Bis (dimethylaluminum) selenide. J Synth Org Chem Jpn 1990, 48, 1064–1065.
[40] Segi, M., Koyama, T., Nakajima, T., Suga, S., Murai, S., and Sonoda, N. Novel route to selenoketones from ketones by the use of bis(dimethylaluminum) selenide. Tetrahedron Lett 1989, 30, 2095–2098.
[41] Segi, M., Kojima, A., Nakajima, T., and Suga, S. A facile synthesis of seleno- and telluroformamides. Synlett 1991, 105–106.
[42] Segi, M., Takahashi, T., Ichinose, H., Li, GM., and Nakajima, T. An efficient construction of a selenocarbonyl unit by the reaction of acetal derivatives with bis(dimethylaluminum) selenide. Terahedron Lett 1992, 33, 7865–7868.

[43] Li, GM., Niu, S., Segi, M., Zingaro, RA., Yamamoto, H., Watanabe, K., Nakajima, T., and Hall, MB. [4 + 2] Dimerization and cycloaddition reactions of α,β-unsaturated selenoaldehydes and selenoketones. J Org Chem 1999, 64, 1565–1575.
[44] Maeda, H., Takashima, M., Sakata, K., Watanabe, T., Honda, M., and Segi, M. One-pot synthesis of selenoureas and selenocarbamates via selenation of isocyanates with bis (dimethylaluminum) selenide. Tetrahedron Lett 2011, 52, 415–417.
[45] O'Neal, S., and Kolis, JW. Convenient preparation and structures of selenometalates $MoSe_4^{2-}$, WSe_4^{2-} and $MoSe_9^{2-}$ from polyselenide anions and metal carbonyls. J Am Chem Soc 1988, 110, 1971–1973.
[46] Saravanan, V., Porhiel, E., and Chandrasekaran, S. Tetraethylammonium tetraselenotungstate: A new and efficient selenium transfer reagent for the chemoselective synthesis of functionalised diselenides. Tetrahedron Lett 2003, 44, 2257–2260.
[47] Saravanan, V., Mukherjee, C., Das, S., and Chandrasekaran, S. Facile conversion of amides and lactams to selenoamides and selenolactams using tetraethylammonium tetraselenotungstate. Tetrahedron Lett 2004, 45, 681–683.
[48] Sivapriya, K., Suguna, P., and Chandrasekaran, S. Novel cyclic tetraselenides of mannose: Synthesis and mechanistic studies. Tetrahedron Lett 2007, 48, 2091–2095.
[49] Sureshkumar, D., Koutha, S., and Chandrasekaran, S. A new selena-aza-payne-type rearrangement of aziridinylmethyl tosylates mediated by tetraselenotungstate. Eur J Org Chem 2007, 4543–4551.
[50] Sureshkumar, D., Ganesh, V., and Chandrasekaran, S. Conformationally locked bridged bicyclic diselenides: Synthesis, structure, Se···O interaction, and theoretical studies. J Org Chem 2007, 72, 5313–5319.
[51] Sivapriya, K., Suguna, P., Banerjee, A., Saravanan, V., Rao, DN., and Chandrasekaran, S. Facile one-pot synthesis of thio and selenourea derivatives: A new class of potent urease inhibitors. Bioorg Med Chem Lett 2007, 17, 6387–6391.
[52] Hofmann, G. Ueber selencyan- und selenazolverbindungen. Justus Liebigs Ann Chem 1889, 250, 294–322.
[53] Młochowski, J., Kloc, K., Lisiak, R., Potaczek, P., and Wójtowicz, H. Developments in the chemistry of selenaheterocyclic compounds of practical importance in synthesis and medicinal biology. ARKIVOC 2007, 14–46.
[54] Kanapickaitė, V., Martynaitis, V., and Šačkus, A. Facile synthesis of novel functionalized 1,3-selenazoles. ARKIVOC 2009, 268–276.
[55] Hanson, RN., and Davis, MA. Selenium-sulfur analogs. 4. Synthesis and characterization of (±)-β-(2-amino-1,3-selenazol-4-yl)alanine. J Heterocycl Chem 1981, 18, 205–206.
[56] Shafiee, A., Ebrahimzadeh, MA., and Maleki, A. Selenium heterocycles. XLIII. Syntheses of 3,5-diaryl-1,2,4-thiadiazoles and 3,5-diaryl-1,2,4-selenadiazoles. J Heterocycl Chem 1999, 36, 901–903.
[57] Hassaneen, HM., Farag, AM., Algharib, MS., and Shawali, AS. A facile Synthesis of arylazoselenazoles and of aroylselenadiazoles. Org Prep Proced Int 1988, 20, 505–510.
[58] Shafiee, A., and Lalezari, I. Selenium heterocycles. XV. Reaction of 2-aminoselenazoles and 2-amino-1,3,4-selenadiazoles with acetylenic compounds. J Heterocycl Chem 1974, 12, 675–681.
[59] Potewar, TM., Ingale, SA., and Srinivasan, KV. An efficient and eco-friendly synthesis of 2-amino-1,3-selenazoles in an ionic liquid/water system under ambient conditions. ARKIVOC 2008, 117–125.
[60] Ramesh, G., Janardhan, B., and Rajitha, B. Green approach: An efficient synthesis of 2,4-disubstituted-1,3-thiazoles and selenazoles in aqueous medium under ultrasonic irradiation. Res Chem Intermed 2015, 41, 8099–8109.

[61] Facchinetti, V., Avellar, MM., Nery, ACS., Gomes, CRB., Vasconcelos, TRA., and de Souza, MVN. An eco-friendly, Hantzsch-based, solvent-free approach to 2-aminothiazoles and 2-aminoselenazoles. Synthesis 2016, 48, 437–440.
[62] Maradolla, M., and Chandramouli, GVP. A convenient synthesis of 2-amino-1,3-selenazoles using ionic liquid and microwave irradiation. Phosphorus Sulfur Silicon Relat Elem 2011, 186, 1650–1654.
[63] Narender, M., Reddy, MS., Kumar, VP., Srinivas, B., Sridhar, R., Nageswar, YVD., and Rao, KR. Aqueous-phase one-pot synthesis of 2-aminothiazole- or 2-aminoselenazole-5-carboxylates from β-keto Esters, thiourea or selenourea, and *N*-bromo-succinimide under supramolecular catalysis. Synthesis 2007, 3469–3472.
[64] Madhav, B., Murthy, SN., Anil Kumar, BSP., Ramesh, K., and Nageswar, YVD. A tandem one-pot aqueous phase synthesis of thiazoles/selenazoles. Tetrahedron Lett 2012, 53, 3835–3838.
[65] Narender, M., Reddy, MS., Kumar, VP., Reddy, VP., Nageswar, YV., and Rao, KR. Supramolecular synthesis of selenazoles using selenourea in water in the presence of β-cyclodextrin under atmospheric pressure. J Org Chem 2007, 72, 1849–1851.
[66] Madhav, JV., Kuarm, BS., and Rajitha, B. Solid-state synthesis of 1,3-selenazoles employing $CuPy_2Cl_2$ as a Lewis acid catalyst. Synthetic Commun 2008, 38, 3514–3522.
[67] Koketsu, M., Kanoh, K., and Ishihara, H. Synthesis of 2-amino-1,3-selenazoles by reaction of *N, N*-unsubstituted selenoureas with α,β-unsaturated aldehydes. Heterocycles 2006, 68, 2647–2652.
[68] Koketsu, M., Kanoh, K., and Ishihara, H. Preparation of 2-amino-1,3-selenazoles by reaction of *N, N*-unsubstituted selenoureas with α,β-unsaturated ketones in alcohol. Heterocycles 2006, 68, 2145–2152.
[69] Kanoh, K., Ishihara, H., and Koketsu, M. Preparation of 2-dialkylamino-1,3-selenazoles by reaction of *N, N*-unsubstituted selenoureas with α-diketones. Heterocycles 2007, 74, 1009–1014.
[70] Koketsu, M., Kanoh, K., Ando, H., and Ishihara, H. A facile synthesis of 2-amino-1,3-selenazole by reaction of *N, N*-unsubstituted selenourea with ketone. Heteroatom Chem 2006, 17, 88–92.
[71] Koketsu, M., Nada, F., and Ishihara, H. Facile preparation of 2-imino-1,3-selenazolidin-4-one derivatives by reaction of *N, N*'-disubstituted selenoureas with α-haloacyl halides. Synthesis 2002, 195–198.
[72] Koketsu, M., Tanaka, H., and Ishihara, H. A facile preparation of 2-amino-1,3-selenazoles by reactions of *N, N*-unsubstituted selenoureas with chloroacetonitrile. Chem Lett 2005, 34, 1260–1261.
[73] Koketsu, M., Kanoh, K., and Ishihara, H. Synthesis of 2-amino-4,5-dihydro-1,3-selenazol-4-ones by reaction of *N, N*-disubstituted selenoureas with acetylenedicarboxylate. Heterocycles 2006, 68, 2627–2633.
[74] Simchen, G. Synthesis of derivatives of isoquinoline, of 1,3-benzoselenazin-4-one, and of 1*H*-2-benzazepin-1-one. Angew Chem Int Ed Engl 1968, 7, 464.
[75] Liebscher, J., and Hartmann, H. Ein einfacher zugang zu neuartigen 2-amino-1,3-selenaziniumsalzen. Tetrahedron Lett 1976, 17, 2005–2006.
[76] Takahashi, M., Watanabe, S., and Kasai, T. The formation of 1,2-diselenol-5-one and 5,6-dihydro-4*H*-1,3-selenazin-4-one derivatives from diphenylcyclopropenone. Heterocycles 1980, 14, 1921–1924.
[77] Koketsu, M., Taura, M., and Ishihara, H. Synthesis and characterization of 2-iminoperhydro-1,3-selenazin-4-ones by reaction of *N, N*'-disubstituted selenoureas with acryloyl chloride. J Heterocycl Chem 2009, 41, 783–785.

[78] Kanoh, K., Ishihara, H., and Koketsu, M. Preparation of 2-amino-4*H*-5,6-dihydro-1,3-selenazin-4-ones by reaction of *N, N*-unsubstituted selenoureas with α,β-unsaturated acid chlorides. Synthesis 2007, 2617–2620.
[79] Proshin, AN., Serkov, IV., Lermontov, AS., Shevtsova, EF., Neganova, ME., and Bachurin, SO. Novel bicyclic derivatives of 1,3-selenazine. Russ Chem Bull 2013, 62, 142–146.
[80] Dupont, L., Pignon, B., and Christiaens, L. 8-Chloro-11-(4-methylpiperazin-1-yl)dibenzo[*b, f*]-1,4-selenazepine. Acta Cryst 1999, C55:IUC9900064.
[81] Sashida, H., and Minamida, H. Thermal ring expansion of 2-azidoselenochromenes: First synthetic examples of 1,3-benzoselenazepines. Chem Pharm Bull 2004, 52, 485–487.
[82] Sommen, GL., Linden, A., and Heimgartner, H. First synthesis of a selenazepane. Tetrahedron Lett 2005, 46, 6723–6725.
[83] Wang, J., Li, H., Leng, T., Liu, M., Ding, J., Huang, X., Wu, H., Gao, W., and Wu, G. Copper-catalyzed *ipso*-selenation of aromatic carboxylic acids. Org Biomol Chem 2017, 15, 9718–9726.
[84] Garud, DR., and Koketsu, M. Synthesis of 3-selena-1-dethiacephems and selenazepines via iodocyclization. Org Lett 2008, 10, 3319–3322.
[85] Amosova, SV., Elokhin, VN., Nakhmanovich, AS., Larina, LI., Martynov, AV., Steele, BR., and Potapov, VA. Reactions of selenourea with benzoyl- and 2-thienoylbromoacetylenes: Synthesis of 1,3-diselenetanes and 1,4-diselenafulvenes. Tetrahedron Lett 2008, 49, 974–976.
[86] Chen, X., Baek, K-H., Kim, Y., Kim, S-J., Shin, I., and Yoon, J. A selenolactone-based fluorescent chemodosimeter to monitor mercury/methylmercury species *in vitro* and *in vivo*. Tetrahedron 2010, 66, 4016–4021.
[87] Koketsu, M., Suzuki, N., and Ishihara, H. Preparation of isoselenocyanate and synthesis of carbodiimide by oxidation of selenourea. J Org Chem 1999, 64, 6473–6475.
[88] Shekharappa, LS., Kumar, LR., and Sureshbabu, VV. Synthesis of carbodiimides from 1,3-disubstituted selenoureas by iodine-mediated deselenation. Indian J Chem 2018, 57B, 1291–1294.
[89] Li, X., Gan, B., Peng, Z., Mi, Z., and Xie, Y. Zinc acetate-mediated regioselective *N*-acylation of 1,3-disubstituted selenoureas. J Chem Res 2018, 42, 251–254.
[90] Xie, Y., and Pan, H. Regioselective acetylate of 1,3-disubstituted selenoureas promoted by recyclable ion-supported hypervalent iodine(III) reagent. Phosphorus Sulfur Silicon Relat Elem 2014, 189, 98–105.
[91] May, SW. Selenium-based drug design: Rationale and therapeutic potential. Expert Opin Investig Drugs 1999, 8, 1017–1030.
[92] May, SW. Selenium-based pharmacological agents: An update. Expert Opin Investig Drugs 2002, 11, 1261–1269.
[93] Nogueira, CW., Zeni, G., and Rocha, JBT. Organoselenium and organotellurium compounds: Toxicology and pharmacology. Chem Rev 2004, 104, 6255–6286.
[94] Soriano-Garcia, M. Organoselenium compounds as potential therapeutic and chemopreventive agents: A review. Curr Med Chem 2004, 11, 1657–1669.
[95] Naithani, R. Organoselenium compounds in cancer chemoprevention. Mini Rev Med Chem 2008, 8, 657–668.
[96] Tiekink, ERT. Therapeutic potential of selenium and tellurium compounds: Opportunities yet unrealized. Dalton Trans 2012, 41, 6390–6395.
[97] Merino-Montiel, P., Maza, S., Martos, S., López, Ó., Maya, I., and Fernández-Bolaños, JG. Synthesis and antioxidant activity of *O*-alkyl selenocarbamates, selenoureas and selenohydantoins. Eur J Pharm Sci 2013, 48, 582–592.

[98] Takahashi, H., Nishina, A., Fukumoto, RH., Kimura, H., Koketsu, M., and Ishihara, H. Selenoureas and thioureas are effective superoxide radical scavengers *in vitro*. Life Sci 2005, 76, 2185–2192.
[99] Ha, SK., Koketsu, M., Lee, K., Choi, SY., Park, JH., Ishihara, H., and Kim, SY. Inhibition of tyrosinase activity by *N, N*-unsubstituted selenourea derivatives. Biol Pharm Bull 2005, 28, 838–840.
[100] Roldán-Peña, JM., Alejandre-Ramos, D., López, Ó., Maya, I., Lagunes, I., Padrón, JM4., Peña-Altamira, LE., Bartolini, M., Monti, B., Bolognesi, ML., and Fernández-Bolaños, JG. New tacrine dimers with antioxidant linkers as dual drugs: Anti-Alzheimer's and antiproliferative agents. Eur J Med Chem 2017, 138, 761–773.
[101] Angeli, A., Abbas, G., Del Prete, S., Carta, F., Capasso, C., and Supuran, CT. Acyl selenoureido benzensulfonamides show potent inhibitory activity against carbonic anhydrases from the pathogenic bacterium *Vibrio cholerae*. Bioorg Chem 2017, 75, 170–172.
[102] Daziano, JP., Günther, WH., Krieg, M., Tsujino, I., Miyagi, K., Anderson, GS., Sampson, RW., Ostrowski, MD., Muir, SA., Bula, RJ., and Sieber, F. Photochemically generated elemental selenium forms conjugates with serum proteins that are preferentially cytotoxic to leukemia and selected solid tumor cells. Photochem Photobiol 2012, 88, 448–460.
[103] Ramisetti, SR., Pandey, MK., Lee, SY., Karelia, D., Narayan, S., Amin, S., and Sharma, AK. Design and synthesis of novel thiobarbituric acid derivatives targeting both wild-type and BRAF-mutated melanoma cells. Eur J Med Chem 2018, 143, 1919–1930.
[104] Hussain, RA., Badshah, A., Tahir, MN., and Tamoor-ul-Hassan, Bano A. Synthesis, chemical characterization, DNA binding, antioxidant, antibacterial, and antifungal activities of ferrocence incorporated selenoureas. J Biochem Mol Toxicol 2014, 28, 60–68.
[105] Hussain, RA., Badshah, A., Pezzuto, JM., Ahmed, N., Kondratyuk, TP., and Park, EJ. Ferrocene incorporated selenoureas as anticancer agents. J Photochem Photobiol B 2015, 148, 197–208.
[106] Sogabe, S., Ando, H., Koketsu, M., and Ishihara, H. A novel de-*O*-chloroacetylation reagent: 1-seleonocarbamoylpiperidine. Tetrahedron Lett 2006, 47, 6603–6606.
[107] Tamai, H., Ando, H., Tanaka, HN., Hosoda-Yabe, R., Yabe, T., Ishida, H., and Kiso, M. The total synthesis of the neurogenic ganglioside LLG-3 isolated from the starfish *Linckia laevigata*. Angew Chem Int Ed Engl 2011, 50, 2330–2333.
[108] Tamai, H., Ando, H., Ishida, H., and Kiso, M. First synthesis of a pentasaccharide moiety of ganglioside GAA-7 containing unusually modified sialic acids through the use of *N*-Troc-sialic acid derivative as a key unit. Org Lett 2012, 14, 6342–6345.
[109] Goto, K., Sawa, M., Tamai, H., Imamura, A., Ando, H., Ishida, H., and Kiso, M. The total synthesis of starfish ganglioside GP3 bearing a unique sialyl glycan architecture. Chem Eur J 2016, 22, 8323–8331.
[110] Casula, A., Llopis-Lorente, A., Garau, A., Isaia, F., Kubicki, M., Lippolis, V., Sancenón, F., Martínez-Máñez, R., Owczarzak, A., Santi, C., Andrea Scorciapino, M., and Caltagirone, C. A new class of silica-supported chromo-fluorogenic chemosensors for anion recognition based on a selenourea scaffold. Chem Commun 2017, 53, 3729–3732.
[111] Amelia, M., Lincheneau, C., Silvi, S., and Credi, A. Electrochemical properties of CdSe and CdTe quantum dots. Chem Soc Rev 2012, 41, 5728–5743.
[112] Lv, W., Wang, X., Qiu, Q., Wang, F., Luo, Z., and Weng, W. Hydrothermal synthesis and characterization of novel PbSe dendritic structures. J Alloy Compd 2010, 493, 358–361.
[113] Khan, ZMSH., Khan, H., and Zulfequar, M. CdSe quantum dots using selenourea as selenium source in polymer matrix. J Mater Sci Mater Electron 2017, 28, 14638–14645.
[114] Campos, MP., Hendricks, MP., Beecher, AN., Walravens, W., Swain, RA., Cleveland, GT., Hens, Z., Sfeir, MY., and Owen, JS. A library of selenourea precursors to PbSe nanocrystals with size distributions near the homogeneous limit. J Am Chem Soc 2017, 139, 2296–2305.

Baitan Chakraborty, Aniruddha Das and Umasish Jana

5 Selenium Compounds as Reagents, Catalysts, and Ligands

5.1 Introduction

Two centuries after the discovery by Barzelius, selenium keeps expanding its spectrum of potential in the field of chemistry and biology, at par with its biological counterpart, sulfur [1]. Initially, the organoselenium chemistry was confined to the aliphatic selenides, selenols, and diselenides, which were difficult to work with. With the emergence of new aromatic selenium compounds, synthetic organoselenium chemistry was flooded with a tide of new methodological developments. In addition, as the versatile nature of organoselenium compounds was explored in the fields of material sciences [2] and biochemistry [3], the concerned chemistry has drawn much attention over the years [4]. Introduction of selenium into organic molecules has also resulted in important biological activities [5], which is exploited in drug designing [6].

The ability of selenium to expand its valency to interact with different atoms (C, H, N, O etc.) enables organoselenium compounds to participate in different types of reactions and redox processes [7]. Organoselenium compounds are manoeuvred into various forms of reagents of nucleophilic, electrophilic, and radical natures. These compounds are extensively used as reagents and catalysts to carry out various types of organic transformations with emphasis on chemo-, regio-, and stereoselectivities [8]. Moreover, the excellent coordinating ability of selenium has been utilized in the development of various selenium ligands. Apart from the classical thermal techniques, alternative activation techniques like microwave [9], ultrasound [9], and electrolysis [9] are applied to expand the scope of the organoselenium chemistry. In this chapter, we will bring out the 2008 onward developments in the chemistry of selenium as reagent, catalyst, and ligand.

5.2 Selenium as reagent

Organoselenium compounds have emerged as versatile reagents, as well as intermediates in organic synthesis. Selenium can be introduced as electrophile, nucleophile, or

Baitan Chakraborty, Aniruddha Das and Umasish Jana, Department of Chemistry, Jadavpur University, Kolkata, India

https://doi.org/10.1515/9783110625110-005

radical in organic synthesis. The reaction of selenium electrophiles with alkene involves seleniranium ion intermediate, which is rapidly opened by external nucleophile giving additional products; or internal nucleophilic attack generates cyclized products. The ability of selenium to stabilize negative charge generates nucleophilic character of selenium compound. The reaction between selenium-stabilized carbanions with electrophiles is to promote functional group transformation and new carbon–carbon bond formation. Due to the stability and the ease of preparation, organoselenium compounds are also extensively used as versatile radical precursors, which offer advantages over organic halides as radical precursors.

5.2.1 Selenium as electrophile

5.2.1.1 Enantioselective selenolactonization

Electrophilic functionalization of unreactive olefins through asymmetric catalysis has been a topic of great importance in the field of organic chemistry in the past decades [10]. In 2015, Niu et al. developed a novel and efficient approach of enantioselective selenolactonization of olefinic acids using $(DHQD)_2PHAL$ as the catalyst in the presence of structurally simple and commercially available N-phenylselenophthalimide (NPSP) as the electrophilic selenium reagent [11]. Finally, enantioselective selenolactonization had been obtained successfully up to 96% enantiomeric excess (ee) (Figure 5.1).

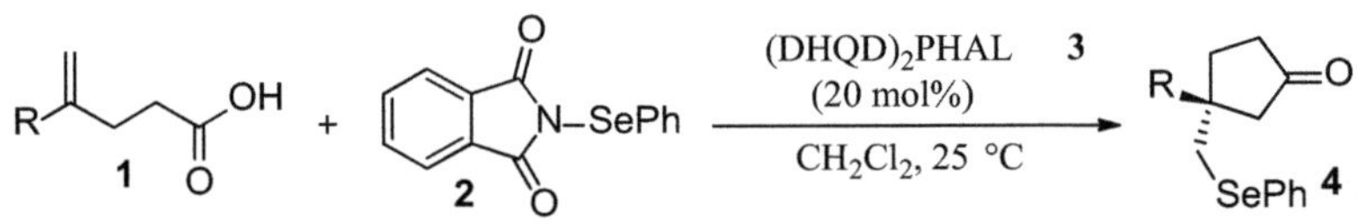

R= 2-naphthyl, 4-CH_3-C_6H_4, 4-CH_3O-C_6H_4, 4-Cl-C_6H_4, 4-F-C_6H_4, 4-CF_3-C_6H_4, 3-CH_3-C_6H_4, 3-CH_3O-C6H_4, 3-Cl-C_6H_4, 3, 5-$(CF_3)_2$-C_6H_4, 2-$CH_3C_6H_4$, 3-thienyl, cyclohexyl

Figure 5.1: Asymmetric selenolactonization.

It was suggested that hydrogen-bonded complex **5** was formed by the interaction of olefinic acid **1** with $(DHQD)_2PHAL$ **3**. Then, quinuclidine–NPSP complex **6** was generated by subsequent interaction of **5** with electrophilic selenium source NPSP **2**. Complex **6** was further activated with another carboxylic acid **1** to afford ionic species **7** along with the elimination of phthalimide molecule **8**. Consequencetly, more electrophilic selenium species **7** smoothly promoted cyclization to produce the desired product **4** (Figure 5.2).

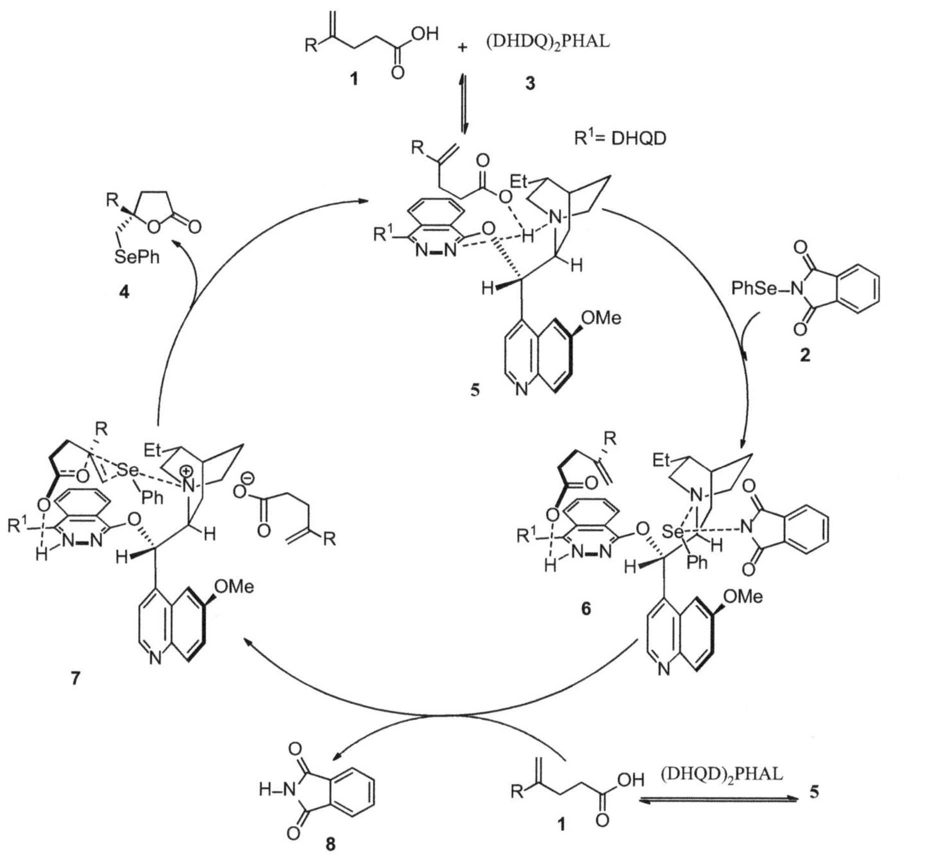

Figure 5.2: Plausible mechanism of the (DHQD)$_2$PHAL catalyzed asymmetric selenolactonization.

5.2.1.2 Catalytic selenium-promoted intermolecular Friedel–Crafts alkylation with simple alkenes

In 2016, Tang et al. reported selenium-promoted intermolecular Friedel–Crafts (F–C) alkylation reactions with simple alkenes using trimethylsilyl trifluoromethanesulfonates as catalyst and NPSP as an efficient selenium source [12]. Electron-rich arenes smoothly underwent F–C alkylation with a variety of alkenes to afford alkylated products in good yields and with high regioselectivity and diastereoselectivity (Figure 5.3).

Mechanistically, it was proposed that TMSOTf activated weak electrophilic organoselenide NPSP by chelating to the amide carbonyl group to facilitate the formation of the episelenonium ion intermediates **12** from the alkenes **10** and the subsequent reaction with the arenes **9** led to the formation of F–C alkylated products **11** and the regeneration of the TMSOTf catalyst (Figure 5.4).

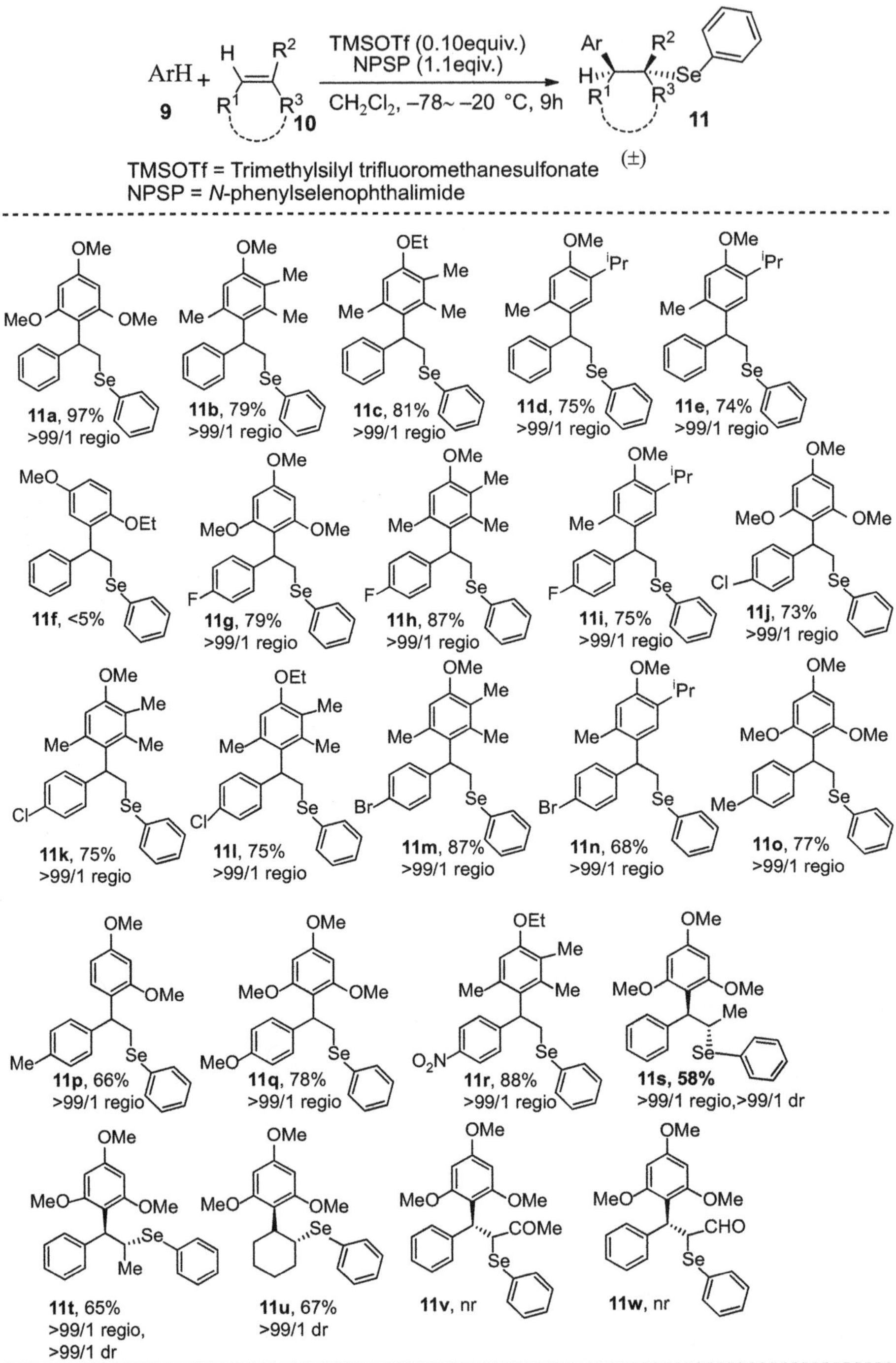

Figure 5.3: Scope of intermolecular selenium-promoted F–C alkylation with alkenes.

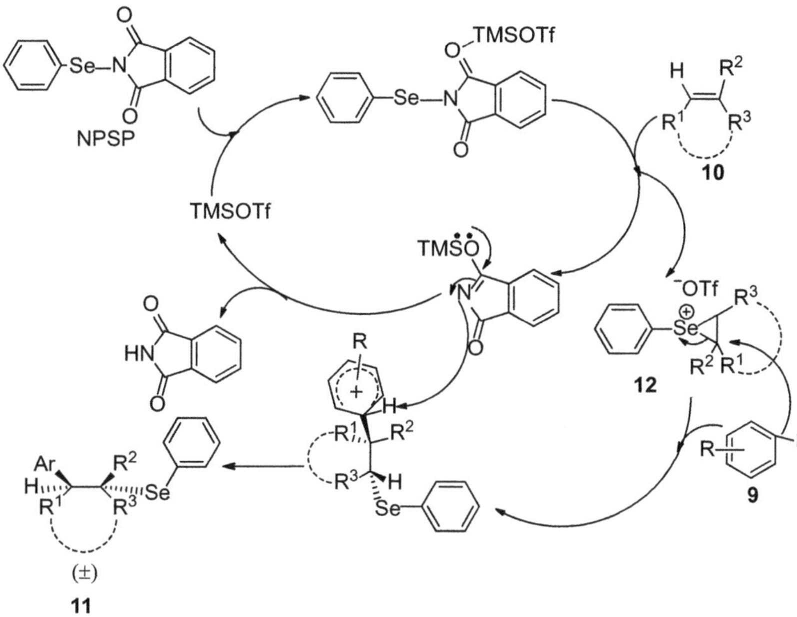

Figure 5.4: Mechanistic hypothesis for the intermolecular NPSP-Promoted F–C alkylation reaction with alkenes.

5.2.1.3 Iodine-mediated vicinal difunctionalization of alkenes for building C–Se and C–S bonds

Due to the wide spectrum of scientific application in the field of organic synthesis as well as medicinal biology, organochalcogen (Se, S) compounds are of growing interest in recent years [13]. Recently, in 2018, Wang et al. developed iodine-mediated vicinal difunctionalization of alkenes **13** with electrophilic selenium species diselenides **14** and nucleophilic thiolating agent carbamodithioates **15** (Figure 5.5). A new kind of compound, β-selanylethyl dithiocarbamates **16** with both C–Se and C–S bonds had been prepared in good yields by this mild reaction condition and simple procedure [14].

Addition of radical scavenger 2,2,6,6-tetramethylpiperidine-1-oxyl (TEMPO) into the reaction mixture had no influence on the reaction, and the difunctionalization proceeded well. This observation clearly indicated that the reaction did not follow radical pathway. The reaction was proposed to follow electrophilic addition mechanism. First, active electrophilic selenium species R^2SeI was generated in situ by the reaction of molecular I_2 with diselenide **14** through the active intermediate **17**. Then, in situ generated electrophilic selenium species R^2SeI **18** reacted with alkene to form unstable cyclic seleniranium intermediate **19**, which was then attacked by nucleophile sodium dithiocarbamate **15** via an S_N1 mechanism to

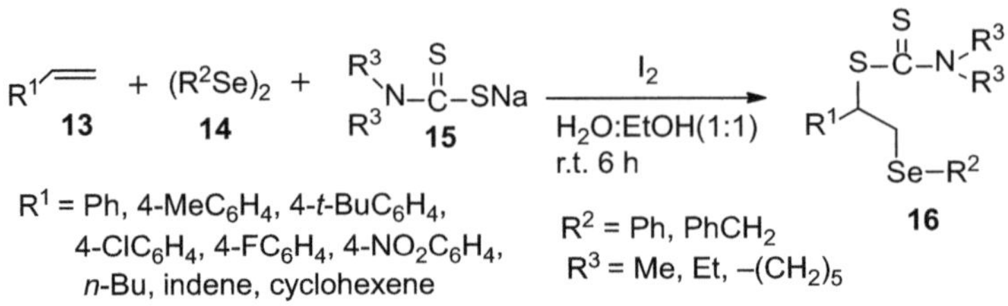

Figure 5.5: Iodine-mediated vicinal difunctionalization of alkenes.

afford the desired product **16** as a single isomer with high regioselectivity. But, S_N2 mechanism was observed in case of aliphatic alkene (Figure 5.6).

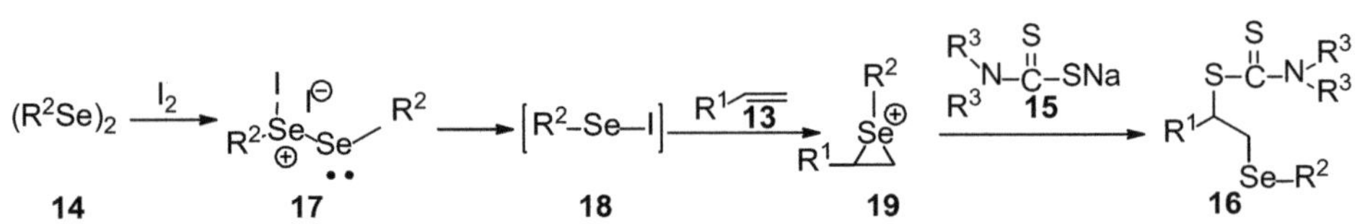

Figure 5.6: Proposed mechanism for the I_2-mediated difunctionalization of alkenes.

5.2.1.4 Redox-neutral synthesis of selenoesters by oxyarylation of selenoalkynes

A mild and efficient approach for the synthesis of selenoesters have been reported by Baldasarri and coworkers in 2018 [15] through an acid-catalyzed, redox-neutral oxyarylation reaction of selenoalkynes **20**. The reaction involved activation of triple bond of selenoalkyne **20** to generate selenium-stabilized vinyl cation **21**, which was trapped by aryl sulfoxide **22** followed by Claisen-type [3,3]-sigmatropic rearrangement to produce α-arylated selenoester product **23** (Figure 5.7). To elucidate the nature of the Se-stabilized carbocation, computational study was used. The superior role of selenium to sulfur in stabilizing the vinyl cation intermediate is established by DFT and NBO analysis.

5.2.1.5 Regio- and stereoselective synthesis of unsaturated compounds with the S–Se bond and their cyclization to 2,3-dihydro-1,4-thiaselenines

The regio- and stereo-selective synthesis of organoselenium compounds by electrophilic selenium reagents are very important and interesting area in the field of organoselenium chemistry [16]. In 2019, Amosova et al. have developed regio- and stereoselective ring-opening reaction of 2-bromomethyl-1,3-thiaselenole **24** with thiols to produce unsaturated selanyl sulfides, (*Z*)-CH_2=CHSCH=CHSeSR **26** by

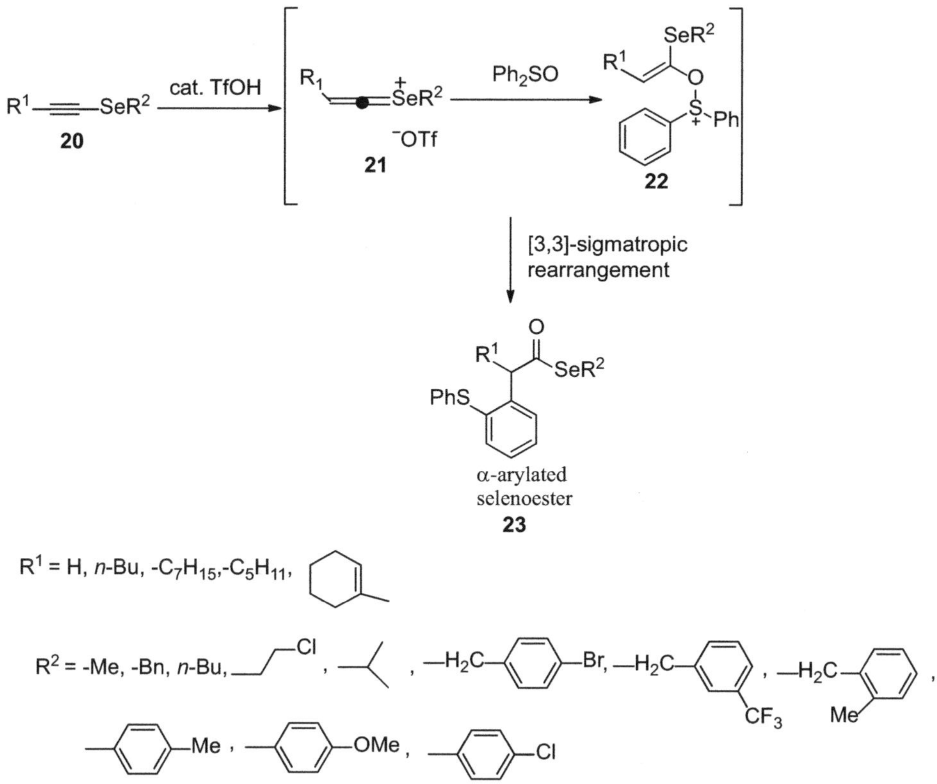

Figure 5.7: Synthesis of selenoesters by acid-catalyzed, redox-neutral oxyarylation reaction of selenoalkynes.

nucleophilic attack at the selenium atom of seleniranium intermediate **25** [17]. Then, acid-catalyzed cyclization of this unsaturated selanyl sulfide was developed to synthesize 2-(organylsulfanyl)-2,3-dihydro-1,4-thiaselenines **28**. They also demonstrated the synthesis of symmetrical polyunsaturated compounds with two S–Se bonds **27** by involving dithiols in this reaction (Figure 5.8). Mild reaction conditions and short reaction times were important features of this methodology.

The anchimeric assistance of selenium atom in thiaselenole **24** was the driving force for the generation of the seleniranium intermediate **25**. Synthesis of symmetrical polyunsaturated compounds **27** with two S–Se bonds involved nucleophilic attacks of two mercapto groups at the selenium atom of the seleniranium cation **25**. Besides, synthesis of 2-(organylsulfanyl)-2,3-dihydro-1,4-thiaselenines **28** was mechanistically rationalized by the formation of thermodynamically more stable heterocycle through acid-catalyzed cyclization pathway (Figure 5.9).

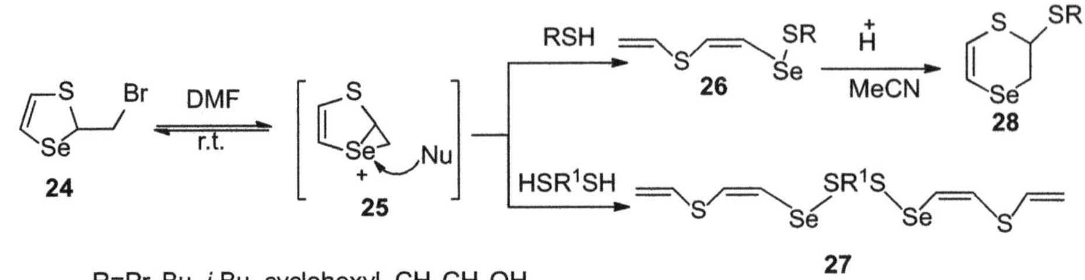

R=Pr, Bu, *i*-Bu, cyclohexyl, CH_2CH_2OH, CH_2Ph, CH_2CO_2H, CH_2CO_2Me,

R^1=CH_2CH_2, $CH_2CH_2OCH_2CH_2$

Figure 5.8: Reactions of thiaselenole **24** with various nucleophiles *via* intermediate seleniranium cation **25**.

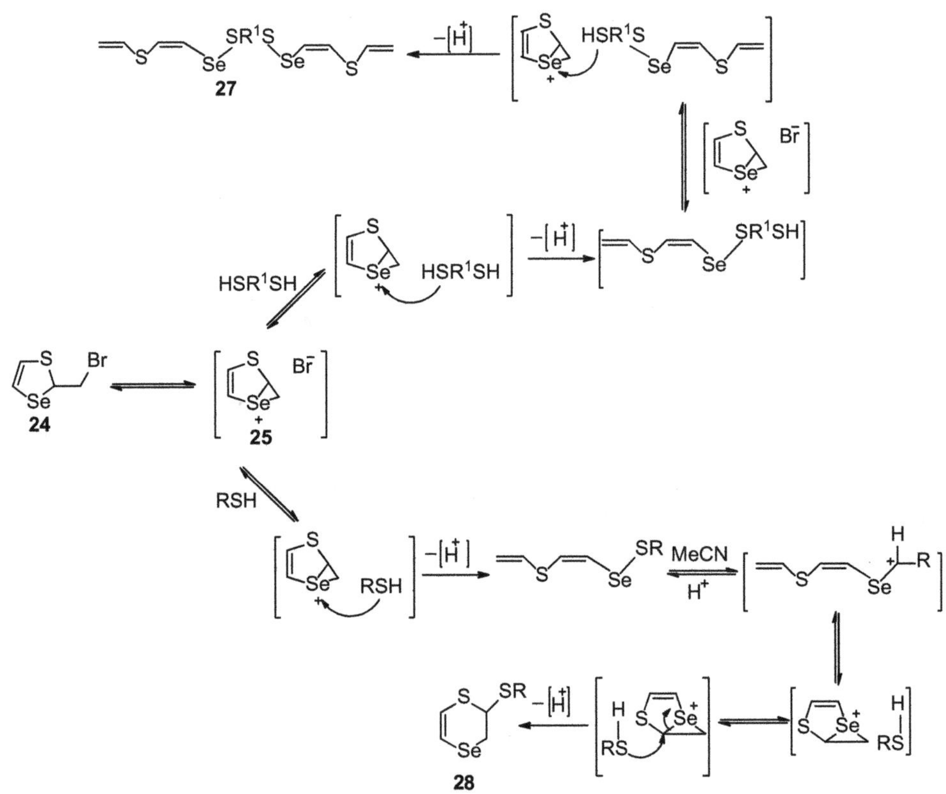

Figure 5.9: Proposed reaction mechanism.

5.2.1.6 Electrochemical oxidative selenylation of imidazo[1,2-*a*]pyridines with diselenides

An efficient protocol for the synthesis of 3-selenylated imidazo[1,2-*a*]pyridine derivatives **31** has been reported by Kim et al. [18] in 2019 via electrochemical oxidative selenylation of imidazo[1,2-*a*]pyridine derivatives **29** with diselenides **30**. The reaction was conducted in an undivided electrochemical cell equipped with glassy carbon plate as electrodes in the presence of $LiClO_4$ as a supporting electrolyte and acetonitrile solvent under constant 7mA current (Figure 5.10). This selenylation

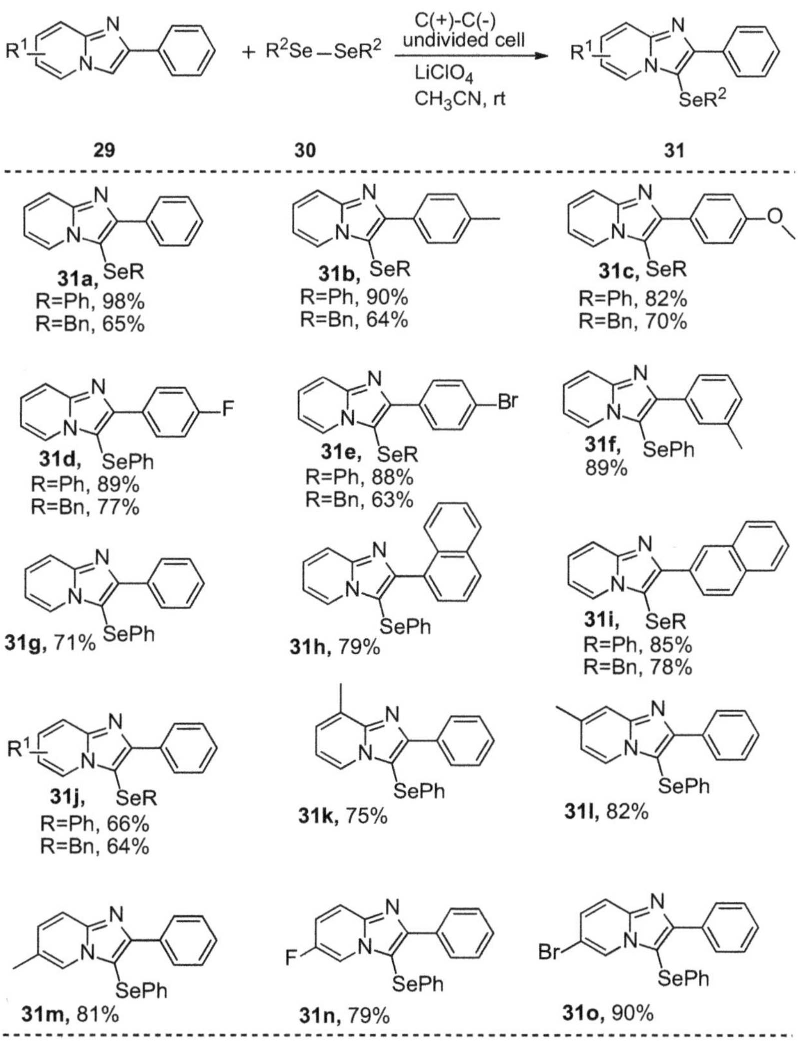

Figure 5.10: Substrate scope.

reaction went smoothly with substituted imidazo[1,2-*a*]pyridines containing halogen and methyl groups, with high yields.

The reaction was not affected by radical scavengers TEMPO and 2,6-di-*tert*-butyl-4-methylphenol (BHT). The desired selenylated products were obtained in high yields, which indicated that the reaction did not follow radical pathway. The new synthetic strategy was environmentally benign by using shelf-stable diselenides as selenium source and electrons as oxidizing reagents. Diphenyl diselenide was oxidized to generate phenylselenium cation **32**, which was then reacted with imidazo[1,2-*a*]pyridine **29** to produce intermediate **33** followed by deprotonation to afford selenylated imidazo[1,2-*a*]pyridine derivatives **31** (Figure 5.11).

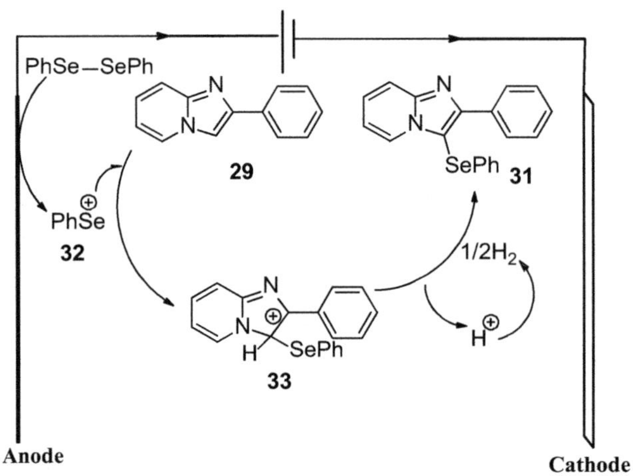

Figure 5.11: Proposed reaction mechanism.

5.2.1.7 Iron(III)-promoted synthesis of 3-(Organoselanyl)-1,2-dihydroquinolines from diorganyl diselenides and N-arylpropargylamines

Iron salts-catalyzed diorganyl diselenides have been reported in literature [19] as useful alternatives to the cyclization of unsaturated substrates and to introduce organoselenium functionality in heterocycles in a one-step reaction. Recently, Goulart et al. have demonstrated the preparation of 3-(organoselanyl)-1,2-dihydroquinolines **35** by iron-promoted tandem cyclization-functionalization reaction of N-arylpropargylamines **34** with diorganyl diselenides [20]. The optimized reaction condition was obtained when the reactions of N-arylpropargylamines have been carried out with the combination of diorganyl diselenides with $FeCl_3 \cdot 6H_2O$ in nitromethane at 70 °C (Figure 5.12).

It was proposed that the iron-seleno complex **36** was generated in situ by the reaction of $FeCl_3$ and diorganyl diselenide $(RSe)_2$. Then, coordination of electrophilic

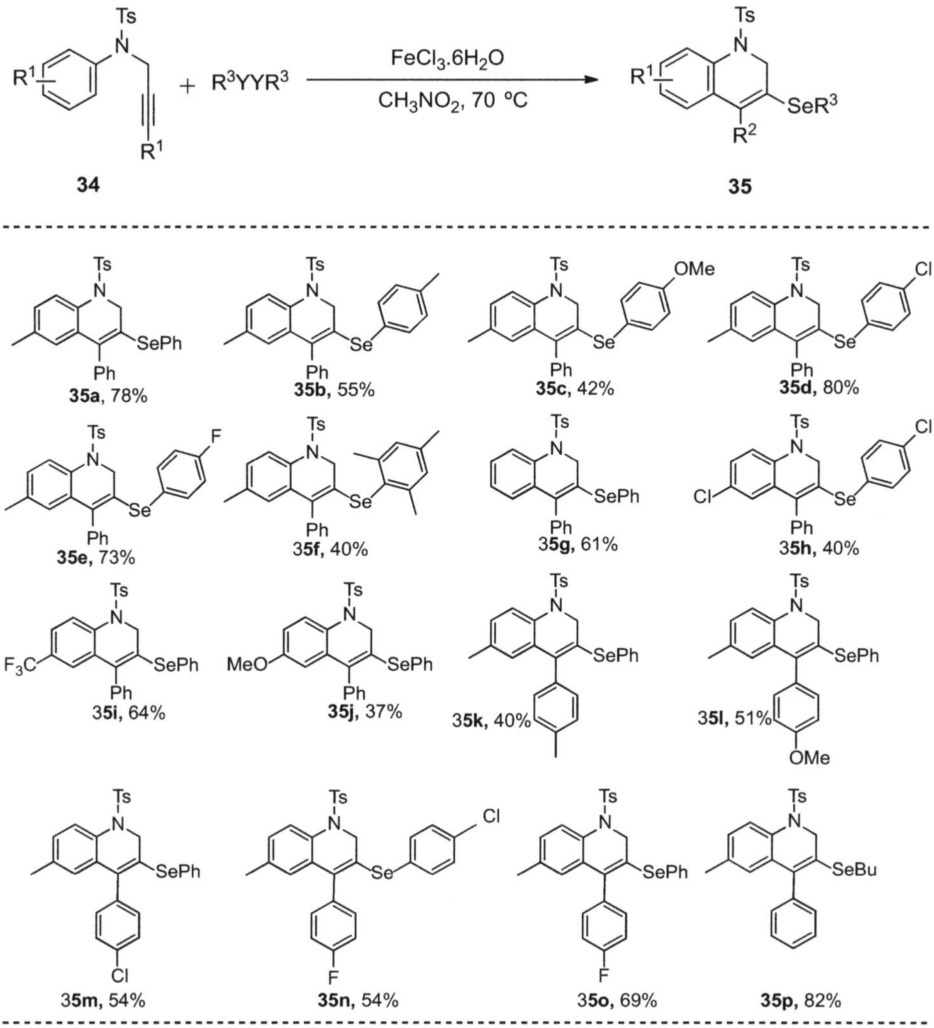

Figure 5.12: Synthesis of 3-(organoselanyl)-1,2-dihydroquinolines **35**.

portion of the selenium species to the carbon–carbon bond of alkyne of **34** generated the seleniranium ion **37**, which facilitated carbon–carbon bond toward nucleophilic attack to afford the intermediate **38** via a selective intramolecular 6-endo-dig cyclization. Removal of the hydrogen by breaking the C–H bond, promoted by the selenolate anion, finally produced the cyclized product **35** (Figure 5.13).

The synthetic application of 3-(organoselanyl)-1,2-dihydroquinolines **35p** has been demonstrated in the transition metal-catalyzed cross-coupling reactions with boronic acid (Figure 5.14).

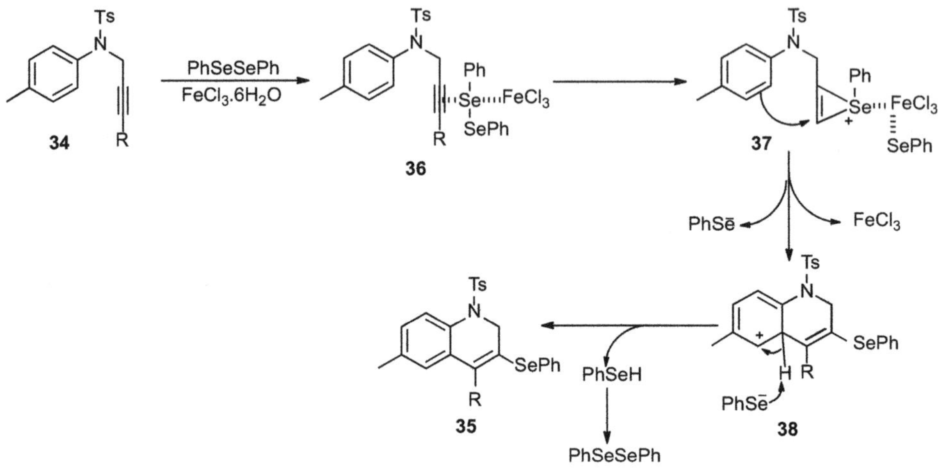

Figure 5.13: Proposed mechanism.

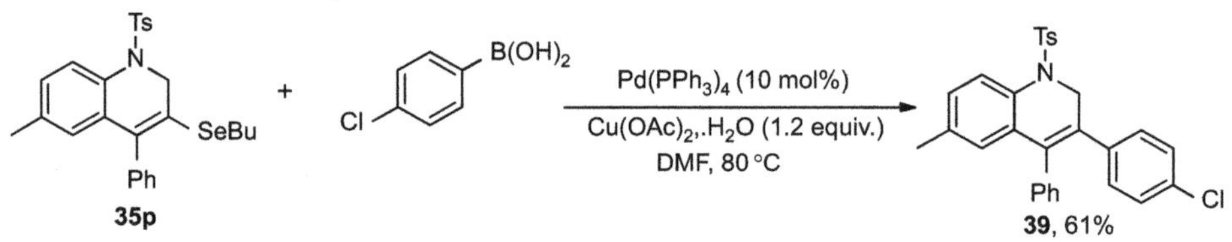

Figure 5.14: Suzuki cross-coupling reaction on 3-(organoselanyl)-1,2-dihydroquinoline **35p.**

5.2.2 Selenium as nucleophile

5.2.2.1 Synthesis of 2-imino-1,3-thiaselenolanes via iodocyclization

Selenium-containing heterocycles attract increasing attention because of their unique reactivity [21] and potential biological activity [22]. The method to introduce selenium into ring is very challenging due to their instability, toxicity and difficulties. In 2012, Toyoda et al. was successful to synthesize selenium-containing heterocycles 2-imino-1,3-thiaselenolanes **44** by the reaction of isoselenocyanates **40** with allyl mercaptan **41** via iodocyclization reaction [23]. Finally, 2-imino-1,3-thiaselenolanes **44** were obtained as *Z/E* mixture at the imine position (Figure 5.15).

Firstly, S-allyl-phenyl-selenothiocarbamate **42** was synthesized by the reaction of phenylisoselenocyanate **40** and allyl mercaptan **41** under the basic conditions [24]. Next, iodocyclization reaction using selenothiocarbamates **42** was carried out using 1 equiv. of iodine in $CH_{2r}Cl_2$ at room temperature. Then both selenium-containing

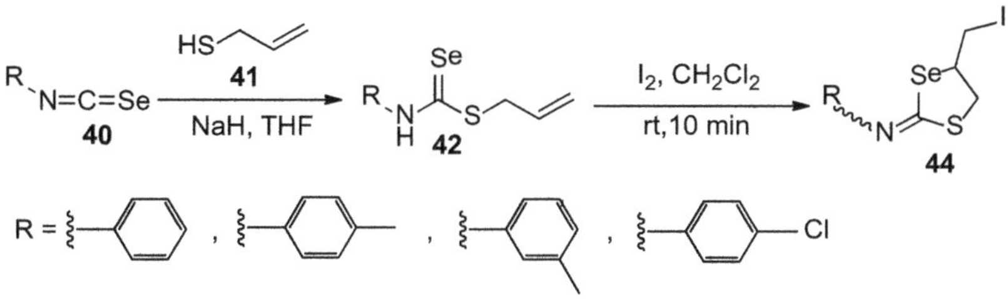

Figure 5.15: Synthesis of 2-imino-1,3-thiaselenolanes 3 *via* iodocyclization of S-allyl-selenothiocarbamate.

heterocyclic ring and exocyclic imines **44** were formed by selenium nucleophilic cyclization from the selenocarbonyl intermediates **43** (Figure 5.16).

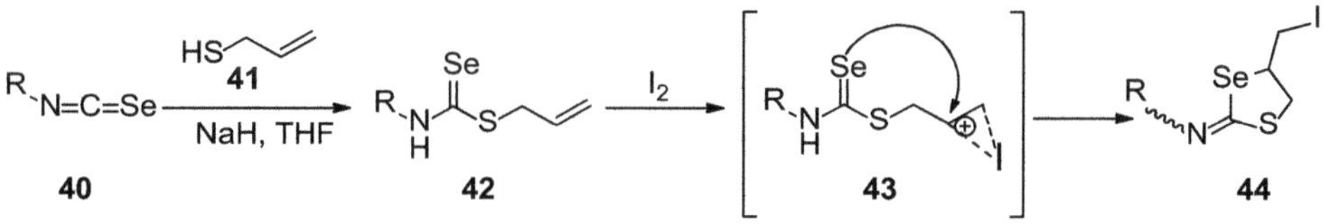

Figure 5.16: Reaction mechanism for the synthesis of 2-imino-1,3-thiaselenolanes *via* iodocyclization.

5.2.2.2 Synthesis of selenophenes and selanyl selenophenes

Selenophenes have attracted great attention due to their diverse biological activities including antitumoral [25], antiinflammatory [26], antihypertensive [27], and anticonvulsant [28] properties. These compounds are also used as versatile building blocks in the synthesis of many biologically active compounds and natural products [29]. Thus synthesis of selenophenes is of great importance. Previously reported synthesis of selenophenes in literature either involved harsh condition and high temperature [30] or multiple reaction steps [31, 32]. So, development of an efficient method for the preparation of selenophenes is very challenging. In 2014, Maity et al. developed a general and efficient protocol for the synthesis of selenophenes **46** in high yields by a simple one-pot CuO nanoparticle-catalyzed coupling of 1,3-dienyl bromides **45** and potassium selenocyanide [33] (Figure 5.17). The reaction with 1,3-dienyl gem-dibromides **45** was also reported to produce selanyl selenophenes **47**, new class of organoselenocycles (Figure 5.18). The use of inexpensive CuO nanoparticles as the catalyst, KSeCN as the selenium source, and an intramolecular nucleophilic cyclization of a selenium moiety to an alkene unit are the significant characteristic features of this reaction.

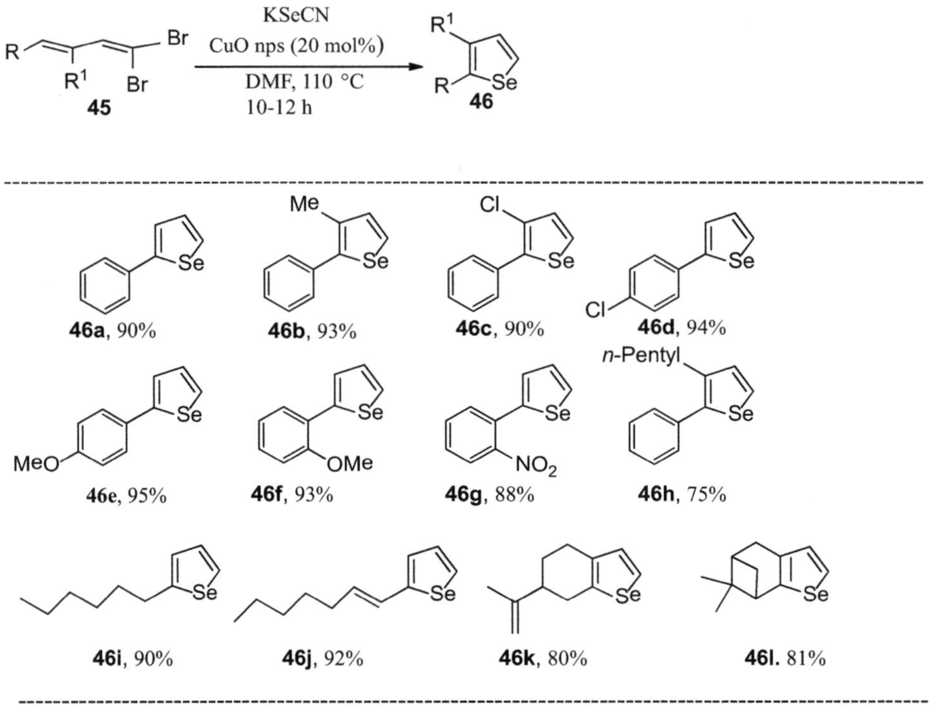

Figure 5.17: Synthesis of substituted selenophenes by copper catalyzed coupling of KSeCN and 1,3-di-enyl bromides.

5.2.2.3 Metal-free synthesis of unsymmetrical organoselenides and selenoglycosides

Organoselenide compounds take part in organic synthesis as a vital substrate [34]. A number of approaches to synthesize organoselenides have been developed. The most established methods for C(sp^2)–Se bond formation mainly involve Fe [35], Pd [36], Cu [37], or Ni [38] as catalyst. In 2017, Guan et al. developed a one-pot metal-free synthesis of organoselenides **52** and selenoglycosides **53**, which involved alkylation, arylation, or alkynylation of selenium anions [39] (Figure 5.19).

In the first step of the reaction, an arylselenocyanate **49** was generated by the arylation of potassium selenocyanate (KSeCN) with an iodonium reagent in the absence of a metal catalyst. By the addition of sodium borohydride ($NaBH_4$) after completing the first arylation reaction, selenol was liberated in a Grieco-type reduction [40]. Then the reaction of selenium nucleophile with aliphatic electrophile, iodonium reagent, or glycosyl halide afforded unsymmetrical organoselenides **52** (Figure 5.20)

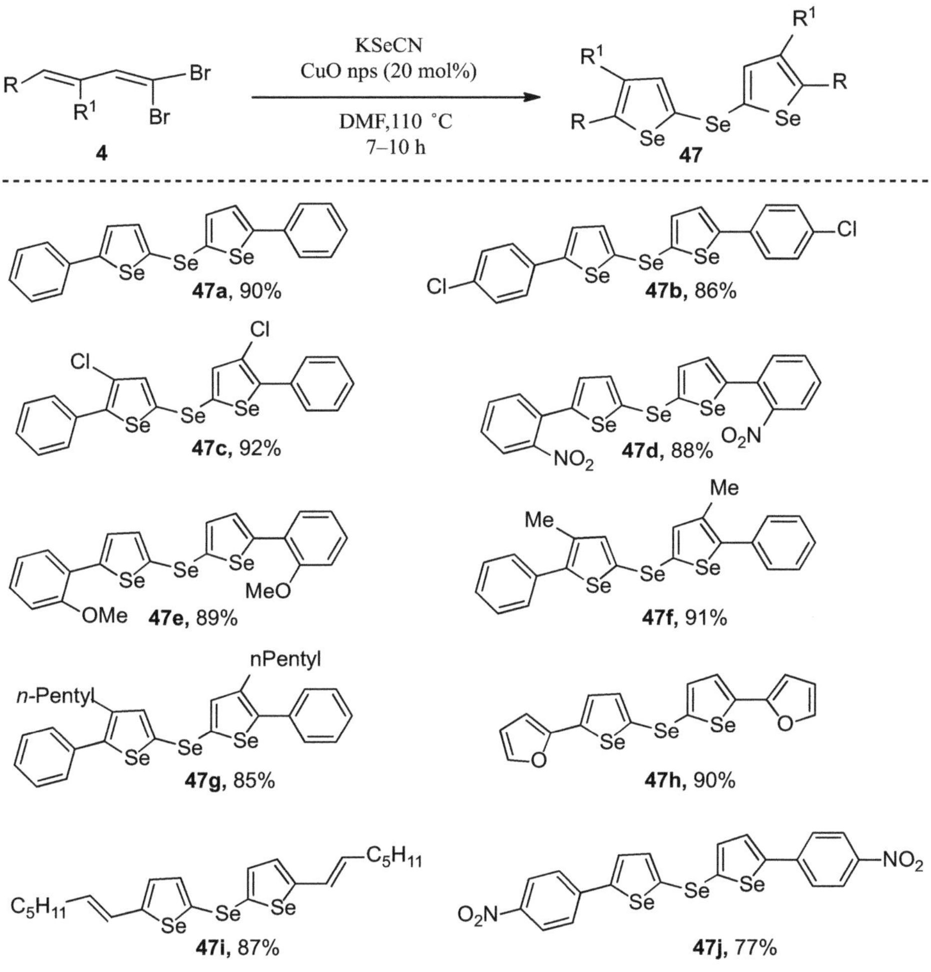

Figure 5.18: Synthesis of selanyl selenophenes by coupling of 1,3-di-enyl-gem-dibromides and KSeCN.

and selenoglycosides **53** (Figure 5.21), respectively. The method represents an umpolung approach to the synthesis of arylselenides.

5.2.2.4 Synthesis of chiral selenazolines

In 2018, Shibahara et al. demonstrated a new synthetic route of chiral selenazolines **55** from readily available N-acyloxazolidinones **54** via a selenative rearrangement of a chiral cyclic skeleton in the presence of elemental selenium, a hydrochlorosilane,

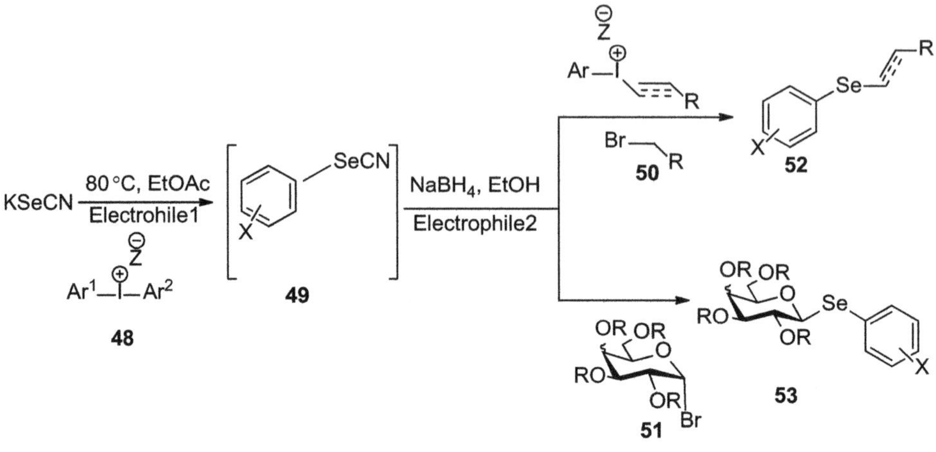

Figure 5.19: Selenium arylation reactions.

and an amine [41]. A wide variety of chiral selenazolines have been prepared by this method (Figure 5.22).

Initially, the selenation occurred randomly at the amide or carbamoyl moiety of **54** to give **56** or **57**. Diselenated intermediate **58** was produced by further selenation process. Then, once generated, **57** was decomposed readily under these conditions via formation of **59** and both reactions were in competition. Highly nucleophilic selenium atom was generated as intermediate **63**. Successive intramolecular S_N2-type cyclizations afforded selenazoline derivatives **55** (Figure 5.23).

5.2.2.5 Se-mediated one-pot synthesis of 2-substituted benzoselenazole derivatives

2-Substituted benzoselenazole derivatives have been of interest due to having unique skeleton for the construction of bioactive compounds [42]. Furthermore, the synthetic approach for the synthesis of 2-substituted benzoselenazoles were rare. In 2018, Gu et al. reported a novel and efficient approach to synthesize 2-substituted benzoselenazoles **59** by three-component reactions of 2-iodoanilines **56**, selenium powder, and arylacetic acids **57** or benzyl chlorides**58** in the presence of copper salt as a catalyst and dimethyl sulfoxide (DMSO) as a solvent (Figure 5.24) [43]. The desired compound was obtained in moderate-to-high yields with good functional group tolerance on the aromatic and heteroaromatic substrates.

The radical mechanism was not involved in the synthesis of 2-substituted benzoselenazole derivatives as the course of the reaction remained unchanged with addition of radical scavenger TEMPO in the reaction mixture. A plausible mechanism was proposed that diselenide **60** was produced initially by the reaction of

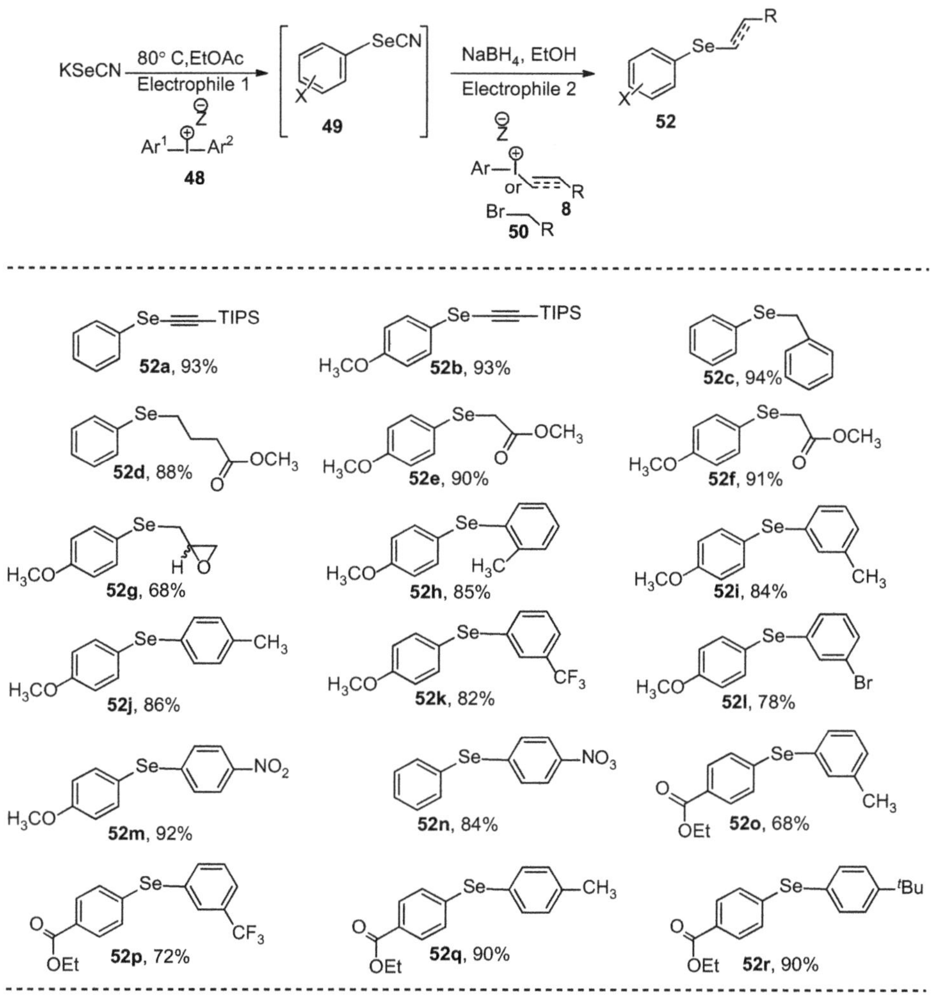

Figure 5.20: One-Pot synthesis of unsymmetrical organoselenides.

2-iodoaniline **56** with selenium powder in the presence of catalyst and base. Meanwhile, the intermediate **61** was formed by the oxidation of phenylacetic acid **57** using selenium powder. Then condensation reaction between **60** and **61** occurred to give the intermediate **62** or **63**, followed by intramolecular cyclization through the nucleophilic attack by selenium on the iminium carbon atom to generate the intermediate **64**. Subsequently, the Se–Se bond was cleaved by the delocalization of an electron pair to generate 1,3-benzoselenazole **59** derivatives through two routes. The route 1 involved the cyclization of **65**, followed by oxidation and release of H^+ and CO_2 to form desired product **59**, and route 2 involved deprotonation of **66** to produce the desired product C (Figure 5.25).

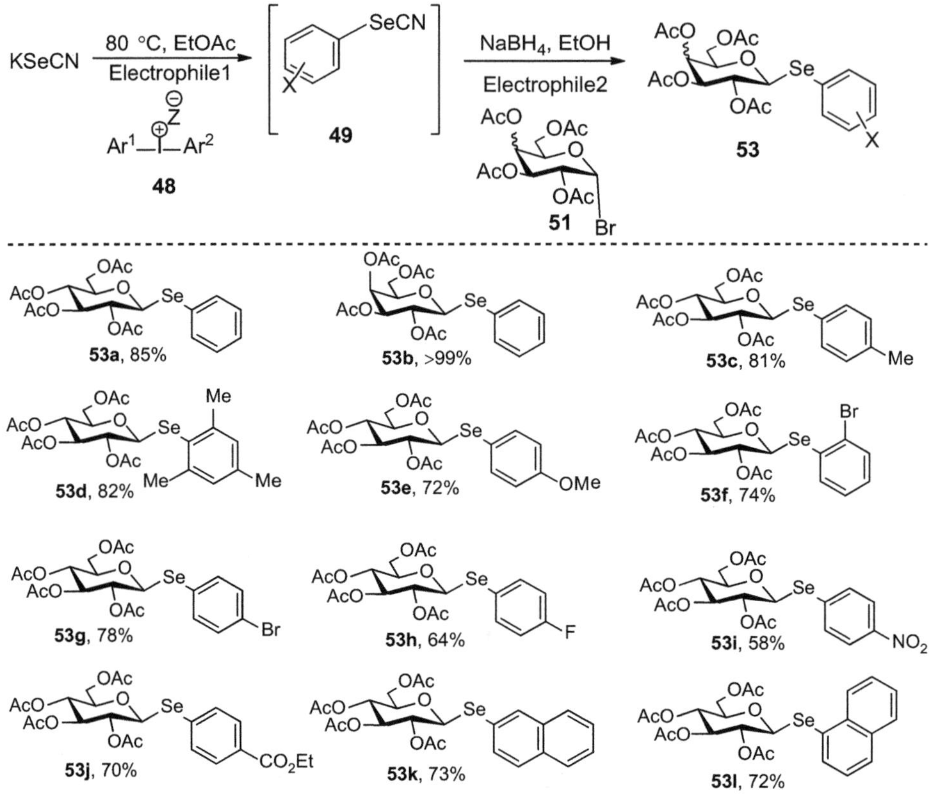

Figure 5.21: One-pot synthesis of glycosylselenides.

5.2.3 Selenium as radical

5.2.3.1 Copper-catalyzed selenylation of imidazo[1,2-*a*]pyridines with selenium powder via a radical pathway

In 2017, Sun et al. reported an efficient and simple approach for construction of-benzo[*b*]selenophene/imidazo[1,2-*a*]pyridine **69** framworks through copper-catalyzed direct selenylation of readily available 2-(2-bromophenyl)imidazo[1,2-*a*]pyridines **68** via functionalization of C(sp^2)–Br and C(sp^2)–H bonds [44]. The reaction was carried out under ligand- and base-free conditions in air using readily available selenium powder as selenylating agent (Figure 5.26). The reaction was successfully carried out with various substituted groups including methoxy, methyl, C–Cl bond, and C–Br bond.

The reaction was suggested to follow radical mechanistic pathway. First, the Cu (I) was oxidized to Cu(II) in the presence of oxygen in air. The single-electron transfer (SET) took place between **68** and Cu(II) to produce radical cation **70.** Proton loss

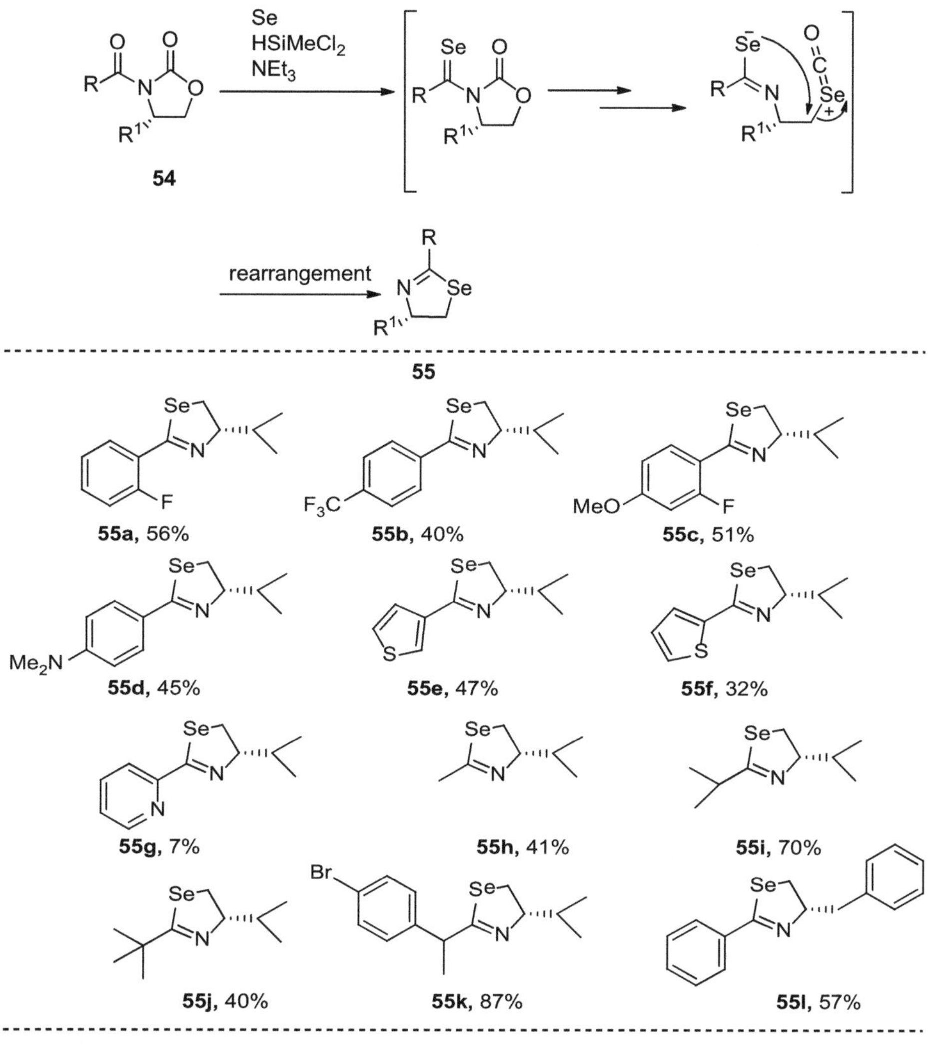

Figure 5.22: Synthesis of chiral selenazolines.

from the intermediate **70** generated vinyl radical intermediate **71**, which was then reacted with elemental selenium to give a selenium-free radical **72**. Subsequently, the intramolecular cyclization of radical **72** was involved to generate radical intermediate **73**. Then, Cu(I)-mediated bromine abstraction of radical intermediate **73** took place and finally the desired product **69** was released along with Br^- and Cu (II) (Figure 5.27).

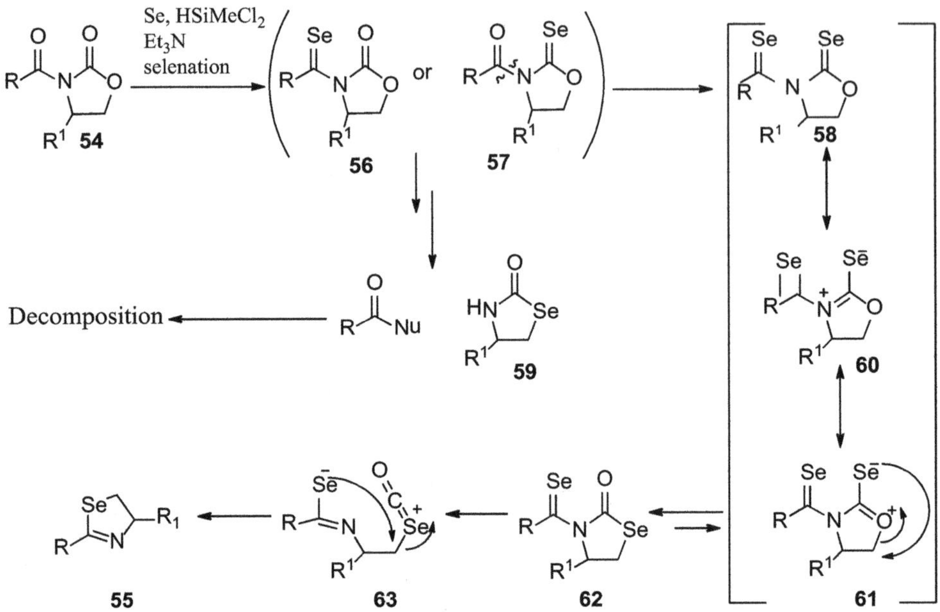

Figure 5.23: Probable reaction pathway.

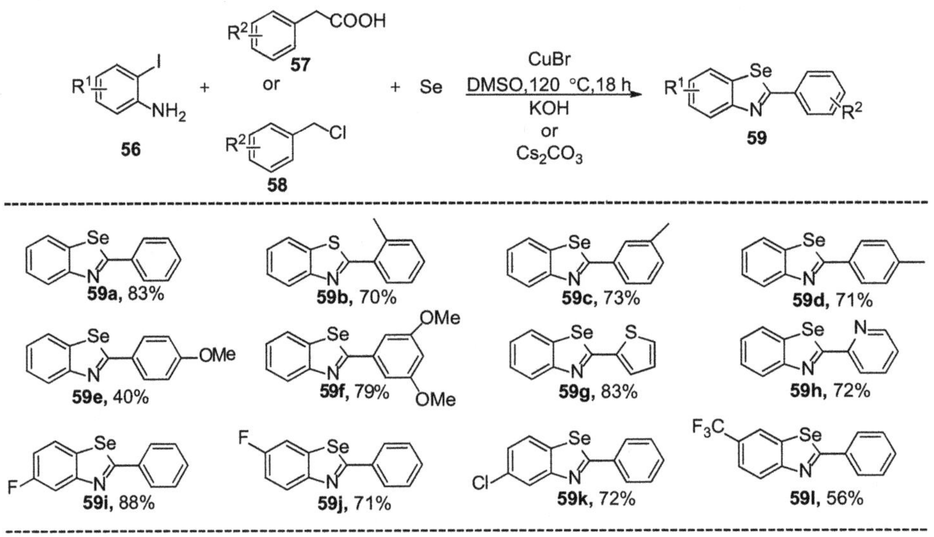

Figure 5.24: Synthesis of 2-substituted benzoselenazoles from iodoanilines, selenium powder, and arylacetic acids or benzyl chloride.

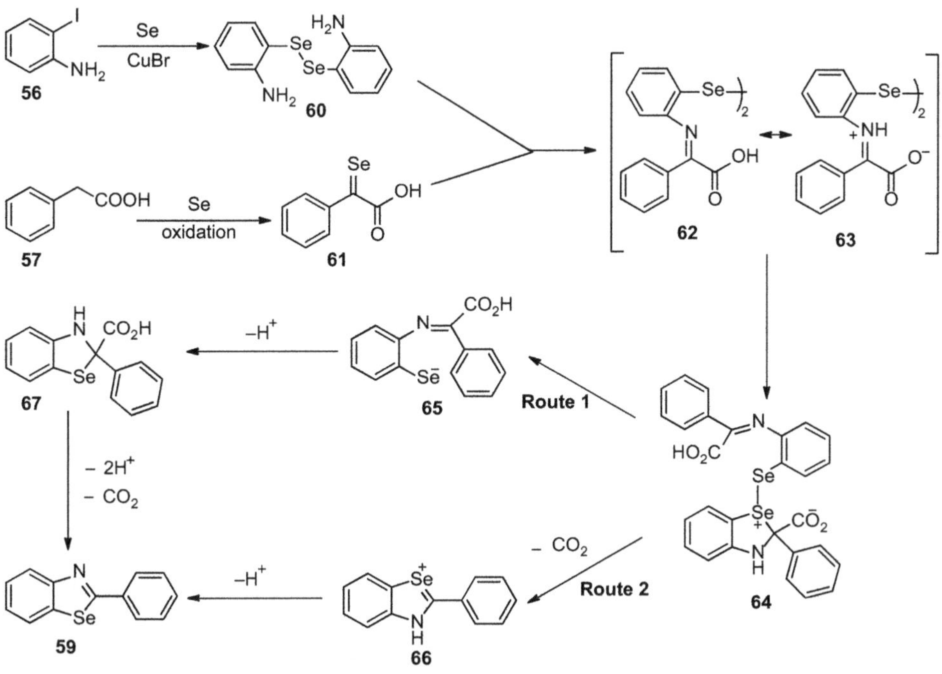

Figure 5.25: Plausible mechanism.

5.2.3.2 Synthesis of 3-selenylindole derivatives by TEMPO-catalyzed aerobic oxidative selenium insertion reaction

In 2018, Liu et al. have developed a novel and efficient approach for selenium functionalization at the C3 position of indoles using selenium powder as the selenium source, catalyzed by TEMPO with O_2 as the green oxidant [45]. With a broad scope of functional group tolerance 3-selenylindoles derivatives were obtained in moderate-to-high yields (Figure 5.28).

The reaction was suggested to follow radical pathway mechanism. Initially, isoselenocyanate **79** was generated in situ by the reaction of isocyanide **75** with elemental selenium **76** under the basic conditions, which was then reacted with an amine to produce selenoate **80**. The equilibrium was established between selenium anion intermediate **80** and nitrogen anion intermediate **81**. Then, nitrogen-centered radical **82** was generated by oxidation of intermediate **81** by O_2/TEMPO. The more active selenium radical intermediate **83** was then generated by the resonance of the nitrogen-centered radical **82**. Simultaneously, another equilibrium between nitrogen anion intermediate **84** and carbon anion intermediate **85** was established by deprotonation of indole in the presence of Cs_2CO_3. Finally, the desired product **78** was obtained by cross-coupling of radical **83** with **85** to generate nitrogen radical cation **86**, followed by SET of intermediate **86** (Figure 5.29).

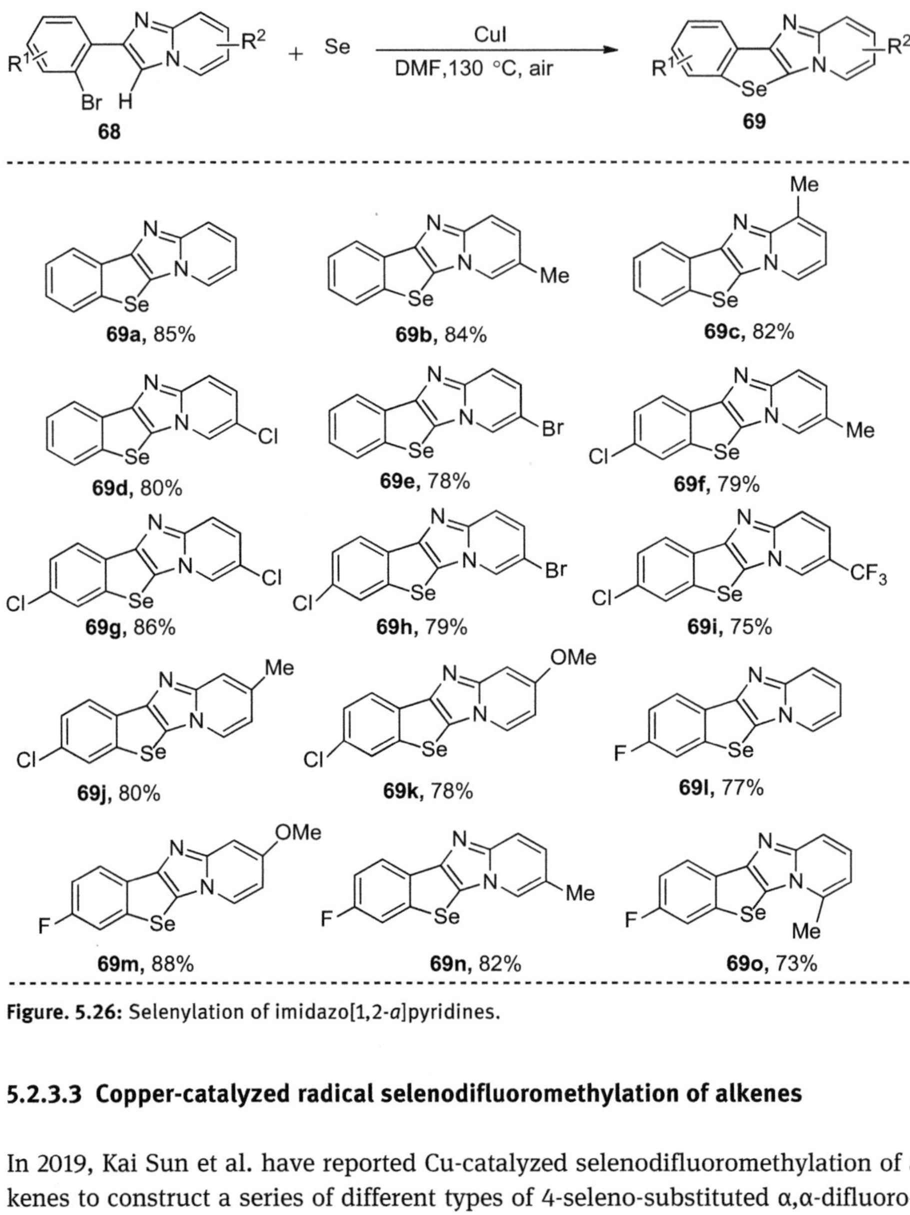

Figure. 5.26: Selenylation of imidazo[1,2-*a*]pyridines.

5.2.3.3 Copper-catalyzed radical selenodifluoromethylation of alkenes

In 2019, Kai Sun et al. have reported Cu-catalyzed selenodifluoromethylation of alkenes to construct a series of different types of 4-seleno-substituted α,α-difluoro-γ-lactams **88** (Figure 5.30) [46]. This attractive strategy has been of much potential for the preparation of other valuable fluorinated γ-lactams due to easy scale-up process with broad substrate scope as well as the scope for product derivatization.

Mechanistic studies suggested that the catalytic system involved a radical cascade cyclization pathway. Initially, the reaction was initiated by the reaction of Cu(I) and N-allyl-2-bromo-2,2- difluoro-N-phenylacetamide **87** to generate radical intermediate **89** along with Cu(II) species via SET process. Then, alkyl radical intermediate **90** has been

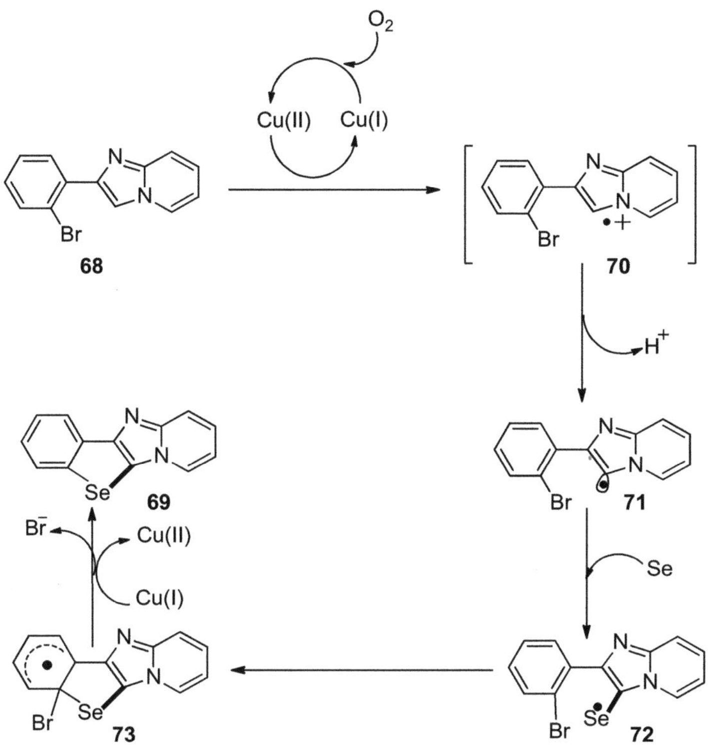

Figure 5.27: Proposed reaction mechanism.

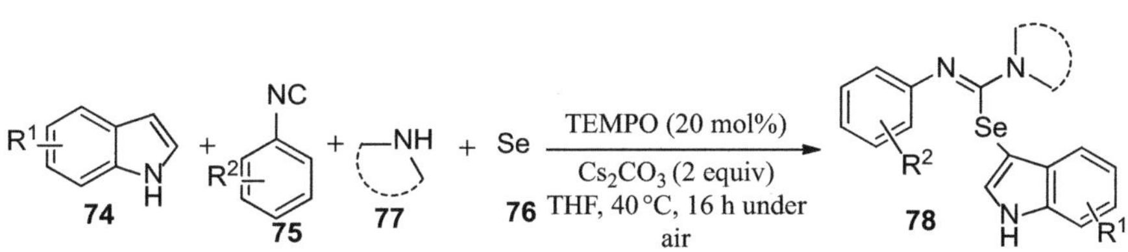

R^1 = 4-CO_2Me, 5-NO_2, 6-CN, 4-Me, 5-Me, 6-Me, 7-Me, 5-OMe, 6-CN, 5-F, 5-Br, 5-I

R^1 = *m*-NO_2, *p*-CN, *p*-$COCH_3$, *p*-$CONEt_2$, *p*-CO_2Et, *p*-CF_3, *p*-Br, *p*-Cl, *p*-OMe

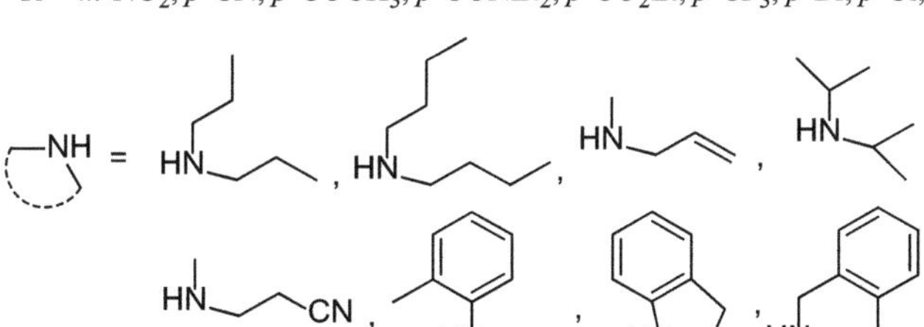

Figure 5.28: Synthesis of 3-selenylindole derivatives from multicomponent reaction of isocyanides, selenium Powder, amines, and indoles by TEMPO-catalyzed aerobic oxidation.

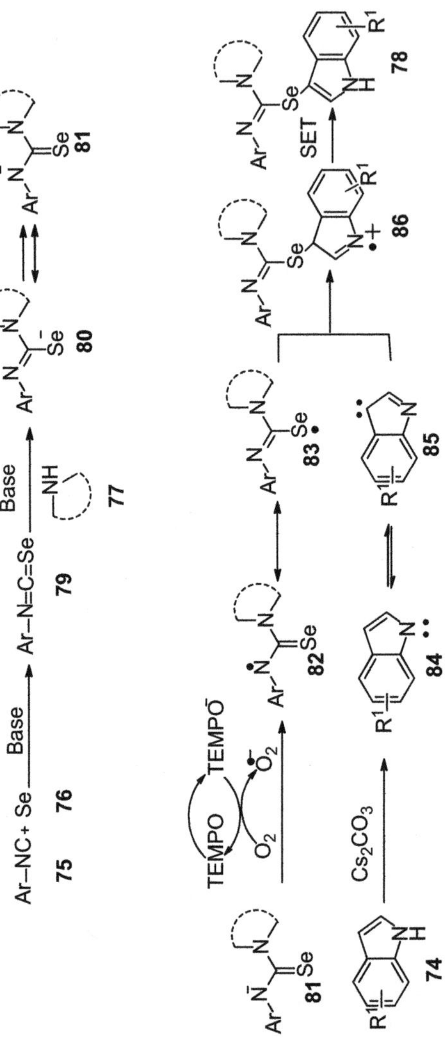

Figure 5.29: Plausible reaction mechanism.

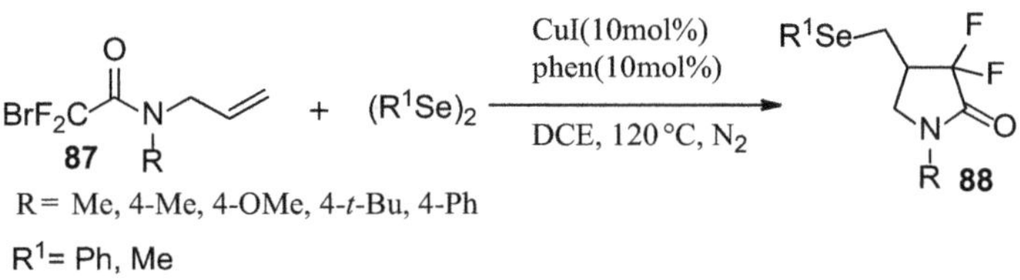

Figure 5.30: Selenodifluoromethylation of alkenes.

produced by rapid 5-exo-trig cyclization by the addition of the fluoroalkyl radical **89** to the unsaturated double bond. Since diselenide is an excellent radical trapping agent, radical intermediate **90** was trapped by the diselenide and afforded desired 4-seleno-substituted α,α-difluoro-γ-lactams **88**. Selenyl radical was converted finally into selenyl anion (PhSeX) with the reduction of Cu(II) to Cu(I) (Figure 5.31).

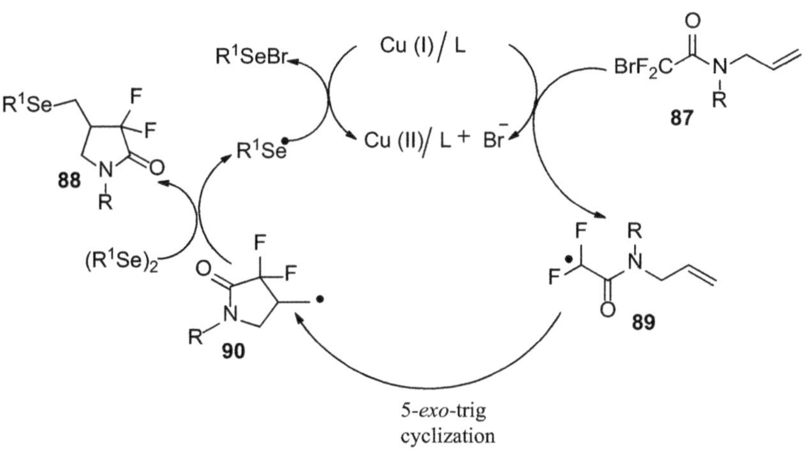

Figure 5.31: Plausible Reaction pathway.

5.2.3.4 Electrochemical radical selenylation/1,2-carbon migration and Dowd–Beckwith-type ring-expansion sequences of alkenylcyclobutanols

Recently, another efficient strategy for the synthesis of cyclohexanone derivatives (**94**) has been reported by Kim et al. via electrochemically induced oxidative radical selenylation/1,2-carbon migration and Dowd–Beckwith-type rearrangement sequences of alkenylcyclobutanol derivatives **91** with diselenides **92** [47]. The reaction was conducted in an undivided electrochemical cell equipped with glassy carbon plate as electrodes in the presence of $n\text{-}Bu_4N^+BF_4^-$ as a supporting electrolyte and acetonitrile solvent under constant 7mA current (Figure 5.32).

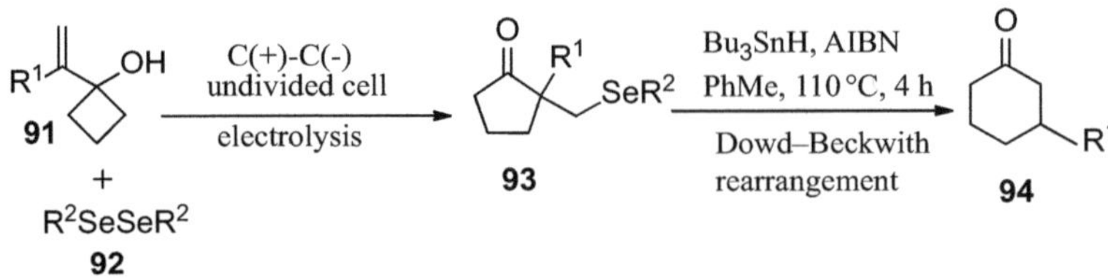

R^1 = Ph, *p*-methylphenyl, *p*-methoxyphenyl, *p*-chlorophenyl, *p*-fluorophenyl, *o*-methylphenyl, *m*-methylphenyl, benzyl, naphthyl
R^2 = Ph, Bn

Figure 5.32: Radical selenation/1,2-carbon migration and Dowd–Beckwith-type ring-expansion sequence.

The reduced yield of the desired products by radical scavengers (TEMPO and BHT) indicated the involvement of radical pathway of the reaction. Diphenyl diselenide **92a** was oxidized to generate cationic radical intermediate **95**, which was decomposed to produce phenyl selenium radical **96** and phenyl selenium cation **97.** Then the reaction was accomplished between phenyl selenium radical **96** with 1-(1-arylvinyl)cyclobutanol **91** to produce intermediate **98**, which was oxidized on anode to afford the cation **99.** 1,2-Carbon migration of cation **99** led to ring expansion producing **93** (path a). In another possibility, phenylselenium cation **97** was attacked by nucleophile 1-(1-arylvinyl)cyclobutanol **91**, followed by 1,2-carbon migration to generate the product **93** (path b) (Figure 5.33).Then Dowd–Beckwith rearrangement of **93** produced the corresponding one-carbon ring-expanded ketones **94.**

5.2.4 Synthesis of diselenophosphinic esters

Recently, synthesis of diselenophosphinic esters attracted increasing attention of researchers because these compounds are broadly used as convenient precursors of nanocrystalline materials [48], anionic ligands for metal complexes [49], and promising building blocks for organic synthesis [50]. A new, efficient, atom-economic synthetic approach of diselenophosphinic esters **104** have been developed by three-component reactions of alkenes **103**, secondary phosphanes **101**, and elemental selenium **102** (Figure 5.34) [51].

Oxidation of secondary phosphane **101** by elemental selenium produced secondary phosphane selenide **105**, which was further oxidized by second equivalent of elemental selenium to produce diselenophosphinic acid **106.** Then the addition of aryl- or hetaryl alkenes **103** to diselenophosphinic acid **106** produced diselenophosphinates **104** (Figure 5.35).

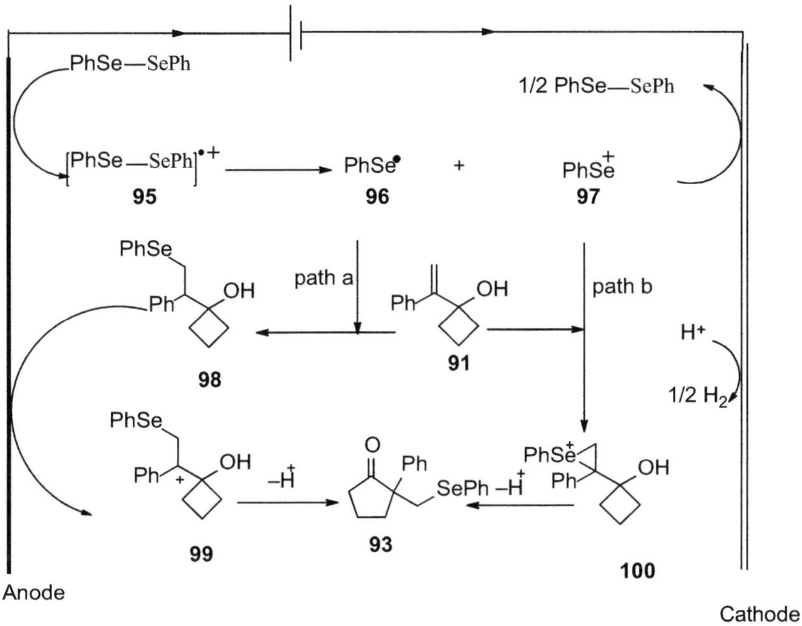

Figure 5.33: Proposed reaction mechanism.

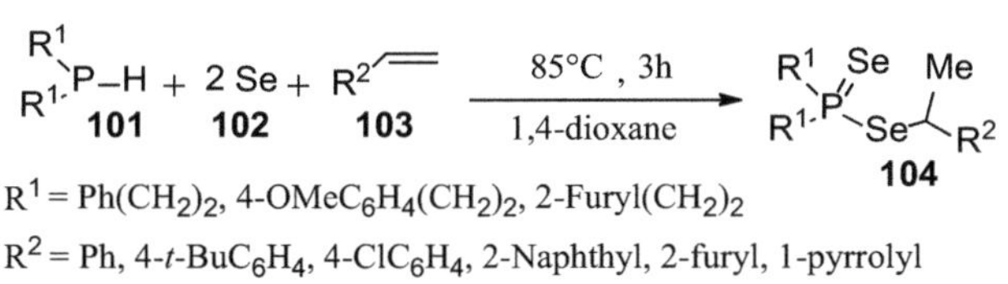

Figure 5.34: One-pot atom-economic synthesis of diselenophosphinic esters from aryl- or hetarylalkenes, secondary phosphanes and elemental selenium.

5.2.5 Markovnikov addition of thioselenophosphinic acids to double bond

The addition of thioselenophosphinic acids, $R_2P(Se)SH$ or $R_2P(S)SeH$, to vinyl ethers was not known in the literature until in 2013 Oparina et al. reported the electrophilic addition of thioselenophosphinic acids **107**, generated from secondary phosphine sulfides **108** and elemental selenium, to a diverse range of vinyl ethers **109** for the first time [52]. Markovnikov adducts, that is, *S*-esters **110** and *Se*-esters **111** of thioselenophosphinic acids, have been obtained in high yields (Figure 5.36). The major isomers were *S*-esters and the ratio depends on the structure of the initial vinyl ethers and thioselenophosphinic acids. This reaction was applicable to secondary phosphine sulfides bearing alkyl, aryl, and heteroaryl groups and aryl

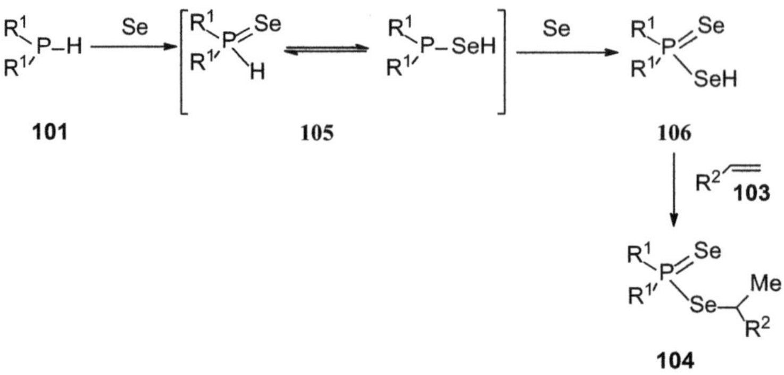

Figure 5.35: Plausible mechanism of synthesis of diselenophosphinic esters for the three-component reaction between secondary phosphanes, elemental selenium and alkenes.

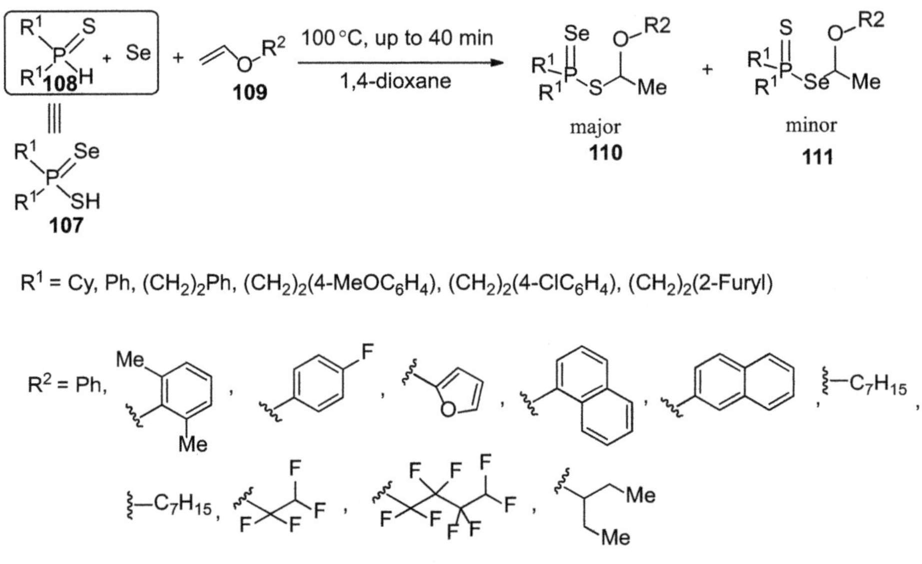

Figure 5.36: Markovnikov addition of thioselenophosphinic acids to double bond.

and alkyl vinyl ethers. Air sensitivity and unstable nature of thioselenophosphinic acids [53] restricted the employment of one-pot protocol for the preparation of thioselenophosphinic esters.

The equimolar mixture of secondary phosphine sulfide **108** and elemental selenium in this three-component reaction was synthetic equivalent of the thioselenophosphinic acids **107**, which was then added to electron-rich double bond of various vinyl ethers **109** in a Markovnikov manner to give a mixture of S-esters **110** and Se-esters **111** of thioselenophosphinic acids where the main isomer was S-ester **110**. This type of addition was electrophilic in nature (Figure 5.37).

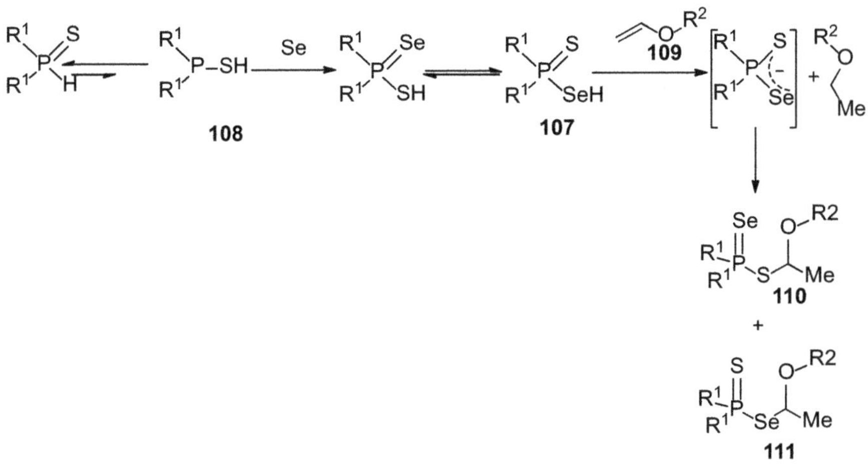

Figure 5.37: Proposed mechanism for the three-component reaction between secondary phosphine sulfides, vinyl ethers and selenium.

5.2.6 Synthesis of 1,4-disubstituted 5-arylselanyl-1,2,3-triazoles

Synthetic methods for constructing multisubstituted 5-selenofunctionalized triazoles are still in high demand. Click reactions [54] and the formation of C–Se bond are well known in the literature [55]. Based on these reactions in 2018, Cui et al. reported a novel and efficient method for the preparation of 1,4-disubstituted 5-arylselanyl-1,2,3-triazoles **115-117** from propiolic acids **112**, diselenides **113**, and azides **114** by Cu-catalyzed decarboxylative/click reaction via intermoleculararazide–alkyne cycloaddition of an alkynyl selenium intermediate (Figure 5.38) [56]. High efficiency and regioselective catalytic reaction featured with mild reaction conditions, easy operational simplicity as well as excellent compatibility with air are main advantages.

The resulting multisubstituted 5-seleno-1,2,3-triazoles were tested for in vitro anticancer activity and compounds **115f**, **115h**, and **115p** showed potent cancer cell-growth inhibition activities.

The plausible mechanism is presented in Figure 5.39. It was proposed that 1,10-phenanthroline coordinated to $Cu(OAc)_2{\cdot}H_2O$ to form the active copper(II) intermediate **118**, and then it reacted with phenylacetic acid **112a** to generate copper(II) intermediate **119**, which produced copper(II) intermediate **120** by decarboxylation through the release of one molecular CO_2 [57]. The intermediate **121** was generated by the reaction of **120** with diphenyl diselenide **113a**. The reductive elimination of **121** resulted in phenyl (phenylethynyl)-selane **125** as well as copper(I) species **122**. Subsequently, the intermediate **123** was produced by the complexation of **122** with phenyl(phenylethynyl)selane **125** and azide **114a**. Then intermediate **124** [58] was generated by oxidative cyclization of intermediate **123**. The reductive elimination of **124** afforded the desired product **115a**

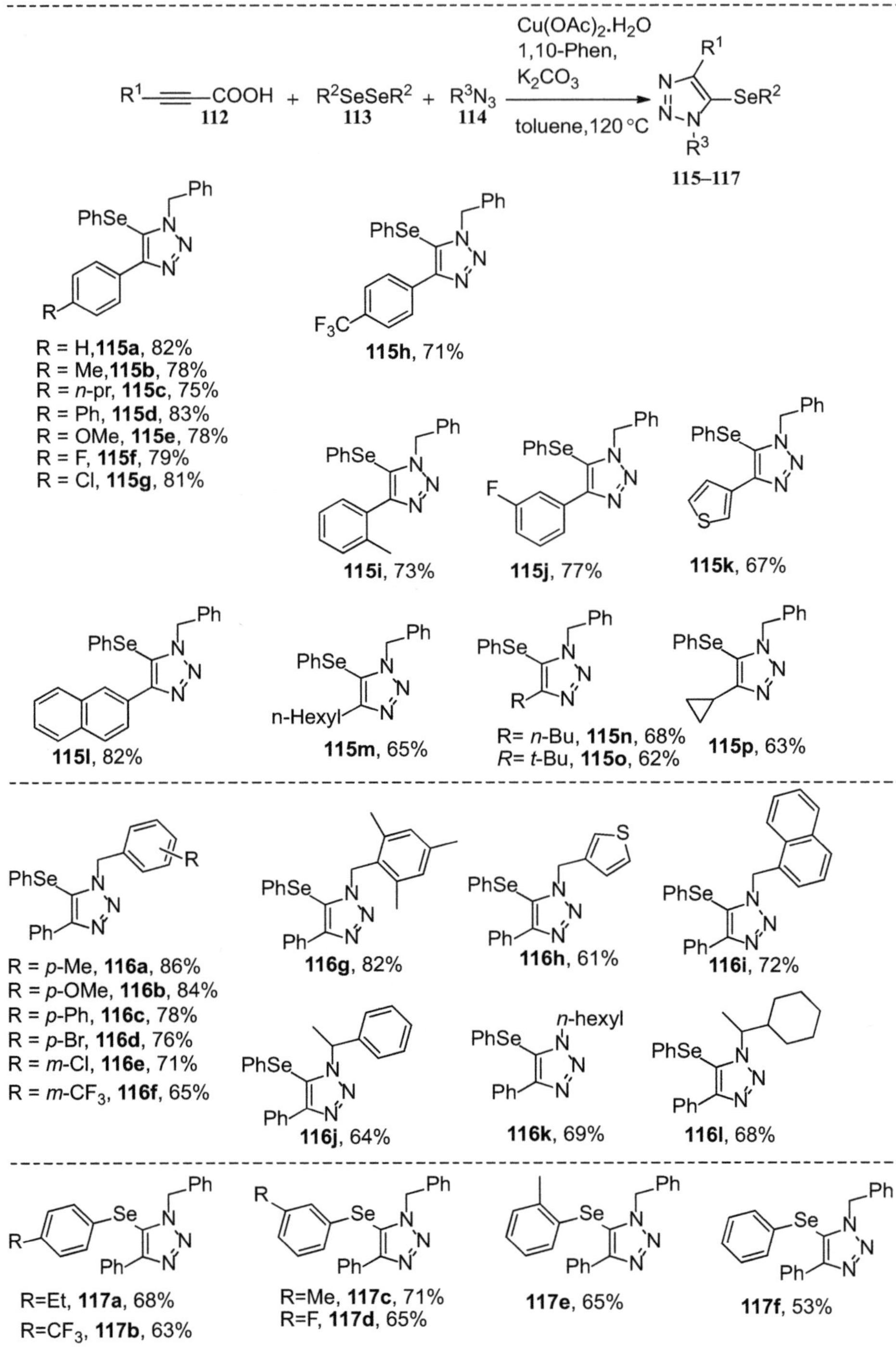

Figure. 5.38: Synthesis of 1,4-disubstituted 5-arylselanyl-1,2,3-triazoles.

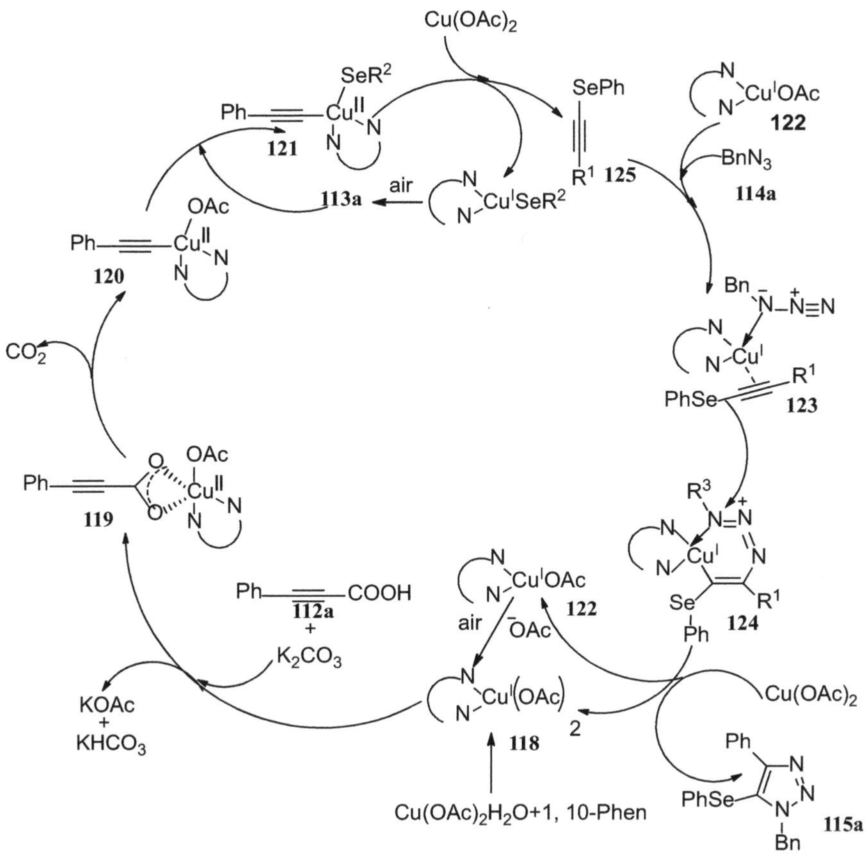

Figure 5.39: Proposed mechanism.

and copper(I) species **122**. Finally, **122** was converted into **118** under air atmosphere (Figure 5.39).

5.2.7 Stereospecific Al-catalyzed tandem C-N/C-Se bond formation of isoselenocyanates with aziridines

The cycloaddition of isothiocyanates with chiral aziridine has been recently reported in literature [59]. Recently, in 2019, Satheesh et al. have reported Al-catalyzed stereospecific tandem C–N/C–Se bond formation of chiral aziridines **127** with isoselenocyanates **126** with 90–99% ee and 77–91% yields (Figure 5.40) [60].

According to experimental and DFT studies, it was suggested that the formation of Al-complex **130** was generated by chelation of isoselenocyanate **126** with Al-salen **128**. Then, Al-complex **130** coupled with aziridine **127** through a concerted S_N2 pathway **131** to produce the target products **129** (Figure 5.41).

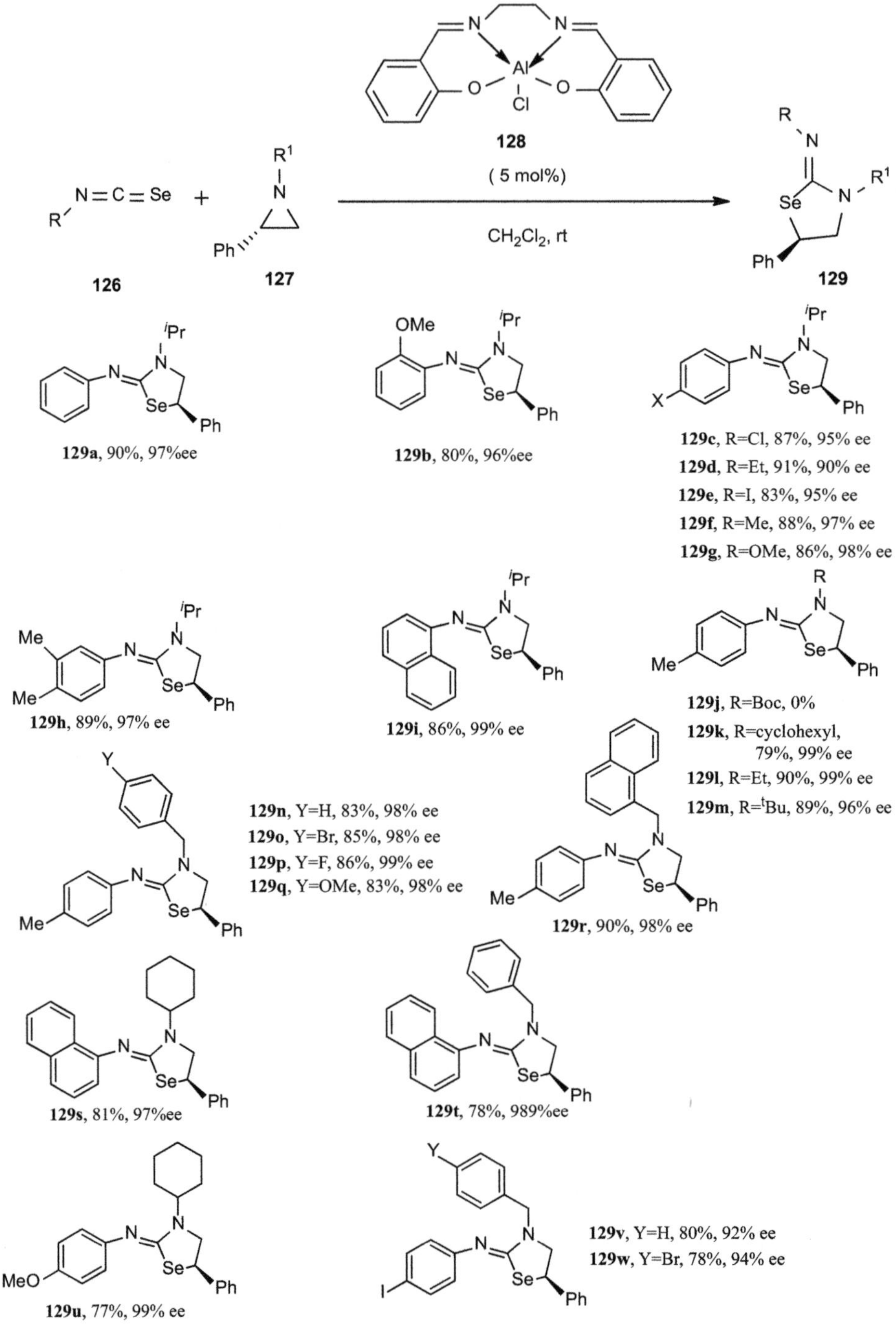

Figure 5.40: Al-catalysed tandem C-N/C-Se bond formation of isoselenocyanates with aziridines.

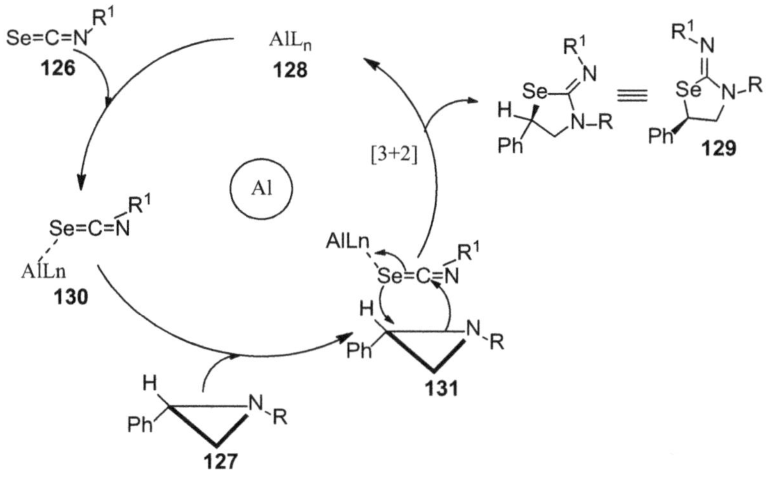

Figure 5.41: Proposed mechanism.

5.3 Selenium as catalyst in organic reactions

Earlier, we have discussed the versatility of organoselenium reagents in organic syntheses. Regarding the toxicity of selenium compounds [61], it is always preferable to shift toward catalytic processes. The use of organoselenium compounds as catalysts has opened a new alley for the synthetic organic chemists, with the first venture by Sonada et al. using elemental Se as a catalyst, in 1975 [62]. Thereafter, the catalytic use of organoselenium reagents gradually increased till the first decade of 21st century, which has been well documented in review articles and book chapters [63]. A considerable progress has taken place in recent years where versatile chemical nature of selenium reagents has been explored thoroughly. This, coupled with newly developed concepts like chiral catalysis and photocatalysis, has enriched the repertoire of organic chemists for processes like oxidations, haloaminations, oxidative cyclizations, enantioselective cyclizations and trifluoromethylthiolations, C–N cross coupling reactions, and so on. A number of such representative works are discussed here, which have been published in the last decade.

5.3.1 Selenium-catalyzed haloamidation reactions

5.3.1.1 Chloroamidation of olefins using Lewis basic selenium catalyst

Selenium-catalyzed allylic chlorination [64], as well as Lewis acid-catalyzed haloamidation of olefins [65], using *N*-halosuccinimide was reported in the literature.

Yeung and coworkers, in 2013, demonstrated a Lewis basic diphenylselenide-catalyzed chloroamidation of olefins **132** to prepare synthetically important *trans*-1,2-chloroamides **133** [66]. *N*-chlorosuccinimide (NCS) and acetonitrile were used as the halogen and the nitrogen source, respectively (Figure 5.42). Yields were good for both alkyl and aryl olefins and excellent regio- and diastereoselectivities were observed for substituted cyclohexenes. Acid-sensitive functional groups are well tolerated in this reaction.

Figure 5.42: Chloroamidation of olefins using Lewis basic diphenylselenide catalyst.

Mechanistically it is proposed that the Lewis basic diphenyl selenide activates the Cl atom of NCS to form an intermediate **134**. This electrophilic Cl atom is intercepted by the olefin **132**. Subsequently, the formed haliranium intermediate **135** undergoes nucleophilic attack by acetonitrile (**135** to **136**), followed by the quenching by water, which leads to the chloroamidation product **133** (Figure 5.43).

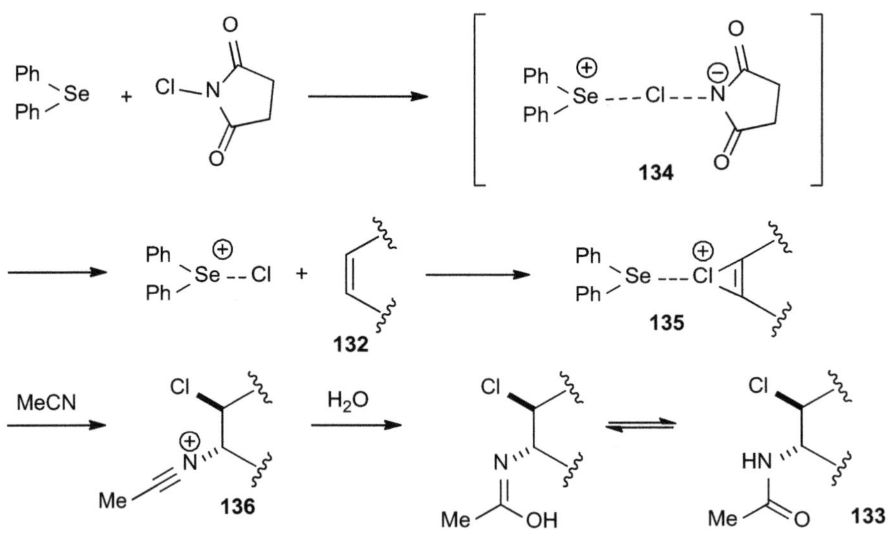

Figure 5.43: Proposed mechanism of the Lewis basic selenium catalyzed chloroamidation.

5.3.1.2 Enantioselective bromoaminocyclization of olefinic amides

Driven by their previous asymmetric bromocyclization [67], catalyzed by bifunctional amino-thiocarbamate, Yeung and coworkers developed a mannitol-derived C_2-symmetric selenium catalyst and employed it successfully in the novel monofunctional Lewis basic selenium-catalyzed enantioselective bromocyclization reaction [68]. Pyrrolidine **138** having two chiral centers was synthesized from trisubstituted olefinic amide using *N*-bromopthalimide as brominating agent and catalyst **139**. Although the reaction needs to be carried out for 5 days at a much lower temperature (−78 °C), yields are excellent (up to 93%) with good-to-excellent enantioselectivity (ee up to 95%). Strong electron-rich and electron-poor groups as R^2 showed poor enantioselectivity (Figure 5.44).

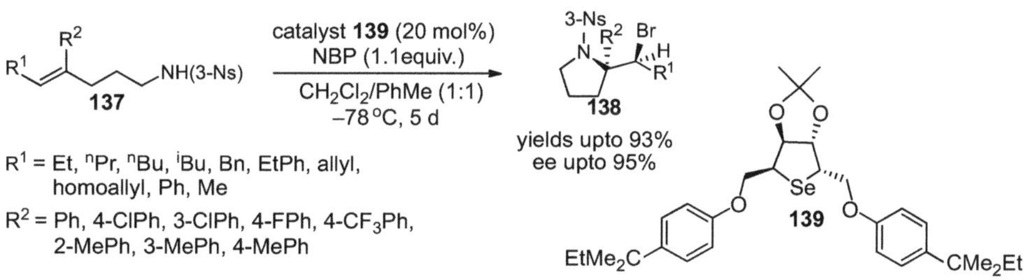

Figure 5.44: **139**-Catalyzed bromocyclization reaction.

Similar to the case of chloroamidation (Figure 5.43), probably, the Lewis basic catalyst **139** transfers the bromonium ion from NBP to the alkene **137** to form a tight ion pair to prevent racemization. Finally, the sulfonated amine attacks intramolecularly to furnish the chiral pyrrolidine **138** enantioselectively [68]. Recently, in 2017, Ishihara and coworkers have reported a diastereoselective chlorocyclizaton of tryptamine derivatives by cooperative catalysis of diphenyl diselenide and iodine [69].

5.3.2 Selenium-catalyzed trifluoromethylthioamination reactions

Due to their significant potential in pharmaceuticals and agrochemicals [70], organic molecules containing SCF_3 groups demand efficient synthetic routes to the molecules containing SCF_3 groups. Through alkene bifunctionalization by acid activation, incorporation of Cl, OTs, CF_3COO, and sulfonyl groups along with SCF_3 groups was reported in 2009 [71]. However, trifuoromethylthioamination under acid activation was not reported, might be due to the protonation of amines. This challenge was met successfully using Lewis basic selenium catalyst, which, like the activation of Cl from NCS in the chloroamidation, activates SCF_3 from N-SCF_3 bonds.

5.3.2.1 Vicinal trifluoromethylthioamination of alkenes

In 2015, research group of Zhao reported the selenium-catalyzed vicinal trifluoromethylthioamination of alkenes **140** using 10 mol% of di(*p*-methoxyphenyl) selenide, 1.3 equiv. of *N*-trifluoromethylthiosaccharin as SCF_3 source, and excess of nitriles **141** as the aminating agent (Figure 5.45). Excess of nitrile might be neutralizing the TfOH used in the reaction. Terminal as well as internal alkenes provided **142** in moderate-to-excellent yields (up to 96%) with good diastereoselectivity in case of the later. The reaction tolerates both alkyl and aryl nitriles well [72].

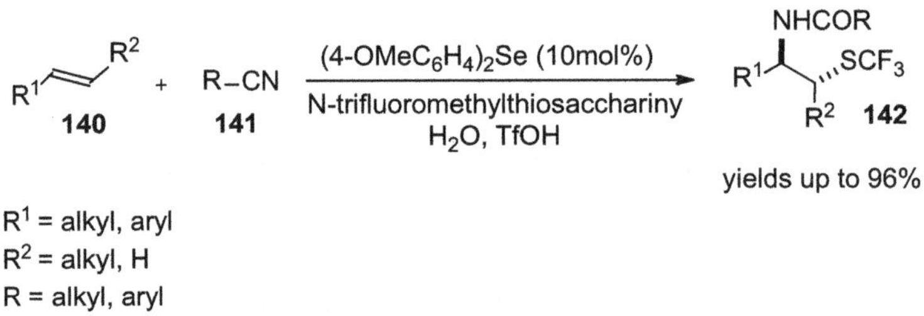

Figure 5.45: Diaryl selenide catalyzed vicinal trifluoromethylthioamination of alkenes.

5.3.2.2 Enantioselective trifluoromethylthiocyclization of olefinic amides

The same group has developed a selenium-catalyzed SCF_3-aminocyclization of olefinicsulfoamides to form various azaheterocycles with high enantioselectivity [73]. An indane-based chiral bifunctional selenide **146** was used as catalyst in the presence of $BF_3.OEt_2$ or NfOH as acid in a mixed solvent of DCE/DCM (1:1), while the source of SCF_3 being $(PhSO_2)_2NSCF_3$ (Figure 5.46). As they have mentioned, the low temperature (-78 °C) was maintained to get the better enantioselectivity (ee up to 97%). With homoallylic amides **143** (n = 1), chiral pyrrolidine derivatives were obtained irrespective of the olefinic substituents, whereas, with an increase of one more C atom (n = 2), two types of products formed. When the terminal substituents were aliphatic, pyrrolidines formed through *5-exo-trig* cyclization. In the case of Ph as terminal substituent, *6-endo-trig* cyclization led to the formation of chiral piperidine derivative. Some of the examples of the products are shown in Figure 5.47.

They have taken this ahead for a trifluoromethylthiocarbocyclization of F–C's alkylation-type reaction, in 2018, with the same catalyst **146** and the same source of SCF_3 as shown in Figure 5.46. Various substituted trifluoromethylthiolated tetrahydronaphthalenes were prepared with high enantioselectivities [74].

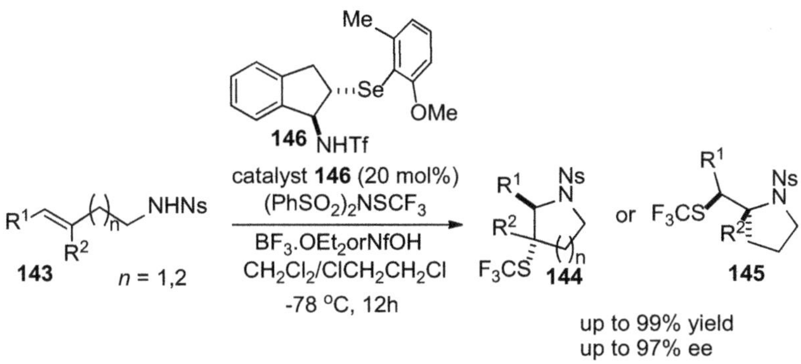

Figure 5.46: Selenide-catalyzed enantioselective synthesis of trifluoromethylthiolated tetrahydronaphthalenes.

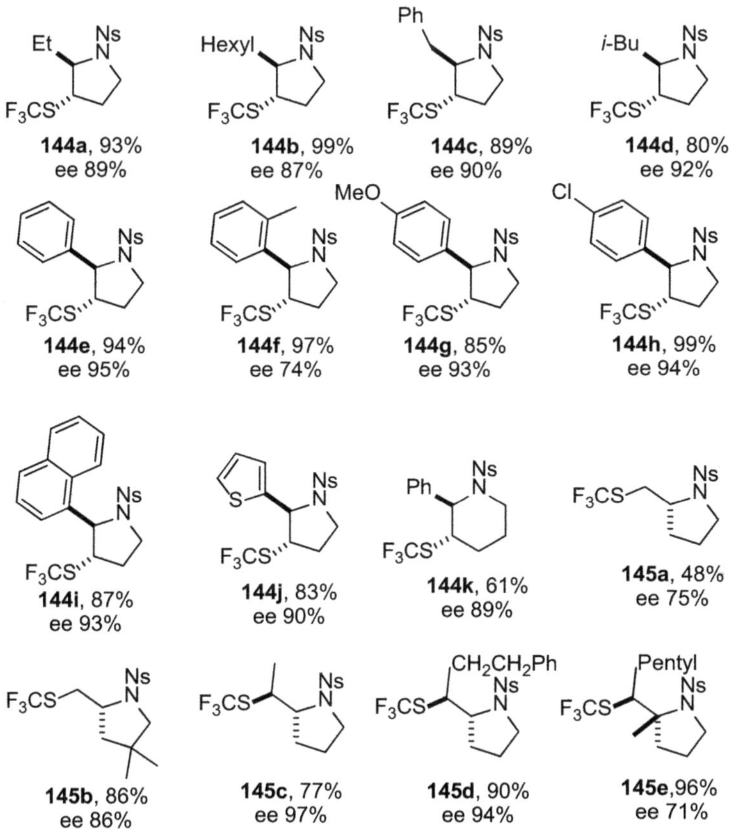

Figure 5.47: Substrate scope.

5.3.3 Selenium-catalyzed oxidative C–H amination reactions

Being one of the most elementary structural foundations of natural products [75], C–N bonds are always hot targets to be achieved by synthetic chemists. Direct oxidative C–H amination is a facile method for C–N bond formation. Among catalytic processes, intramolecular C–H amination reactions using palladium catalysts are well known [76]. However, intermolecular reactions of this type suffer from lack of regioselectivity [77]. A metal-free selenium-catalyzed approach proved to be very successful in this regard.

5.3.3.1 Direct oxidative allylic and vinylic C–H amination

In 2013, Breder and coworkers published the first selenium-catalyzed direct C–H amination reaction of alkenes using *N*-fluorobenzenesulfonimide (NFSI) as the N-source and terminal oxidant [78]. The allylic amination of the alkenes **147** was carried out with NFSI under the catalysis of diphenyl diselenide to get allylic imides **149** (Figure 5.48). As the presence of water leads to a by-product by allylic oxidation, molecular sieves were used to improve the yields of **149**. The presence of an electron-withdrawing group (EWG) at the allylic position appears to be essential for the reaction. Ester, amide, phosphonate, sulfone, cyano and keto groups as EWG, alkyl and benzyl groups as olefinic substituent are tolerated well (yields up to 89%). The same procedure was, then, successfully applied to cyclic alkenes **148** to get vinylic imides **150**, mostly, in excellent yields (up to 95%).

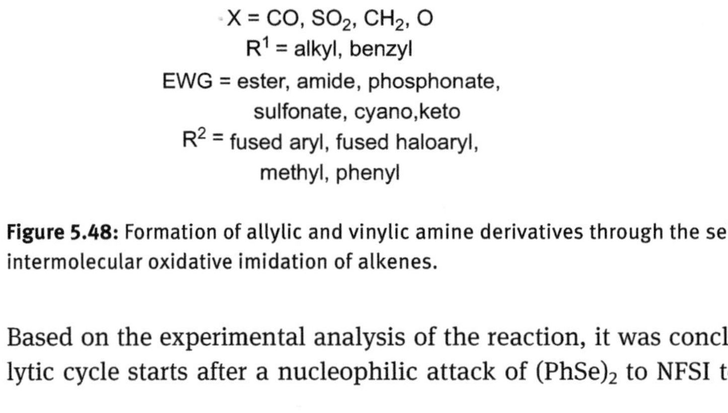

Figure 5.48: Formation of allylic and vinylic amine derivatives through the selenium-catalyzed intermolecular oxidative imidation of alkenes.

Based on the experimental analysis of the reaction, it was concluded that the catalytic cycle starts after a nucleophilic attack of $(PhSe)_2$ to NFSI to form the cationic species **151** (Figure 5.49). After the cationic adduct **152** is formed by the addition of **151** to the alkene **147** or **148**, elimination takes place to lead to the formation of the product **149** or **150** [78].

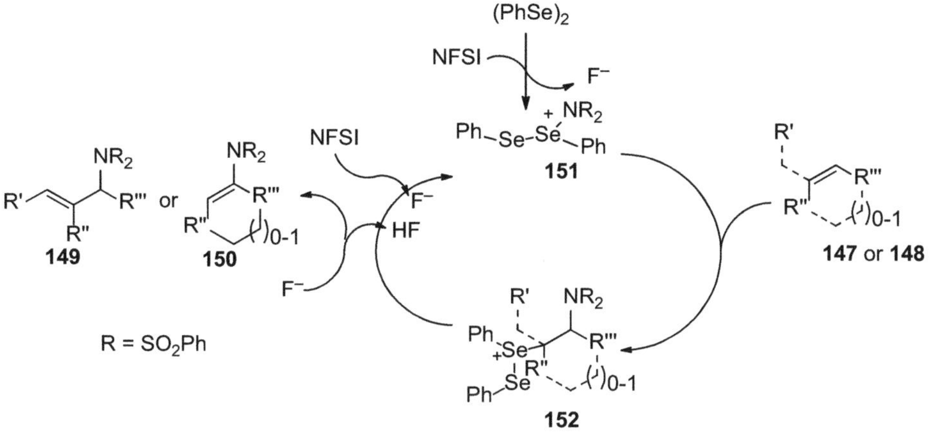

Figure 5.49: Tentative catalytic cycle for the oxidative imidation of alkenes **147** and **148**.

5.3.3.2 Direct regioselective oxidative C–H amination of simple terminal alkenes

In spite of the above-mentioned work and the work by Zhao's group [79], for the terminal amination of allyl alcohols, there was no method for a regioselective direct C–H amination of simple terminal alkenes. Very recently, Michael and coworkers introduced a new phosphine selenide catalyst in the metal-free aza-Heck reaction of simple terminal alkenes for a regioselective C–H amination [80]. Terminal alkenes **153** were converted to terminal *E*-enamides **154** by using NFSI as the terminal oxidant and the N-source and tri(*o*-tolyl)phosphine selenide as the catalyst (Figure 5.50). Very good regio- and steroselectivities were found with varieties of substituents such

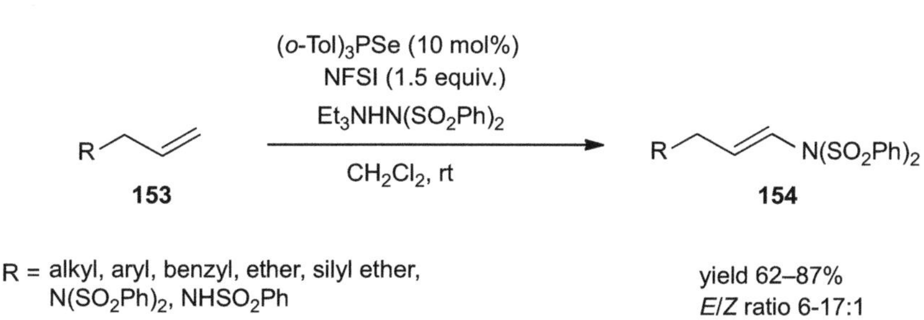

Figure 5.50: Regioselective metal-free Aza-Heck reactions of terminal alkenes catalyzed by phosphine selenides.

as alkyl, aryl, esters, ethers, silyl ethers, sulfonamides, nitriles, and so on. Moderate-to-high yields (62–87%) were obtained along with *E*/*Z* ratios between 6:1 and 17:1.

The proposed catalytic cycle is similar to that of the diphenyl diselenide-catalyzed reaction shown in Figure 5.49. In this work, deuterium-labeled experiments suggested that *E*/*Z* ratio of product increases with *Z*-deuterated alkene, whereas the *E*/*Z* ratio decreases with *E*-deuterated alkene as starting material. Based on this observation, antiaddition/syn-elimination mechanism is explained in Figure 5.51 [80].

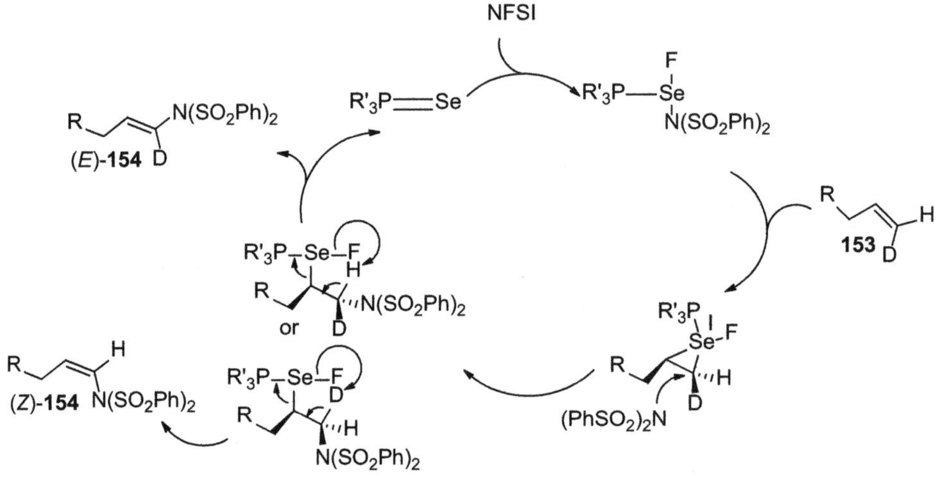

Figure 5.51: Proposed catalytic cycle.

5.3.3.3 Intramolecular oxidative C–H amination in the synthesis of pyrazoloquinazolinones

Among the latest progress in this area, an intramolecular oxidative C–H amination of alkene, catalyzed by selenium, is noticeable. Breder's group was the first to exploit the carbophilicity of selenium electrophiles in the intramolecular C(sp^2)-H amination in 2015 [81]. In the next year, Zhao et al. reported a similar method for the synthesis of oxygen- and nitrogen-containing heterocycles [82]. This year, Chen and coworkers synthesized well-known biologically active [83] pyrazolo[5,1-*b*]quinazolinone derivatives **156** using only 5 mol% of diphenyldislenide from the 2-aryl-3-(arylamino)quinazolinones **155** (Figure 5.52) [84]. Here, NFSI acts only as the oxidizing agent. The reaction goes well with different substituents R^1, R^2, and R^3 at the three aryl rings, except with highly electron-deficient difluoro and dichloro groups as R^3 (Figure 5.52). Yield was highly satisfactory even at the gram scale (78%).

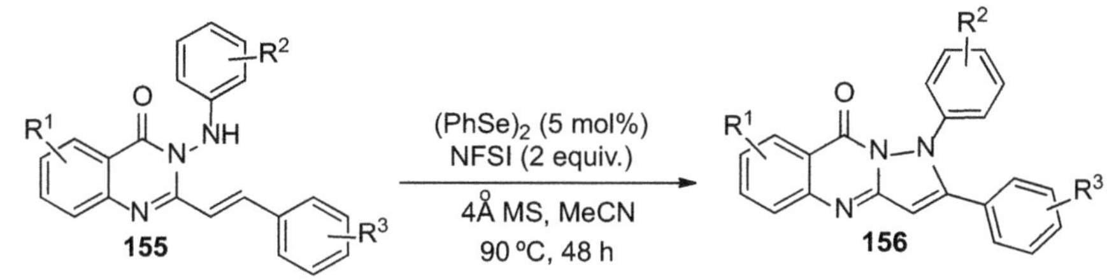

R^1 = Me, OMe, di-Me, F, Cl, Br, I
CF_3
R^2 = Me, OMe, F, Cl, CF_3
R^3 = OMe, di-F

yields up to 87%

Figure 5.52: Selenium-catalyzed oxidative C-H amination of 2-aryl-3-(arylamino)quinazolinones **155**.

5.3.4 Selenium-catalyzed oxidative lactonization reactions

Catalytic use of selenium reagents in cyclization processes dates back to 2007, when Wirth's group reported the cyclization of 3-butenoic acids into their corresponding butenolides using selenium electrophiles as catalyst [85]. This work opened a new route for cyclization reactions, especially for the lactonization of unsaturated carboxylic acids.

5.3.4.1 Regioselective cyclization of γ,δ- unsaturated carboxylic acids

Wirth and coworkers, subsequently, used the same strategy for the cyclization of stilbene-2-carboxylic acids to isocoumarin derivatives [86]. A year later, in 2011, they reported the more challenging aliphatic analogue. Cyclization of γ,δ-pentenoic acids **157** to the corresponding 3,6-dihydro-2*H*-pyran-2-ones **158** was catalyzed by diphenyl diselenide with [bis(trifluoroacetoxy)iodo]benzene as oxidant (Figure 5.53).

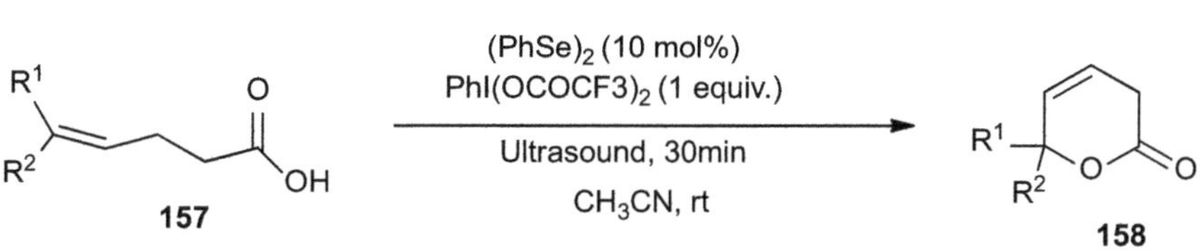

R^1 = H, aryl
R^2 = aryl, alkyl

yields up to 87%

Figure 5.53: Regioselective cyclization of γ,δ-pentenoic acids **157** catalyzed by diphenyl diselenide.

The yields were improved when performed in ultrasonic bath. The products were formed regioselectively only in 30 min, with yields varying from 51% to 87% [87].

They have proposed the mechanism, shown in Figure 5.54, on the basis of their earlier proof by NMR that the reaction is initiated by the formation of phenylselenenyl trifluoroacetate **159** that reacts with the substrate **157** to form the selenolactone **160**, which is again activated by [bis(trifluoroacetoxy)iodo]benzene to the intermediate **161**. Finally, the elimination of the selenylated hypervalent iodine compound, which goes back to the catalytic cycle, leads to the product **158**.

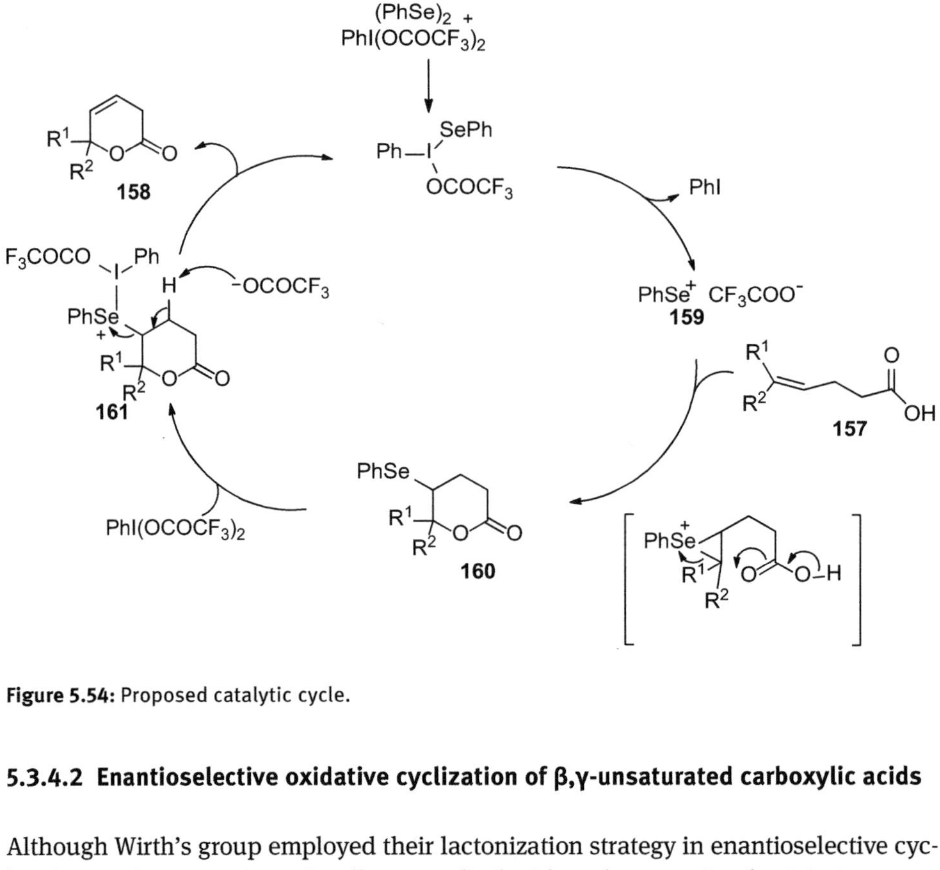

Figure 5.54: Proposed catalytic cycle.

5.3.4.2 Enantioselective oxidative cyclization of β,γ-unsaturated carboxylic acids

Although Wirth's group employed their lactonization strategy in enantioselective cyclizations using enantiomeriacally pure diselenides, the enantioselectivity was not very high [85, 88]. Much later, in 2016, Maruoka and coworkers established a highly enantioselective lactonization process. Success of this effort relied on the development of an indanol-based chiral electrophilic selenium catalyst **165**, synthesis of which is shown in Figure 5.55. The synthesis involves a crucial optical resolution of *rac*-**162** using phenylalanine derivative **163** to get the vital enantiopure indanol (*S*)-**162**. About 10 mol % of the catalyst **165** enantioselectively converted β,γ-unsaturated butenoic acids **166** to the corresponding butenolides **167** with 1.1 equiv. of NFSI

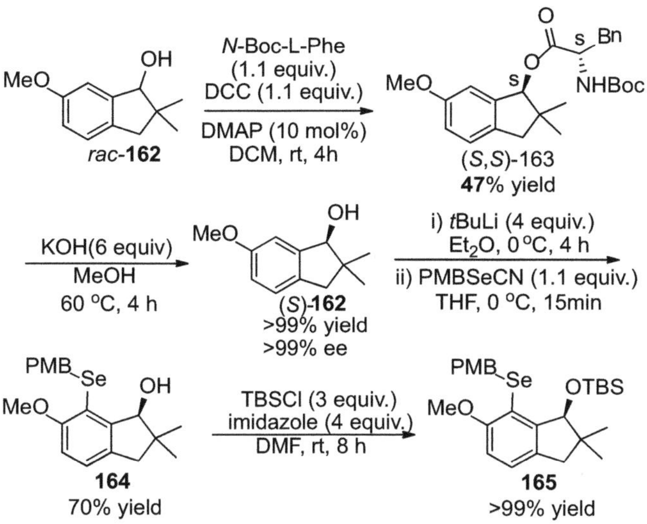

Figure 5.55: Catalyst synthesis.

acting as oxidizing agent and $CaCO_3$ or $TMSOCOCF_3$ as additive (Figure 5.56). Varieties of aliphatic and aromatic substituents containing terminal alkenes were shown to produce products in good-to-excellent yields, with ee up to 97% [89].

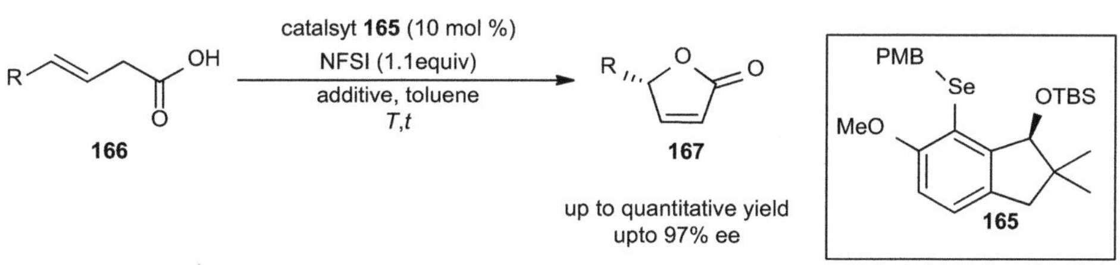

R	Additive	*T*	*t*
Alkyl	$CaCO_3$(3 equiv)	rt	2 h
Aryl	$TMSOCOCF_3$ (1.2 equiv)	10°C	12 h

Figure 5.56: Selenium-catalyzed oxidative cyclization of β,γ-unsaturated carboxylic acid.

5.3.4.3 Oxidative C(sp^3)-H acyloxylation of *o*-allylic benzoic acids

In 2015, Breder's group reported an intramolecular oxidative cyclization of carboxylic acid with an allylic sp^3 carbon atom. *o*-Allylic benzoic acids **168** were cyclized to the isobenzofuranones **169** using 10 mol% of diphenyl diselenide (Figure 5.57).

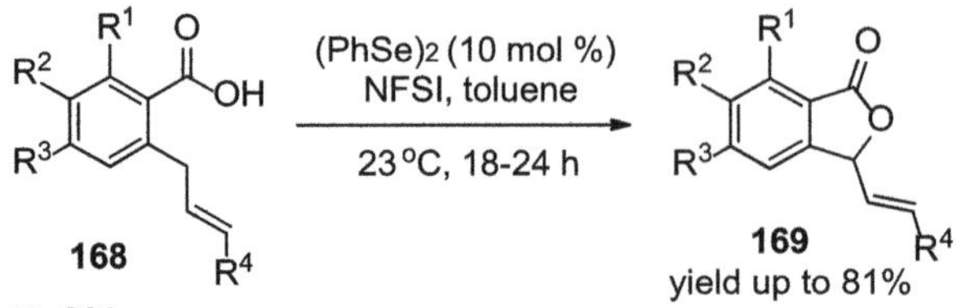

168 169
yield up to 81%

R^1 = H, OMe
R^2 = H, Me, OMe, PivNH, CF_3
R^3 = H, CF_3
R^4 = Et, Ph, 4-MeC_6H_4, 4-FC_6H_4, 4-$CF_3C_6H_4$, 1-Nap, 2-thienyl

Figure 5.57: Selenium-catalyzed C(sp^3)-H acyloxylation.

NFSI acts as the terminal oxidant in this case too. The reaction seems to be a little sluggish with EWGs at the aromatic ring whereas unwanted addition reaction takes place with alkyl groups at the olefinic double bond [90].

5.3.5 Selenium-catalyzed oxidative cyclization in the total synthesis of (+)-Greek tobacco lactone

A new and efficient approach for the total synthesis of a rarely found C_{11}-homoterpenoid (+)-Greek tobacco lactone **171** [91] was developed in 2017, where the final step involved a selenium-catalyzed oxidative cyclization of chiral hydroxyalkene **170** (Figure 5.58) [92]. Christmann and coworkers used a $(PhSe)_2$-catalyzed photoredox protocol where air is the terminal oxidant and 2,4,6-tri(4-methoxyphenyl)pyrylium tetrafuoroborate (5 mol %) is the photocatalyst. Compared to the vanadium- or palladium-catalyzed reaction, this selenium-catalyzed reaction gave better result, with the yield being 83% and the diastereomeric ratio of 84:16.

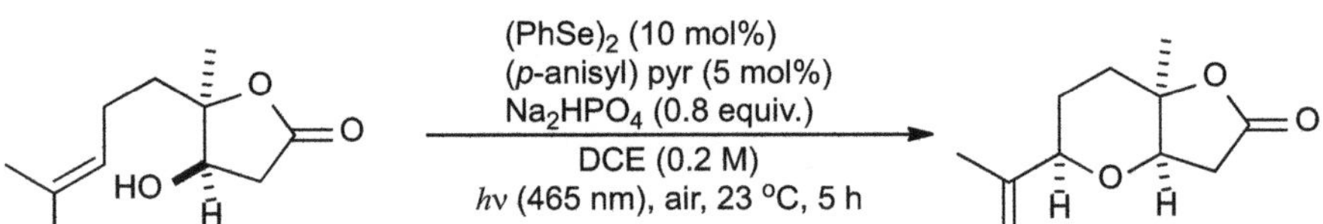

170 171

83%
dr 84:16

Figure 5.58: Selenium-catalyzed oxidative cyclization.

5.3.6 Selenium-catalyzed direct oxidation reactions

One of the earliest uses of selenium in organic synthesis was its potential behavior as an oxidizing agent. Initially, Se at its elemental form was used in many dehydrogenation reactions [93]. The next name that comes in mind in this regard is selenium dioxide. Those early reactions are well documented in some of the reviews [94]. Oxidation of alcohols by Barton et al. [95], allylic oxidation of alkenes using catalytic SeO_2 or seleninic acids by Sharpless [96] and by Barton [97], and epoxidation of alkenes by Sharpless [98] are some of the earliest oxidation reactions catalyzed by selenium. There has been a significant development in organoselenium catalyzed oxidation reactions in the first decade of this century [99]. In the last decade, significant works have been contributed by many organic chemists. Representative reactions are reported in this section.

5.3.6.1 Oxidation of alkenes to carbonyl compounds by C=C bond cleavage

Oxidative cleavage of C=C bonds usually requires metal catalysts or strong oxidizing agents [100]. Yu and coworkers have presented a new method for this cleavage by H_2O_2 using a relatively green selenium catalysis protocol [101]. As shown in Figure 5.59, di- or trisubstituted alkenes **172** were oxidized to the carbonyls **173** using 5 mol % of dialkyl diselenides as catalyst. Notably, less used catalysts such as $(PhCH_2Se)_2$, (n-$C_4H_9Se)_2$, and (c-$C_6H_{11}Se)_2$ were proved to be more efficient in this reaction. Up to 74% yields were obtained with geminal disubstituted alkenes whereas the yield was 30% when two substituents were Et and Me. Average yields were lower in case of trisubstituted alkenes. The reaction did not take place at all with four phenyl groups as substituents.

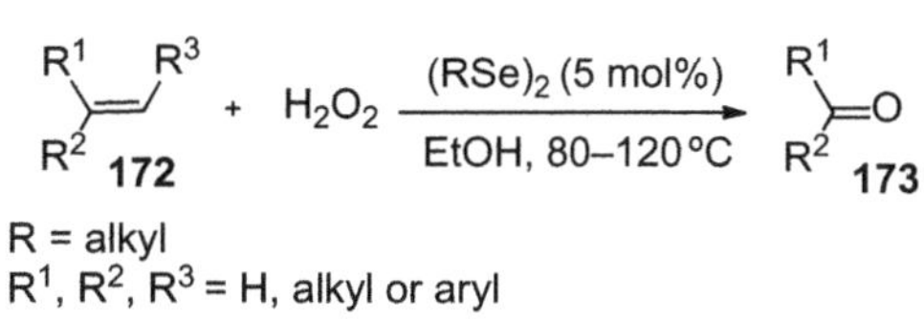

Figure 5.59: Oxidation of alkenes **172** into carbonyl compounds **173** by organoselenium-catalyzed oxidative C=C bond cleavage.

5.3.6.2 Catalytic oxidation of sulfides to sulfoxides using cyclic seleninate esters

In 2012, Back and coworkers reported a method for the oxidation of sulfides to sulfoxides by hydrogen peroxide using cyclic seleninate esters **176** instead of seleninic acids (Figure 5.60) [102]. The oxidation was clean in a mixed solvent of

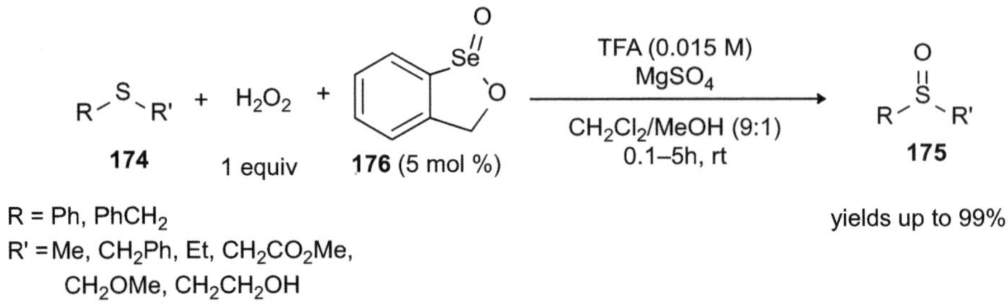

Figure 5.60: Preparation of sulfoxides.

DCM/MeOH (9:1) at room temperature, with TFA and $MgSO_4$ as additives. Both aryl and alkyl groups are tolerated well for the formation of sulfoxides in high yields.

According to the proposed mechanism (Figure 5.61), the cyclic seleninate ester gets protonated by TFA and, subsequently, it reacts with hydrogen peroxide to form the required peroxyselenurane **177** or peroxyseleninate **178**. These Se(IV) intermediates **177** or **178** are presumed to be the active oxygen transfer species for the catalytic oxidation of sulfides [102].

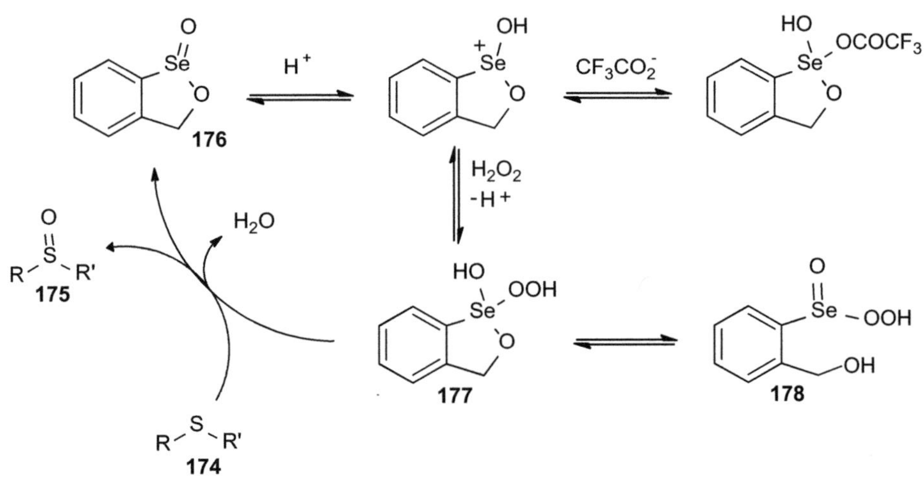

Figure 5.61: A plausible mechanism for the sulfide oxidation.

5.3.6.3 Selective C(sp^3)-H oxidation of benzylpyridines with molecular oxygen

Although there are metal-catalyzed C(sp^3)-H oxidations of benzylpyridines with molecular oxygen [103], a metal-free alternative is always desirable. Recently, Law and coworkers reported a selenium-catalyzed approach for the selective C(sp^3)-H oxidation of benzylpyridines **179** (Figure 5.62) with molecular oxygen as the oxidant. About 5 mol% of PhSeBr as catalyst and 1 equiv. of AcOH was used to convert **179** into the corresponding benzoylpyridines **180** in good-to-excellent yields. Both EDGs and EWGs at the aryl and heteroaryl ring underwent smooth reactions. Only, *p*-CHO at the aryl ring gave a yield of 28% [104].

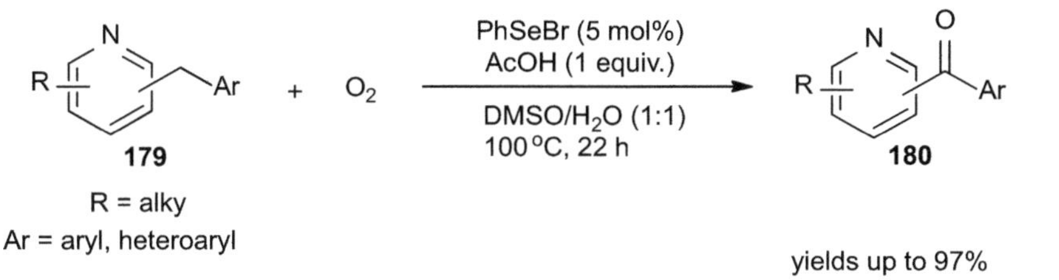

Figure 5.62: Selenium-catalyzed selective C(sp^3)-H oxidation of benzylpyridines with molecular oxygen.

Based on the analytical experiments, they have proposed a radical mechanism (Figure 5.63). Protonated benzylpyridine **181** reacts with PhSe radical, which comes from PhSeBr, to form the key radical intermediate **182**. Then, **182** interacts with molecular oxygen to form **183**, which on protonation becomes the hydroperoxidate intermediate **184**. This, on elimination of water and HOAc, leads to the formation of the product **180**. The final step might also go through the oxidation of the alcohol **185** [104].

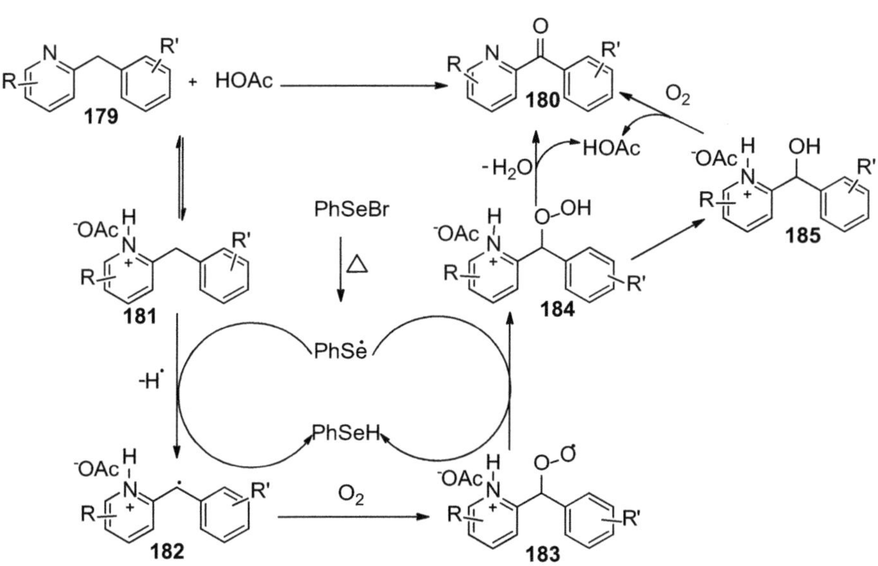

Figure 5.63: A proposed mechanism.

5.3.6.4 Selenide ion catalyzed homo- and crossed-Tishchenko reaction

Selenide ions were presented as a better alternative to the metals and thiolate anions for a practically useful Tishchenko reaction. Connon and coworkers, in 2012, established a homo- and crossed-Tishchenko reaction with improved catalyst efficiency and broader substrate scope [105]. Benzaldehydes **186** were converted to the corresponding benzyl esters **187** using benzyl selenide anion (Figure 5.64) generated by the the addition of commercially available dibutyl magnesium to dibenzyl diselenide in THF at room temperature. Most of the substrates including the aliphatic cyclohexyl aldehyde underwent reactions efficiently. The simple benzyl selenide ions were not effective in case of crossed-Tishchenko reactions involving the benzaldehydes **188** and α,α,α-trifluoromethylacetophenones **189**. *M*-CF_3-substituted diphenyl diselenide acted as an efficient catalyst for this conversion to the product **190**. It has been proposed that this modification is credited to the formation of highly electrophilic acylating agent **191**. In addition, the heteroaryl aldehydes were tolerated in case of both homo- and crossed Tishchenko reactions [105].

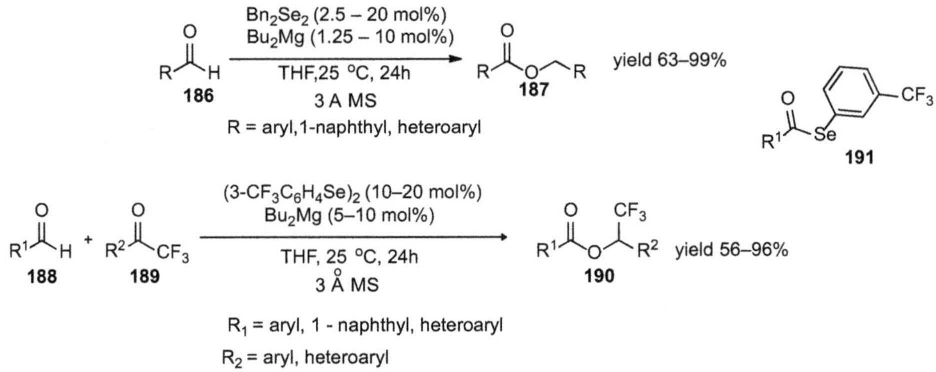

Figure 5.64: Selenide ions as catalysts for homo- and crossed-Tishchenko reactions.

5.3.6.5 Baeyer–Villiger oxidation of α,β-unsaturated ketones

The combination of stoichiometric amount of hydrogen peroxide with catalytic organoselenium reagents has been applied successfully in the Baeyer–Villiger oxidation reactions. In 2014, Yu and coworkers developed a greener protocol compared to the earlier methods [106] for Baeyer–Villiger oxidation of α,β-unsaturated ketones. Α,β-Unsaturated ketones **192** were oxidized by H_2O_2 to the vinylic esters **193** using 5 mol % of dibenzyl diselenide in acetonitrile at room temperature (Figure 5.65). Both alkyl and aryl groups as R^2 and aryl and alkenyl groups as R^1 were tolerated well. However, EWG such as a keto substituent was not suitable for this reaction [107].

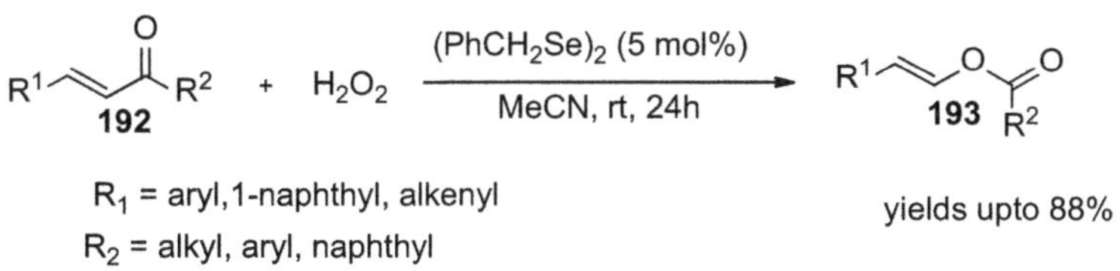

Figure 5.65: Organoselenium-catalyzed Baeyer–Villiger oxidation for vinyl ester **193** synthesis.

As shown in Figure 5.66, selective nucleophilic O–H addition of the organoseleninoperoxoic acid **194**, generated in situ by the oxidation of the diselenide catalyst by H_2O_2, to the C=O of the ketone **192** is probably the initiation of the reaction. This intermediate **195** would undergo the alkenyl migration in a usual fashion of Baeyer–Villiger oxidation to produce the vinyl ester **193** [107].

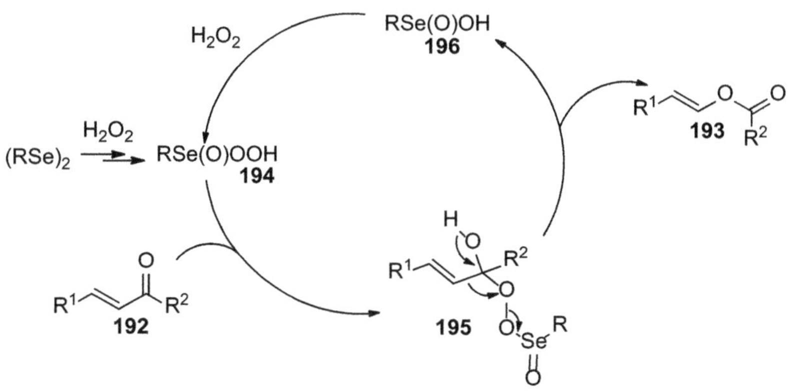

Figure 5.66: Possible mechanism for the organoselenium-catalyzed Baeyer–Villiger oxidation.

5.3.6.6 Dehydration of aldoximes for the synthesis of nitriles

Encouraged by the observation from a failed attempt to oxidize benzaldoximes **197** to nitrobenzenes by organoselenium catalysis, Yu et al. explored a different path of oxidation of the aldoximes that leads to benzonitriles **198** by dehydration of **197** (Figure 5.67). The diselenide catalyst $(3\text{-}FC_6H_4Se)_2$ as well the H_2O_2 were used in a catalytic load of just 4 mol%. The yields were moderate to very good in the case of aromatic and alkenylic aldoximes. But only long chain alkyl substrates underwent the reaction successfully. The catalyst showed its activity up to four to six cycles [108].

According to the proposed mechanism (shown in Figure 5.68), diselenide $(ArSe)_2$ gets oxidized to ArSeOH (**199**), which was probably converted to the more effective seleninic anhydride **200**. This might form a mixed anhydride **201** with the aldoxime **197**. This, again, could rearrange to the selenoxide **202**. Finally, this might undergo syn-elimination of **199** to give the product **198** [108].

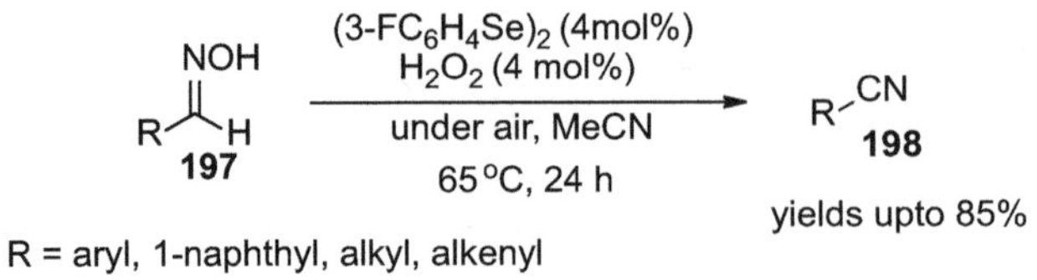

Figure 5.67: Organoselenium-catalyzed aldoxime dehydration reaction for organonitrile synthesis.

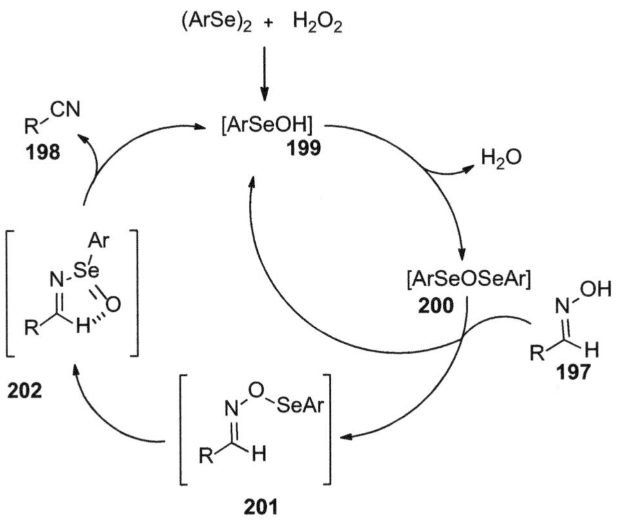

Figure 5.68: Possible mechanism for organoselenium catalyzed aldoxime dehydration reaction.

5.3.7 Allylic C–O bond formation by oxidative coupling reaction

5.3.7.1 Oxidative allylic esterification by cooperative catalysis

Alkenes were oxidized to β-selenenylated ethers and esters using stoichiometric quantities of diphenyl diselenide with 1,4-dicyanonaphthalene as photosensitizers by Pandey et al. [109]. This concept of activation of diselenides with photosensitizer was used by Breder and his coworkers in 2016. Catalytic amounts of both diphenyl diselenide and 2,4,6-tris(4-methoxyphenyl)pyrylium tetrafuoroborate (**206**) were combined to carry out the allylic esterification of the alkenes **203** with carboxylic acids **204** to form the allyl esters **205** (Figure 5.69). Air played the role of the oxidant, whereas the irradiation at 465 nm in acetonitrile at room temperature facilitated photosensitization. The reaction provided good-to-excellent yields with high regioselectivity for both functionalized as well as nonfunctionalized alkenes. Both alkyl and aryl groups at the olefin terminus increased the substrate scope. No aryl variant is reported in the carboxylic acids [110].

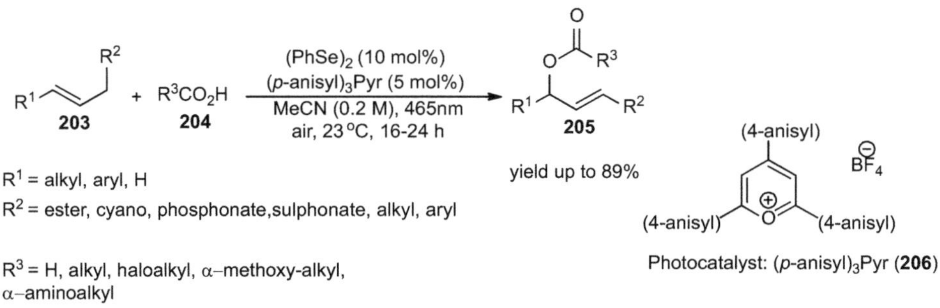

Figure 5.69: Selenium-catalyzed oxidative allylic esterification of alkenes **203** using air as the sole oxidant.

5.3.7.2 Allylic phosphatation of alkenes by cooperative catalysis

Driven by the above-mentioned method, Breder and coworkers explored the cooperative catalysis protocol for the aerobic phosphatation of simple alkenes [111]. This dual photoredox/ selenium π-acid catalysis proved to be highly efficient for the transformation of the alkenes **207** to the allylic phosphates **209** using simple phosphoric acids **208** as phosphate source and air or molecular oxygen as terminal oxidant (Figure 5.70). The same set of catalysts, diphenyl diselenide and the photocatalyst **206**, was used in this case also. For an improved yield, Na_2HPO_4 was used as base in DCE with irradiation at 465 nm. Simple aliphatic and alicyclic hydrogen phosphates were successfully used in the phosphatation with moderate-to-high yields (up to 89%). No diastereoselectivity was observed with chiral enantiopure phosphates. When performed with different alkenes, both cyclic and acyclic substrates showed moderate-to-good results.

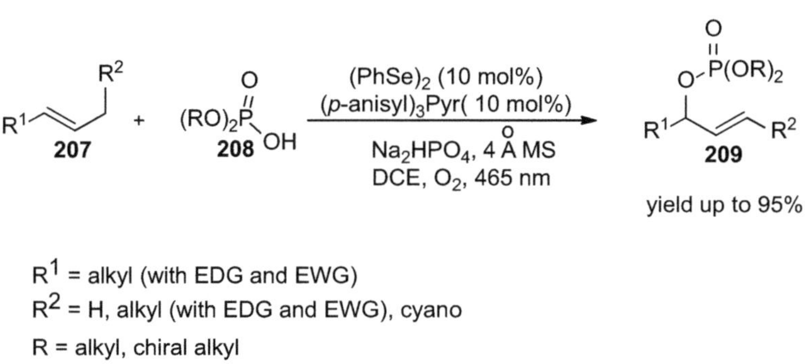

Figure 5.70: Photocatalytic aerobic phosphatation of alkenes.

5.4 Selenium as ligands in organic reactions

As discussed in the previous section, the catalytic activity of organoselenium compounds in metal-free processes has been highlighted. Although metal-free processes are highly desirable for moving toward a green synthetic world, the synthetic advantages of certain metal-catalyzed processes cannot be ignored [112]. To extract the maximum activity from minimum amount of metal, judicious use of ligands is very important. In this context, the demand of selenium is increasing as many selenium compounds having suitable donor sites have been proven as effective ligands [113]. Uemura and coworkers were the first to report the use of selenium as ligand in a rhodium-catalyzed asymmetric hydrosilylation of ketones, in 1994 [114]. Since then, the use of achiral and chiral ligands in organic transformations has gained momentum [4, 115]. Some of the recent developments in the applications of selenium ligands in elementary organic reactions are presented below.

5.4.1 Pd-catalyzed asymmetric allylic alkylation

Chiral organoselenium ligands showed their utility in palladium-catalyzed asymmetric allylic alkylation reactions in the past [116]. Andrade et al. have prepared a new set of chiral selenium–amine ligands **210** and **211** and used them for asymmetric allylic alkylation processes [117]. To synthesize the ligands **210** and **211**, they started with the stereoselective amination of organoselenium acetophenones (Figure 5.71). Then, the kinetic resolution of racemic organoselenium amines catalyzed by transaminases provided the chiral amine in enantiopure form, which was then easily converted to the desired ligands **210** and **211** by simple reactions.

These ligands **210** and **211** were then used in the allylic alkylation of allylic acetates **212** with various malonate nucleophiles **213** to get the alkylated products **214** (Figure 5.72). About 10 mol% of the ligand and 5 mol% of the catalyst $[Pd(\eta^3\text{-}C_3H_5)Cl]_2$ were used under basic conditions. Although the substrate scopes shown are very limited, high yield of 94% was achieved with Me as R^1 and H as R^2, with an enantiomeric excess of 89%. The highest enantioselectivity was obtained with diethyl phenylmalonate, but the yields were very low (yield 18% and ee > 99%) [117].

5.4.2 Asymmetric Darzens reaction catalyzed by organoselenium–lithium complex

Due to their importance in organic synthesis as useful chiral intermediates [118], enantiopure oxiranes have featured as useful synthetic targets [119]. In 2010, Watanabe et al. prescribed a new kind of asymmetric Darzens reaction using aorganoselenide–lithium hydroxide complex for the synthesis of enantiopure *trans*-oxiranes [120].

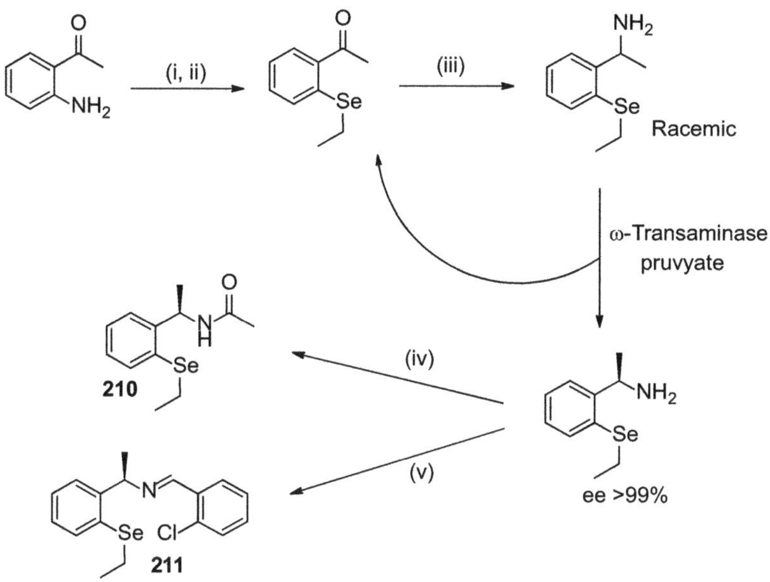

Figure 5.71: Synthesis of new chiral selenium-amine ligands for Pd-catalysis.

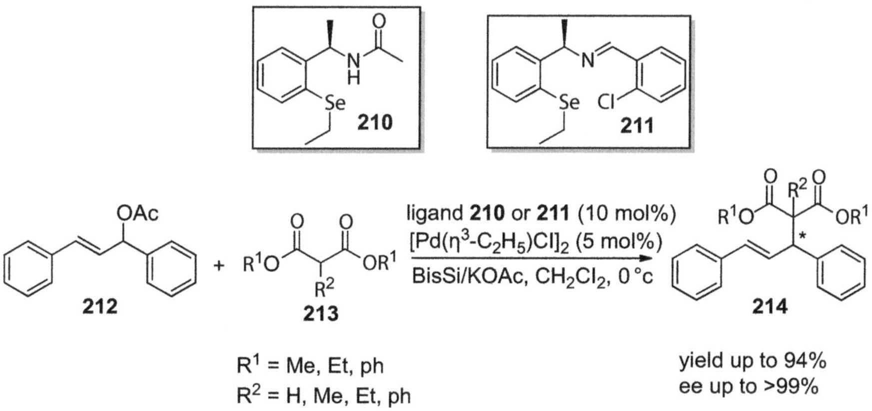

Figure 5.72: Evaluation of novel chiral organoselenium compounds **210, 211** as ligands in the palladium-catalyzed asymmetric allylic alkylation.

As shown in Figure 5.73, catalytic amounts of the C_2-symmetric chiral selenide **218** was promoted by 4 equiv. of LiOH to carry out the reaction between phenacyl bromide **215** and aldehydes **216** for the formation of the oxiranes **217**. It was observed that the selection of solvent was very crucial to get better enantioselectivity. In addition, aliphatic aldehydes failed to provide any enantioselectivity. In case of aromatic aldehydes, yields were better for electron-deficient substrates.

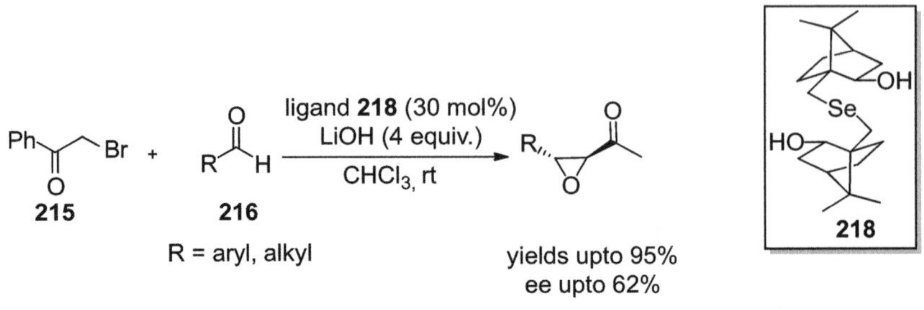

Figure 5.73: Asymmetric Darzens reaction with aldehydes using chiral chalcogenides.

5.4.3 Enantioselective cross-coupling reaction of arylboronic acids and aryl aldehydes

Based on the previous methods [121], Braga's group developed a new ephedrine-based diselenide ligand **220** for the organozinc reagent to perform the arylation of aryl aldehydes **222** using arylboronic acids **221** (Figure 5.74 and 5.75). The chiral diselenide **220** was easily prepared from the chiral amino alcohol **219** by two elementary reactions (Figure 5.73) in an affordable yield of 53%. About 2.5 mol% of ligand **220** provided the best result for the cross-coupling reaction.

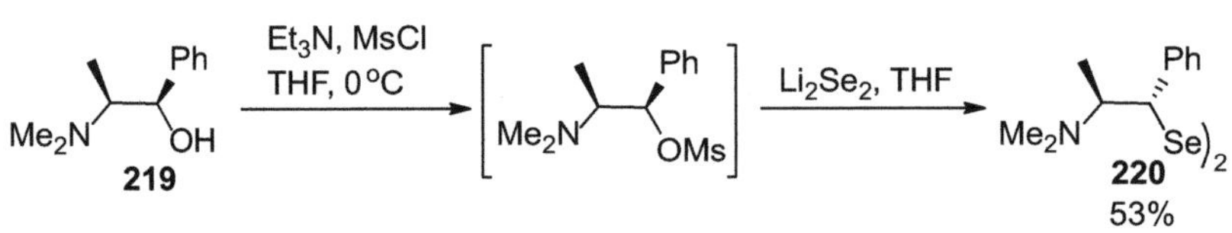

Figure 5.74: Synthesis of Chiral selenium ligand **220**.

The substrate scopes with different arylboronic acids **221** and aryl aldehydes **222** suggest the generality of the reaction with good to excellent yields and enantioselectivities. In the absence of the arylboronic acid, asymmetric addition of diethyl zinc takes place [122].

5.4.4 Palladium-catalyzed Suzuki–Miyaura coupling using selenium ligands

Among all the palladium-catalyzed cross-coupling reactions to form C–C bonds, Suzuki–Miyaura coupling is one of the widely used protocols. The most widely used ligands for this purpose are the phosphorous ligands. But the phosphorous ligands

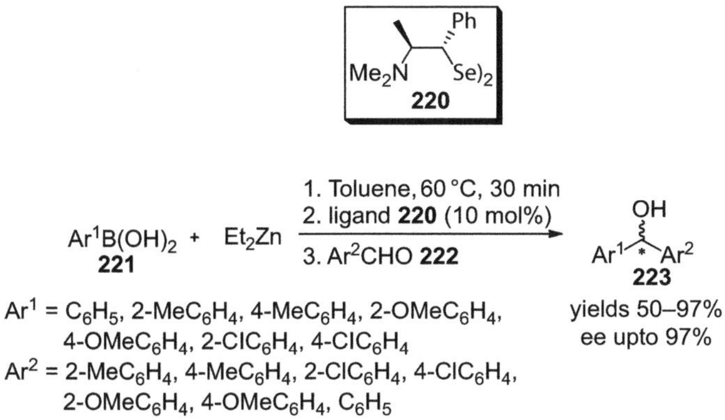

Figure 5.75: Boronic acid addition to aldehydes catalyzed by **220**.

are not that easy to handle due to their sensitivity toward air and moisture [123]. Singh and coworkers showed that these problems can be overcome by replacing phosphorous ligands by suitable chalcogen ligands that are quite stable in air and moisture. In their consistent effort in the synthesis and application of new selenium ligands, they developed two organoselenium ligands to form the palladium complexes **224** [124] and **225** [125], which performed well in the Suzuki–Miyaura coupling of aryl bromides **226** with phenylboronic acid to prepare the biaryls **227** (Figure 5.76). The substrate scope was very limited with the complex **225** where both the donor

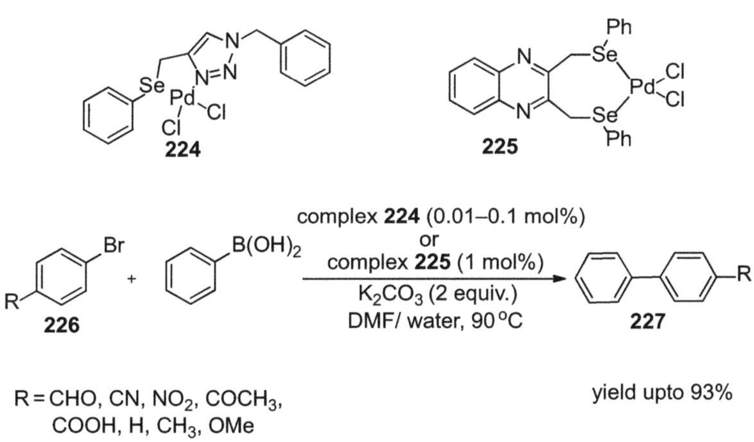

Figure 5.76: Suzuki Miyaura coupling reaction catalyzed by complex **224** or **225**.

atoms are Se. A broader substrate scope was established with complex **224** where one N atom acts as a donor along with the Se atom being the other. Nanoparticles composed of palladium and selenium and protected with the ligand or its fragment were found to be formed during the reaction. They also performed the coupling reaction with isolated nanoparticles, which gave very good results [124, 125].

5.4.5 Transition metal-catalyzed base free transfer hydrogenation of carbonyls and *N*-alkylation of anilines

In continuation of their work with organoselenium ligands, Singh and coworkers standardized a protocol for efficient base free transfer hydrogenation of carbonyl compounds and *N*-alkylation of anilines with various benzyl alcohols [126, 127]. Initially in 2018, they have synthesized the vital bidentate organoselenium ligand **229** by the Schiff base condensation of anthracene-9-carbaldehyde **228** (Figure 5.76). Thereafter, this ligand **229** delivered all the positive results for the above-mentioned two class of reactions. At first, Ir(III) complexes of the ligand **229**, **232a**, and **232b** were synthesized as shown in Figure 5.77. These complexes very efficiently catalyzed the transfer hydrogenation of carbonyls **233** to **234** and the *N*-alkylation of anilines **235** with alcohols **236** to the secondary amines **237** (Figure 5.78) [126]. Next year, the same protocol was applied in case of Ru(II)- and Rh(III)-catalyzed reactions. While the rhodium complexes **231a** and **231b** carried out both the reactions

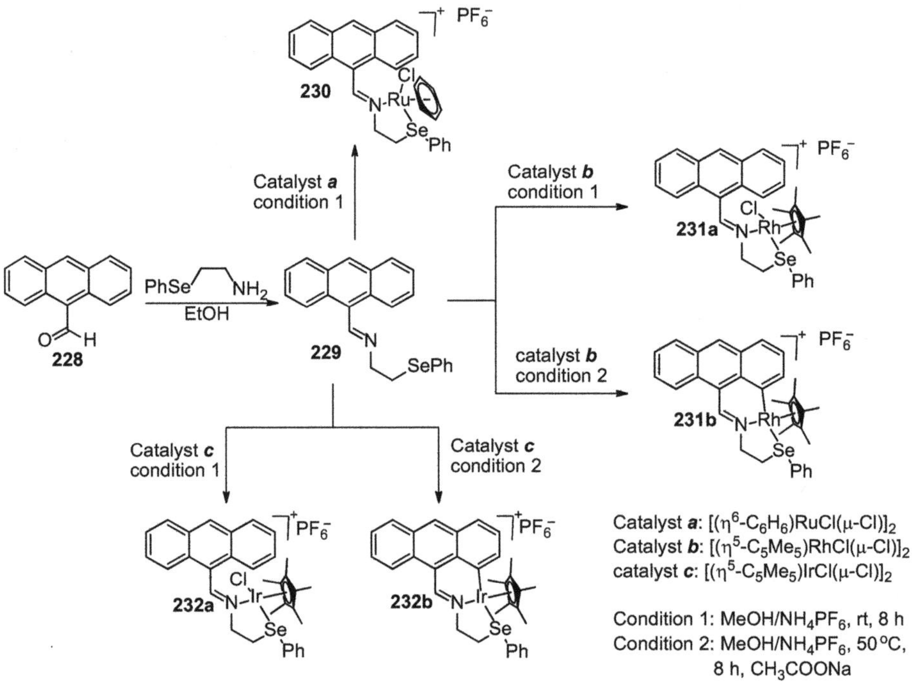

Figure 5.77: Synthesis of ligand **230–232**.

efficiently (Figure 5.78), the ruthenium complex **230** could execute only the base free *N*-alkylation reaction [127].

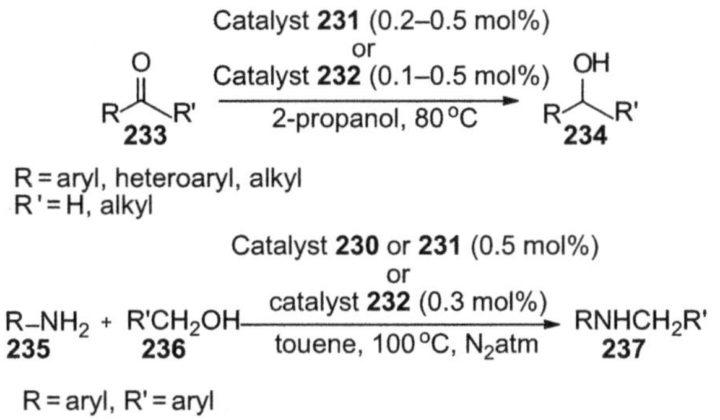

Figure 5.78: Transfer hydrogenation of carbonyl compounds catalyzed with **231–232** as well as N-Alkylation of aromatic benzyl alcohol and aniline catalyzed with **230–232**.

5.4.6 Palladium-catalyzed aminocarbonylation using organoselenium ligands

Palladium-catalyzed CO insertion is a well-known technique for aminocarbonylation of aryl halides [128]. As stated earlier, there is always a demand for ligands other than phosphines for the palladium-catalyzed processes. Bhanage and coworkers contributed to this objective by performing a palladium-catalyzed aminocarbonylation reaction of aryl iodides using organoselenium ligand [129]. Aminocarbonylation of the aryl iodides **238** with primary amines **239** was carried out using the selenium-ligated palladium complex **241** under CO atmosphere (Figure 5.79). Although this complex contained PPh_3 ligands as well, the presence of the 4-pyridylselenolate ligand might stabilize the catalyst to a greater extent. Electronic nature of the aryl iodide and

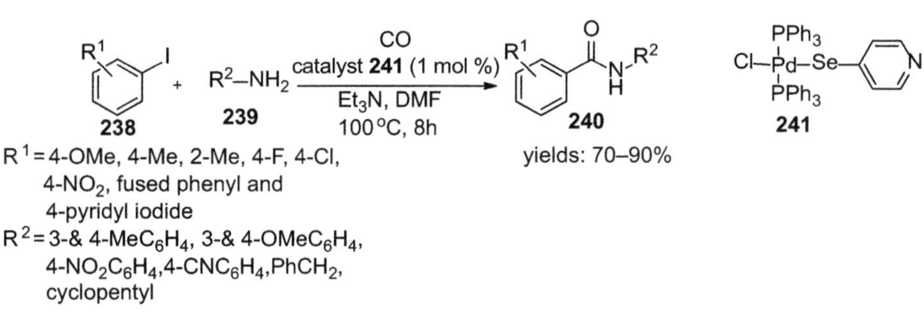

Figure 5.79: Palladium-catalyzed aminocarbonylation reaction using organoselenium ligand **241**.

amine does not affect the reaction significantly. Good-to-excellent yields (up to 90%) were obtained with merely 1 mol% of the catalyst [129].

5.4.7 Ruthenium-catalyzed oxidation of alcohols

Selenium-ligated transition metal complexes not only catalyze reduction processes via transfer hydrogenation but also have potential in catalyzing oxidation processes. It has been demonstrated by Singh's group that primary and secondary alcohols can be oxidized to corresponding carbonyl compounds in an organoselenium-ligated ruthenium-catalyzed reaction [130]. Primary and secondary alcohols **242** were oxidized to the corresponding aldehydes and ketones **243** with *N*-methylmorpholine *N*-oxide under the catalysis of the ruthenium(II) complex **244** (Figure 5.80). Only 0.01 mol% of catalyst load was sufficient for the transformation. Yields were very good with both electron-rich and electron-deficient substrates. The advantage of this reaction is that no further oxidation of the product takes place. In addition, the complex **244** can be synthesized from the corresponding triazole-based ligands in a single easy step [130].

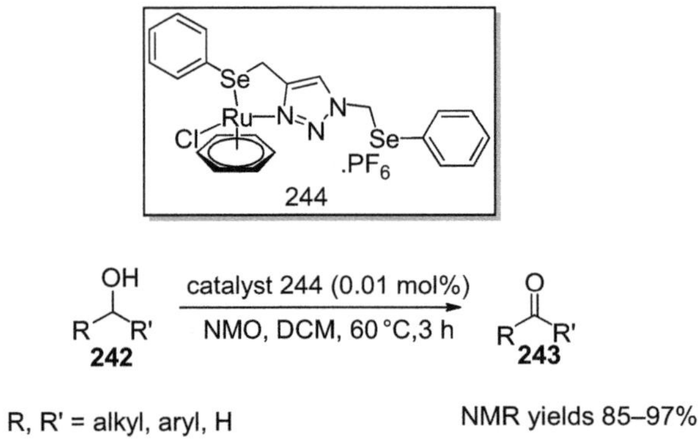

Figure 5.80: Oxidation of alcohol with NMO.

5.4.8 Selenium containing pincer ligands in palladium-catalyzed copper-free Sonogashira reaction

5.4.8.1 Unsymmetrical (O^-, N, Se) pincer ligands in palladium-catalyzed Sonogashira reaction

There are many symmetrical pincer ligands known to be used in Sonogashira coupling reaction [131], where CuI is required as a cocatalyst. Very few unsymmetrical pincer ligands containing (P, N, F) and (N, N, C) were reported in palladium-catalyzed

Sonogashira reactions [132]. Singh et al. developed a unique oxine-based unsymmetrical (O^-, N, Se) pincer ligand **245** from a simple quinoline derivative (shown in Figure 5.81) [133]. After complexation of **245** with palladium, complex **246** catalyzed a copper and amine-free Sonogashira reaction with a very low catalyst load. Various aryl halides **247** coupled with terminal alkynes **248** under the catalysis of **246** to give the Sonogashira products **249** mostly with very good yields (Figure 5.82). Aryl chlorides are not much reactive and the reaction was even poorer with silylacetylene.

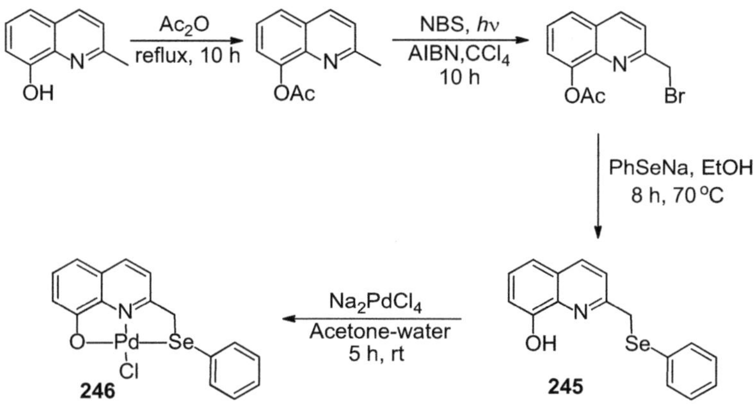

Figure 5.81: Synthesis of **245** and its Pd(II) complex **246**.

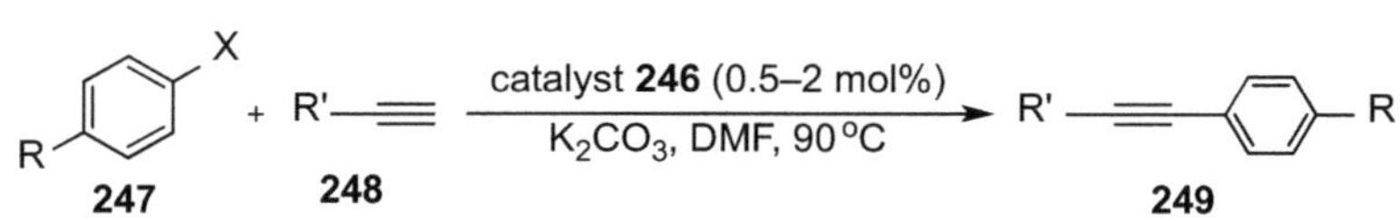

R = H, CH_3, CHO, CN, $COCH_3$, NO_2, OMe
R' = Ph, 4-$CF_3C_6H_4$, 3-thiophenyl, $Si(^iPr)_3$
X = Br, I, Cl

yields: 14–95 %

Figure 5.82: Sonogashira coupling of various alkynes catalyzed with **246**.

5.4.8.2 Polystyrene-supported (Se, C, Se) pincer ligands in palladium-catalyzed Sonogashira reaction

Although palladium-catalyzed C–C cross-coupling reaction was achieved by the symmetrical (SeCSe) pincer ligand developed in 2004 [134], this type of reaction is based on homogeneous catalysis. Homogeneous catalysis suffers from the difficulty in separation of the products from the catalyst and lack of practicality on the reuse of catalyst. Inspired by the previous method of heterogeneous catalysis using PEG-supported selenium pincer ligand [135], Movassagh and coworkers developed a polystyrene supported ligand **250** from easily available 5-hydroxyisophthalic acid by some elementary organic reactions (Figure 5.83). This symmetrical pincer **250** formed the complex **251** with $PdCl_2$ to be used in the copper-free Sonogashira reaction. TBAF was used as the base in the coupling reaction of the aryl halides **252** with the alkynes **253** (Figure 5.84). The internal alkynes **254** were obtained mostly in good to excellent yields (up to 95%). Moreover, this protocol was successfully applied in the Sonogashira coupling of vinyl bromides to get the desired products stereoselectively [136].

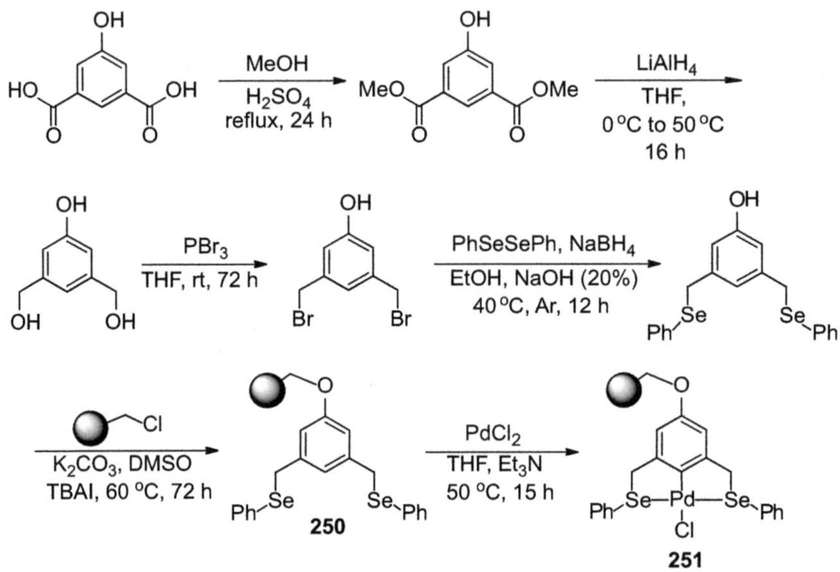

Figure 5.83: Synthetic pathway of PS-[PdCl(SeCSe)] **251**.

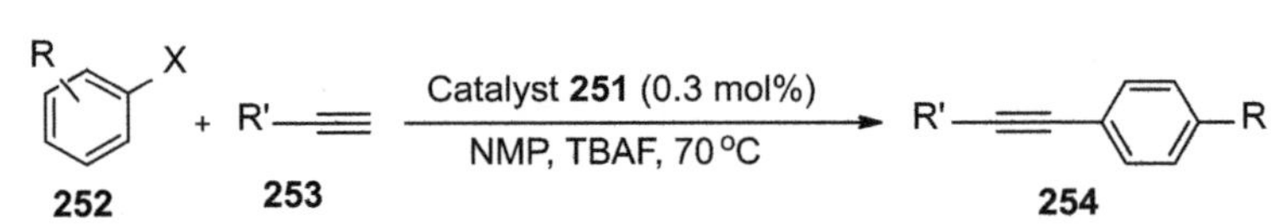

R = H, 4-OMe, 4-CN, 4-$COCH_3$, 4-NO_2
R' = Ph, 4-MeC_6H_4, 4-$OMeC_6H_4$, n-C_4H_9, n-C_6H_{13}
X = I, Br, Cl

yields: 23-95 %

Figure 5.84: Sonogashira cross-coupling reaction of aryl halides with terminal alkynes catalyzed with **251**.

5.4.9 Palladium-catalyzed regioselective arylation of imidazoles using selenium containing NHC ligands

In 2017, Singh and coworkers showed that Pd(II) complex of selenated NHC ligands can catalyze the Heck reaction to form C–C bonds [137]. One year later, the group of Singh developed two palladium complexes of NHC amidate from selenated acetamide-functionalized 1*H*-benzimidazolium salts and applied them in a regioselective arylation of imidazoles [138]. About 0.5–1 mol% of catalysts **258a** or **258b** were sufficient for the arylation of imidazoles **255** with aryl halides **256** in the presence of pivalates in polar *N*,*N*-dimethylacetamide at high temperature (Figure 5.85). The C-5 arylated imidazoles **257** were obtained with good-to-excellent regioselectivity for both aryl and heteroaryl halides.

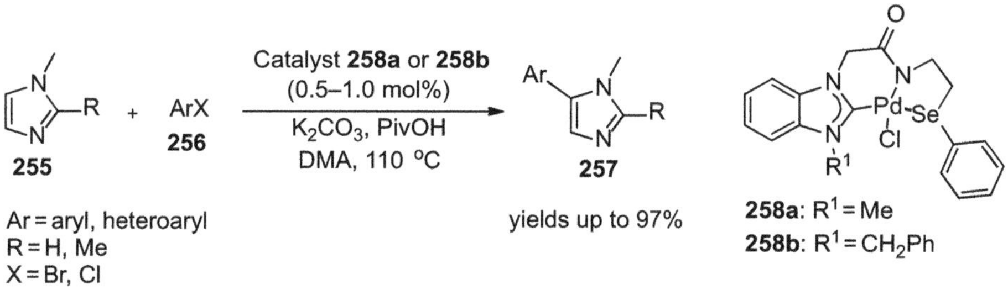

Figure 5.85: Direct Arylation of 1-Methyl-1*H*-imidazole **255** with Aryl Halides **256**.

5.5 Conclusions

In this chapter, we have brought together the significant developments in the versatile use of organoselenium compounds over the last decade. Various types of organoselenium reagents have been discussed in detail in the first section. These reagents give access to a plethora of selenium compounds, for example, selenides, selenophenes, other selenocycles, thiaselenocycles, selenazoles, and so on. Then, we have described the various synthetically important organic transformation facilitated by variety of achiral and chiral organoselenium catalysts. Their catalytic activity spans from the primitive oxidation reactions to the contemporary fields of asymmetric synthesis and C–H functionalization. In the last part, we have discussed the significant participation of selenium ligands in metal-catalyzed organic transformations. All these applications suggest that selenium may have more contributions to make in the field of organic chemistry in the coming days.

References

[1] Reich, HJ., and Hondal, RJ. Why nature chose selenium. ACS Chem Biol 2016, 11, 821–41.

[2] (a) Bryce, MR. Recent progress on conducting organic charge-transfer salts. Chem Soc Rev 1991, 20, 355–90; (b) Kanatzidis MG, Huang SP. Coordination chemistry of heavy polychalcogenide ligands. Coord Chem Rev 1994, 130, 509–621; (c) Hitchman ML, Jenson KF. Chemical Vapour Deposition: Principles and Applications; Academic Press: London, 1993, 677; (d) Stringfellow GB. Organometallic Vapour Phase Epitaxy: Theory and Practice; Academic Press: New York, 1989. € Wei C, Zhang Y, Xu H, Xu Y, Xu Y, Lang M. Well-defined labile diselenide-centered poly(ε-caprolactone)-based micelles for activated intracellular drug release. J Mater Chem B 2016, 4, 5059–67.

[3] (a) Klayman DL, Gu¨nther WHH Eds, Organic Selenium Compounds: Their Chemistry and Biology, Wiley Interscience, New York, 1973, (b) Shamberger RJ. In: Biochemistry of Selenium; EDs. Plenum Press: New York, 1983; (c) Nicolaou KC, Petasis NA. In: Selenium in Natural Product Synthesis; CIS: Philadelphia, PA, 1984; (d) Mugesh G, du Mont W-W, Sies H. Chemistry of biologically important synthetic organoselenium compounds. Chem Rev 2001, 101, 2125–79; (e) Chasteen TG, Bentley R. Biomethylation of selenium and tellurium: Microorganisms and plants. Chem Rev 2003, 103, 1–26; (f) Nogueira CW, Zeni G, Rocha JBT. Organoselenium and organotellurium compounds: toxicology and pharmacology. Chem Rev 2004, 104, 6255–85; (g) Zeni G, Lüdtke DS, Panatieri RB, Braga AL. Vinylic tellurides: from preparation to their applicability in organic synthesis. Chem Rev 2006, 106, 1032–76; (h) Kedarnath G, Jain VK. Pyridyl and pyrimidyl chalcogen (Se and Te) compounds: A family of multi utility molecules. Coord Chem Rev 2013, 257, 1409–35; (i) Bhasin AKK, Jacob C, Sasse F, Karaghiosoff K, Bhasin KK, Mehta, SK, Sharma A, Rishu. Design, synthesis and characterization of picoline based organoselenium compounds, 1,2-*bis*(pyridine-2/3/4-yl) methyl diselanes: X-ray crystal structure of 1,2-*bis*(pyridine-3-yl)methyldiselane. J Organomet Chem 2015, 785, 19–25.

[4] (a) Krief, A., EW, Abel., FGA, Stone., and Wilkinson, G. Selenium, Comprehensive Organometallic Chemistry-II, ed., Krief A, Oxford, UK, Elsevier, 1995, 515; (b) Wirth T, Selenium. In: Comprehensive Organometallic Chemistry-III, volume ed. Knochel P, ed. CrabtreeRH, Mingos DMP. Oxford, UK, Elsevier, 2007, 457; (c) Wirth T. Organoselenium Chemistry, Cardiff, UK, Wiley, 2012.

[5] (a) Klayman, DL., and Günter, HH. Organoselenium Compounds: Their Chemistry and Biology, New York, USA, Wiley, 1973; (b) Back TG, Moussa Z. Remarkable activity of a novel cyclic seleninate ester as a glutathione peroxidase mimetic and its facile in situ generation from allyl 3-hydroxypropyl selenide. J Am Chem Soc 2002, 124, 12104–5; (c) Piovan L, Alves MFM, Juliano L, Bromme D, Cunha RLOR, Andrade LH. Chemoenzymatic synthesis of organoselenium(IV) Compounds and their evaluation as cysteine protease inhibitors. J Braz Chem Soc 2010, 21, 2108–18; (d) Hort MA, Straliotto MR, Netto PM, da Rocha JBT, de Bem AF, Ribeiro-do-Valle RM. Diphenyl diselenide effectively reduces atherosclerotic lesions in LDLr −/− mice by attenuation of oxidative stress and inflammation. J. Cardiovasc Pharmacol 2011, 58, 91–101; (e) Seng H, Tiekink ERT. Anti-cancer potential of selenium- and tellurium-containing species: opportunities abound!. Appl Organomet Chem 2012, 26, 655–62; (f) Gai BM, Bortolatto CF, Brüning CA, Zborowski VA, Stein AL, Zeni G, Nogueira CW. Depression-related behavior and mechanical allodynia are blocked by 3-(4-fluorophenylselenyl)-2,5-diphenylselenophene in a mouse model of neuropathic pain induced by partial sciatic nerve ligation. Neuropharmacology 2014, 79, 580–9; (g) Sahu PK, Umme T, Yu JH, Nayak, A, Kim G, Noh M, Lee JY, Kim DD, Jeong LS. Selenoacyclovir and selenoganciclovir: Discovery of a new template for antiviral agents. J Med Chem 2015, 58, 8734–8; (h) Luo DP, Wu G, Yang H, Liu

MC, Gao WX, Huang XB, Chen JX, Wu HY. Copper-catalyzed three-component reaction for regioselective aryl- and heteroarylselenation of indoles using selenium powder. J Org Chem 2016, 81, 4485–93; (i) Gao C, Wu G, Min L, Liu MC, Gao WX, Ding JC, Chen JX, Huang XB, Wu HY. Copper-catalyzed three-component coupling reaction of azoles, Se powder and aryl iodides. J Org Chem 2017, 82, 250–5.
[6] (a) Gutierrez-Hernandez, AI., JG, Lopez-Cortes., Ortega-Alfaro, MC., Ramirez-Apan, MT., Cazares-Marinero, JJ., and Toscano, RA. Ferrocenylselenoamides: Synthesis, characterization and cytotoxic properties. J Med Chem 2012, 55, 4652–63; (b) Yoshimoto N, Itoh T, Inaba Y, Ishii H, Yamamoto K. Structural basis for inhibition of carboxypeptidase B by selenium-containing inhibitor: selenium coordinates to zinc in enzyme. J Med Chem 2013, 56, 7527–35; (c) Martins IL, Charneira C, Gandin V, Ferreira da Silva J, Justino GC, Telo JP, Vieira AJSC, Marzano C, Antunes AM. Selenium-containing chrysin and quercetin derivatives: attractive scaffolds for cancer therapy. J Med Chem 2015, 58, 4250–65; (d) Macegoniuk K, Grela E, Palus J, Rudzińska-Szostak E, Grabowiecka A, Biernat M, Berlicki L. 1,2-Benzisoselenazol-3 (2*H*)-one derivatives as a new class of bacterial urease inhibitors. J Med Chem 2016, 59, 8125–33; (e) Iwasaki M, Miki N, Tsuchiya Y, Nakajima K,Nishihara Y. Synthesis of benzoisoselenazolone derivatives by nickel-catalyzed dehydrogenative direct selenation of C (sp^2)-H bonds with elemental selenium in air. Org Lett 2017, 19, 1092–5; (f) Pang YQ, An BJ, Lou L L, Zhang JS, Yan J, Huang L, LiXS, Yin S. Design, synthesis, and biological evaluation of novel selenium-containing *iso*combretastatins and phenstatins as antitumor agents. J Med Chem 2017, 60, 7300–14.
[7] Mukherjee, AJ., Zade, SS., Singh, HB., and Sunoj, RB. Organoselenium chemistry: Role of intramolecular interactions. Chem Rev 2010, 110, 4357–416.
[8] (a) Ogawa, A., Ogawa, I., Obayashi, R., Umezu, K., Doi, M., and Hirao, T. Highly selective thioselenation of vinyl cyclopropanes with a (PhS)(2)-(PhSe)(2) binary system and its application to thiotelluration. J Org Chem 1999, 64, 86–92; (b) Ogawa A, Doi M, Ogawa I, Hirao T. Highly selective three-component coupling of ethyl propiolate, alkenes and diphenyl diselenide under visible-light irradiation. Angew Chem Int Ed 1999, 38, 2027–9; (c) Tsuchii K, Tsuboi Y, Kawaguchi S-I, Takahashi J, Sonoda N, Nomoto A, Ogawa A. Highly selective double chalcogenation of isocyanides with disulfide–diselenide mixed systems. J Org Chem 2007, 72, 415–23; (d) Mitamura T, Iwata K, Nomoto A,Ogawa A. Photochemical intramolecular cyclization of *o*-alkynylaryl isocyanides with organic dichalcogenides leading to 2,4-bischalcogenated quinolines. Org Biomol Chem 2011, 9, 3768–75.
[9] (a) Perin, G., Guimar˜aes, RG., de Azambuja, F., Dos Santos, GFF., and Lenard˜ao, EJ. Addition of chalcogenolate anions to terminal alkynes using microwave and solvent-free conditions: Easy access to bis-organochalcogen alkenes. Tet Lett 2006, 47, 935–8; (b) Kumar S, Engman L. Microwave-assisted copper-catalyzed preparation of diaryl chalcogenides. J Org Chem 2006, 71, 5400–3; (c) Kundu D, Ahammed S, Ranu BC. Microwave-assisted reaction of aryl diazonium fluoroborate and diaryl dichalcogenides in dimethyl carbonate: a general procedure for the synthesis of unsymmetrical diaryl chalcogenides. Green Chem 2012, 14, 2024–30; (d) Botteselle GV, Godoi M, Galetto FZ, Bettanin L, Singh D, Rodrigues OED, Braga AL. Microwave-assisted one-pot synthesis of symmetrical diselenides, ditellurides and disulfides from organoyl iodides and elemental chalcogen catalyzed by CuO nanoparticles. J Mol Catal A Chem 2012, 365, 186–93; (e) Jafari M, Salavati-Niasari M, Sobhani A. Silver selenide nanoparticles: synthesis, characterisation and effect of preparation conditions under ultrasound radiation. Micro & Nano Letters 2013, 8, 508–11; (f) Gautheron B, Tainturier G, Degrand C. Ultrasound-induced electrochemical synthesis of the anions selenide (Se22-, Se2-), and telluride (Te22-, and Te2-). J Am Chem Soc 1985, 107, 5579–81; (g) Thompson DP, Boudjouk P. A convenient synthesis of alkali metal selenides and diselenides in

tetrahydrofuran and the reactivity differences exhibited by these salts toward organic bromides. Effect of ultrasound. J Org Chem 1988, 53, 2109–12; (h) Vukievic R, Konstantinovic S, Mihailovic ML. Electrochemical cyclization of unsaturated hydroxy compounds. Phenylselenoetherification and phenylselenolactonization. Tetrahedron 1991, 47, 859–65.
[10] Z-X, (a) Wang., Tu, Y., Frohn, M., J-R, Zhang., and Shi, Y. An efficient catalytic asymmetric epoxidation method. J Am Chem Soc 1997, 119, 11224–35; (b) Müller P, Fruit C. Enantioselective catalytic aziridinations and asymmetric nitrene insertions into CH bonds. Chem Rev 2003, 103, 2905–20; (c) Zhu YG, Wang Q, Cornwall RG, Shi Y. Organocatalytic asymmetric epoxidation and aziridination of olefins and their synthetic applications. Chem Rev 2014, 114, 8199–256.
[11] Niu, W., and Yeung, Y-Y. Catalytic and highly enantioselective selenolactonization. Org Lett 2015, 17, 1660–63.
[12] Tang, E., Zhao, Y., Li, W., Wang, W., Zhang, M., and Dai, X. Catalytic selenium-promoted intermolecular Friedel–Crafts alkylation with simple alkenes. Org Lett 2016, 18, 912–5.
[13] Y-X, (a) Gao., Tang, G., Cao, Y., and Y-F, Zhao. A novel and general method for the formation of *S*-aryl, *Se*-aryl, and *Te*-aryl phosphorochalcogenoates. Synthesis 2009, 1081–6; (b) Lauer AM, Mahmud F, Wu J. Cu(I)-catalyzed, α-selective, allylic alkylation reactions between phosphorothioate esters and organomagnesium reagents. J Am Chem Soc 2011, 133, 9119–23; (c) Zhang A, Sun J, Lin C, Hu X, Liu W. Enantioselective interaction of acid α-naphthyl acetate esterase with chiral organophosphorus insecticides. J Agric Food Chem 2014, 62, 1477–81; (d) Zhu Y, Chen T, Li S, Shimada S, Han L-B. Efficient Pd-Catalyzed Dehydrogenative Coupling of P(O)H with RSH: A Precise Construction of P(O)–S Bonds. J Am Chem Soc 2016, 138, 5825–8.
[14] Wang, J., Sheng, W., and Yan, J. Iodine-mediated vicinal difunctionalization of alkenes: A convenient method for building C–Se and C–S bonds. Synlett 2018, 29, 1654–8.
[15] Baldassari, LL., Mantovani, AC., Senoner, S., Maryasin, B., Maulide, N., and Lüdtke, DS. Redox-neutral synthesis of selenoesters by oxyarylation of selenoalkynes under mild conditions. J Org Chem 2018, 20, 5881–5.
[16] (a) Freudendahl, DM., and Wirth, T. In: New Selenium Selenium and Tellurium Chemistry. From Small Molecules to Biomolecules and Materials. Electrophiles and Their Reactivity, Woollins JD,Laitinen RS, ed., Springer-Verlag, Heidelberg, 2011, 41; (b) Santi C, Santoro S. In: Organoselenium Chemistry: Synthesis and Reactions, Wirth, T., ed. Wiley-VCH: Weinheim, 2012, 1;(c) Freudendahl DM, Iwaoka M, Wirth T. Synthesis of new sulfoxide-containing diselenides and unexpected cyclization reactions to 2,3-dihydro-1,4-benzoselenothiine 1-oxides. Eur J Org Chem 2010, 3934–44; (d) Singh FV, Wirth T. Selenium-catalyzed regioselective cyclization of unsaturated carboxylic acids using hypervalent iodine oxidants. Org Lett 2011, 13, 6504–7.
[17] Amosova, SV., Filippov, AS., Potapov, VA., Penzik, MV., Makhaeva, NA., and Albanov, AI. Regio- and stereoselective synthesis of a novel family of unsaturated compounds with the S–Se bond and their cyclization to 2,3-dihydro-1,4-thiaselenines. Synthesis 2019, 51, 1832–40.
[18] Kim, YJ., and Kim, DY. Electrochemical oxidative selenylation of imidazo[1,2–*A*]pyridines with diselenides. Tet Lett 2019, 60, 739–42.
[19] (a) Mantovani, AC., Goulart, TA., Back, DF., Menezes, PH., and Zeni, G. Iron(III) chloride and diorganyl diselenides-mediated 6-endo-dig cyclization of arylpropiolates and arylpropiolamides leading to 3-organoselenyl-2H-coumarins and 3-organoselenyl-quinolinones. J Org Chem 2014, 79, 10526–36; (b) Neto JSS, Iglesias BA, BackDF, Zeni G. Iron-Promoted Tandem Cyclization of 1,3-Diynyl Chalcogen Derivatives with Diorganyl Dichalcogenides for the Synthesis of Benzo[*b*]furan-Fused Selenophenes. Adv Synth Catal 2016, 358, 3572–85; (c) Prochnow T, Back DF, Zeni G. Iron(III) Chloride and diorganyl

diselenide-promoted nucleophilic closures of 1-benzyl-2-alkynylbenzenes in the preparation of 9-(organoselanyl)-5*H*-benzo[7]annulenes. Adv Synth Catal 2016, 358, 1119–29; (d) Recchi AMS, Back DF, Zeni G. Sequential carbon–carbon/carbon–selenium bond formation mediated by iron(III) chloride and diorganyl diselenides: synthesis and reactivity of 2-organoselenyl-naphthalenes. J Org Chem 2017, 82, 2713–23.

[20] Goulart, TAC., Kazmirski, JAG., Back, DF., and Zeni, G. Iron(III)-promoted synthesis of 3-(organoselanyl)-1,2-dihydroquinolines from diorganyl diselenides and *N*-arylpropargylamines by sequential carbon-carbon and carbon-selenium bond formation. Adv Synth Catal 2019, 361, 96–104.

[21] (a) Imakubo, T., Shirahata, T., and Kibune, M. Synthesis, crystal structure and electrochemical properties of bis(ethylenedioxy)tetraselenafulvalene (BEDO-TSeF). Chem Commun 2004, 1590–1; (b) Shirahata T, Kibune M, Yoshino H, Imakubo T. Ambient-Pressure Organic Superconductors κ-(DMEDO-TSeF)$_2$[Au(CN)$_4$](solv.): T_c Tuning by Modification of the Solvent of Crystallization. Chem-Eur J 2007, 13, 7619–30.

[22] (a) Mugesh, G., Du Mont, WW., and Sies, H. Chemistry of biologically important synthetic organoselenium compounds. Chem Rev 2001, 101, 2125–79; (b) Back TG, Moussa Z. Diselenides and allyl selenides as glutathione peroxidase mimetics. Remarkable activity of cyclic seleninates produced in situ by the oxidation of allyl ω-hydroxyalkyl selenides. J Am Chem Soc 2003, 125, 13455–60; (c) Nogueira CW, Zeni, G, Rocha JB. Organoselenium and organotellurium compounds: toxicology and pharmacology. Chem Rev 2004, 104, 6255–85; (d) Ninomiya M, Garud DR, Koketsu M. Biologically significant selenium-containing heterocycles. Coord Chem Rev 2011, 255, 2968–90.

[23] Toyoda, Y., and Koketsu, M. Synthesis and Z/E isomerization of 2-imino-1,3-thiaselenolanes *via* iodocyclization. Tetrahedron 2012, 68, 10496–501.

[24] GL, (a) Sommen., Linden, A., and Heimgartner, H. Synthesis of 2-selenoxo-1,3-thiazolidin-4-ones and 2-selenoxo-1,3-thiazinan-4-ones from isoselenocyanates. Heterocycles 2005, 65, 1903–15; (b) Koketsu M, Fukuta Y, Ishihara H. Reaction of *N*, *N*-dimethylselenocarbamoyl chloride with nucleophiles. preparation of diselenocarbamates, selenothiocarbamates and selenoureas. J Org Chem 2002, 67, 1008–11; (c) MuraiT. Organoselenium Chemistry In Synthesis and Reactions, Wirth T, ed., Wiley-VCH: Weinheim, Germany, 2012, 257.

[25] Shiah, H-S., Lee, W-S., Juang, S-H., Hong, P-C., Lung, C-C., Chang, C-J., Chou, K-M., and Chang, J-Y. Mitochondria-mediated and p53-associated apoptosis induced in human cancer cells by a novel selenophene derivative, D-501036. Biochem Pharmacol 2007, 73, 610–9.

[26] Zeni, G., Nogueira, CW., Panatieri, RB., Silva, DO., Menezes, PH., Braga, AL., Silveira, CC., Stefani, HA., and Rocha, JBT. Synthesis and anti-inflammatory activity of acetylenic thiophenes. Tet Lett 2001, 42, 7921–3.

[27] May, SW., Wang, L., Gill-Woznichak, MM., Browner, RF., Ogonowski, AA., Smith, JB., and Pollock, SH. An orally active selenium-based antihypertensive agent with restricted CNS permeability. J Pharm Exp Ther 1997, 283, 470–7.

[28] Wilhelm, EA., Jesse, CR., Bortoletto, CF., Nogueira, CW., and Savegnago, L. Anticonvulsant and antioxidant effects of 3-alkynyl selenophene in 21-day-old rats on pilocarpine model of seizures. Brain Res Bull 2009, 79, 281–7.

[29] Lam, LKT., and Ahamed, NUS. Organoselenium compounds for cancer chemoprevention. US Pat Appl Publ 2002, 20020165215.

[30] Rhoden, CRB., and Zeni, G. New development of synthesis and reactivity of seleno- and tellurophenes. Org Biomol Chem 2011, 9, 1301–13.

[31] RF, (a) Schumacher., AR, Rosario., ACG, Souza., PH, Menezes., and Zeni, G. Synthesis of 2,3-dihydroselenophene and selenophene derivatives by electrophilic cyclization of homopropargyl selenides. Org Lett 2010, 12, 1952–5; (b) Schumacher RF, Rosário AR, Leite

MR, Zeni G. Cyclization of homopropargyl chalcogenides by copper(II) salts: selective synthesis of 2,3-dihydroselenophenes, 3-arylselenophenes and 3-haloselenophenes/ thiophenes. Chem Eur J 2013, 19, 13059–64.
[32] AL, (a) Stein., Alves, D., JT, da Rocha., CW, Nogueira., and Zeni, G. Copper iodide-catalyzed cyclization of (Z)-chalcogenoenynes. Org Lett 2008, 10, 4983–6; (b) Barancelli DA, Schumacher RF, Leite MR, Zeni G. Copper(II)-mediated intramolecular cyclization of (*Z*)-chalcogenoenynes: synthesis of 3-halochalcogenophene derivatives. Eur J Org Chem 2011, 6713–8.
[33] Maity, P., Kundu, D., Roy, R., and Ranu, BC. A direct synthesis of selenophenes by Cu-catalyzed one-pot addition of a selenium moiety to (E,E)-1,3-dienyl bromides and subsequent nucleophilic cyclization. Org Lett 2014, 16, 4122–5.
[34] Sharpless, KB., Gordon, KM., Lauer, RF., Patrick, DW., Singer, SP., and Young, MW. The utility of selenium reagents in organic synthesis. Chem Scr (Sweden) 1975, 8A, 9–13.
[35] Yang, ZJ., Hu, BL., Deng, CL., and Zhang, XG. Iron-promoted electrophilic annulation of aryl enynes with disulfides or diselenides leading to polysubstituted naphthalenes. Adv Synth Catal 2014, 356, 1962–66.
[36] Beletskaya, IP., Sigeev, AS., Peregudov, AS., and Petrovskii, PV. New approaches to the synthesis of unsymmetrical diaryl selenides. J Organomet Chem 2000, 605, 96–101.
[37] (a) Taniguchi, N. Tetrahedron 2016, 72, 5818; (b) Taniguchi N. J Org Chem 2015, 80,1764; (c) Gao C, Wu G, Min L, Liu MC, Gao WX, Ding JC, Chen JX, Huang XB, Wu HY. J Org Chem 2017, 82, 250.
[38] Fang, WY., Dong, T., Han, JB., Zha, GF., and Zhang, CP. Expeditious trifluoromethylthiolation and trifluoromethylselenolation of alkynyl(phenyl)iodoniums by $[XCF_3]^-$ (X = S, Se) anions. Org Biomol Chem 2016, 14, 11502–9.
[39] Guan, Y., and Townsend, SD. Metal-free synthesis of unsymmetrical organoselenides and selenoglycosides. Org Lett 2017, 19, 5252–55.
[40] KB, (a) Sharpless., and MW, Young. Olefin synthesis. Rate enhancement of the elimination of alkyl aryl selenoxides by electron-withdrawing substituents. J Org Chem 1975, 40, 947–9; (b) Grieco PA, Gilman S, Nishizawa M. Organoselenium chemistry. A facile one-step synthesis of alkyl aryl selenides from alcohols. J Org Chem 1976, 41, 1485–6.
[41] Shibahara, F., Fukunaga, T., Kubota, S., Yoshida, A., and Murai, T. Synthesis of chiral selenazolines from N-acyloxazolidinones *via* a selenative rearrangement of chiral cyclic skeletons. Org Lett 2018, 20, 5826–30.
[42] DA, (a) Stoyanovsky., Jiang, J., MW, Epperly., Zhang, X., Li, S., JS, Greenberger., Kagan, VE., and Bayir, H. Design and synthesis of a mitochondria-targeted mimic of glutathione peroxidase, mitoEbselen-2, as a radiation mitigator. ACS Med Chem Lett 2014, 5, 1304–7; (b) Yan J, Guo Y, Wang Y, Mao F, Huang L, Li X. Design, synthesis, and biological evaluation of benzoselenazole-stilbene hybrids as multi-target-directed anti-cancer agents. Eur J Med Chem. 2015, 95 220–9; (c) Sundaram GSM, Dhavale DD, Prior JL, Yan P, Cirrito JR, Rath NP, Laforest R, Cairns NJ, Lee JM, Kotzbauer P, Sharma V. Fluselenamyl: A novel benzoselenazole derivative for PET detection of amyloid plaques (Aβ) in Alzheimer's disease. Sci Rep 2016, 6, 35636–47.
[43] Gu, R., Wang, X., Yang, Z., and Han, S. Se-mediated one-pot synthesis of 2-substituted benzoselenazole derivatives from 2 to iodoanilines and arylacetic acids/arylmethyl chlorides. Tet Lett 2018, 59, 2835–8.
[44] Sun, P., Jiang, M., Wei, W., Min, Y., Zhang, W., Li, W., Yang, D., and Wang, H. Copper-catalyzed selenylation of imidazo[1,2-*a*]pyridines with selenium powder *via* a radical pathway. J Org Chem 2017, 82, 2906–13.
[45] Liu, H., Fang, Y., Wang, S-Y., and Ji, S-J. TEMPO-catalyzed aerobic oxidative selenium insertion reaction: Synthesis of 3-selenylindole derivatives by multicomponent reaction of isocyanides, selenium powder, amines, and indoles under transition-metal-free conditions. Org Lett 2018, 20, 930–3.

[46] Sun, K., Wang, S., Feng, R., Zhang, Y., Wang, X., Zhang, Z., and Zhang, B. Copper-catalyzed radical selenodifluoromethylation of alkenes: access to CF_2-containing γ-lactams. Org Lett 2019, 21, 2052–5.
[47] Kim, YJ., and Kim, DY. Electrochemical radical selenylation/1,2-carbon migration and Dowd–Beckwith-type ring-expansion sequences of alkenylcyclobutanols. Org Lett 2019, 21, 1021–25.
[48] (a) Panneerselvam, A., CQ, Nguyen., MA, Malik., O'Brien, P., and Raftery, J. The CVD of silver selenide films from dichalcogenophosphinato and imidodichalcogenodiphosphinatosilver(I) single-source precursors. J Mater Chem 2009, 19, 419–27; (b) Maneeprakorn W, Malik MA, O'Brien P. The preparation of cobalt phosphide and cobalt chalcogenide (CoX, X = S, Se) nanoparticles from single source precursors. J Mater Chem 2010, 20, 2329–35; (c) Cossairt BM, Owen JS. CdSe clusters: at the interface of small molecules and quantum dots. Chem Mater 2011, 23, 3114–9.
[49] TS, (a) Lobana., J-C, Wang., and CW, Liu. Recent advances in the coordination chemistry of diselenophosphates and allied ligands. Coord Chem Rev 2007, 251, 91–110; (b) Artem'ev AV, Malysheva SF, Gusarova NK, Trofimov BA. New synthesis of diselenophosphinates of heavy metals. Russ J Gen Chem 2011, 81, 1449–52.
[50] (a) Song, E., Jo, Y., Bae, G., IK, Oh., HM, Jung., and Lee, S. Synthesis of phosphinodiselenoic acid ester derivatives and their application in the controlled radical polymerization of styrene. Bull Korean Chem Soc 2009, 30, 2129–31; (b) Artem'ev AV, Chernysheva NA, Gusarova NK,Malysheva SF, Yas'ko SV, Albanov AI, Trofimov BA. An efficient and general synthesis of Se-esters of diselenophosphinic acids via reaction of alkali metal diselenophosphinates with organic halides. Synthesis 2011, 1309–13.
[51] Gusarova, NK., Artem'ev, AV., Malysheva, SF., Tarasova, OA., and Trofimov, BA. A three-component reaction between alkenes, secondary phosphanes, and elemental selenium: A novel, efficient, atom-economic synthesis of diselenophosphinic esters. Tet Lett 2011, 52, 6985–7.
[52] Oparina, LA., Artem'ev, AV., Vysotskaya, OV., Kolynanov, NA., Bagryanskaya, IY., EP, Doronina., and NK, Gusarova. Three-component reaction between secondary phosphine sulfides, elemental selenium and vinyl ethers: The first examples of Markovnikov addition of thioselenophosphinic acids to double bond. Tetrahedron 2013, 69, 6185–95.
[53] (a) Kimura, T., Murai, T., Miwa, A., Kurachi, D., Yoshikawa, H., and Kato, S. Phosphinoselenothioic acids and their salts: Synthesis, characterization, and reaction with electrophiles. J Org Chem 2005, 70, 5611–7; (b) Artem'ev AV, Oparina LA, Gusarova NK, Kolyvanov NA, Vysotskaya OV, Bagryanskaya IY, Trofimov BA. Chemoselective synthesis of first representatives of bis(diorganothiophosphinyl)selenides, $(R_2P = S)_2Se$, from secondary phosphine sulfides and elemental selenium. Inorg Chem Comm 2013, 30, 124–7.
[54] YC, (a) Wang., YY, Xie., HE, Qu., HS, Wang., YM, Pan., and FP, Huang. $Ce(OTf)_3$-catalyzed [3 + 2] cycloaddition of azides with nitroolefins: Regioselective synthesis of 1,5-disubstituted 1,2,3-triazoles. J Org Chem 2014, 79, 4463–9; (b) Xie YY, Wang YC, Qu HE, Tan XC, Wang HS, Pan YM. Regioselective synthesis of β-aryl enaminones and 1,4,5- trisubstituted 1,2,3-triazoles from chalcones and benzyl azides. Adv Synth Catal 2014, 356, 3347–55; (c) Wang Y, Li J, He Y, Xie Y, Wang H, Pan Y. Synthesis of polysubstituted imidazoles and pyridines *via* samarium(III) triflate-catalyzed [2+2+1] and [4+2] annulations of unactivated aromatic alkenes with azides. Adv Synth Catal 2015, 357, 3229–41; (d) Li JL, Wang YC, Li WZ, Wang HS, Mo DL, Pan YM. A facile synthesis of 2,5-disubstituted oxazoles *via* a copper-catalyzed cascade reaction of alkenes with azides. Chem Commun 2015, 51, 17772–4; (e) Xie YY, Wang YC, He Y, Hu DC, Wang HS, Pan YM. Catalyst-free synthesis of fused 1,2,3-triazole and isoindoline derivatives *via* an intramolecular azide–alkene cascade reaction. Green Chem, 2017, 19, 656–9.

[55] Cui, FH., Chen, J., Su, SX., Xu, YL., Wang, HS., and Pan, YM. Regioselective synthesis of selenide ethers through a decarboxylative coupling reaction. Adv Synth Catal 2017, 359, 3950–61.
[56] Cui, FH., Chen, J., Mo, ZY., Su, SX., Chen, YY., Ma, XL., Tang, HT., Wang, HS., Pan, YM., and Xu, YL. Copper-catalyzed decarboxylative/Click cascade reaction: Regioselective assembly of 5-selenotriazole anticancer agents. Org Lett 2018, 20, 925–9.
[57] Li, X., Yang, F., Wu, Y., and Wu, Y. Copper-mediated oxidative decarboxylative coupling of arylpropiolic acids with dialkyl H-phosphonates in water. Org Lett 2014, 16, 992–5.
[58] Stefani, HA., Vasconcelos, SNS., Manarin, F., Leal, DM., Souza, FB., Madureira, LS., Zukerman-Schpector, J., Eberlin, MN., Godoi, MN., and de Souza Galaverna, R. Synthesis of 5-organotellanyl-1*H*-1,2,3-triazoles: Functionalization of the 5-position scaffold by the Sonogashira cross-coupling reaction. Eur J Org Chem 2013, 3780–85.
[59] (a) Craig, RA., O'Connor, NR., Goldberg, AFG., and Stoltz, BM. Stereoselective Lewis acid mediated (3 + 2) cycloadditions of *N*–H- and *N*–Sulfonylaziridines with heterocumulenes. Chem Eur J 2014, 20, 4806–13; (b) Sengoden M, Irie R, Punniyamurthy T. Enantiospecific aluminum-catalyzed (3+2) cycloaddition of unactivated aziridines with isothiocyanates. J Org Chem 2016, 81, 11508–13.
[60] Satheesh, V., Srivastava, HK., Kumar, SV., Sengoden, M., and Punniyamurthy, T. Stereospecific Al-catalysed tandem *C–N*/*C–Se* bond formation of isoselenocyanates with aziridines: Synthesis and DFT study. Adv Synth Catal 2019, 361, 55–8.
[61] (a) Nogueira, CW., Zeni, G., and Rocha, JBT. Organoselenium and organotellurium compounds: Toxicology and pharmacology. Chem Rev 2004, 104, 6255–85; (b) Mugesh G, Singh HB. Synthetic organoselenium compounds as antioxidants: glutathione peroxidase activity. Chem Soc Rev 2000, 29, 347–57; (c) Nogueira CW, Meotti FC, Curte E, Pilissao C, Zeni G, Rocha JBT. Investigations into the potential neurotoxicity induced by diselenides in mice and rats. Toxicology 2003, 183, 29–37.
[62] Sonoda, N., Yamamoto, G., Natsukawa, K., Kondo, K., and Murai, S. A new synthesis of heterocycles by the reaction of amino alcohols or amino thiols with carbon monoxide in the presence of selenium. Tetrahedron Lett 1975, 24, 1969–72.
[63] (a) Back, TG. Organoselenium Chemistry, Oxford, UK, Oxford University Press, 1999; (b) Wirth T. Organoselenium Chemistry: Modern Developments in Organic Synthesis, Top Curr Chem, vol. 208. Berlin, Germany, Springer, 2000; (c) Beaulieu PL, Déziel R. In Organoselenium Chemistry: A Practical Approach. ed. Back TG. Oxford, UK, Oxford University Press, 1999, 35; (d) Freudendahl DM, Wirth T. In Selenium and Tellurium Chemistry. ed. Woolins JD, Laitinen RS. Berlin, Germany, Springer, 2011, 41; (e) Singh FV and Wirth T. In Patai Series: Organic Selenium and Tellurium Compounds, Vol. 3. ed. Rappoport Z. John Wiley & Sons, 2012, 303; (f) Nogueira CW, Rocha JBT. In Patai Series: Organic Selenium and Tellurium Compounds, Vol. 3. ed. Rappoport Z. John Wiley & Sons, 2012, 1277; (g) Braga AL, Rafique J. in Patai Series: Organic Selenium and Tellurium Compounds, Vol. 4. ed. Rappoport Z. John Wiley & Sons, 2014, 1175. (h)Santoro S, Azeredo JB, Nascimento V, Sancineto L, Braga AL, Santi C. The green side of the moon: ecofriendly aspects of organoselenium chemistry. RSC Adv 2014, 4, 31521–35; (j) Singh FV, Wirth T. Selenium reagents as catalysts. Catal Sci Technol 2019, 9, 1073–91.
[64] (a) Tunge, JA., and SR, Mellegaard. Selective selenocatalytic allylic chlorination. Org Lett 2004, 6, 1205–7; (b) Barrero AF, Quílez del Moral JF, Herrador MM, Cortés M, Arteaga P, Catalán JV, Sánchez EM, Arteaga JF. Solid-phase selenium-catalyzed selective allylic chlorination of polyprenoids: Facile syntheses of biologically active terpenoids. J Org Chem, 2006, 71, 5811–4.
[65] (a) Yeung, YY., Gao, X., and EJ, Corey. A general process for the haloamidation of olefins. Scope and mechanism. J Am Chem Soc 2006, 128, 9644–5; (b) Yadav JS, Subba Reddy BV,

NarasimhaChary D, Chandrakanth D. InX_3-catalyzed haloamidation of vinyl arenes: a facile synthesis of α-bromo- and α-fluoroamides.Tet Lett 2009, 50, 1136–8.
[66] Tay, DW., Tsoi, IT., Er, JC., Leung, GYC., and Yeung, YY. Lewis basic selenium catalyzed chloroamidation of olefins using nitriles as the nucleophiles. Org Lett 2013, 15, 1310–3.
[67] (a) Zhou, L., CK, Tan., Jiang, X., Chen, F., and YY, Yeung. Asymmetric bromolactonization using amino-thiocarbamate catalyst. J Am Chem Soc 2010, 132, 15474–6; (b) Tan CK, Zhou L, Yeung YY. Aminothiocarbamate-catalyzed asymmetric bromolactonization of 1,2-disubstituted olefinic acids. Org Lett. 2011, 13, 2738–41; (c) Zhou L, Chen J, Tan CK, Yeung YY. Enantioselective bromoaminocyclization using amino-thiocarbamate catalysts. J Am Chem Soc 2011, 133, 9164–7; (d) Chen J, Zhou L, Tan CK, Yeung YY. An enantioselective approach toward 3,4-dihydroisocoumarin through the bromocyclization of styrene-type carboxylic acids. J Org Chem 2012, 77, 999-1009; (e) Chen J, Zhou L, Yeung YY. A highly enantioselective approach towards 2-substituted 3-bromopyrrolidines. Org Biomol Chem 2012, 10, 3808–11; (f) Tan CK, Le C, Yeung YY. Enantioselective bromolactonization of *cis*-1,2-disubstituted olefinic acids using an amino-thiocarbamate catalyst. Chem Commun 2012, 48, 5793–5; (g) Jiang X, Tan CK, Zhou L, Yeung YY. Enantioselective bromolactonization using an S-alkyl thiocarbamate catalyst. Angew Chem Int Ed 2012, 51, 7771–5.
[68] Chen, F., Tan, CK., and Yeung, YY. C_2-Symmetric cyclic selenium-catalyzed enantioselective bromoaminocyclization. J Am Chem Soc 2013, 135, 1232–5.
[69] Horibe, T., Ohmura, S., and Ishihara, K. Selenium–Iodine cooperative catalyst for chlorocyclization of tryptamine derivatives. Org Lett 2017, 19, 5525–8.
[70] (a) Leroux, F., Jeschke, P., and Schlosser, M. α-Fluorinated ethers, thioethers, and amines: Anomerically biased species. Chem Rev 2005, 105, 827–56; (b) Muller K, Faeh C, Diederich F. Fluorine in pharmaceuticals: looking beyond intuition. Science 2007, 317, 1881–6; (c) Hagmann WK. The many roles for fluorine in medicinal chemistry. J Med Chem 2008, 51, 4359–69; (d) Purser S, Moore PR, Swallow S, Gouverneur V. Fluorine in medicinal chemistry. Chem Soc Rev 2008, 37, 320–30; (e) Manteau B, Pazenok S, Vors JP, Leroux FR. New trends in the chemistry of α-fluorinated ethers, thioethers, amines and phosphines. J Fluorine Chem 2010, 131, 140–58.
[71] Ferry, A., Billard, T., Langlois, BR., and Bacqué, E. Trifluoromethanesulfanylamides as easy-to-handle equivalents of the trifluoromethanesulfanyl cation (CF(3)S(+)): Reaction with alkenes and alkynes. Angew Chem Int Ed 2009, 48, 8551–5.
[72] Luo, J., Zhu, Z., Liu, Y., and Zhao, X. Diaryl selenide catalyzed vicinal trifluoromethylthioamination of alkenes. Org Lett 2015, 17, 3620–3.
[73] Luo, J., Liu, Y., and Zhao, X. Chiral selenide-catalyzed enantioselective construction of saturated trifluoromethylthiolated azaheterocycles. Org Lett 2017, 19, 3434–7.
[74] Luo, J., Cao, Q., Cao, X., and Zhao, X. Selenide-catalyzed enantioselective synthesis of trifluoromethylthiolated tetrahydronaphthalenes by merging desymmetrization and trifluoromethylthiolation. Nat Commun 2018, 9, 527–35.
[75] Hili, R., and Yudin, AK. Making carbon-nitrogen bonds in biological and chemical synthesis. Nat Chem Biol 2006, 2, 284–7.
[76] (a) Inamoto, K., Saito, T., Katsuno, M., Sakamoto, T., and Hiroya, K. Palladium-catalyzed C–H activation/intramolecular amination reaction: A new route to 3-aryl/alkylindazoles. Org Lett 2007, 9, 2931–4; (b) Youn SW, Bihn JH, Kim BS. Pd-catalyzed intramolecular oxidative C-H amination: synthesis of carbazoles. Org Lett 2011, 13, 3738–41; (c) Mei TS, Leow D, Xiao H, Laforteza BN, Yu JQ. Synthesis of indolines via Pd(II)-catalyzed amination of C-H bonds using PhI(OAc)2 as the bystanding oxidant. Org Lett 2013, 15, 3058–61; (d) Park Y, Kim Y, Chang S. Transition metal-catalyzed C-H amination: Scope, mechanism, and applications. Chem Rev 2017, 117, 9247–301.

[77] (a) Liu, G., Yin, G., and Wu, L. Palladium-catalyzed intermolecular aerobic oxidative amination of terminal alkenes: Efficient synthesis of linear allylamine derivatives. Angew Chem Int Ed 2008, 47, 4733–6; (b) Pattillo CC, Strambeanu II, Calleja P, Vermeulen NA, Mizuno T, White MC. Aerobic linear allylic C–H amination: Overcoming benzoquinone inhibition. J Am Chem Soc 2016, 138, 1265–72; (c) Timokhin VI, Anastasi NR, Stahl SS. Dioxygen-coupled oxidative amination of styrene. J Am Chem Soc 2003, 125, 12996–7; (d) Brice JL, Harang JE, Timokhin VI, Anastasi NR, Stahl SS. Aerobic oxidative amination of unactivated alkenes catalyzed by palladium. J Am Chem Soc 2005, 127, 2868–9.
[78] Trenner, J., Depken, C., Weber, T., and Breder, A. Direct oxidative allylic and vinylic amination of alkenes through selenium catalysis. Angew Chem Int Ed 2013, 52, 8952–6.
[79] Deng, Z., Wei, J., Liao, L., Huang, H., and Zhao, X. Organoselenium-catalyzed, hydroxy-controlled regio- and stereoselective amination of terminal alkenes: Efficient synthesis of 3-amino allylic alcohols. Org Lett 2015, 17, 1834–7.
[80] Zheng, T., Tabor, JR., Stein, ZL., and Michael, FE. Regioselective metal-free aza-Heck reactions of terminal alkenes catalyzed by phosphine selenides. Org Lett 2018, 20, 6975–8.
[81] Ortgies, S., and Breder, A. Selenium-catalyzed oxidative $C(sp^2)$–H amination of alkenes exemplified in the expedient synthesis of (aza-)indoles. Org Lett 2015, 17, 2748–51.
[82] Guo, R., Huang, J., Huang, H., and Zhao, X. Organoselenium-catalyzed synthesis of oxygen- and nitrogen-containing heterocycles. Org Lett 2016, 18, 504–7.
[83] (a) Sircar, JC., Capiris, T., Kesten, SJ., and Herzig, DJ. Pyrazolo[5,1-b]quinazolin-9-ones: A new series of antiallergic agents. J Med Chem 1981, 24, 735–42; (b) Vidule RR, Shirodkar SG. Synthesis and antimicrobial studies of few new substituted 2-methyl-3-(aryldiazenyl) pyrazolo[5,1-b]quinazolin-9(3H)-ones. Res J Chem Sci 2013, 3, 60–8.
[84] Zhang, Y., Shao, Y., Gong, J., Zhu, J., Cheng, Y., and Chen, J. Selenium-catalyzed oxidative C-H amination of (E)-3-(arylamino)-2-styrylquinazolin-4(3H)-ones: A metal-free synthesis of 1,2-diarylpyrazolo[5,1- *b*]quinazolin-9(1H)-ones. J Org Chem 2019, 84, 2798–807.
[85] Browne, DM., Niyomura, O., and Wirth, T. Catalytic use of selenium electrophiles in cyclizations. Org Lett 2007, 9, 3169–71.
[86] Shahzad, SA., Venin, C., and Wirth, T. Diselenide- and disulfide-mediated synthesis of isocoumarins. Eur. J Org Chem 2010, 3465–72.
[87] Singh, FV., and Wirth, T. Selenium-catalyzed regioselective cyclization of unsaturated carboxylic acids using hypervalent iodine oxidants. Org Lett 2011, 13, 6504–7.
[88] Browne, DM., Niyomura, O., and Wirth, T. Catalytic addition-elimination reactions towards butenolides. Phosphorus Sulfur Silicon Relat Elem 2008, 183, 1026–35.
[89] Kawamata, Y., Hashimoto, T., and Maruoka, K. A chiral electrophilic selenium catalyst for highly enantioselective oxidative cyclization. J Am Chem Soc 2016, 138, 5206–9.
[90] Krtzschmar, F., Kaßel, M., Delony, D., and Breder, A. Selenium-catalyzed $C(sp^3)$-H acyloxylation: Application in the expedient synthesis of isobenzofuranones. Chem Eur J 2015, 21, 1–6.
[91] (a) Shi, YS., Liu, YB., Ma, SG., Li, Y., Qu, J., Li, L., SP, Yuan., Hou, Q., YH, Li., Jiang, JD., and Yu, SS. Bioactive sesquiterpenes and lignans from the fruits of *Xanthium sibiricum*. J Nat Prod 2015, 78, 1526–35; (b) Pettersson T, Eklund AM, Wahlberg I. New lactones from tobacco. J Agric Food Chem, 1993, 41, 2097–103.
[92] Leisering, S., Riaño, I., Depken, C., Gross, LJ., Weber, M., Lentz, D., Zimmer, R., Stark, CBW., Breder, A., and Christmann, M. Synthesis of (+)-greek tobacco lactone *via* a diastereoablative epoxidation and a selenium-catalyzed oxidative cyclization. Org Lett 2017, 19, 1478–81.
[93] (a) Diels, O., and Karstens, A. Über Dehydrierungen mit Selen (II. Mitteilung). Chem Berichte 1927, 60B, 2323–5; (b) Diels O, Gädke W, Körding P. Uber die Dehydrierung des Cholesterins. Annalen 1927, 459, 1–26; (c) Clemo GR, Dickenson HG. The action of selenium on compounds

containing angular methyl groups. J Chem Soc 1935, 735–8; (d) House WT, Orchin M. A study of the selenium dehydrogenation of guaiol and related compounds. Selenium as a hydrogen transfer agent. J Am Chem Soc 1960, 82, 639–42; (e) Silverwood HA, Orchin M. Aromatization reactions with selenium and aryl diselenides. J Org Chem 1962, 27, 3401–4.
[94] (a) Waitkins, GR., and Clark, CW. Selenium dioxide: Preparation, properties, and use as oxidizing agent. Chem Rev 1945, 36, 235–89; (b) Rabjohn N. Selenium dioxide oxidation. Org React 1949, 5, 331–86; (c) Rabjohn N. Selenium dioxide oxidation. Org React 1976, 24, 261–415; (d) Jerussi RA. In Selective Organic Transformations. Ed. Thyagarajan BS. New York, USA, Wiley, 1970, 301; (e) Fieser LF, Fieser M. Reagents for Organic Synthesis, Vol. 1. New York, USA, Wiley, 1967; (f) Reich HJ. In Organic Chemistry, Vol. 5C. ed. Trahanovsky WS. New York, USA, Academic Press, 1978, Chapter 1; (g) Nicolaou KC, Petasis NA. Selenium in Natural Products Synthesis. Philadelphia, USA, CIS, 1984, Chapters 2 and 3. (h) Ley SV. In Organoselenium Chemistry. Ed. Liotta D. New York, USA, Wiley, 1987, 163.
[95] Barton, DHR., Brewster, AG., Hui, AHF., Lester, DJ., Ley, SV., and Back, TG. Oxidation of alcohols using benzeneseleninic anhydride. J Chem Soc, Chem Commun 1978, 952–4.
[96] Umbreit, MA., and Sharpless, KB. Allylic oxidation of olefins by catalytic and stoichiometric selenium dioxide with tert-butyl hydroperoxide. J Am Chem Soc 1977, 99, 5526–8.
[97] Barton, DHR., and Crich, D. Oxidation of olefins with 2-pyridineseleninic anhydride. Tetrahedron 1985, 41, 4359–64.
[98] Hori, T., and Sharpless, KB. Synthetic applications of arylselenenic and arylseleninic acids. Conversion of olefins to allylic alcohols and epoxides. J Org Chem 1978, 43, 1689–97.
[99] (a) Brink, GJT., Fernandes, BCM., van Vliet, MCA., Arends, IWCE., and Sheldon, RA. Selenium catalysed oxidations with aqueous hydrogen peroxide. Part I: Epoxidation reactions in homogeneous solution. J Chem Soc, Perkin Trans 2001, 1, 224–8; (b) Santi C, Lorenzo RD, Tidei C, Bagnoli L, Wirth T. Stereoselective selenium catalyzed dihydroxylation and hydroxymethoxylation of alkenes. Tetrahedron 2012, 68, 10530–5; (c) Santoro S, Santi C, Sabatini M, Testaferri L, Tiecco M. Eco-friendly olefin dihydroxylation catalyzed by diphenyl diselenide. Adv Synth Catal 2008, 350, 2881–4; (d) Carrera I, Brovetto MC, Seoane GA. Selenium-catalyzed iodohydrin formation from alkenes. Tet Lett 2006, 47, 7849–52; (e) Mellegaard SR, Tunge JA. Selenium-catalyzed halolactonization: nucleophilic activation of electrophilic halogenating reagents. J Org Chem 2004, 69, 8979–81; (f) Ichikawa H, Usami Y, Arimoto M. Synthesis of novel organoselenium as catalyst for Baeyer–Villiger oxidation with 30% H_2O_2. Tet Lett 2005, 46, 8665–8; (g) Brink GJT, Vis JM, Arends IWCE, Sheldon RA. Selenium catalysed oxidations with aqueous hydrogen peroxide. Part 3: Oxidation of carbonyl compounds under mono/bi/triphasic conditions. Tetrahedron 2002, 58, 3977–83; (h) Brink GJT, Vis JM, Arends IWCE, Sheldon RA. Selenium-catalyzed oxidations with aqueous hydrogen peroxide. 2. Baeyer–Villiger reactions in homogeneous solution. J Org Chem 2001, 66, 2429–33; (i) Singh FV, Wirth T. Selenium-catalyzed regioselective cyclization of unsaturated carboxylic acids using hypervalent iodine oxidants. Org Lett 2011, 13, 6504–7; (j) Bennett SM, Tang Y, McMaster D, Bright FV, Detty MR. A xerogel-sequestered selenoxide catalyst for brominations with hydrogen peroxide and sodium bromide in an aqueous environment. J Org Chem 2008, 73, 6849–52; (k) Niyomura O, Cox M, Wirth T. Electrochemical generation and catalytic use of selenium electrophiles. Synlett 2006, 251–4; (l) Goodman MA, Detty MR. Substituent effects in arylseleninic acid-catalyzed bromination of organic substrates with sodium bromide and hydrogen peroxide. Organometallics 2004, 23, 3016–20; (m) Barrero AF, delMoral JFQ, Herrador MM, Cortés M, Arteaga P, Catalán JV, Sánchez EM, Arteaga JF. Solid-phase selenium-catalyzed selective allylic chlorination of polyprenoids: Facile syntheses of biologically active terpenoids. J Org Chem 2006, 71, 5811–4; (n) Mellegaard-Waetzig SR, Wang C, Tunge JA. Selenium-catalyzed oxidative halogenation.

Tetrahedron 2006, 62, 7191–8; (o) Drake MD, Bateman MA, Detty MR. Substituents effects in arylseleninic acid-catalyzed bromination of organic substrates with sodium bromide and hydrogen peroxide. Organometallics 2003, 22, 4158–62; (p) Gebhardt C, Priewisch B, Irran E, Rück-Braun K. Oxidation of anilines with hydrogen peroxide and selenium dioxide as catalyst. Synthesis 2008, 1889–94; (q) Zhao D, Johansson M, Bäckvall J-E. In situ generation of nitroso compounds from catalytic hydrogen peroxide oxidation of primary aromatic amines and their one-pot use in hetero-Diels–Alder reactions. Eur J Org Chem 2007, 4431–6.

[100] (a) Gonzalez-de-Castro, A., and J-L, Xiao. Green and efficient: iron-catalyzed selective oxidation of olefins to carbonyls with O_2. J Am Chem Soc 2015, 137, 8206–18; (b) Moorthy JN, Parida KN. Oxidative cleavage of olefins by in situ-generated catalytic 3,4,5,6-tetramethyl-2-iodoxybenzoic acid/oxone. J Org Chem 2014, 79, 11431–9; (c) Daw P, Petakamsetty R, Sarbajna A, Laha S, Ramapanicker R, Bera JK. A highly efficient catalyst for selective oxidative scission of olefins to aldehydes: Abnormal-NHC-Ru(II) complex in oxidation chemistry. J Am Chem Soc 2014, 136, 13987–90; (d) Hossain MM, Huang W-K, Chen H-J, Wang P-H, Shyu SG. Efficient and selective copper-catalyzed organic solvent-free and biphasic oxidation of aromatic *gem*-disubstituted alkenes to carbonyl compounds by *tert*-butyl hydroperoxide at room temperature. Green Chem 2014, 16, 3013; (e) Wang G-Z, Li X-L, Dai J-J, Xu H-J. AIBN-catalyzed oxidative cleavage of gem-disubstituted alkenes with O_2 as an oxidant. J Org Chem 2014, 79, 7220–5; (f) Wang T, Jiao N. TEMPO-catalyzed aerobic oxygenation and nitrogenation of olefins *via* C—C double-bond cleavage. J Am Chem Soc 2013, 135, 11692–5; (g) Lin R, Chen F, Jiao N. Metal-free, NHPI catalyzed oxidative cleavage of C–C double bond using molecular oxygen as oxidant. Org Lett 2012, 14, 4158–61.

[101] Wang, T., Jing, X., Chen, C., and Yu, L. Organoselenium-catalyzed oxidative C—C bond cleavage: A relatively green oxidation of alkenes into carbonyl compounds with hydrogen peroxide. J Org Chem 2017, 82, 9342–9.

[102] Mercier, EA., Smith, CD., Parvez, M., and Back, TG. Cyclic seleninate esters as catalysts for the oxidation of sulfides to sulfoxides, epoxidation of alkenes, and conversion of enamines to α-hydroxyketones. J Org Chem 2012, 77, 3508–17.

[103] (a) Sterckx, H., JD, Houwer., Mensch, C., Caretti, I., KA, Tehrani., WA, Herrebout., SV, Doorslaer., and BUW, Maes. Mechanism of the Cu^{II}-catalyzed benzylic oxygenation of (aryl)(heteroaryl)methanes with oxygen. Chem Sci 2016, 7, 346–57; (b) Sterckx H, Houwer JD, Mensch C, Herrebout W, Tehrani KA, Maes BUW. Base metal-catalyzed benzylic oxidation of (aryl)(heteroaryl)methanes with molecular oxygen. Beilstein J Org Chem 2016, 12, 144–53; (c) Houwer JD, Tehrani KA, Maes BUW. Synthesis of aryl(di)azinyl ketones through copper- and iron-catalyzed oxidation of the methylene group of aryl(di)azinylmethanes. Angew Chem 2012, 124, 2799–802; Angew Chem Int Ed 2012, 51, 2745–8; (d) Pieber B, Kappe CO. Direct aerobic oxidation of 2-benzylpyridines in a gas–liquid continuous-flow regime using propylene carbonate as a solvent. Green Chem 2013, 15, 320–4; (e) Liu JM, Zhang X, Yi H, Liu C, Liu R, Zhang H, Zhuo KL, Lei AW. Chloroacetate-promoted selective oxidation of heterobenzylic methylenes under copper catalysis. Angew Chem 2015, 127, 1277–81; Angew Chem Int Ed 2015, 54, 1261–5; (f) Itoh M, Hirano K, Satoh T, Miura M. Copper-catalyzed α-methylenation of benzylpyridines using dimethylacetamide as one-carbon source. Org Lett 2014, 16, 2050–3; (g) Ren, LH, Wang LY, Lv Y, Li GS, Gao S. Synergistic H_4NI–AcOH Catalyzed Oxidation of the $C_{sp}{}^3$–H Bonds of Benzylpyridines with Molecular Oxygen. Org Lett 2015, 17, 2078–81.

[104] Jin, W., Zheng, P., Wong, W-T., and Law, G-L. Efficient selenium-catalyzed selective $C(sp^3)$–H oxidation of benzylpyridines with molecular oxygen. Adv Synth Catal 2017, 359, 1–7.

[105] Curran, SP., and Connon, SJ. Selenide ions as catalysts for homo- and crossed-Tishchenko reactions of expanded scope. Org Lett 2012, 14, 1074–7.

[106] Poladura, B., Martinez-CastaÇeda, A., Rodriguez-Solla, H., Llavona, R., Concellon, C., and Amo, Vd. General metal-free Baeyer-Villiger-type synthesis of vinyl acetates. Org Lett 2013, 15, 2810–3.
[107] Zhang, X., Ye, J-Q., Yu, L., Shi, X-K., Zhang, M., Xu, Q., and Lautens, M. Organoselenium-catalyzed Baeyer–Villiger oxidation of α,β-unsaturated ketones by hydrogen peroxide to access vinyl esters. Adv Synth Catal 2015, 357, 955–60.
[108] Yu, L., Li, H., Zhang, X., Ye, J., Liu, J., Xu, Q., and Lautens, M. Organoselenium-catalyzed mild dehydration of aldoximes: An unexpected practical method for organonitrile synthesis. Org Lett 2014, 16, 1346–9.
[109] (a) Pandey, G., BBV, SomaSekhar., and UT, Bhalerao. Photoinduced single electron transfer initiated heterolytic carbon-selenium bond dissociation. Sequential one-pot selenenylation and deselenenylation reaction. J Am Chem Soc 1990, 112, 5650–1; (b) Pandey G, SomaSekhar BBV. In situ generation and utilization of electrophilic selenium species (PhSe+) by photooxidative (single electron transfer) cleavage of diphenyl diselenide (PhSeSePh). J Org Chem 1992, 57, 4019–23.
[110] Ortgies, S., Depken, C., and Breder, A. Oxidative allylic esterification of alkenes by cooperative selenium-catalysis using air as the sole oxidant. Org Lett 2016, 18, 2856–9.
[111] Depken, C., Krätzschmar, F., Rieger, R., Rode, K., and Breder, A. Photocatalytic aerobic phosphatation of alkenes. Angew Chem Int Ed 2018, 57, 2459–63.
[112] (a) Wencel-Delord, J., Dröge, T., Liu, F., and Glorius, F. Towards mild metal-catalyzed C–H bond activation. Chem Soc Rev 2011, 40, 4740–61; (b) Seechurn CCJ, Kitching MO, Colacot TJ, Snieckus V. Palladium-catalyzed cross-coupling: a historical contextual perspective to the 2010 Nobel Prize. Angew Chem Int Ed 2012, 51, 5062–85; (c) Bolm C, Legros J, Paih JL, Zani L. Iron-catalyzed reactions in organic synthesis. Chem Rev 2004, 104, 6217–54; (d) Blanco-Urgoiti J, Añorbe L, Pérez-Serrano L, Domínguez G, Pérez-Castells J. The Pauson-Khand reaction, a powerful synthetic tool for the synthesis of complex molecules. Chem Soc Rev 2004, 33, 32–42.
[113] (a) Nishibayashi, Y., Segawa, K., JD, Singh., Fukuzawa, S., Ohe, K., and Uemura, S. Novel chiral ligands, diferrocenyl dichalcogenides and their derivatives, for rhodium- and iridium-catalyzed asymmetric hydrosilylation. Organometallics 1996, 15, 370–9; (b) Braga AL, Paix˜ao MW, Lüdtke DS, Silveira CC, Rodrigues OED. Synthesis of new chiral aliphatic amino diselenides and their application as catalysts for the enantioselective addition of diethylzinc to aldehydes. Org Lett, 2003, 5, 2635–8; (c) Braga AL, Silva SJN, Lüdtke DS, Drekener RL, Silveira CC, Rocha JBT, Wessjohann LA. Chiral diselenide ligands for the asymmetric copper-catalyzed conjugate addition of Grignard reagents to enones. Tet Lett 2002, 43, 7329–31; (d) Adams E, Anderson JC, Cubbon R, James DS, Mathias JP. Origins of Enantioselectivity with Nitrogen-Sulfur Chelate Ligands in Palladium-Catalyzed Allylic Substitution. J Org Chem 1999, 64, 8256–62; (e) Braga AL, Lüdtke DS, Sehnem JA, Alberto EE. Modular chiral selenium-containing oxazolines: synthesis and application in the palladium-catalyzed asymmetric allylic alkylation.Tetrahedron 2005, 61, 11664–71; (f) Braga AL, Vargas F, Sehnem JA, Braga RC. Efficient synthesis of chiral beta-seleno amides via ring-opening reaction of 2-oxazolines and their application in the palladium-catalyzed asymmetric allylic alkylation. J Org Chem, 2005, 70, 9021–4.
[114] Nishibayashi, Y., Singh, JD., Segawa, K., Fukuzawaand, S-I., and Uemura, S. Rhodium(I)-catalysed asymmetric hydrosilylation of ketones using new diferrocenyl dichalcogenides (*R*, *S*)-{[$EC_5H_3CHMe(NMe_2)$]$Fe(C_5H_5)$}$_2$(E = S, Se, Te), as chiral ligands. J Chem Soc Chem Commun 1994, 1375–6.
[115] Godoi, M., Paix˜ao, MW., and Braga, AL. Chiral organoselenium-transition-metal catalysts in asymmetric transformations. Dalton Trans 2011, 40, 11347–55.

[116] (a) Sprinz, J., Kiefer, M., Helmchen, G., Reggelin, M., Huttner, G., Walter, O., and Zsolnai, L. Catalysis of allylic substitutions by Pd complexes of oxazolines containing an additional P, S, or Se Center. X-ray crystal structures and solution structures of chiral π-allyl palladium complexes of phosphinoaryloxazolines. Tet Lett 1994, 35, 1523–6; (b) Hiroi K, Suzuki Y, Abe I. (*S*)-Proline-derived new chiral ligands with phosphino, organosulfur or organoselenenyl functionality as an enantiocontrollable coordinating element. Tetrahedron: Asymmetry 1999, 10, 1173–88; (c) Braga AL, Paix˜ao MW, Marin G. Seleno-imine: A new class of versatile, modular N, Se ligands for asymmetric palladium-catalyzed allylic alkylation. Synlett 2005, 1675–8.

[117] Andrade, LH., Silva, AV., Milani, P., Koszelewski, D., and Kroutil, W. ω-Transaminases as efficient biocatalysts to obtain novel chiral selenium-amine ligands for Pd-catalysis. Org Biomol Chem 2010, 8, 2043–51.

[118] (a) Smith, JG. Synthetically useful reactions of epoxides. Synthesis 1984, 629–56; (b) Berkessel A, Gröger H. In: Asymmetric Organocatalysis; Wiley-VCH: Weinheim, 2005; (c) Yudin AK. In Aziridines and Epoxides in Organic Synthesis; Wiley-VCH: Weinheim, 2006; (d) Xia Q-H, Ge H-Q, Ye C-P, Liu Z-M, Su K-X. Advances in homogeneous and heterogeneous catalytic asymmetric epoxidation. Chem Rev 2005, 105, 1603–62; (e) McGarrigle EM, Myers EL, Illa O, Shaw MA, Riches SL, Aggarwal VK. Chalcogenides as organocatalysts. Chem Rev 2007, 107, 5841; (f) Wong OA, Shi Y. Organocatalytic oxidation. Asymmetric epoxidation of olefins catalyzed by chiral ketones and iminium salts. Chem Rev 2008, 108, 3958–87.

[119] (a) Rosen, T. In: Comprehensive Organic Synthesis, Trost BM, Fleming I, ed., Pergamom, Oxford, 1991, 409; (b) Newman MS, Magerlein BJ. In: Organic Reactions; Adams, R., Ed.; Wiley: New York, 1949, 413.

[120] Watanabe, S-i., Hasebe, R., Ouchi, J., Nagasawa, H., and Kataoka, T. Enantioselective Darzens reaction using organoselenide–Lithium hydroxide complexes. Tet Lett 2010, 51, 5778–80.

[121] (a) Braga, AL., Rodrigues, OED., Paixão, MW., Appelt, HR., Silveira, CC., and Bottega, DP. One-pot synthesis of new chiral sulfides and selenides containing oxazolidines: Catalyst in the enantioselective addition of diethylzinc to benzaldehyde. Synthesis 2002, 2338–40; (b) Bolm CM, Kesselgruber A, GrenzHermmanns N, Hildebrand JP. A novel ferrocenyl diselenide for the catalytic asymmetric aryl transfer to aldehydes. New J Chem 2001, 25, 13–5; (c) Schwab RS, Soares LC, Dornelles L, Rodrigues OED, Paixão MW, Godoi M, Braga AL. Chiral chalcogen peptides as ligands for the catalytic enantioselective aryl transfer reaction to aldehydes. Eur J Org Chem 2010, 3574–8.

[122] Soares, LC., Alberto, EE., Schwab, RS., Taube, PS., Nascimento, V., Rodriguesa, OED., and Braga, AL. Ephedrine-based diselenide: A promiscuous catalyst suitable to mimic the enzyme glutathione peroxidase (GPx) and to promote enantioselective C–C coupling reactions. Org Biomol Chem 2012, 10, 6595–9.

[123] (a) Bedford, RB., and Cazin, CSJ. Highly active catalysts for the Suzuki coupling of aryl chlorides. Chem Commun 2001, 1540–1; (b) Bedford RB, Hazelwood SL, Limmert ME. Extremely high activity catalysts for the Suzuki coupling of aryl chlorides: the importance of catalyst longevity. Chem Commun 2002, 2610–1.

[124] Saleem, F., Rao, GK., Kumar, A., Kumar, S., Singh, MP., and Singh, AK. Palladium(II) complexes bearing the 1,2,3-triazole based organosulfur/ selenium ligand: Synthesis, structure and applications in Heck and Suzuki–Miyaura coupling as a catalyst *via* palladium nanoparticles. RSC Adv 2014, 4, 56102–11.

[125] Saleem, F., Rao, GK., Singh, P., and Singh, AK. Chalcogen-dependent palladation at the benzyl carbon of 2,3-bis[(phenylchalcogeno)methyl]quinoxaline: Palladium complexes catalyzing Suzuki–Miyaura coupling *via* palladium–Chalcogen nanoparticles. Organometallics 2013, 32, 387–95.

[126] Dubey, P., Gupta, S., and Singh, AK. Base free *N*-alkylation of anilines with $ArCH_2OH$ and transfer hydrogenation of aldehydes/ketones catalyzed by the complexes of η^5-Cp*Ir(III) with chalcogenated Schiff bases of anthracene-9-carbaldehyde. Dalton Trans 2018, 47, 3764–74.

[127] Dubey, P., Gupta, S., and Singh, AK. Complexes of Pd(II), η^6-C_6H_6Ru(II), and η^5-Cp*Rh(III) with chalcogenated schiff bases of anthracene-9-carbaldehyde and base-free catalytic transfer hydrogenation of aldehydes/ketones and *N*-alkylation of amines. Organometallics 2019, 38, 944–61.

[128] (a) Schoenberg, A., and RF, Heck. Palladium-catalyzed amidation of aryl, heterocyclic, and vinylic halides. J Org Chem 1974, 39, 3327–31; (b) Martinelli JR, Clark TP, Watson DA, Munday RH, Buchwald SL. Palladium-catalyzed aminocarbonylation of aryl chlorides at atmospheric pressure: the dual role of sodium phenoxide. Angew Chem Int Ed 2007, 46, 8460–3; (c) Skoda-Foldes R, Takacs E, Horvath J, Tuba Z, Kollar L. Palladium-catalysed aminocarbonylation of steroidal 17-iodo-androst-16-ene derivatives in *N,N'*-dialkyl-imidazolium-type ionic liquids. Green Chem 2003, 5, 643–5; (d) Mane RS, Bhanage BM. Pd/C-catalyzed aminocarbonylation of aryl iodides *via* oxidative C-N bond activation of tertiary amines to tertiary amides. J Org Chem 2016, 81, 1223–8; (e) Tambade PJ, Patil YP, Bhanushali MJ, Bhanage BM. $Pd(OAc)_2$-catalyzed aminocarbonylation of aryl iodides with aromatic or aliphatic amines in water. Synthesis 2008, 2347–52.

[129] Chavan, SP., Dey, S., Jain, VK., and Bhanage, BM. Palladium(II) complex of 4-pyridylselenolate ligand: An efficient catalyst for aminocarbonylation of aryl and hetero aryl iodides with primary amines. Proc Natl Acad Sci India Sect A Phys Sci 2016, 86, 581–7.

[130] Saleem, F., Rao, GK., Kumar, S., Singh, MP., and Singh, AK. Complexes of (η^6-benzene) ruthenium(II) with 1,4-bis(phenylthio/seleno-methyl)-1,2,3-triazoles: Synthesis, structure and applications in catalytic activation of oxidation and transfer hydrogenation. Dalton Trans 2015, 44, 19141–52.

[131] (a) Gallego, D., Bruck, A., Irran, E., Meier, F., Kaupp, M., Driess, M., and JF, Hartwig. From bis (silylene) and bis(germylene) Pincer-type nickel(II) complexes to isolable intermediates of the nickel-catalyzed Sonogashira cross-coupling reaction. J Am Chem Soc 2013, 135, 15617–26; (b) Eberhard MR, Wangand Z, Jensen CM. Investigations into the Pd-catalysed cross-coupling of phenylacetylene with aryl chlorides: simple one-pot procedure and the effect of $ZnCl_2$ co-catalysis. Chem Commun 2002, 818–9; (c) Bolligera JL, Frecha CM. Highly convenient, clean, fast and reliable Sonogashira coupling reactions promoted by aminophosphine-based Pincer complexes of palladium performed under additive- and amine-free reaction conditions. Adv Synth Catal 2009, 351, 891–902; (c) Zhang J-H, Li P, Hub W-P, Wang H-X. Substituent effect of diimino-palladium (II) pincer complexes on the catalysis of Sonogashira coupling reaction. Polyhedron 2015, 96, 107–12; (d) Tyson GE, Tokmic K, Oian CS, Rabinovich D, Valle HU, Hollis TK, Kelly JT, Cuellar KA, McNamara LE, Hammer NI, Webster CE, Oliver AG, Zhang M. Synthesis, characterization, photophysical properties, and catalytic activity of an SCS bis(N-heterocyclic thione) (SCS-NHT) Pd pincer complex. Dalton Trans 2015, 44, 14475–82.

[132] (a) Scharf, A., Goldberg, I., and Vigalok, A. Evidence for metal–Ligand cooperation in a Pd–PNF pincer-catalyzed cross-coupling. J Am Chem Soc 2013, 135, 967–70; (b) Gu S, Chen W. Pincer complexes of palladium- and nickel-containing 3-butyl-1-(1,10-phenanthrolin-2-yl) imidazolylidene as efficient aqueous Sonogashira and Kumada coupling reactions. Organometallics 2009, 28, 909–14.

[133] Kumar, S., Saleem, F., Mishra, MK., and Singh, AK. Oxine based unsymmetrical (O^-, N, S/Se) pincer ligands and their palladium(II) complexes: Synthesis, structural aspects and applications as a catalyst in amine and copper-free Sonogashira coupling. New J Chem 2017, 41, 2745–55.

[134] Yao, Q., Kinney, EP., and Zheng, C. Selenium-ligated palladium(II) complexes as highly active catalysts for carbon-carbon coupling reactions: The Heck reaction. Org Lett 2004, 6, 2997–9.
[135] Sigeev, AS., Peregudov, AS., Cheprakov, AV., and Beletskaya, IP. The palladium slow-release pre-catalysts and nanoparticles in the "phosphine-free" Mizoroki–Heck and Suzuki–Miyaura reactions. Adv Synth Catal 2015, 357, 417–29.
[136] Mohammadi, E., and Movassagh, B. A polystyrene supported [PdCl–(SeCSe)] complex: A novel, reusable and robust heterogeneous catalyst for the Sonogashira synthesis of 1,2-disubstituted alkynes and 1,3-enynes. New J Chem 2018, 42, 11471–9.
[137] Rishu, Prashanth B., Bawari, D., Mandal, U., Verma, A., Choudhury, AR., and Singh, S. Hg(II) and Pd(II) complexes with a new selenoether bridged biscarbene ligand: Efficient mono- and bis-arylation of methyl acrylate with a pincer biscarbene Pd(II) precatalyst. Dalton Trans 2017, 46, 6291–302.
[138] Bhaskar, R., Sharma, AK., and Singh, AK. Palladium (II) complexes of N-heterocyclic carbene amidates derived from chalcogenated acetamide-functionalized 1*H*-benzimidazolium salts: Recyclable catalyst for regioselective arylation of imidazoles under aerobic conditions. Organometallics 2018, 37, 2669–81.

Eder J. Lenardão, Laura Abenante, Angelita M. Barcellos, Gelson Perin, Diego Alves, Marcio S. Silva, Daniela Hartwig and Raquel G. Jacob

6 Synthesis of organoselenium compounds using nonconventional reaction media

Keywords

green chemistry, organoselenium, solvent effects, microwaves, ultrasound, mechanochemistry, solvent-free, water

6.1 Introduction

Considering that solvents are responsible for up to 90% of mass utilization in the pharmaceutical and fine chemical industries [1], the use of nonvolatile and renewable solvents is a hot topic for a greener and sustainable organic synthesis [2–19]. To reduce the impact of chemical processes on the environment, another prevailing trend is the development of efficient reaction conditions by using alternative sources of energy such as microwave irradiation, sonochemistry, mechanochemistry, and electrochemistry [20].

In the last decade, these concepts in modern chemistry have been widely explored for the synthesis of organochalcogen compounds [21, 22]. The synthetic applicability and the biological activities of organochalcogen compounds, mainly those containing selenium atoms, are well documented in several review articles [23–35] and books [36–43]. Besides, aspects regarding the application of organoselenium compounds as reagents and catalysts with a focus on the development of green protocols have been discussed in three chapters of this book [44–46]. In this chapter, a comprehensive and updated review on recent green alternative methods available for the synthesis of organoselenium compounds is presented and discussed. We focus on the use of safe solvents, such as water, polyethylene glycol, glycerol, ionic liquids (ILs) or solvent-free conditions, as well as alternative sources of energy to promote the reactions.

Considering the large number of synthetic methodologies described, for a better discussion this chapter was divided into 14 sections, according to the chemical class of the synthesized organoselenium compounds: (1) diorganyl diselenides; (2) selenoethers; (3) β-seleno amines; (4) β-oxy selenides; (5) seleno ketones; (6) selenoesters; (7) vinyl selenides; (8) bis-organoselanyl alkenes; (9) selenoalkynes;

Eder J. Lenardão, Laura Abenante, Angelita M. Barcellos, Gelson Perin, Diego Alves, Marcio S. Silva, Daniela Hartwig and Raquel G. Jacob, *Laboratório de Síntese Orgânica Limpa – LASOL – CCQFA, Universidade Federal de Pelotas,* Pelotas, Brazil

https://doi.org/10.1515/9783110625110-006

(10) seleno-functionalized heterocycles; (11) selenium-containing heterocycles; (12) selenoxides and selenones; (13) organophosphorus selenides; and (14) diselenocarbamates.

6.2 Diorganyl diselenides

Diorganyl diselenides have attracted special interest due to the large number of chemical and biological applications of these compounds. They act as possible intermediates in some biologically important process, such as the catalytic activity of glutathione peroxidase, a selenium-containing enzyme (Figure 6.1) [27, 47]. From a synthetic point of view, the ease of reduction of the Se–Se bond in diselenides by homo- or heterolytic cleavage and the high reactivity of the generated species (radical, electrophile, and nucleophile) are reasons for a wide use of diselenides to introduce the selenium moiety in organic compounds or to catalyze organic transformations [43, 44, 48, 49]. In general, dichalcogenides are relatively more stable in organic reactions such as oxidation, acylation, and alkylation, compared to the corresponding free species [50]. Besides these, environmental and safety aspects were responsible to encourage the development of various methods for the preparation of dichalcogenides.

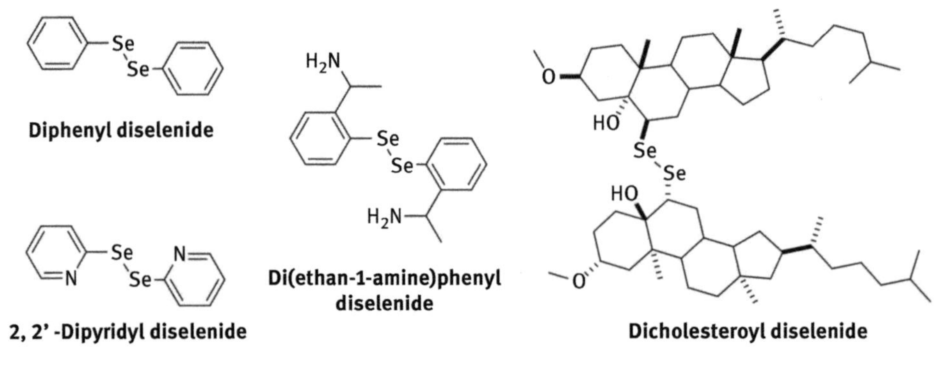

Figure 6.1: Some examples of diselenides with antioxidant activity.

Despite the various traditional approaches for the synthesis of diselenides, Hu et al. [51] described water as a solvent in the alkylation of metal diselenides. By this method, a variety of dialkyl and dialkenyl diselenides **1** were obtained in good yields and short reaction times. First, hydrazine hydrate was used as a reducing agent to generate the diselenide dianion from elemental selenium in basic solution. Then, alkyl halides **2** and tetrabutyl ammonium bromide (TBAB) were added to the reaction flask under nitrogen atmosphere. Further, the authors reported that the phase-transfer catalyst TBAB was essential in this procedure, since no reaction occurred in its absence (Scheme 6.1).

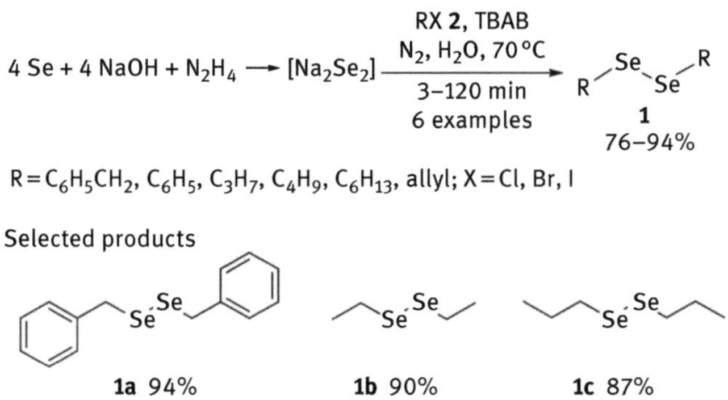

Scheme 6.1: Synthesis of dialkyl and dialkenyl diselenides **1** [51].

A new methodology to prepare symmetrical disulfides **3** and diselenides **1** was proposed by Li et al. [52]. This procedure involves a copper-catalyzed coupling reaction in water between aryl halides **2** and elemental sulfur or selenium. As described earlier, the reaction was inefficient without the addition of phase transfer agent [$(^n\text{Bu})_4$NF]. Generally, aryl iodides gave slightly higher yields than their bromo analogs; however, this protocol allowed the preparation of the corresponding disulfides **3** and diselenides **1** in good to excellent yields (76–96%) after 24 h at 100 °C (Scheme 6.2).

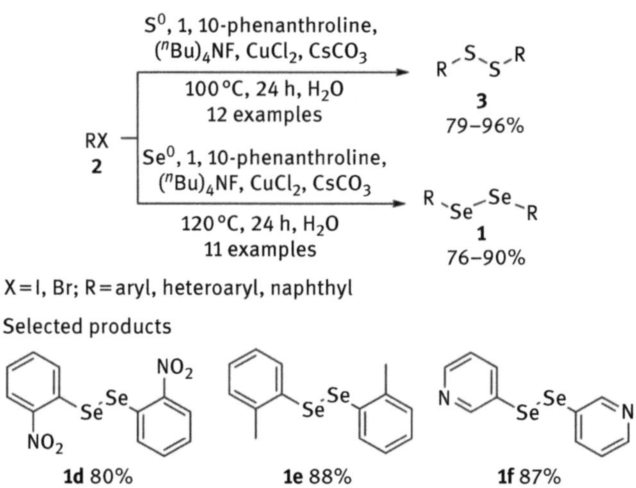

Scheme 6.2: Synthesis of disulfides **3** or diselenides **1** using elemental chalcogenium [52].

Recently, another method to synthesize dialkyl dichalcogenides (S, Se) was described in aqueous media under catalyst-free conditions [53]. This one-pot and efficient procedure transformed benzylic, allylic, and primary halides **2** and tosylates **4** with elemental sulfur and selenium into organochalcogen compounds. Dialkyl dichalcogenides (Se **1**, S **3**) were obtained as the only products in good yields and short reaction times from primary, benzylic, and allylic halides after easy workup (Scheme 6.3). Due to steric hindrance, trisulfides (S–S–S) and triselenides (Se–Se–Se) were obtained from secondary and tertiary halides under the same conditions.

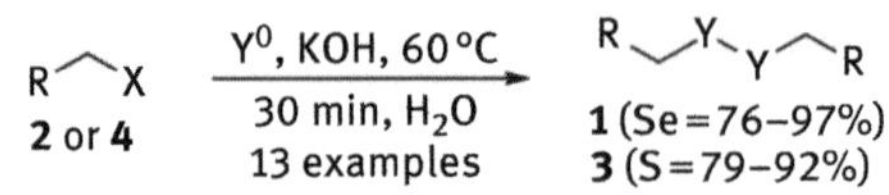

R = phenyl, alkyl, allyl; X = Br, Cl, OTs

Selected products

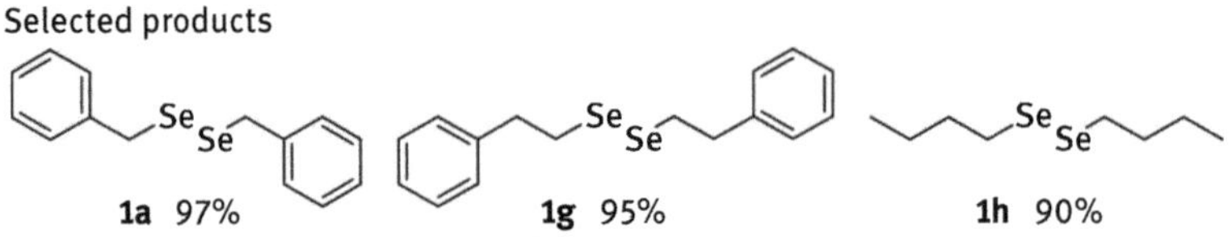

Scheme 6.3: Synthesis of dialkyl and dialkenyl diselenides **1** and disulfides **3** [53].

Kholshin et al. [54] reported a convenient method for the synthesis of 3-(4-hydroxyaryl)propyl selenosulfates and the corresponding diselenides **1** by the reaction between Na_2SeSO_3 and bromopropyl-substituted phenol **2** in 50% aqueous ethanol (Scheme 6.4). First, the selenium reactive species is formed in situ from selenium and Na_2SO_3 and the nucleophilic selenium species attacks bromide **2**. The hydrolysis of the alkyl selenosulfate intermediate **5** under the reaction conditions affords the corresponding diselenides **1**. To improve the yield of compound **1**, the authors proposed an azeotropic removal of ethanol in the final step. Additionally, to explore the reduction of the Se–Se bond, symmetrical and unsymmetrical selenides (derivatives of alkylated phenols and pyrocatechol) were synthesized.

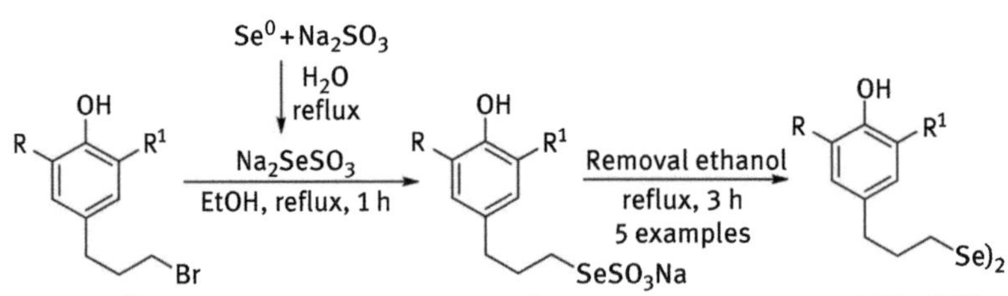

$R = {}^tC_4H_9, CH_3, Br, H; R^1 = {}^tC_4H_9, CH_3, H$

Scheme 6.4: Synthesis of selenosulfates **5** and diselenides **1** [54].

6.3 Selenoethers

Organoselenium and organosulfur compounds are part of a range of biologically active molecules, participating in multiple therapeutic functions of great importance, including as anticancer and antiviral agents and in a variety of situations where free radicals are involved [30]. Specifically, diorganyl chalcogenides are versatile tools in organic synthesis due to their applicability in asymmetric catalysis [55] and their usefulness in the generation of reactive organometallics, which are used in the preparation of bioactive compounds [56]. Their high versatility in synthesis is linked to the ease of introduction of the organochalcogen moiety as either electrophiles or nucleophiles in other organic molecules [37, 38].

In this context, a number of green procedures for the synthesis of a variety of seleno- and thioethers have been developed. In general, the transition metal-mediated aryl–chalcogen bond formation is the more common protocol, which mainly includes metals such as copper, palladium, and zinc, as well as methods involving substitution reactions [23]. In the last years, some alternatives to the traditional synthetic strategies to access chalcogenoethers have been reported, and they will be discussed below.

In 2009, Singh et al. [57] reported the cross-coupling of aryl or alkyl bromides **2** with diaryl diselenides **1**, catalyzed by copper oxide nanopowder (CuO NPs, 0.5 mol%), using [bmim][BF_4] as the solvent and KOH (2.0 equiv.) as a base (Scheme 6.5). In this article, unsymmetrical diorganyl selenoethers **6** were isolated from moderate to good yields (70–82%) after 1 h at room temperature. The results show that when alkyl bromides were used as starting materials, lower yields were obtained compared to the aryl ones. However, the reaction was not sensitive to electronic effects in the diaryl diselenide. Furthermore, to verify the possibility of reusing [bmim][BF_4], after the reaction between 1-bromo-4-methylbenzene and diphenyl diselenide was complete, the IL was removed by filtration and used directly in the following four coupling reactions, affording the respective selenoether **6** in 82%, 82%, 80%, and 78% yields.

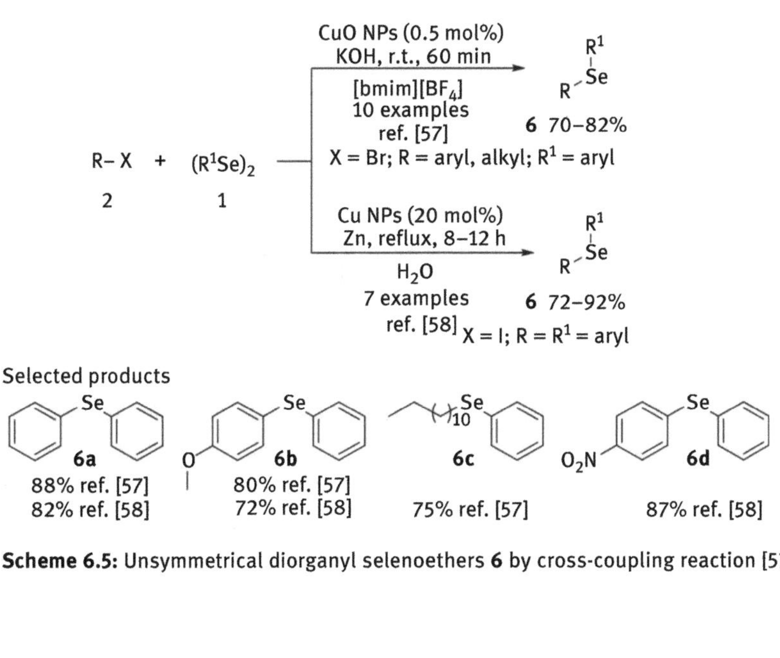

Scheme 6.5: Unsymmetrical diorganyl selenoethers **6** by cross-coupling reaction [57, 58].

Similarly, Saha et al. [58] described the use of Cu(0) NPs (20 mol%) in the presence of zinc dust (1.5 equiv.) to access selenoethers **6** by cross-coupling of diphenyl diselenide **1i** with aryl iodides **2** (Scheme 6.5) or vinyl bromides (described in Section 6.8). In this green method, besides the ligand-free conditions, the reaction was promoted using water as the solvent. The phenylselenylation reaction shows general applicability and compatibility with different functionalities in the aryl iodides, such as OCH_3, CO_2H, CO_2CH_3, CF_3, NO_2, and Cl, affording the desired compounds in 72–92% yields under reflux. Interestingly, due to the agglomerating tendency of the catalyst under the reaction conditions, the remaining Cu NPs were reused with good performance in only three successive runs (88%, 79%, and 78% yields).

Regardless of the low atom economy, the substitution of alkyl and benzyl halides with chalcogenolate anions is an easy and simple strategy to prepare selenoethers. Zinc organoselenolates were used in efficient substitution reactions of alkyl and benzyl halides **2** (Cl, Br, and I) (Scheme 6.6) [59, 60]. The reactive species of selenium was generated in situ from the reaction between diphenyl diselenide **1i** and Zn dust using [bmim][BF_4] as the solvent. Through this Lewis acid-free procedure, a variety of unsymmetrical diorganyl selenides were prepared from good to excellent yields in a few minutes. The procedure shows a large tolerance to different functional groups, such as protected aminoester, nitrile, ester, and allyl, affording the corresponding products **6** in acceptable yields. When the sterically hindered 2-methoxybenzenoselenolate reacted with benzyl chloride, however, a lower yield of the desired selenoether was obtained (45% yield) after 45 min of reaction. Moreover, the IL was recovered and reused in five successive reactions between diphenyl diselenide and benzyl chloride, giving the product in good yields, with a little decrease only after the fourth reaction.

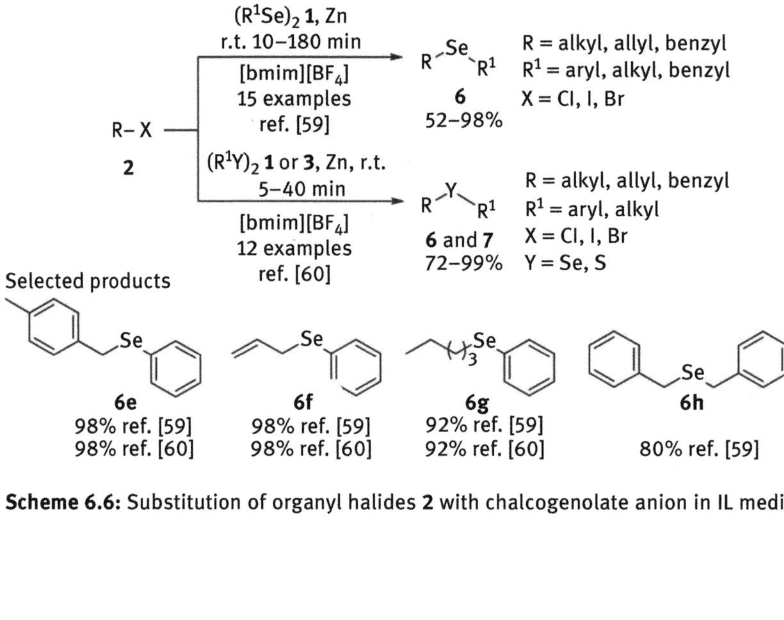

Scheme 6.6: Substitution of organyl halides **2** with chalcogenolate anion in IL media [59, 60].

Under the same experimental conditions described above to prepare selenoethers **6**, the Zn/[bmim][BF_4] system was successfully applied in the synthesis of structurally diverse diorganyl sulfides **7** (Scheme 6.6) [60]. Several alkyl and arylthiolates reacted with benzyl, alkyl, and allyl halides to give the respective unsymmetrical thioethers in 72–99% yields at room temperature in short reaction times.

In the search for new procedures to prepare unsymmetrical diorganyl selenides **6** using an alternative and recyclable solvent, Braga et al. [61] developed a procedure that uses InI as a reducing agent for the Se–Se bond cleavage of diaryl or dialkyl diselenides **1** in [bmim][BF_4] as the solvent (Scheme 6.7). Several diaryl and alkyl arylselenoethers **6** were obtained in 52–97% yields by the reaction between alkyl, allyl, and benzyl halides (Cl, I, and Br) **2** and diselenides **1**. In this study, no reaction occurred when tertiary halides **2** were used, and a significant decrease in yield was observed when the electron-rich 1,2-bis(2-methoxyphenyl)diselenide was used. Similar to the other procedures using IL, it was successfully reused in five successive reactions of diphenyl diselenide with benzyl chloride, giving the desired selenide in 90%, 85%, 82%, 77%, and 76% yields, consecutively.

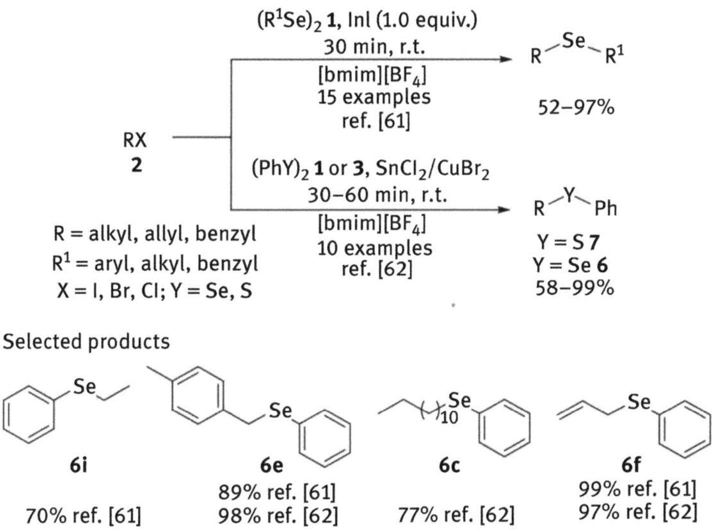

Scheme 6.7: Synthesis of unsymmetrical diorganyl selenides **6** and sulfides **7** [61, 62].

The same group [62] published the use of a mixture of tin(II) and copper(II) salts ($SnCl_2/CuBr_2$) in [bmim][BF_4] in the efficient synthesis of diorganyl selenides **6** and sulfides **7** from the corresponding dichalcogenides **1** or **3** and aryl or alkyl halides **2** (Scheme 6.7). This bimetallic system (1.2 equiv. of $SnCl_2$ and 0.1 equiv. of $CuBr_2$) was responsible for the reductive cleavage of the Y–Y bond (Y = S or Se) in the dichalcogenide. After the chalcogenolate anion formation and their reaction with the halides, the unsymmetrical selenoethers **6** and thioethers **7** were isolated in 65–99% and 58–98% yields after 30 and 60 min, respectively. Additionally, by this procedure it was possible to synthesize more complex molecules, such as chiral β-sulfur and β-selenoamines, in good yields (62–75%) (described in Section 6.4). As expected, the IL was reused in successive reactions without significant loss in yields after four successive reactions.

Due to the considerable interest in the development of new efficient strategies to prepare chalcogenoethers **6**, **7** or **8** with diverse patterns of substitution, the use of substrates alternative to organyl halides has been investigated. For example, Alves and coworkers [63] reported the use of boronic acids in the copper-catalyzed cross-coupling reaction to prepare selenoethers **6** (Scheme 6.8). This protocol worked well with several diaryl diselenides **1** and aryl boronic acids **9**, affording the unsymmetrical diaryl selenides **6** from good to excellent yields under green conditions. Specifically, 4-methoxyphenylboronic acid efficiently reacted with a range of diaryl diselenides **1** containing electron-withdrawing and electron-donating groups at the aromatic ring to give the respective products in good yields (76–89%). Also, *ortho*- and *para*-bromo-substituted arylboronic acids were evaluated and the corresponding selenoethers **6** were isolated in 82% and 86% yields, respectively. These

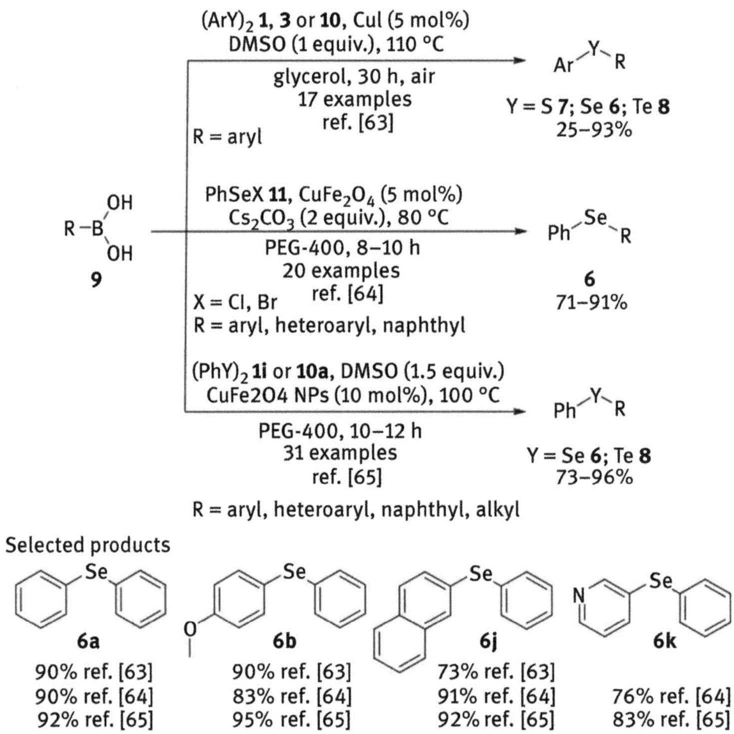

Scheme 6.8: Preparation of chalcogenoethers **6**, **7** or **8** [63–65].

results highlight that there is a differentiation on the reactivity between boron and bromine atoms, because only coupling products at boron moiety were formed, without by-products. In addition, the reaction between 4-methoxyphenylboronic acid and diphenyl ditelluride afforded the telluroether **8** in 93% yield.

The magnetically separable $CuFe_2O_4$ NPs were used as catalyst in the cross-coupling reaction between heteroaryl or arylboronic acids **9** and phenylselenyl bromide and chloride **11**, using polyethylene glycol-400 (PEG-400) as the solvent (Scheme 6.8) [64]. By this procedure, unsymmetrical diaryl selenides **6** were prepared in good yields (71–91%), with a little influence of electronic effects. In this work, the reuse of $CuFe_2O_4$ was easily performed up to four new reactions with good activity (82–88% yields). After each reaction of phenylselenyl bromide **11** with phenylboronic acid **9**, the catalyst was magnetically separated, washed with solvents, dried under vacuum, and reused in another reaction.

In a closely related work, the $CuFe_2O_4$ NPs/PEG-400 system was employed in the synthesis of seleno- and telluroethers **6** and **8** by coupling reaction of diphenyl dichalcogenides (Se and Te) **1i** or **10a** with a range of organoboronic acids **9** (Scheme 6.8) [65]. This atom-economic and base-free procedure was carried out with several functionalized organoboranes **9** containing alkyl, styryl, and phenylethynyl groups. The authors also extended the scope of the reaction to aryl trifluoroborates (80–92% yields after 10–12 h) and boronic acid pinacol esters (71–92% yields after 16–18 h) with similar efficiency. Interested in the biological potential as an antioxidant agent [66], the authors have prepared the telluride derivative of 4,4′-biphenyl diboronic acid, which was isolated in 85% yield. At the end of the reactions, the catalyst was easily collected by a magnetic rod from the reaction medium and washed with solvents and was reused. $CuFe_2O_4$ was used for up to seven additional reactions without any considerable loss of activity (the yields remained above 80%).

In parallel to these studies, Kumar et al. [67] have prepared symmetrical selenoethers and thioethers by the copper-catalyzed cross-coupling of aryl halides with potassium chalcogenocyanates in the presence of a base. This green alternative involves the reaction of aryl iodides or bromides **2** with potassium selenocyanate **12** in water, using CuI (10 mol%) as a catalyst, $CsCO_3$ as a base, and *trans*-1,2-diaminocyclohexane as a ligand (Scheme 6.9). Several symmetrical diaryl selenides **6** were prepared from moderate to good yields, with aryl iodides **2** giving better yields than the aryl bromides **2** analogues. Electron-rich aryl halides were slightly more reactive than the electron-poor ones.

In 2011, Zhao et al. [68] reported a clean procedure to prepare selenoethers **6**, by the Pd-catalyzed cross-coupling reaction of aryl- or alkyl halides **2** with aryltributylstannyl selenides **13** using the IL [bmim][PF_6] as the solvent (Scheme 6.10). In this work, aryl, alkyl, and benzyl bromides or iodides were efficiently used. However, in reactions with bromides, long reaction times, higher temperature (110 vs. 80 °C), and catalyst loading (10 mol%) were required. While 1-chlorooctane was not reactive

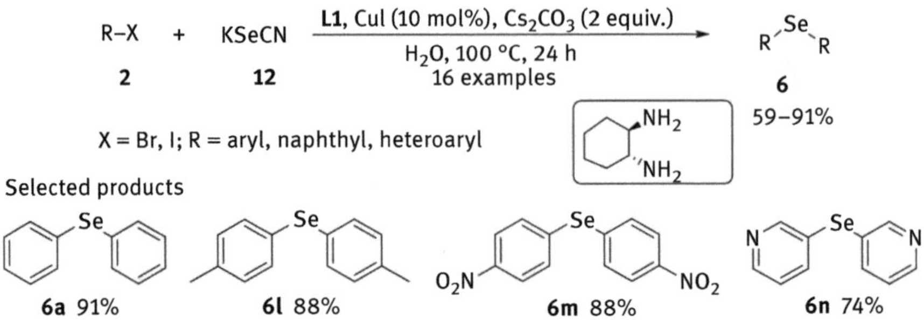

Scheme 6.9: Selenoethers **6** by the reaction between potassium selenocyanate **12** and aryl halides **2** [67].

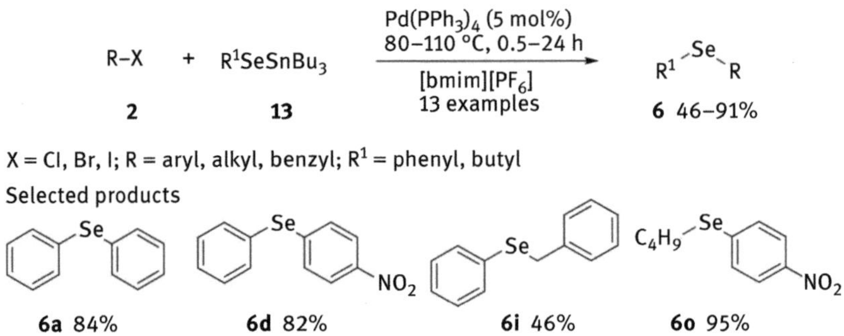

Scheme 6.10: Unsymmetrical diorganyl selenoethers **6** [68].

under these conditions, the benzyl and phenyl chlorides afforded poor yields of the desired compounds, even in reactions at 110 °C (46% and 55% yields, respectively). Butyl tributylstannyl selenide reacted with aryl or benzyl halides giving excellent yields of the isolated product (80–95%) in reaction times up to 2 h at 80 °C. Further, the IL/$Pd(PPh_3)_4$ system was recycled and used again in the reaction between 3-iodotoluene and $PhSeSnBu_3$. Selenoethers were obtained in excellent yields (89–91%) after six successive reactions, without adding more catalyst.

In 2011, Freitas et al. [69] developed a new catalyst- and metal-free strategy to obtain unsymmetrical diaryl selenides **6** (Scheme 6.11). The reaction was conducted at room temperature using electrophilic species of selenium and aryl boronic acids **9** or potassium aryltrifluoroborates **14** as nucleophiles in ILs, such as [bmim][PF_6] and [bmim][BF_4]. Arylboronic acids **9** and arylselenium chlorides **11** were used in the presence of [bmim][PF_6] as the solvent, affording the respective diarylselenides **6** in 75–96% yields. The reaction of potassium aryltrifluoroborates **14** and ArSeBr in [bmim][BF_4] afforded a series of diaryl selenides **6** in 73–88% yields after 3 h of reaction. In addition, the ILs can be recovered and reused for five successive reactions after a simple extraction of the product with hexanes.

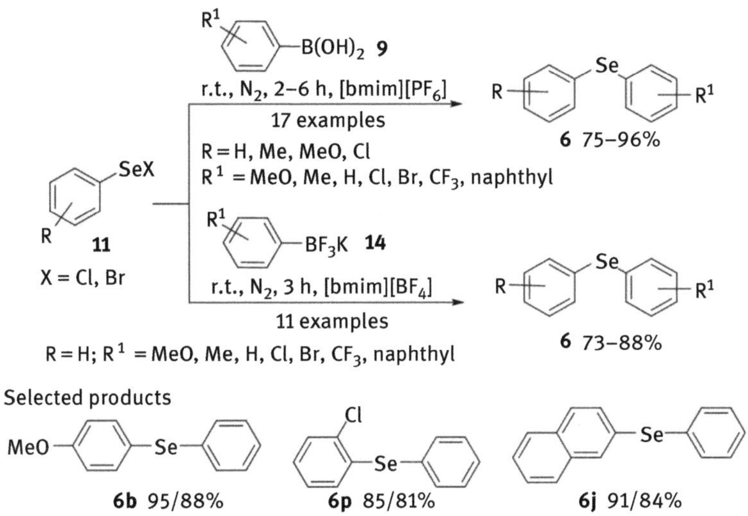

Scheme 6.11: Synthesis of diaryl selenides in ionic liquids [69].

In 2012, Banerjee et al. [70] described the metal-free synthesis of unsymmetrical selenides **6** using the IL [pmim][Br] (Scheme 6.12). The unsymmetrical diorganyl selenides **6** were obtained after 2–6 h, at 75 °C in 70–86% yields. The unsymmetrical sulfide analogues **7** were formed in 70–88% yields at 75 °C after 1.5–5.5 h of reaction. The cleavage of the chalcogen–chalcogen bond of diphenyl diselenide or disulfide was promoted by [pmim]Br/PPh_3 system and the resulting chalcogenolate anions reacted efficiently with various benzyl and allyl halides **2**.

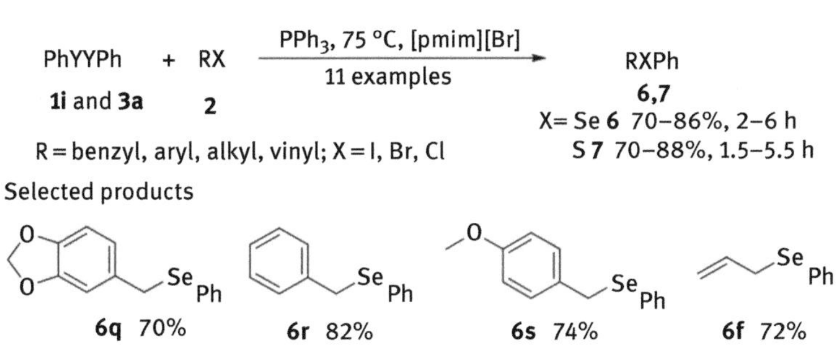

Scheme 6.12: Synthesis promoted by PPh_3/ionic liquid [70].

The authors proposed an addition–elimination mechanism for the reaction, in which PPh_3 reacts with diphenyl diselenide or disulfide to form the intermediate **I**

(Scheme 6.13). The intermediate **I** is cleaved generating the selenolate or thiolate anion, which on reaction with the halide **2** afforded the respective products **6** (Y = Se) or **7** (Y = S).

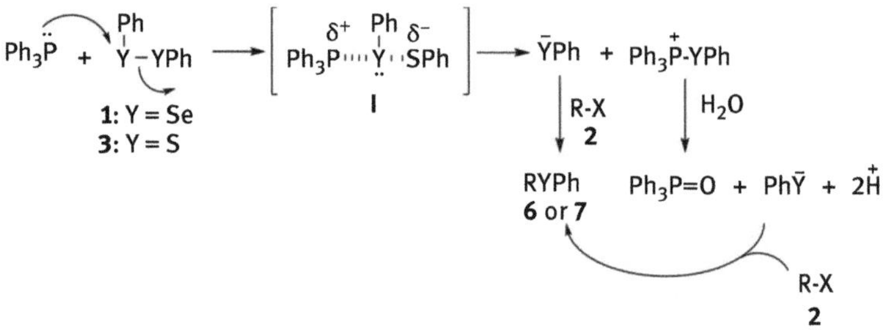

Scheme 6.13: Mechanism proposed by Banerjee et al. [70].

Mukherjee et al. [71] developed a new solvent-, ligand-, and metal-free protocol to prepare unsymmetrical diaryl chalcogenides through the reaction between diazonium tetrafluoroborates **15** and diaryl dichalcogenides **1, 3,** and **10**, in the presence of KOH (0.75 equiv.) in neutral alumina under ball-milling (6 balls, 600 rpm, Scheme 6.14). Unsymmetrical diaryl selenides **6** were prepared after 15–20 min at room temperature in 73–78% yields (Scheme 6.14). Also, unsymmetrical diaryl sulfides **7** and diaryl tellurides **8** were prepared in similar yields.

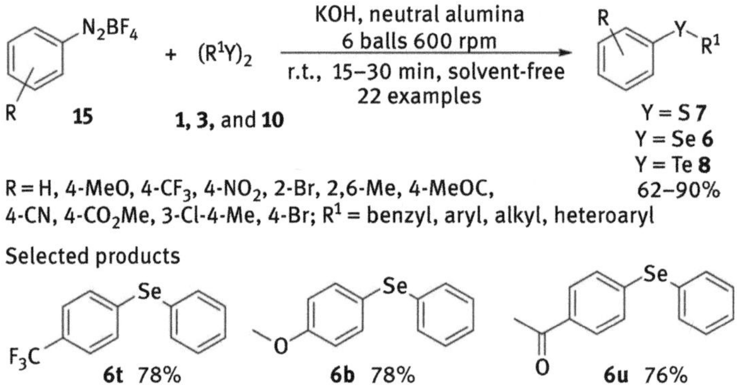

Scheme 6.14: Synthesis of unsymmetrical diaryl chalcogenides under ball-milling [71].

More recently, in 2016, Kumar et al. [72] reported a metal-free methodology to obtain symmetrical diaryl selenides **6** in moderate to good yields (67–82%) using SeO_2 and boronic acids **9** in PEG-400. The reaction was conducted in the presence of a base, K_2CO_3, for 3 h at 110 °C and various selenides **6** were prepared (Scheme 6.15).

This C–Se cross-coupling was effective for all the tested aromatic boronic acid substrates, including aryl bearing electron-withdrawing and electron-donating groups, as well as heteroaryl ones.

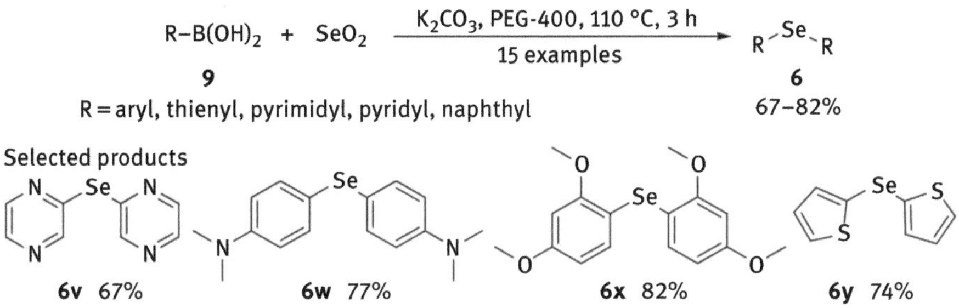

Scheme 6.15: Synthesis proposed by Kumar et al. [72].

In the same year, Navarro and coworkers [73] reported the preparation of symmetrical organochalcogenides **6**, **7**, and **8** through an electrochemical synthesis in aqueous NaOH medium (Scheme 6.16). The reaction proceeded in two steps. In the first step, selenide (Se^{2-}), sulfide (S^{2-}), and telluride (Te^{2-}) anions were generated in aqueous medium in an electrochemical cell under argon atmosphere. After the mixture became colorless, the halogenated compounds **2** were added and the reaction was left under argon, at room temperature for 12 h. Moreover, the formation of the respective dichalcogenides as by-products was also observed, but the monochalcogenides **6**, **7**, and **8** remained as the major products. When the telluride ion was employed, the respective products **8** were obtained in moderate to high yields (64–95%), while the selenide and sulfide were less reactive, giving lower yields of the respective chalcogenides (5–96% and 5–59%, respectively).

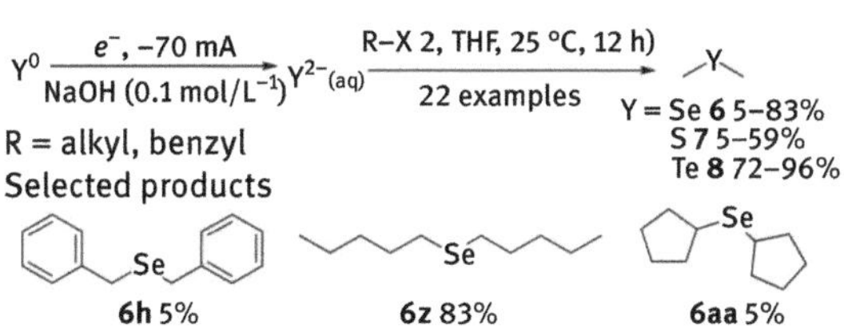

Scheme 6.16: Electrochemical synthesis [73].

In 2018, an alternative solvent- and metal-free protocol to obtain diaryl chalcogenides in good to excellent yields was reported. Rodrigues et al. [74] described the use of nontoxic and easily available KIO_3 to catalyze the chalcogenations (S and Se) of several (hetero)arenes **16** in the presence of ethylene glycol (4 equiv.) as an additive. Through this direct C(sp^2)-H bond chalcogenation, several diaryl selenides **6** and diaryl sulfides **7** were obtained in 52–99% and 30–80% yields, respectively, after 3 h at 110 °C (Scheme 6.17). Moreover, the synthesized compounds have a pharmaceutical interest as anti-Alzheimer agents.

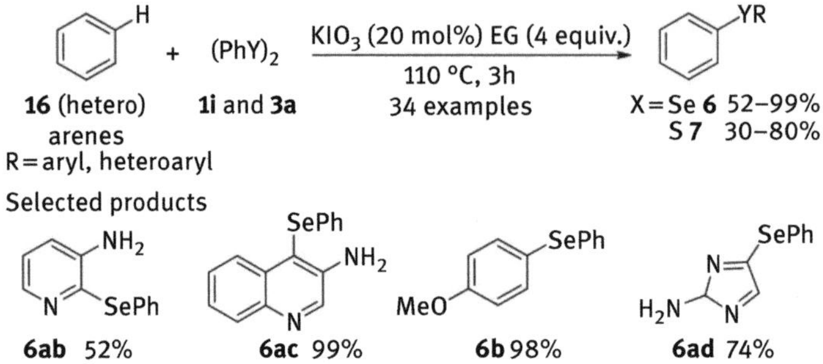

Scheme 6.17: Synthesis catalyzed by KIO_3 [74].

In addition, the authors reported two possible mechanisms for the reaction, which could be contemporary. In the first proposal (Scheme 6.18), the species **I** attacks KIO_3 to form the intermediate **II,** from which is generated the species **III.** This intermediate loses ^{-}OH to form the intermediate **IV,** which undergoes a homolytic cleavage forming the radical **V.** In the sequence, **V** reacts with diselenide, forming the product **6** and the species **VI** that, after reaction with KOH, regenerates the catalyst KIO_3.

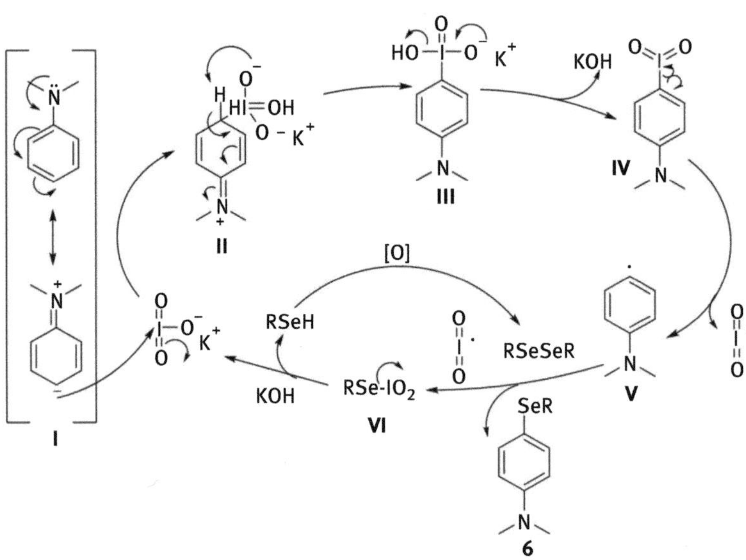

Scheme 6.18: Possible mechanism of the reaction – suggestion 1 [74].

The second proposed mechanism differs from the previous one in that intermediate **IV'** reacts with diselenide to give species **V** and selenolate anion. In the sequence, **V** reacts with KOH to form product **6**, regenerating the catalyst KIO_3 (Scheme 6.19).

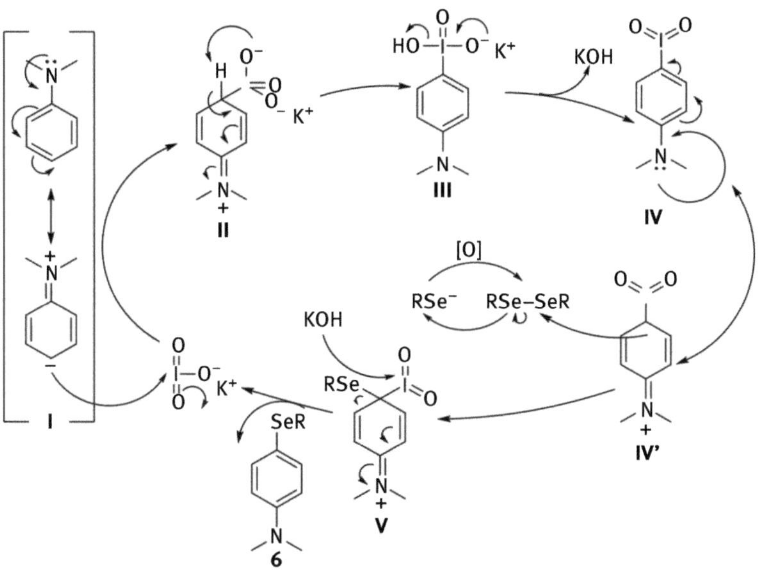

Scheme 6.19: Possible mechanism of the reaction – suggestion 2 [74].

More green and alternative methods to prepare densely functionalized selenoethers were also reported in the literature. For example, Silveira et al. [75] developed a procedure for the functionalization of nitrogen-containing heterocycle, generating the selenoether **6ae** (Scheme 6.20). In this study, the use of morpholine **17a** in a solvent-free reaction with phenylseleno acrylonitrile **18a**, provided the corresponding phenylseleno-β-amino nitriles **6ae** in 82% yield after 5 min.

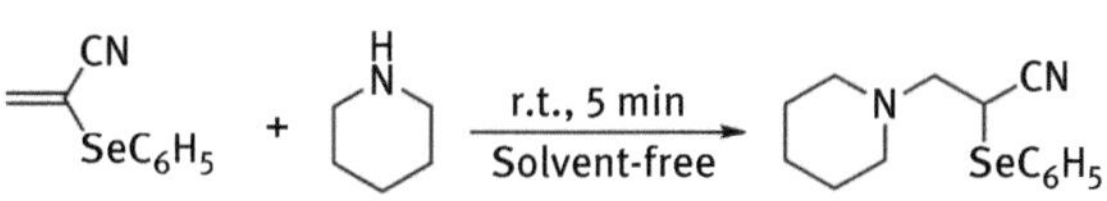

18a **17a** **6ae** 82%

Scheme 6.20: Synthesis of selenoether **6ae** under solvent-free conditions [75].

More recently, a new strategy to obtain selenides from a natural and eco-friendly source was reported by Jacob et al. [76]. Castor bean oil was used to prepare compound **19**, which was used as a starting material in the synthesis of (*Z*)-12-organylselenooctadec-9-enoates **20** in good yields (Scheme 6.21). These semisynthetic selenoethers were formed by the nucleophilic substitution of the tosylate derived from methyl ricinoleate by arylselenolate anions, using PEG-400 as the solvent.

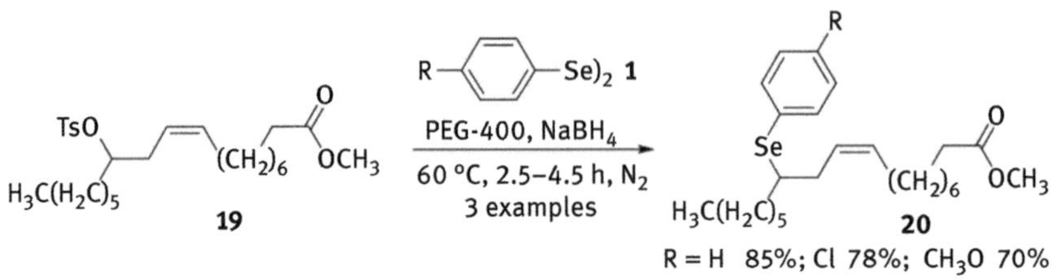

Scheme 6.21: Synthesis proposed by Jacob et al. [76].

In 2018, an ultrasound-promoted cyclization procedure to obtain 2-organoselanylnaphthalenes **6** was described by Perin et al. (Scheme 6.22) [77]. Using water as the solvent, selenoethers **6** were prepared by the reaction between diorganyl diselenides **1** and alkynols **21** in the presence of Oxone®. Specifically, Oxone® promoted the oxidative cleavage of the Se–Se bond of diselenides to form the electrophilic species of selenium in situ. In this paper, the best reaction condition was extended to different substrates, using electron-donor and electron-withdrawing groups attached to the aromatic ring of the diselenides and the alkynols. Aliphatic diselenides **1** and alkynols **21** were also good substrates for the reaction, and the respective 2-organoselanylnaphthalenes **6** were isolated in moderate to excellent yields (56–94%).

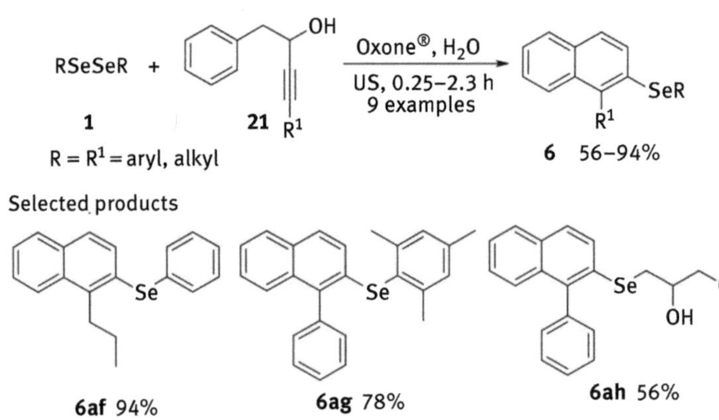

Scheme 6.22: Synthesis promoted by US, using Oxone® [77].

Further, the authors proposed a possible mechanism for the reaction, in which the oxidative cleavage of diphenyl diselenide **1i** to form the intermediates **I** and **II** was the first step (Scheme 6.23). Then, the electrophilic species **I** reacts with the alkynol **21**, forming the intermediate **III**. At this point, a 6-*endo-dig* cyclization takes place, generating the species **IV**. After having restored the aromaticity and losing a molecule of water, the product **6ai** is formed.

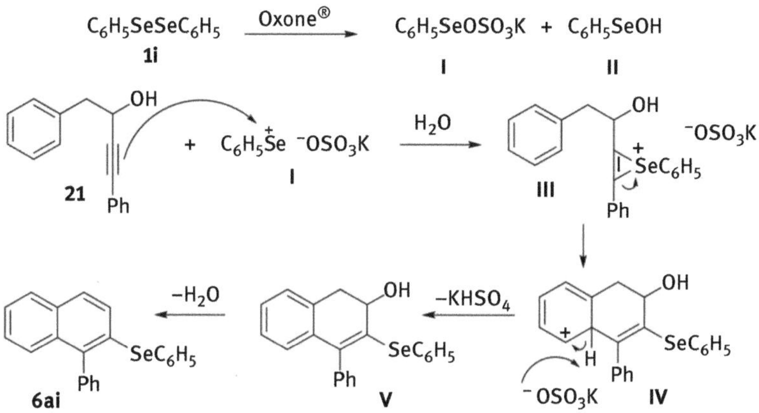

Scheme 6.23: Mechanism proposed by Perin et al. [77].

The same group [78] reported the synthesis of selenoether derivatives **6** of glycerol, by the reaction of the diselenide derived from glycerol **1j** and organyl halides **2** (Scheme 6.24). At first, bis(2,2-dimethyl-1,3-dioxolanylmethyl) diselenide **1j** was reduced by the sodium borohydride ($NaBH_4$)/PEG-400 system under Ar atmosphere, followed by the addition of a variety of alkyl, allyl, and benzyl halides. In all cases, the products **6** were obtained in acceptable yields (53–90%) after short reaction times.

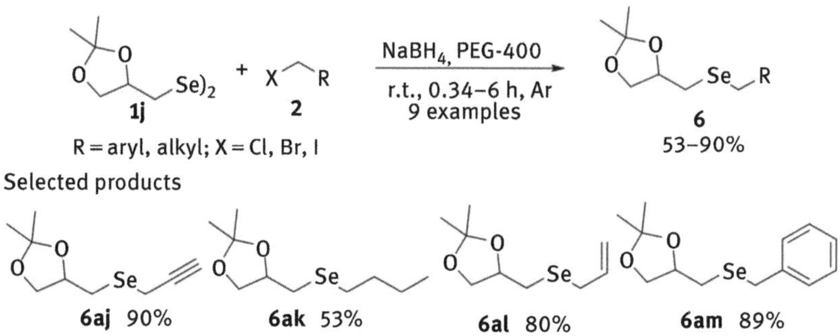

Scheme 6.24: Glycerol-derivatives selenoethers [78].

Further, the authors reported glycerol derivatives containing pyridyl selenides **6**, starting from 2,2′-dipyridyl diselenide **1k** [79] and tosyl solketal **4a** or carbonate **4b**. Selenoethers **6** were isolated in acceptable yields after 2–3 h of reaction at 50 °C. Compounds **6an**–**6ap** showed antioxidant and anticholinesterase properties (Scheme 6.25).

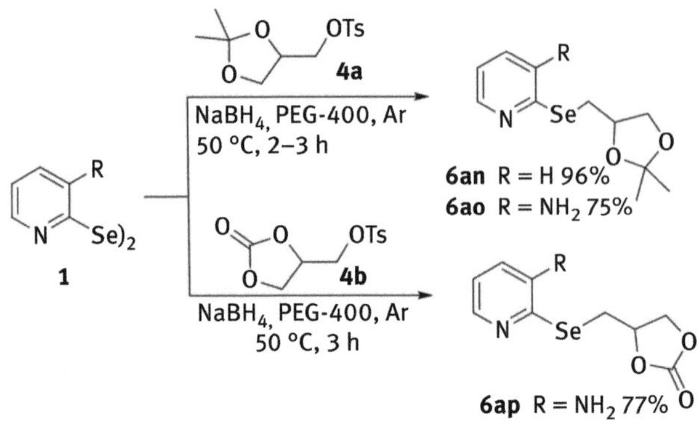

Scheme 6.25: Selenoether glycerol derivatives [79].

6.4 β-Seleno amines

The studies of the chiral β-chalcogen amines has increased due to the versatility of these compounds as chiral ligands in enantioselective reactions and also as important synthetic targets, such as cysteine and selenocysteine derivatives [80, 81]. Additionally, synthetic analogs of this class exhibit interesting biological activities, including antioxidant, antimicrobial and antitumor ones [28, 82]. The nucleophilic ring-opening reaction of aziridines is the conventional method for the straightforward preparation of chiral β-substituted amines [83, 84]. By this procedure, a range of biologically important molecules can be conveniently prepared, such as amines, amino acids, amino alcohols, and nitrogen-containing building blocks [85]. The approach to prepare β-seleno amines involves in a first step the reductive cleavage of Se–Se bonds with reducing agents, such as $NaBH_4$, Zn, and Zn/$InCl_3$, followed by the selective ring-opening reaction with the selenolate nucleophiles generated in situ. The mechanism for the carbon–nitrogen bond cleavage on the C-2 or on C-3 positions in the aziridine is presented in Scheme 6.26 [86].

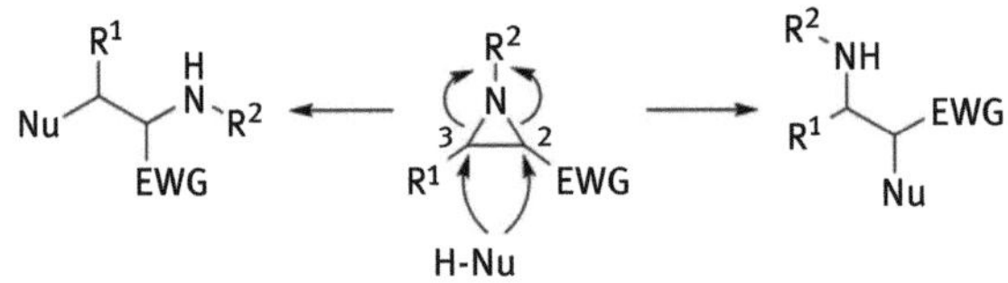

Scheme 6.26: The general mechanism of nucleophilic ring-opening reaction of aziridines.

Salman et al. [87] described an environmentally benign approach to prepare β-seleno amines **22** by the nucleophilic ring-opening reaction of several aziridines **23** under neutral conditions, employing a stable phenylselenolate species (PhSeZnBr **24**) as a nucleophile in [bmim][BF_4]. The respective products **22** were obtained in short reaction times from a broad range of unprotected and protected aziridines **23** (Scheme 6.27). Generally, better yields were obtained using *N*-Ts-protected aziridines compared to the *N*-Boc and unprotected ones (85–99% vs. 60–81% vs. 52–70% yields, respectively). Further, considering economic and environmental aspects, the possibility of recycling the IL was investigated, and a good level of efficiency was observed for up to four reactions.

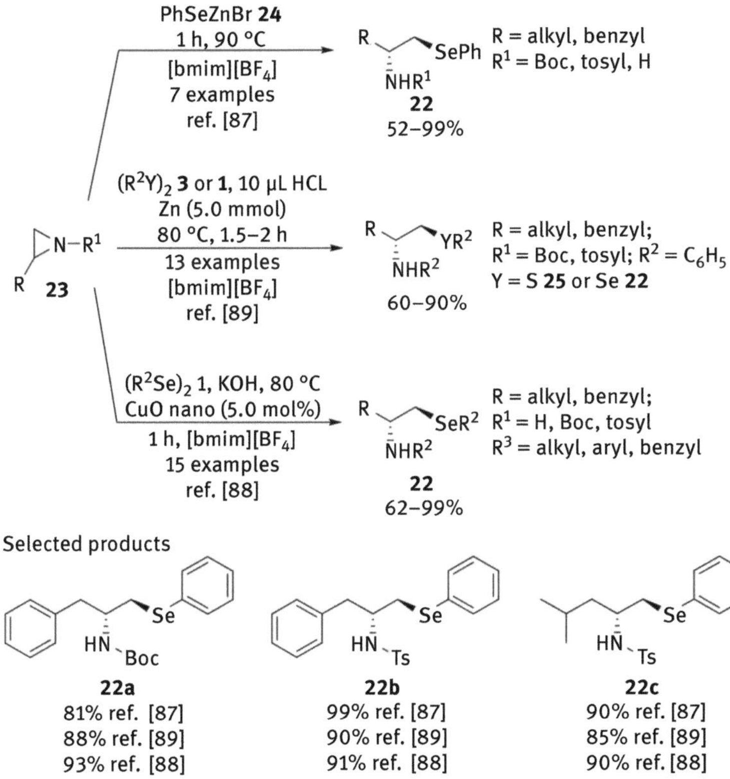

Scheme 6.27: Synthesis of chiral β-chalcogen amines **22** or **25** [87–89].

The success in the use of [bmim][BF_4] as solvent for the synthesis of β-seleno amines led the same group to develop a new procedure using IL [88]. They reported the reaction between different protected and unprotected aziridines **23** and diorganyl diselenides **1** mediated by CuO NP and using [bmim][BF_4] as solvent (Scheme 6.27). The reactions proceeded in 60 min to give the expected products **22** in good to excellent yields (62–99%). However, when diaryl diselenides **1** containing electron-donating groups in the aromatic ring or unprotected aziridine **23** were used, lower yields were obtained. Furthermore, the CuO/IL system was recovered and after a simple pre-treatment, it was used in a new reaction, maintaining its good level of efficiency for additional four successive reactions.

The same group investigated the possibility of generating the nucleophilic species in situ, by the reductive cleavage of diphenyl diselenide **1i** and disulfide **3a** in $[bmim][BF_4]$ medium (Scheme 6.27) [89]. This new zinc- and HCl-mediated route provides an easy access to chiral β-seleno **22** or β-sulfur amines **25** in a stereospecific and regioselective manner in good to excellent yields. The chiral β-chalcogen amines **22** and **25** were obtained by opening *N*-protected aziridines derived from L-phenyl alanine, L-leucine, L-valine, and L-isoleucine and the better yields were obtained when R was a small group in the *N*-Ts aziridines **23**. Further, the IL was further reused in four additional reactions, affording the desired product in a close range of yield. Additionally, some tests were conducted on the antimicrobial activity against Gram-positive bacterial strains. Moderate activities were found using chiral *N*-protected β-chalcogen amines derived from L-isoleucine against *Listeria monocytogenes*, *Bacillus cereus*, and *Paenibacillus* species.

Rodrigues and coworkers [90] described the synthesis of chiral β-seleno amines **22** and selenocysteine derivatives from *N*-protected β-amino mesylates or tosylates **26** (Scheme 6.28). The reactions were performed using the catalytic system ZnO_{NPs}/$[bmim][BF_4]$ and Zn dust, which is able to reduce PhSeSePh, allowing the formation of the reactive zinc selenolate (PhSeZnSePh). In this interesting study, the method was successfully applied to different diaryl diselenides **1** bearing electron-withdrawing or -donating groups, as well as to β-amino mesylates **26** derived from L-leucine, L-isoleucine and L-valine. For instance, the biologically active selenocysteine derivative **22d** was obtained from the corresponding β-amino mesylate in 78% yield. The authors reported that the effect of the leaving group in the protected β-amino alcohols was not so pronounced when mesylate or tosylate was used. Additionally, in the same work it was observed that the catalyst cannot be recycled due to the contamination of ZnO_{NPs} with Zn dust. However, the IL was reused in two additional reactions without significant decrease in the yields of the product (84%, 82%, and 75%).

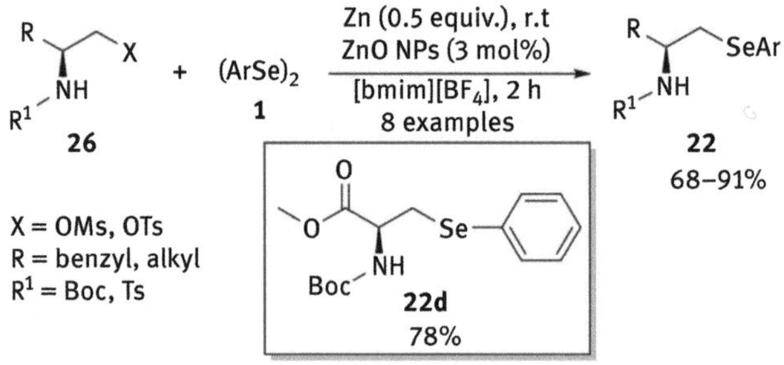

Scheme 6.28: Accessing chiral β-seleno amines **22** from *N*-protected β-amino mesylates or tosylates **26** [90].

Similarly, chiral β-sulfur **25** and β-seleno amines **22** were obtained from the reaction of β-amino mesylate **26** and bromo derivatives with diphenyl diselenide/disulfide **1i** and **3a** using bimetallic reagent Sn(II)/Cu(II) in [bmim][BF_4] (Scheme 6.29) [62]. In this work, four β-chalcogen amines were synthesized under mild conditions and with good yields. More specifically, the seleno-compounds needed more reaction time than sulfur ones.

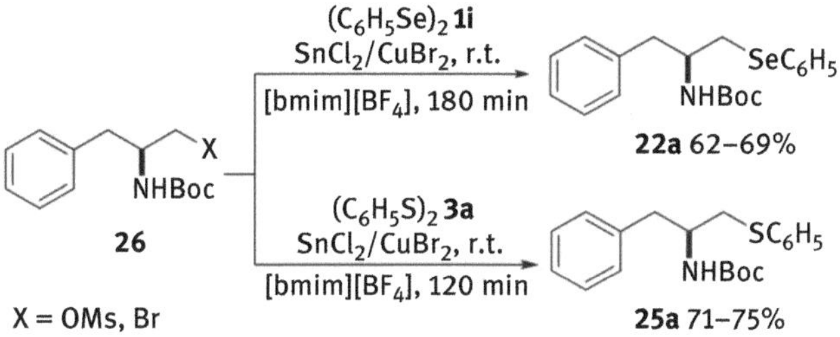

Scheme 6.29: Accessing chiral β-seleno amines **22** from *N*-protected β-amino mesylates or bromo derivatives **26** [62].

6.5 β-Oxy selenides

The β-oxy selenides are important organic compounds in pharmaceutical and natural products chemistry. They play a key role in the synthesis of natural products, such as sphingosine [91], siastatin, and their analogues [92].

For the synthesis of β-hydroxy selenides, in 2008 Yang et al. [93] reported a mild, simple, and environmentally friendly approach, through a regioselective ring opening of 1,2-epoxides **27** with aryl selenols **28** promoted by the IL [bmim][BF_4]. A large range of β-hydroxy selenides **29** was prepared in excellent yields (91–98%) after 1–2 h at 50 °C. Also, the thiocompounds **30** were prepared and were obtained in excellent yields after 3–6 h at 60 °C (81–99%, Scheme 6.30). The IL was used in five successive reactions between epoxide and 4-chlorophenylthiol, allowing the desired product in high yields.

In 2015, Santi and Braga [94] developed an ecofriendly, solvent-, and metal-free procedure to form β-aryl and β-alkyloxy selenides **29**. The desired products **29** were prepared by the reaction of olefins **32** and diorganoyl diselenides **1** in modest to excellent yields (21–96%) after 10 min microwave irradiation using I_2/DMSO as an oxidant catalytic system (Scheme 6.31). Further, the authors performed the reaction between styrene **32** and different diorganoyl disulfides **3**, obtaining the products **30** in satisfactory yields (75–81%). In the case of the reaction between styrene **32** and diphenyl ditelluride **10**, the respective product **33** was formed in only 14% yield. Noteworthy, the formation of side products from the reaction between the nucleophile and iodine, generated in situ, was not observed. With this methodology a

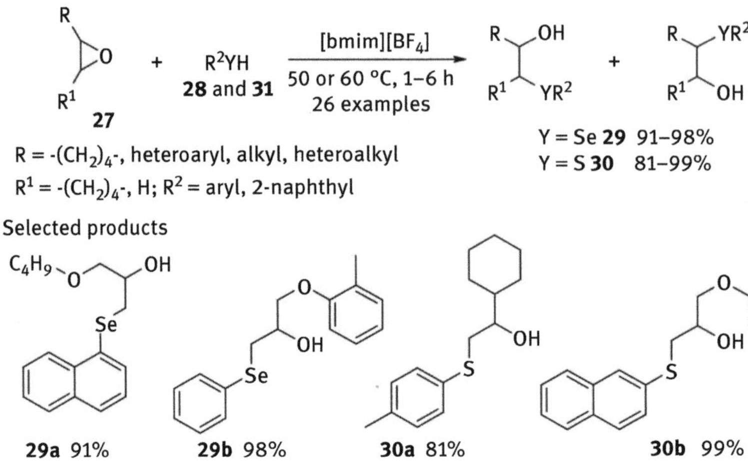

Scheme 6.30: Synthesis promoted by recyclable ionic liquid [93].

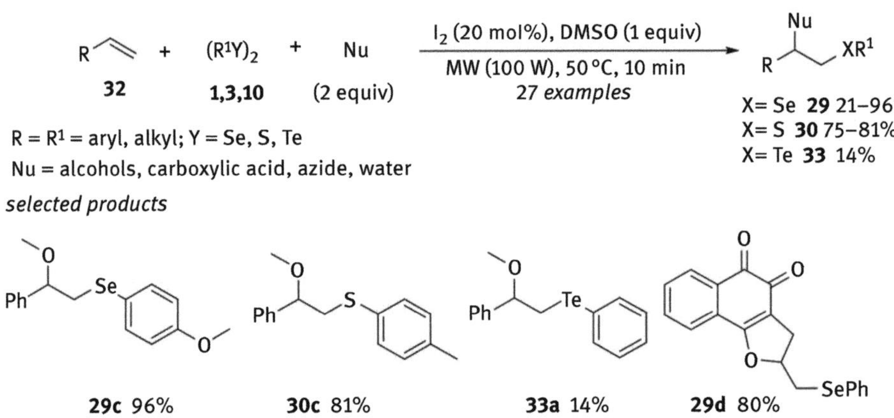

Scheme 6.31: Catalytic selenylation via I_2/DMSO oxidant system [94].

seleno-cyclofunctionalization was performed leading to the compounds with potential biological activity, such as derivatives of lapachol **29d** and lawsone, using olefins with a pendent nucleophile.

The mechanism proposed by the authors involves the formation of the electrophile **I**, generated by the reaction of the catalyst with the diorganyl dichalcogenides **1**, **3**, and **10**. Once formed, **I** reacts with olefins **32** forming the intermediate **II**, which is then attacked by the nucleophile, allowing the formation of the products **29**, **30**, or **33** (Scheme 6.32). The catalyst is regenerated in the reaction media avoiding competition with the nucleophile [94].

In 2018, a new protocol to obtain α-alkoxyl selenides from terminal olefins **32**, diselenides **1**, and different alcohols **34** was reported by Liu et al. [95] (Scheme 6.33).

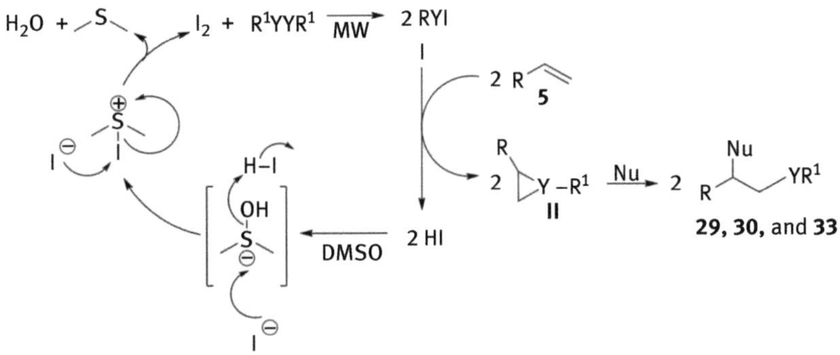

Scheme 6.32: Proposed mechanism [94].

This procedure involves the visible light irradiation for 20 h at room temperature, iodine as catalyst, the use of air as a supplemental mild oxidant, and only 0.5 equiv. of H_2O_2. The desired products **29** were formed in good to excellent yields (40–99%, Scheme 6.33). To obtain the complete conversion of olefins **32** and diselenides **1** and to avoid the nonselective overoxidation of diselenides **1**, the amount of H_2O_2 is crucial. The authors performed several control experiments, which demonstrated that the reaction did not occur in the dark under N_2. Instead, the absence of light, I_2 and H_2O_2, or air are mandatory to afford good yields of the expected products.

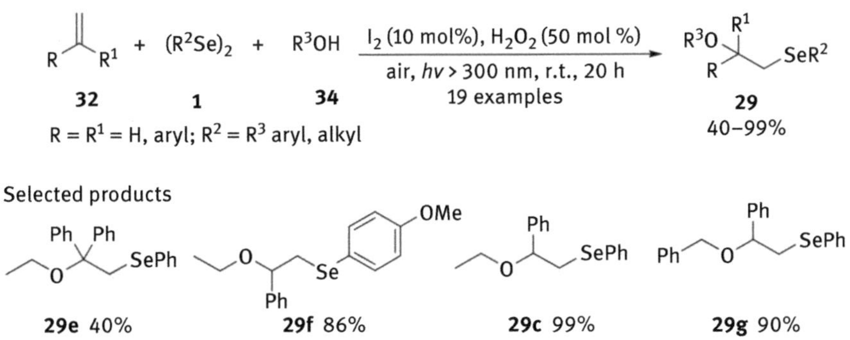

Scheme 6.33: Selenoalkoxylation promoted by visible light [95].

The first step in the mechanism is the oxidation of $(R^2Se)_2$ to the PhSeOH, which reacts with I_2 forming the intermediate R^2SeI **I**. This intermediate rapidly reacts with alkene **32** to form the seleniranium species **II**, which undergoes a nucleophilic attack by the iodide, generating the intermediate **III**. After another nucleophilic attack by the alcohol, **III** is converted to product **29**, releasing HI, which is re-oxidized for a new reaction (Scheme 6.34) [95].

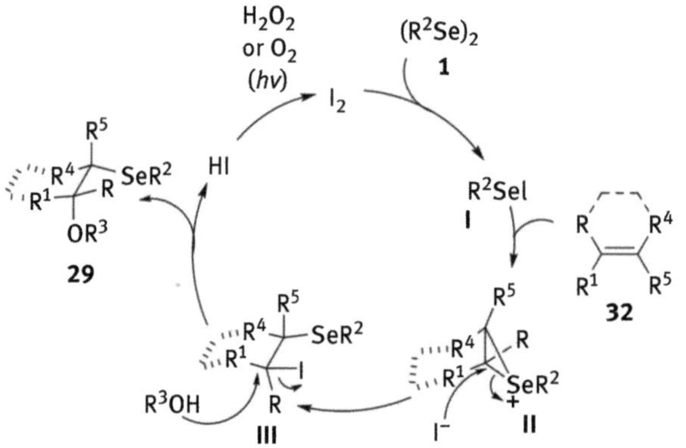

Scheme 6.34: Possible mechanism of selenoalkoxylation [95].

6.6 Seleno ketones

Other class of selenium ether widely studied are the α- or β-seleno carbonyl derivatives. Generally, the methods to access α-seleno carbonyl compounds consist of the reaction of an enolate with an electrophilic organoselenium species or by substitution reactions involving nucleophilic organoselenium. The β-seleno carbonyl derivatives, in turn, are prepared by 1,4-addition reactions to α,β-unsaturated carbonyl compounds. Some studies were reported in the last years reporting the antioxidant properties of the seleno ketones (Figure 6.2) [96–98].

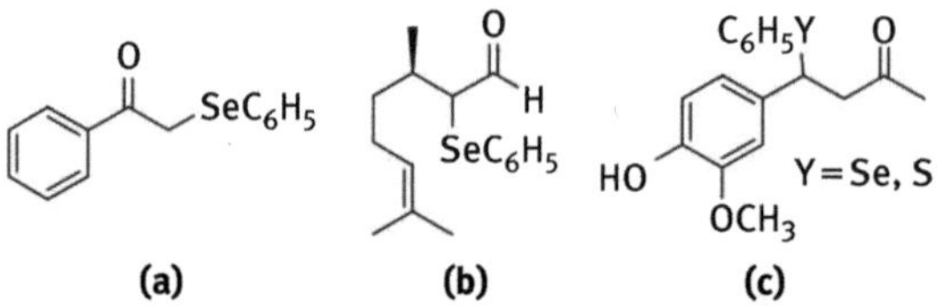

Figure 6.2: Chemical structures of (a) α-(phenylselanyl)acetophenone, (b) α-(phenylselanyl) citronellal, and (c) selenium and sulfur-containing zingerone derivatives.

Lenardão and coworkers [99] described the synthesis of α-seleno aldehydes and ketones employing the solid-supported base KF/Al_2O_3 associated with PEG-400 as the solvent (Scheme 6.35). This clean α-selenylation of carbonyl compounds was successfully applied to cyclic and aliphatic ketones **35** and aldehydes **36**, giving the respective α-phenylseleno derivatives **37** and **38** in 67–96% yields after few hours. It was observed that the reaction was faster using aldehydes (3 h) than using ketones (6–21 h). The solid-supported base was reused for additional four reactions just by washing it with hexanes and drying under vacuum after each reaction. The

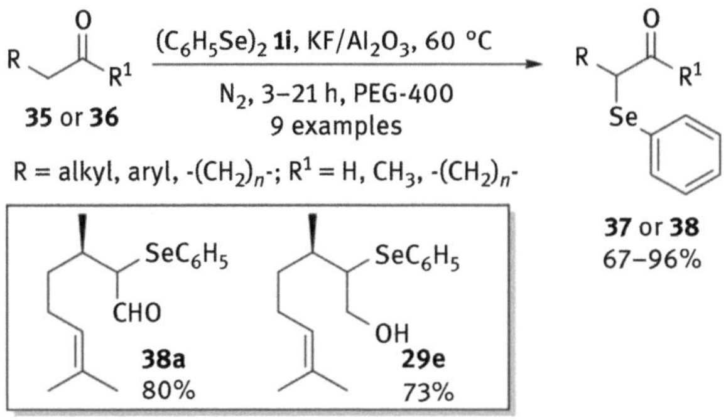

Scheme 6.35: α-Selenylation of ketones **35** and aldehydes **36** [99].

semisynthetic α-phenylseleno citronellal **38a**, derived from (*R*)-citronellal, as well as the alcohol derivative **29e**, showed antibacterial activity against *Staphylococcus aureus*, *Listeria monocytogenes*, and *Salmonella enteritidis*. The results obtained in these assays indicated that the semisynthetic compound **38a** is more bioactive than the natural precursor.

In the last years, some greener methodologies to these Michael-type reactions have been described. For example, Yao and coworkers used a catalytic amount of ceric(IV) ammonium nitrate (CAN) to promote the 1,4-addition of benzeneselenol **28** or thiols **31** to various α,β-unsaturated ketones **39** (Scheme 6.36) [100]. Specifically, CAN accelerates the solvent-free addition of benzeneselenol **28** to enones, which occurred in short reaction times (5–15 min), affording the respective β-phenylselenyl ketones **40** and **41** in almost quantitative yields. Additionally, it was observed that exclusive 1,4-addition of thiol **31** to the conjugated double bond is detriment of the nonactivated one, when dienyl esters are used. The authors proposed a mechanism of the CAN-promoted 1,4-addition reaction. At first, the generation of the selanyl radical cation **I** from selenol **28** occurs, which subsequently is fragmented to radical **II**. The selanyl radical **II** undergoes a radical-chain addition to the enone to form adduct **40** and a new selanyl radical (Scheme 6.36).

A convenient Michael-type addition reaction of unsaturated ketones leading to synthetically useful β-seleno derivatives was proposed by Santi and coworkers (Scheme 6.37) [101]. In this work, the Santi's reagent (PhSeZnCl **24**) was applied as an excellent nucleophile "on water" conditions. The desired β-phenylselanyl ketones **40** were isolated from moderate to quantitative yields. Alternatively, comparable yields were found when the reactions were performed using THF (140 vs. 24 h).

In 2013, Perin et al. [102] reported the in situ generation of nucleophilic species of selenium by the reaction of diorganyl diselenides **1** with $NaBH_4$ in PEG-400 as solvent. These nucleophiles reacted with electron-deficient alkenes **39** to afford β-organylseleno carbonyl compounds **40** (Scheme 6.38). The desired compounds were isolated in 60–94% yields by reaction of diaryl/dibutyl diselenides and methyl

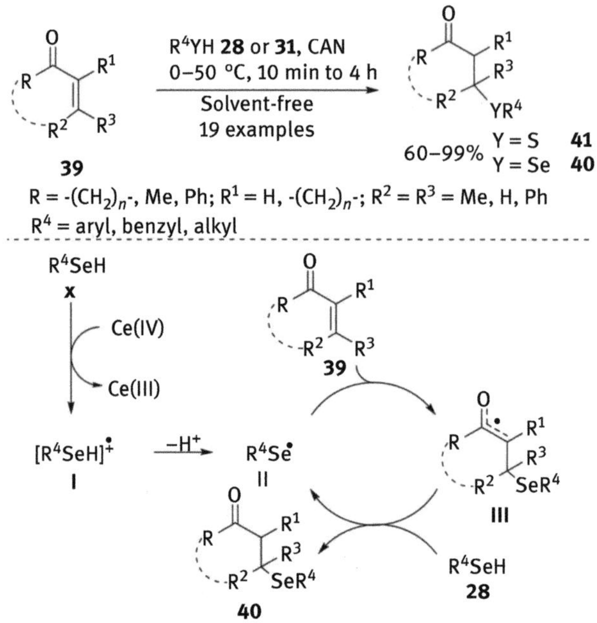

Scheme 6.36: Mechanism of the CAN-promoted 1,4-addition [100].

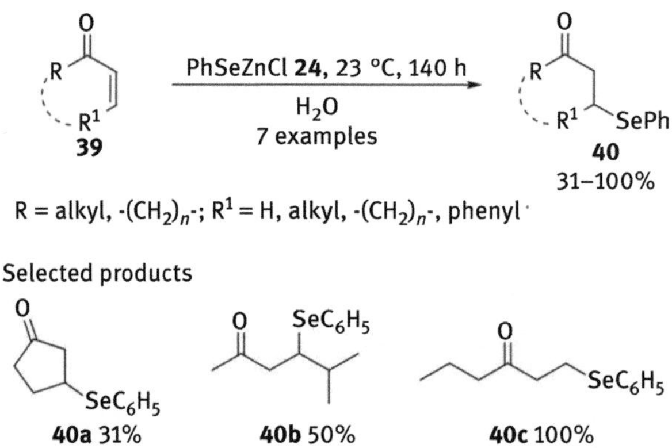

Scheme 6.37: Synthesis of β-phenylselanyl ketones **40** [101].

acrylate, acrylonitrile, enones or acrylic acid in short reaction times. Acrylic acid reacted with the nucleophile obtained from diphenyl diselenide **1i** to give 3-(phenylselanyl)propanoic acid in only 20% yield after 2 h of reaction, while methyl acrylate worked well, giving 94% yield of the respective product.

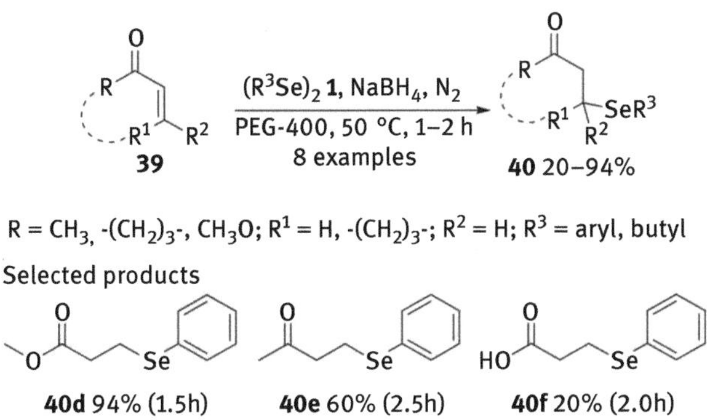

Scheme 6.38: Michael-type addition reaction to electron-deficient alkenes **39** [102].

6.7 Selenoesters

Selenoesters are a class of organic compounds worthy of interest as they exhibit important physical and biological properties and they are useful intermediates in organic synthesis. These compounds can be used as liquid-crystalline materials [103], with applications in optical devices. Moreover, they present antioxidant [104], antiproliferative [105], and cytotoxic activities [106]. They are used in the synthesis of natural compounds [107–109] and are specific reagents in the *trans*-acylation reaction [110]. In the literature, several methods are reported to obtain selenoesters, highlighting their relevance in organic synthesis and medicinal chemistry.

In 2010, Singh et al. [111] developed an efficient synthesis of selenoesters **42** mediated by CuO NP by the reaction between acyl chlorides **43** and diaryl diselenides **1** using the IL [bmim][PF_6] as the solvent. By this method, several selenoesters **42** were isolated in good to excellent yields (57–91%, Scheme 6.39). Additionally, CuO NP and the solvent were recyclable and were used three times without relevant

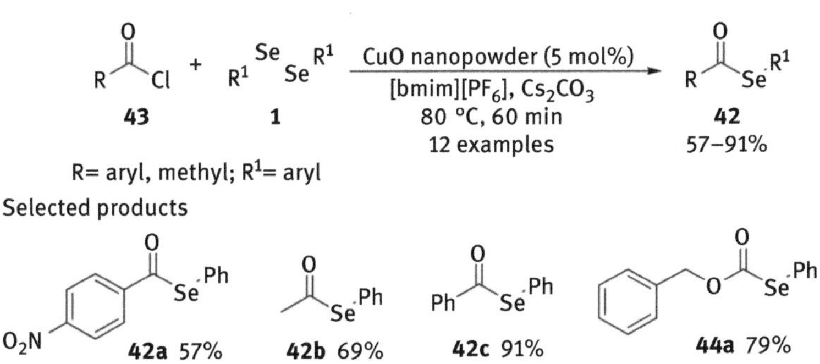

Scheme 6.39: Synthesis of selenoesters catalyzed by CuO nanopowder [111].

loss of the activity. The same procedure was successfully used to prepare different functionalized selenocarbonates **44**.

In the same year, Braga and coworkers [112] published another protocol, which employed the same IL [bmim][PF_6] as solvent, to prepare chalcogenoesters **42** and **45** from acyl chlorides **43** and a variety of stable diaryl chalcogenides **1** and **3** (Scheme 6.40). In the presence of indium metal as a mild reducing agent, selenoesters **42** were isolated in 13–98% yields and thioesters **45** in 36–78% yields in 1 h; the lower yields were obtained using aliphatic acyl chlorides **43**. In addition, the IL was recovered and reused in successive reactions, maintaining its good level of efficiency for two new reactions.

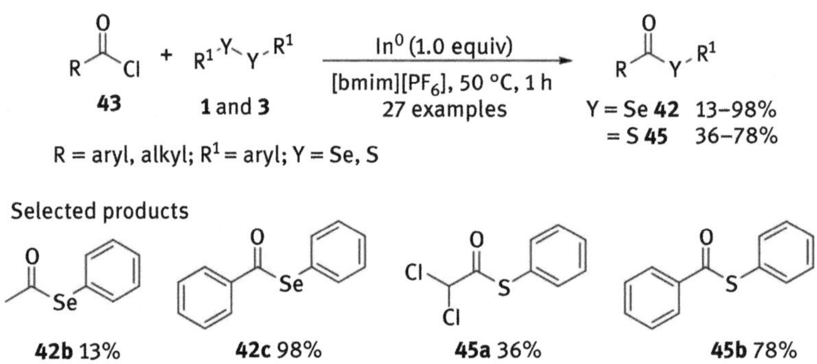

Scheme 6.40: Synthesis proposed by Braga and coworkers [112].

In 2011, Braga and coworkers [60] reported an approach to synthesize selenoesters **42** and thioesters **45** promoted by zinc dust in the presence of IL. Selenoesters **42** and thioesters **45** were achieved in 3 min at room temperature in 39–95% and 42–99% yields, respectively (Scheme 6.41). As mentioned in Section 6.3, this method was applied to obtain diorganyl selenides and sulfides. Moreover, this methodology takes advantage of the reuse of the solvent, maintaining excellent performance in five consecutive reactions.

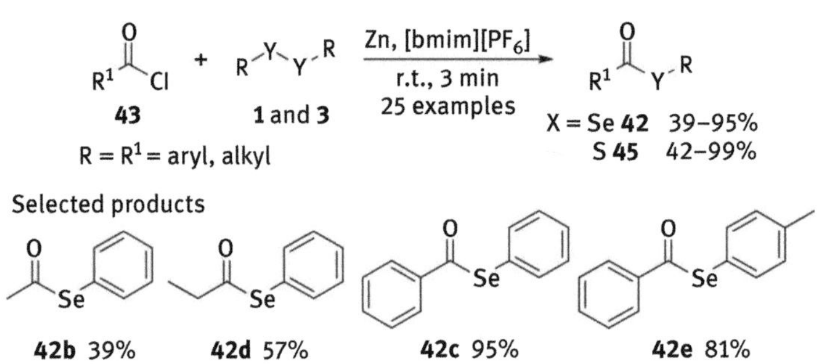

Scheme 6.41: Synthesis of selenoesters promoted by Zn [60].

The IL [bmim][BF_4] was used in combination with the bimetallic system Sn(II)/Cu (II) by Braga and coworkers [62] to generate selenoesters **42** and thioesters **45** (Scheme 6.42). In this chapter, diphenyl diselenide **1** or disulfide **3** reacted with acyl chlorides **43.** The higher yields were obtained starting from aryl chlorides substituted with electron-withdrawing group in the aromatic ring, due to the increase in electrophilicity of the carbonyl center. Further, the authors observed a drastic decrease in yields when acetyl chloride was used (10–17% yield) compared to aroyl ones (52–84% yield). Moreover, diorganyl chalcogenides and chiral β-chalcogen amines were also prepared by this procedure (Sections 6.3 and 6.4).

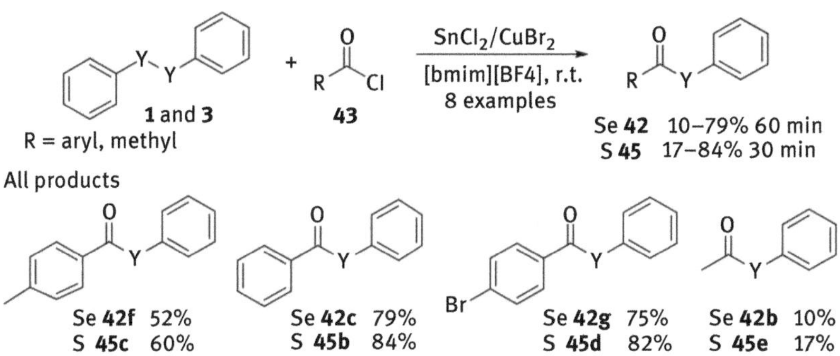

Scheme 6.42: Synthesis of selenoesters and thioesters using a bimetallic system [62].

A solvent-free, microwave-accelerated synthesis of selenoesters **42** was described by Braga and coworkers in 2012 [113]. In this study, the reductive cleavage of the selenium–selenium bond of diorganyl diselenides **1** was promoted by Zn. The nucleophilic zinc selenolate species reacted with several aromatic and aliphatic acyl chlorides **43** in the absence of solvent, affording the desired products **42** in short reaction times and in 40–95% yields (Scheme 6.43).

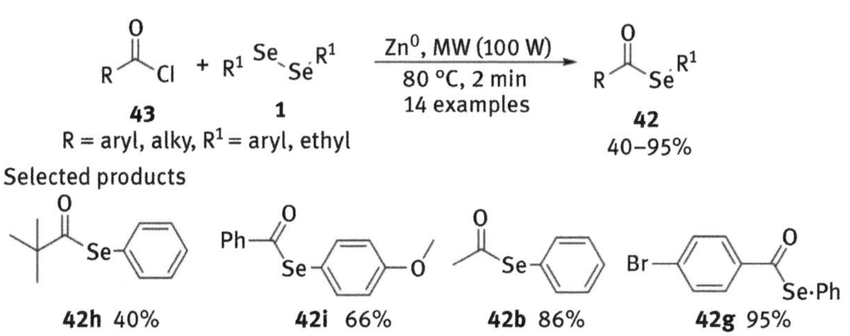

Scheme 6.43: Synthesis of selenoesters **42** under solvent-free conditions [113].

In the proposed reaction mechanism, initially zinc inserts in the Se–Se bond, forming species **I**, which reacts with acyl chloride **43** to give selenoester **42** and species **II** (the Santi's reagent). Once formed, **II** can react with other acyl chloride, forming more product **42** (Scheme 6.44).

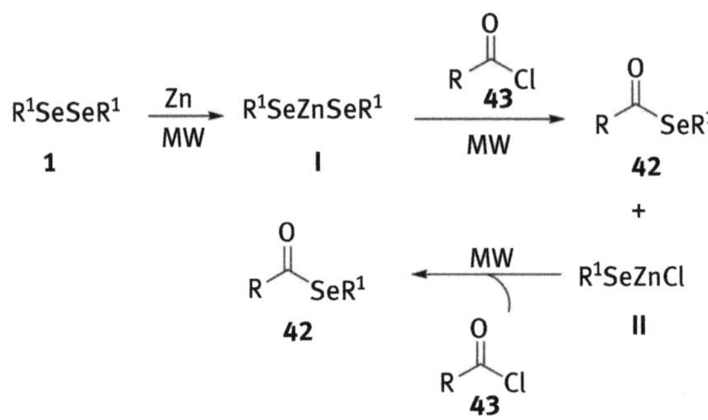

Scheme 6.44: Mechanism proposed by Braga and coworkers [113].

In the same year, Santi and coworkers [114] reported a new, green procedure to form aromatic and aliphatic selenoesters. These compounds are prepared from phenylchalcogeno zinc halides **24** (the Santi's reagent) and acyl chlorides **43** under "on water" conditions (Scheme 6.45). The selenoesters **42** were obtained in 20–97% yields after 3 h of reaction. In this work, it was observed that PhSeZnBr was more reactive than PhSeZnCl in all the tested examples. 4-Butylbenzoyl chloride was the unique exception, and in both cases the corresponding selenoester was obtained in 60% yield. Also, thioesters **45f** and **45g** were prepared in 37% and 55% yields, respectively, under these "on water" conditions (Scheme 6.45). The aqueous medium can be reused for three times in the reaction between benzoyl chloride and PhSeZnBr **24**, after neutralizing the pH (70%, 60%, and 55% yields).

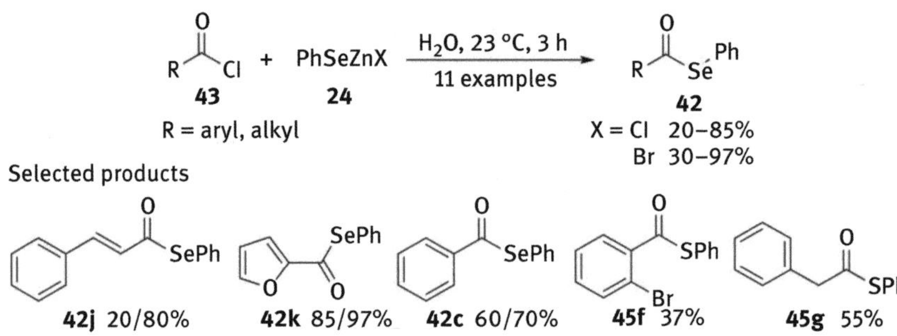

Scheme 6.45: Synthesis of selenoesters **42** "on water" [114].

The reaction involves a concerted mechanism, in which it first generates intermediate **I**, following the activation of the carbonyl group and the nucleophilic attack of the selenium. This intermediate evolves in species **II**, which affords the desired products **42** (Scheme 6.46). The collateral product is formed by the nucleophilic attack of water, which happens slowly.

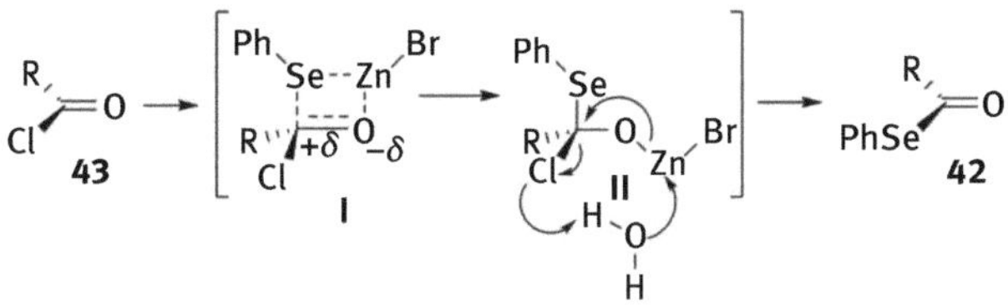

Scheme 6.46: Concerted mechanism proposed by Santi and coworkers [114].

More recently, in 2017, Santi et al. [115] developed a new strategy for the synthesis of selenoesters **42** under "on water conditions," which involved the reaction between acyl chlorides **43** and zinc selenates, such as zinc bis-phenylselenate **46** or TMEDA-stabilized zinc selenate **47** (Scheme 6.47). The desired products **42** were formed in 53–90% yields when compound **46** was used, and in 35–80% using complex **47**. The authors also performed a "one-pot" synthesis, followed by a direct chromatographic purification of the Se-phenyl benzoselenoate **42c**, thus avoiding the workup of the reaction and making the process greener.

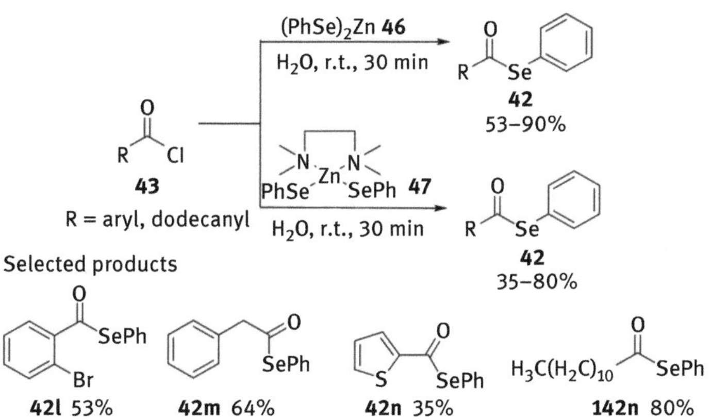

Scheme 6.47: Synthesis of selenoesters **42** with zinc selenates [115].

Perin and coworkers described the synthesis of selenoesters **42** by reacting acyl chlorides **43** and arylselenol **28**, which was generated in situ from the reductive cleavage of the Se–Se bond of different diaryl diselenides **1** with hypophosphorous acid (Scheme 6.48) [116]. The reactions were carried out using PEG-400 as solvent

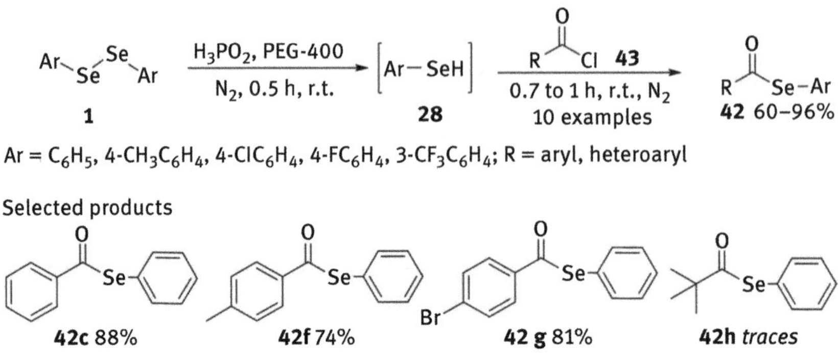

Scheme 6.48: Synthesis of selenoesters **42** in the presence of H_3PO_2 [116].

and they proceeded efficiently with a range of acyl chlorides and diaryl diselenides, both containing electron-donating and electron-withdrawing groups, giving the corresponding selenoesters in good to excellent yields (60–96%). However, when pivaloyl chloride or dibutyl diselenide were used, only traces of the desired products were obtained. Additionally, PEG-400 can be easily recovered and directly reused in the new reactions between diphenyl diselenide and benzoyl chloride. A moderate level of efficiency was maintained for three times.

The same group [117] prepared new glycerol-derived selenoesters **42** from the reaction between anhydrides **48** and bis-(2,2-dimethyl-1,3-dioxolanylmethyl)diselenide **1j** in a reducing system of Rongalite®/K_2CO_3 and PEG-400 as the solvent (Scheme 6.49). The products **42** were prepared in 55–85% yields, after 20–180 min at room temperature. Good results were obtained starting from aromatic anhydrides containing electron-withdrawing and electron-donating groups, as well as heteroaromatic and aliphatic ones and diselenide **1j,** to furnish the respective products **42** in

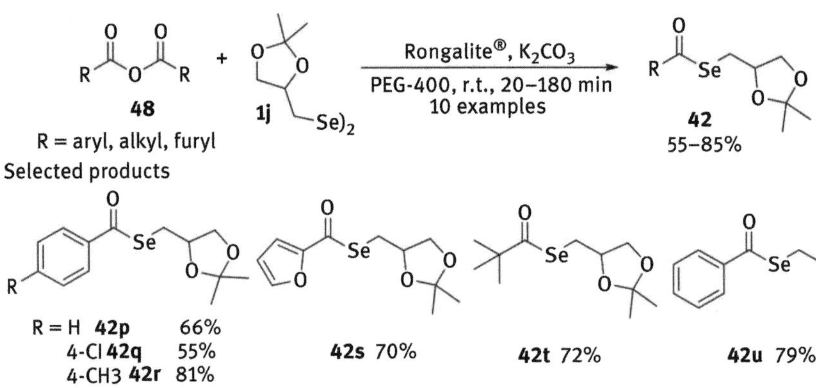

Scheme 6.49: Synthesis of new glycerol-derived selenoesters **42** [117].

acceptable yields. Also, an interesting water-soluble selenoester **42u** was formed, from a ketal deprotection reaction.

The mechanism of the reaction starts by the decomposition of Rongalite® in the presence of K_2CO_3 to formaldehyde and anion **I** (Scheme 6.50A). This anion reacts with the diselenide **1j** through a single electron transfer reaction, forming the intermediates **II** and **III** (Scheme 6.50B). Subsequently, radical **II** undergoes a reduction to anion **III** (Scheme 6.50C). Finally, intermediate **III** attacks anhydride **48,** leading to elenoester **42** (Scheme 6.50D).

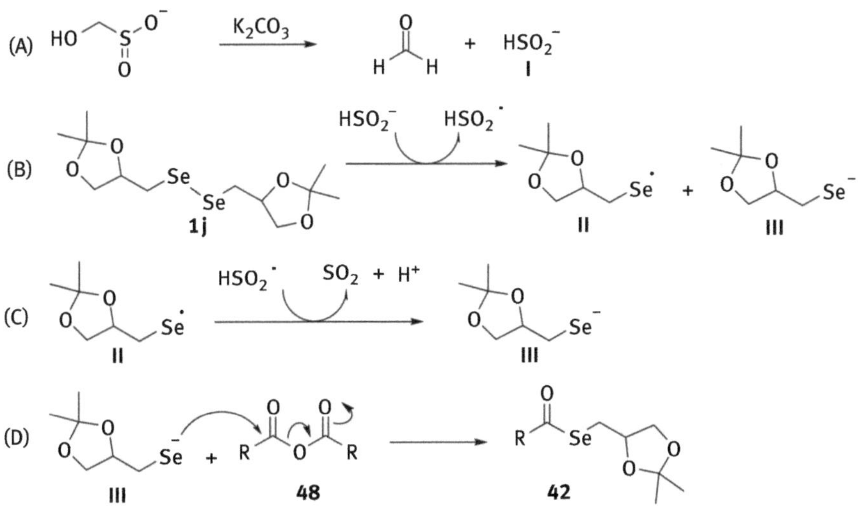

Scheme 6.50: Possible mechanism with Rongalite® [117].

6.8 Vinyl selenides

The vinyl chalcogenides are versatile intermediates in organic synthesis, being useful tools for the selective construction of conjugated or isolated olefins [32, 49]. Besides that, vinyl sulfides are also present in natural occurring compounds, such as griseoviridin and benzylthiocredillidone [118].

Several protocols for the synthesis of vinyl selenides have been reported, and the most common and atom-economic procedures involve the hydrochalcogenation of internal or terminal alkynes using organoselenols or the respective selenolate anions generated in situ [25, 41]. The plausible mechanism of this reaction may involve either an external nucleophilic attack to the triple bond coordinated with metals with high Lewis acidity, which leads to the product of the *anti*-addition of the nucleophile, or an insertion reaction into the metal–chalcogen bond in the chalcogenolate, by formation of a four-membered transition state resulting in the *syn*-addition (Scheme 6.51) [119, 120].

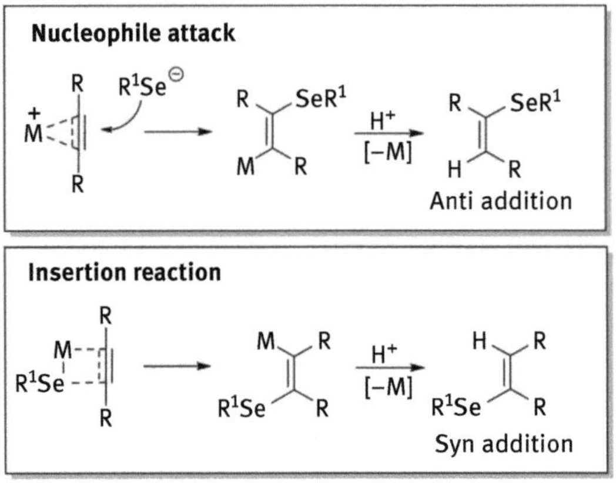

Scheme 6.51: General mechanism for the hydrochalcogenation of alkynes.

The development of safer reaction conditions for the hydroselenation of terminal alkynes has been increased over the years. Additionally, some studies reported the substitution reaction of haloalkenes by nucleophilic selenium species using nonvolatile solvents.

The hydroselenation of alkynes **49** using diorganyl diselenides **1**, $NaBH_4$, and [bmim][BF_4] as the solvent was proposed by Lenardão et al. in 2007 [121]. In this chapter, several alkynyl alcohols **49** were evaluated and in all the cases, a mixture of *gem*-vinyl selenides **50** and (*Z*)-**50** was obtained. The amount of *gem*-product **50** depends on the volume of the substituent at the propargyl alcohol. Still, when using the homo-propargyl alcohol **49**, the lowest selectivity was observed. Interestingly, (*E*)-1,2-bis-phenylselenostyrene was formed from phenylacetylene (63% yield, 3 h), as observed and discussed in Section 6.9 [122]. Moreover, the IL was reused several times without aqueous workup and no significant decrease in the yields of successive reactions was observed (Scheme 6.52) [123].

PEG-400 was also successfully employed as solvent in the synthesis of vinyl selenides **50**. The nucleophilic species of selenium were easily generated in situ by the reaction of diphenyl diselenide **1** with $NaBH_4$ as a reducing agent in PEG-400 (Scheme 6.52) [124]. In this chapter, the hydroselenation of several alkynes **49** promoted by $NaBH_4$/PEG-400 was very efficient, and a mixture of Markovnikov and *anti*-Markovnikov adducts **50** was obtained in good to excellent yields (50–95%) in short reaction times. The recycling of PEG-400 was performed in reactions between diphenyl diselenide and phenylacetylene as starting materials; the product was isolated in similar yields (85–71%) after four runs using the same solvent. When glycerol was used as the solvent in the reaction of phenylacetylene with diphenyl diselenide, the reaction behavior was similar to that using IL, giving a mixture of (*E*)- and (*Z*)-1,2-bis-phenylseleno styrene **52** [121, 123].

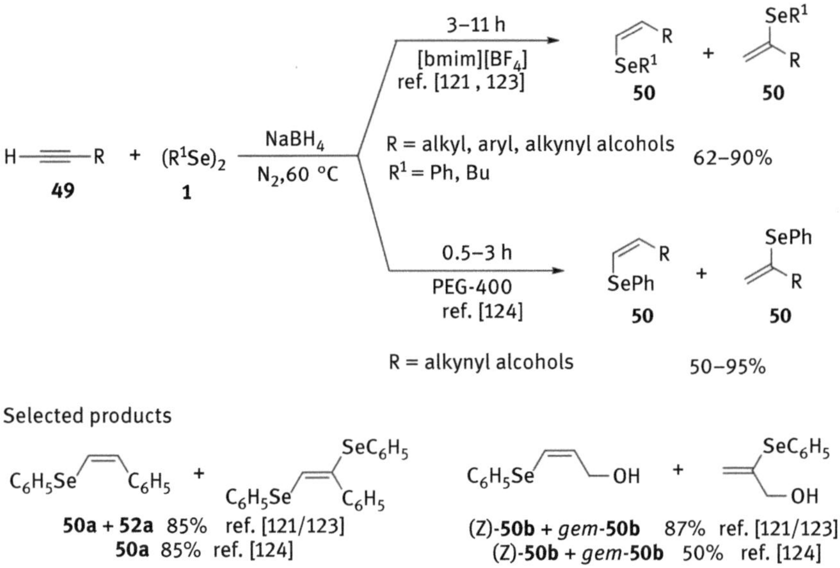

Scheme 6.52: Hydroselenation of alkynes **49** using diorganyl dichalcogenides **1** [121, 123, 124].

Functionalized vinyl selenides **50** have been synthesized from diverse substrates, such as electron-deficient alkynes **49**, using green conditions. For instance, Perin et al. developed a solvent-free hydroselenation of methyl propiolate derivatives to obtain various β-phenylseleno-α,β-unsaturated esters **50** preferentially with *Z*-configuration (*Z*:*E* ratio of 73:27 to 98:2) (Scheme 6.53) [125]. The sodium borohydride supported on alumina ($NaBH_4/Al_2O_3$) system was used in the generation in situ of the reactive species of selenium ($PhSe^-$) under three different conditions: at room temperature, under microwave irradiation (in a domestic microwave oven) and under heating (oil bath). The results indicated that there is no substantial difference in the reaction performance among the three procedures. The vinyl selenides **50** were obtained in good yields (50–83%) with comparable selectivity to the protocols that use volatile organic solvents. Further, the reaction times were reduced from several hours to only few minutes under microwave irradiation.

The method described earlier was extended to the preparation of (*Z*)-β-phenylseleno-α,β-unsaturated ketone **50** starting from 3-organyl-3-butyn-2-ones **49** (Scheme 6.53) [126]. A remarkable feature of this green protocol is the possibility of using $NaBH_4$ as a reducing agent in the presence of the carbonyl group of the ketone without parallel reactions, besides no need to using inert atmosphere. The selenides were obtained in a high selectivity (*Z*:*E* ratio of 85:15 to 96:04). As shown in the previous work [125], the irradiation with microwaves allowed the synthesis of compounds **50** in good yields (62–82%) and short reaction times. For instance, the reaction between diphenyl diselenide and 4-phenyl-3-butyn-2-one gave the

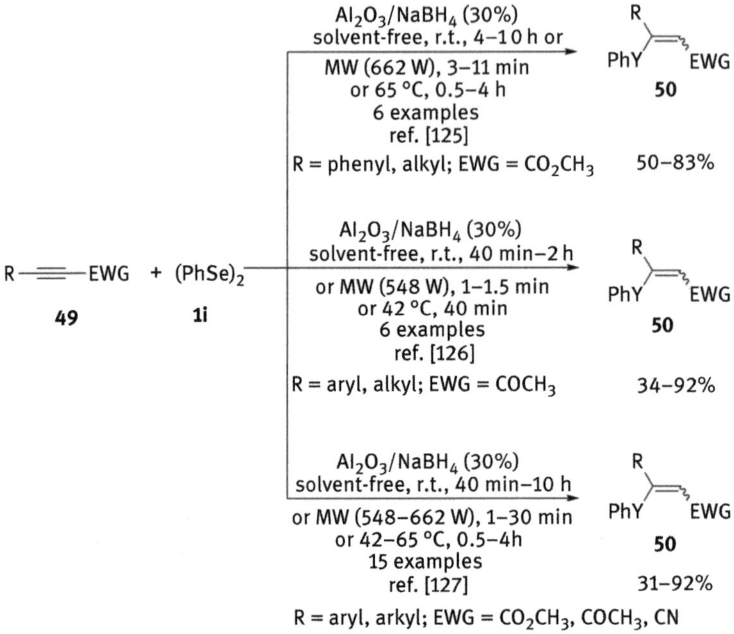

Scheme 6.53: Hydroselenation of electron-deficient alkynes **49** [125–127].

respective β-phenylseleno ketone after 40 min under conventional heating (42 °C, 51% yield), *versus* 1.5 min under microwave irradiation (64% yield). Following this study, the $NaBH_4/Al_2O_3$ system was successfully employed in the solvent-free hydroselenation of several other Michael-acceptor alkynes **49** (esters, ketones, and nitrile) using diphenyl diselenide **1i** (Scheme 6.53) [127].

The Santi's reagent **24** was used as a source of selenium nucleophilic to prepare functionalized vinyl selenides **50** from electron-deficient alkynes **49** on water. After stirring an aqueous suspension of an equimolar mixture of alkyne **49** and the bench-stable PhSeZnCl **24**, the respective Michael adducts **50** were isolated from good to excellent yields with *Z*-configuration, preferentially (83–100% yield; *Z*:*E* ratio = 66:34–100:0) (Scheme 6.54) [101]. A comparison between water and THF as the solvent showed a remarkable acceleration of the reaction in aqueous medium. The authors described that water should play a role in activating the selenium reagent **24**, with no influence in the reactivity of the substrate. The method tolerates a range of electron-deficient alkynes **49**, such as esters, ketones, and aldehydes. As mentioned in Section 6.6, the same Santi's reagent **24** was used in Michael-type additions to electron-deficient alkenes, affording the respective diorganyl selenides in excellent yields.

The vinyl substitution with a nucleophilic or electrophilic organochalcogenium species is often another strategy used to synthesize vinyl chalcogenides. Through this approach, vinyl chalcogenides were obtained with total control of the stereochemistry,

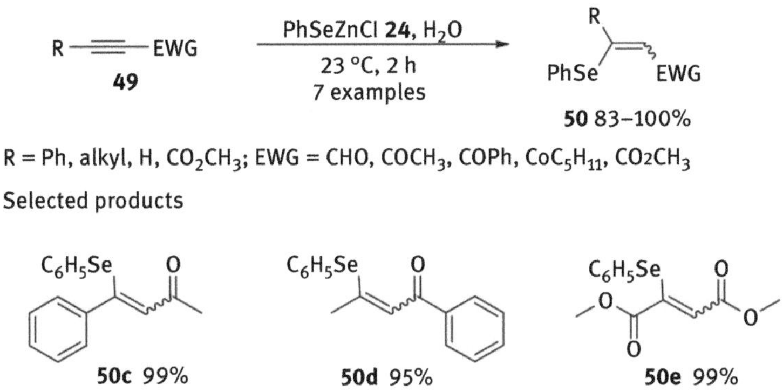

Scheme 6.54: Vinyl selenides **50** from electron-deficient alkynes **49** and Santi's reagent [101].

which depends totally of the purity of starting material. Despite the generally "stereo-conservative" reaction, this procedure is less atom-economic than the direct hydrochalcogenation of alkynes. By using this useful protocol, Kabalka and Venkataiah [128] described the access to (*E*)- and (*Z*)-vinyl selenides **50** by the reaction between vinylboronic acids or vinylboronic esters **54** with phenylselanyl chloride **11** in IL (Scheme 6.55). After stirring equimolar amounts of PhSeCl **11** and the vinyl boronic reagent **54** in [bmim][BF_4] as the solvent for 2 h, the (*E*)- and (*Z*)-products **50** were prepared from the correspondent (*E*)- and (*Z*)-isomers of **50.** In addition, the solvent could be recycled by successive reactions without significant reduction in the product yield.

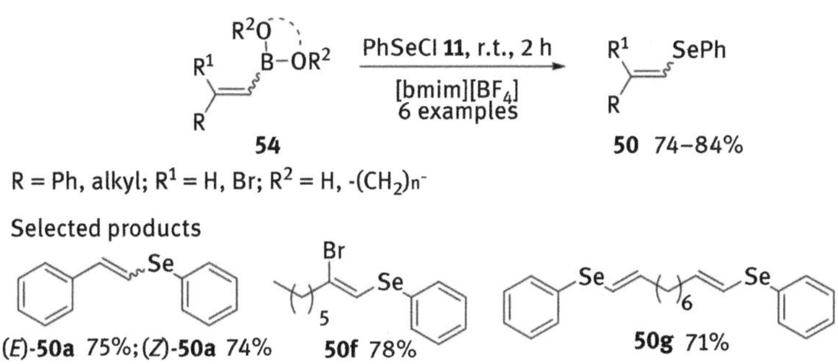

Scheme 6.55: Synthesis of vinyl selenides **50** from vinylboronic acids and esters **54** [128].

An alternative protocol to prepare (*Z*)-vinyl chalcogenides **50** or **53** was reported by Bao and coworkers [129], which involves a highly stereoselective copper-catalyzed coupling reaction. Several alkenyl and styryl bromides **55** reacted with diaryl diselenides **1** or thiols **31** in the presence of zinc powder and using an IL as solvent (Scheme 6.56). The IL based on the amino acid *N,N*-dimethylglycine acts both as a base and as a solvent in the reaction. Different vinyl selenides **50** were isolated in good to excellent yields with high stereoselectivity (*Z*:*E* = 94:6 > 98:2). Vinyl bromides **55** afforded lower yields in the reaction with benzenethiol and diphenyl diselenide, of 70% and 76%, respectively. Additionally, the IL was recycled in the reaction between benzenethiol and (*Z*)-β-bromostyrene. After the product extraction, the IL was concentrated in vacuo and reused for three times with comparable yields of the vinyl sulfide. To further new reactions, a pretreatment to remove HBr from IL with potassium carbonate was necessary.

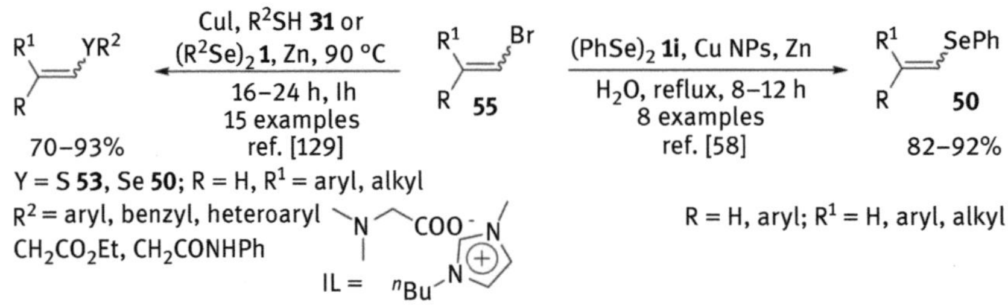

Scheme 6.56: Synthesis of vinyl selenides **50** from vinyl bromides **55** [58, 129].

A similar strategy was described by Ranu and coworkers [58], which used Cu(0) NPs (20 mol%) in the presence of zinc dust (1.5 equiv.) to promote the coupling of vinyl bromides **55** with diphenyl diselenides **1** in water under ligand-free conditions (Scheme 6.56). Interestingly, when (*E*)-vinyl bromide **55** was used, the (*E*)-vinyl selenide **50** was obtained exclusively, while (*Z*)-vinyl bromides **55** gave a mixture of (*E*)- and (*Z*)-vinyl selenides **50** (*E*:*Z* = 54:46 to 20:80 ratios). The reaction was not sensible to electronic effects in the aromatic ring of the styryl bromide and both product yield and stereoselectivity were not affected by the presence of substituents in the phenyl ring. In addition, the efficiency of the Cu(0) NPs as catalyst was remarkable when compared with metallic Cu in the reaction between diphenyl diselenide and styryl bromides (87–88% vs. 35–43% yields). The authors described the

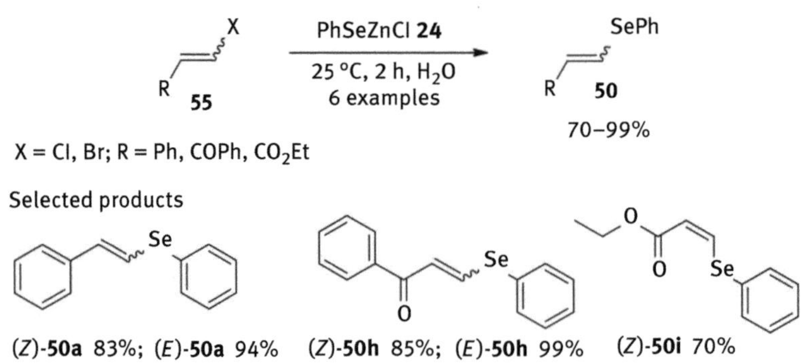

Scheme 6.57: Synthesis of vinyl selenides **50** from vinyl halides **55** [130].

agglomerating tendency of the Cu NPs after various cycles of reusing, resulting in the loss of efficiency and a decrease in yields from 90 to about 75% after three successive reactions. This procedure was used also to prepare diaryl selenides by the cross-coupling of diphenyl diselenide with aryl iodides, as reported in Section 6.3.

As mentioned earlier, PhSeZnCl **24** is a good nucleophile in the vinylic substitution of functionalized vinyl bromides and chlorides **55** to prepare the different vinyl selenides **50** (Scheme 6.57) [130]. By this procedure, the vinyl selenide **50** was obtained strictly with the same stereochemistry of the substrate utilized as a starting material. Only when (*Z*)-3-chloro-1-phenylprop-2-en-1-one was used as a substrate, a little amount of (*E*)-vinyl selenide (5–9%) was isolated together with the (*Z*)- one. In this work, several (*Z*)- and (*E*)-vinyl selenides were synthetized in good to excellent yields starting from (*Z*)- and (*E*)-β-bromostyrenes, as well as (*Z*)- and (*E*)-β-chloro enones and α,β-unsaturated esters. Additionally, a comparative study on the use of water or THF as the solvent at room temperature showed that the "on water" reaction afforded products **50** in higher yields and less time than in THF medium (2 h vs. 24 h).

Encouraged by the results obtained "on water," the reaction between 4-chloro-3-nitro-2H-chromen-2-one **56** and PhSeZnCl **24** was studied, yielding the respective (*Z*)-vinyl selenide **57** in 78% yield, showing that the reaction tolerates a sort of functionalities in the substrate (Scheme 6.58).

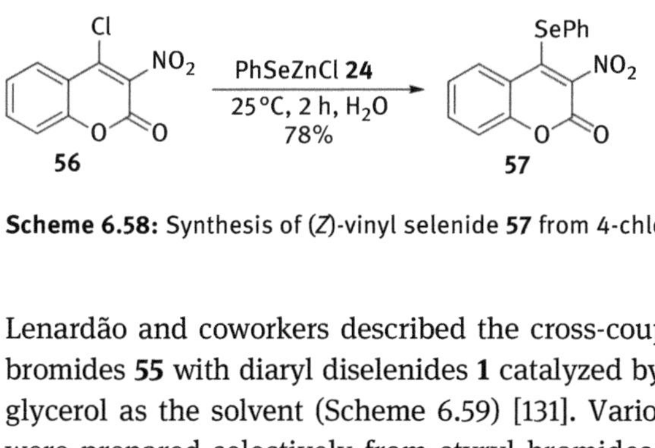

Scheme 6.58: Synthesis of (*Z*)-vinyl selenide **57** from 4-chloro-3-nitro-2H-chromen-2-one **56** [130].

Lenardão and coworkers described the cross-coupling reaction of (*Z*)- and (*E*)-vinyl bromides **55** with diaryl diselenides **1** catalyzed by CuI in the presence of zinc, using glycerol as the solvent (Scheme 6.59) [131]. Various (*Z*)- and (*E*)-vinyl selenides **50** were prepared selectively from styryl bromides **55** bearing electron-withdrawing and electron-donating groups, with retention of the configuration of the starting materials **55**. After completing the reaction between β-bromostyrene and diphenyl diselenide, the recyclability of the CuI/Zn/glycerol mixture was explored. The catalyst/solvent system was directly reused for further reactions, simply by adding more reagents. The recyclable system showed a good level of efficiency in five consecutive reactions; after the fourth run, the yields were slightly decreased. The mechanism of these coupling reactions involves the initial reduction of Cu(I) to Cu(0) by metal zinc [132]. Next, Cu(0) undergoes an oxidative addition to diaryl diselenide **1**, giving the intermediate $(ArSe)_2Cu(II)$ (Scheme 6.59). After reduction by Zn, this intermediate leads to ArSe–Cu(I), which reacts with aryl vinyl bromides **55** to give the respective vinyl selenide **50** via a "transitory" Cu(III) intermediate. $Zn(SeAr)_2$ formed after the reduction of $Cu(SeAr)_2$ by Zn reacts with CuI to form more ArSeCu(I), with both ArSe moieties of ArSeSeAr being used in the overall reaction.

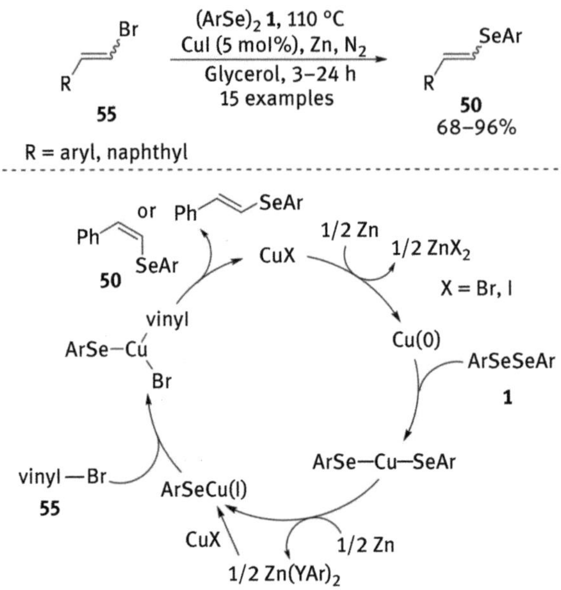

Scheme 6.59: Synthesis of vinyl selenides **50** and mechanism involving cross-coupling reactions [131].

In 2016, Perin and coworkers [133] described the monosubstitution of bromine atom from 1,1-dibromo-alkenes **58** by a diverse array of nucleophilic selenium species, generated in situ from the corresponding diaryl diselenides **1** by reaction with $NaBH_4$ as a reducing agent in PEG-400 (Scheme 6.60). This metal-free method employed diaryl diselenides **1** and 1,1-dibromo-alkenes **58** to prepare (*E*)-1-bromo-1-selenoalkenes **59**. Nine alkenes **59** were formed in good yields (49–95%) at 50 °C using conventional heating for 0.5–2.0 h. Additionally, by simple stoichiometry and temperature-controlling reaction, five ketene selenoacetals **60** were efficiently obtained in a range of 42–76% yield, after 1.0–3.5 h of reaction. The proposed mechanism to give the (*E*)-1-bromo-1-selenoalkenes **59** starts by the generation of the saturated intermediate **C**, after addition of the selenolate anion **A** to the double bond, with the involvement of the solvent to stabilize the transition state **B**. The elimination of HBr, through an *anti*-periplanar conformation, results in the formation of the desired compound **59**. A sequential addition of a second phenylselenyl group to **59** affords the ketene selenoacetal **60**.

Functionalized vinyl selenides **50** were prepared by the Knoevenagel reaction of phenylselenoacetonitrile or ethyl (phenylseleno)acetate **61** with aldehydes **36** in the presence of a solid-supported catalyst (KF/Al_2O_3) under solvent-free conditions

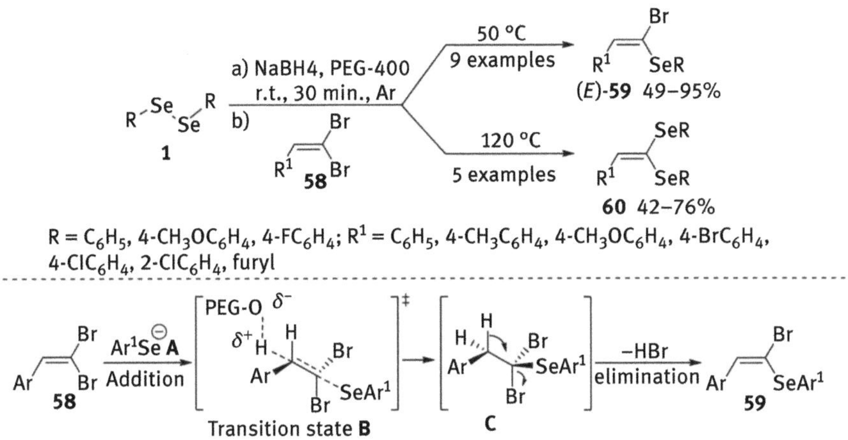

Scheme 6.60: Synthesis of (*E*)-1-bromo-1-selenoalkenes **59** and ketene selenoacetal **60** and the proposed mechanism [133].

(Scheme 6.61) [134]. By this protocol, β-phenylselenoacrylonitriles and β-phenylseleno-α,β-unsaturated esters were prepared in 40–80% yields. In all cases, a mixture of (*Z*)- and (*E*)-alkenes **50** was formed, with preference for the (*Z*)-isomer. The Knoevenagel approach using KF/Al_2O_3 is a straightforward alternative to access functionalized compounds. Despite the modest yields, the reaction is versatile and the use of aromatic, heteroaromatic, and aliphatic α,β-unsaturated aldehydes was suitable. In general, acrylonitriles were obtained in better yields than the ester analogues. Still, the solid support was reused in new reactions with comparable yield, after addition of more KF.

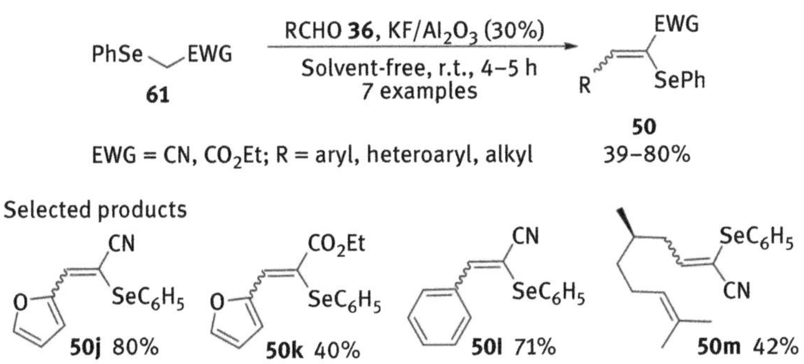

Scheme 6.61: Synthesis of functionalized vinyl selenides **50** by the Knoevenagel reaction [134].

Perin and coworkers [135] reported in 2016 the use of PEG-400 as solvent to prepare selectively enynes **62–64** and dienes **65–66** substituted with chalcogenyl groups, through the hydrochalcogenation of 1,4-diorganyl-1,3-butadiynes **67** under mild reaction conditions (Scheme 6.62). The first step of this temperature-controlled strategy was the generation of nucleophilic species of selenium, tellurium and sulfur in situ, by the reaction of the respective diorganyl dichalcogenides with $NaBH_4$. These reactive species reacted with diynes **64** at 30 °C affording the respective (*Z*)-chalcogenynes **62**–**64** (Se, S, and Te), while at 90 °C, (*Z,Z*)-bis-chalcogen-1,3-butadienes **65**–**66** (Se and S) were produced in good to excellent yields. In general, this stereoselective method works better with aliphatic dichalcogenides, due to their higher nucleophilicity, giving the products in higher yields. The irradiation with microwaves as an alternative energy source to the conventional oil bath provides the expected products in lower reaction times.

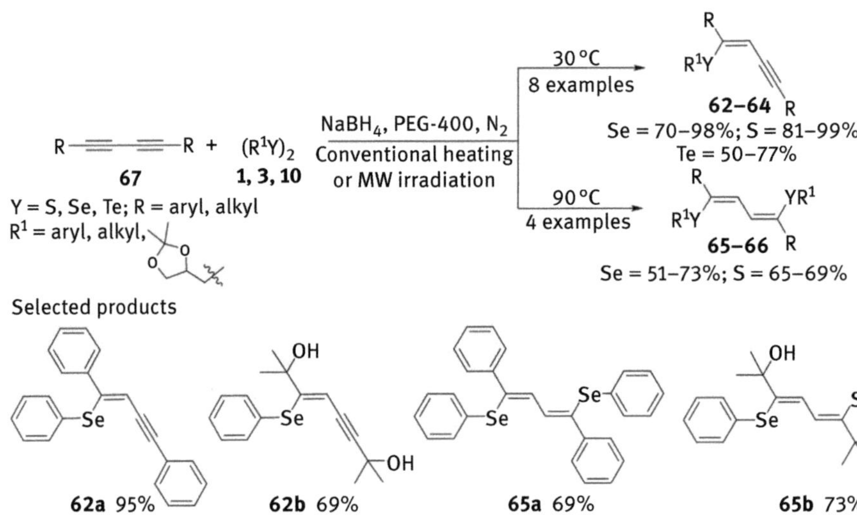

Scheme 6.62: Synthesis of (*Z*)-chalcogenynes (Se, S, and Te) **62–64** and (*Z,Z*)-bis-chalcogen-1, 3-butadienes (Se and S) **65–66** [135].

A new class of vinyl selenides **68** was prepared by Perin and coworkers, by the regio- and stereoselective addition of sodium selenide species to two equivalents of aryl alkynes **49** under green conditions (Scheme 6.63) [136]. In this work, the nucleophilic species of selenium was generated in situ, from the reaction of elemental selenium with $NaBH_4$, utilizing PEG-400 as the solvent. By this one-pot procedure, some divinyl selenides **68** were synthesized in moderate to excellent yields with high selectivity for the (*Z,Z*)-isomer. These reactions proceeded under gentle heating at 60 or 90 °C in short reaction times. Additionally, the reactivity of this class of compounds was explored in the Fe-catalyzed cross-coupling reaction between properly substituted divinyl selenide **68b** and a Grignard reagent leading to the valuable resveratrol trimethyl ether in 57% yield.

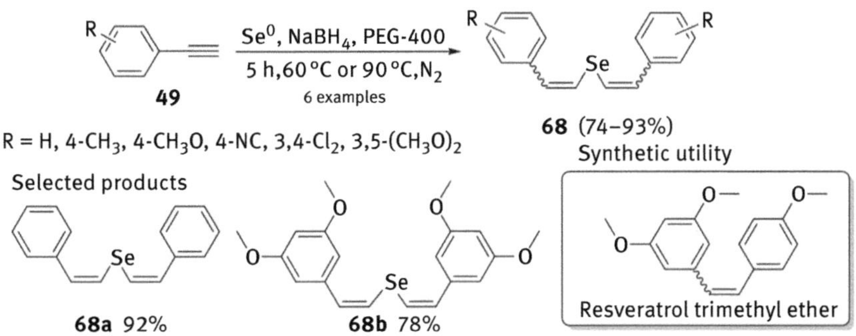

Scheme 6.63: Synthesis of divinyl selenides [136].

6.9 Bis-organoselanyl alkenes

The class of the bis(selanyl)-alkenes is of much importance due to its pharmacological properties, such as antioxidant and antinociceptive ones [22, 35]. They are also versatile tools in organic synthesis, being used as a precursor of enediynes and other functionalized olefins [137, 138]. Several green methods are reported to form this class of compounds, which are alternatives to the classical ones.

In 1991, Ogawa et al. [139] published a photocatalyzed reaction, between diphenyl or dibutyl diselenides **1** and different internal and terminal alkynes **49** using a tungsten lamp (500 W). The expected products **52** were isolated in 18–91% yields with moderate selectivity E:Z = 28:72 to 95:5 (Scheme 6.64).

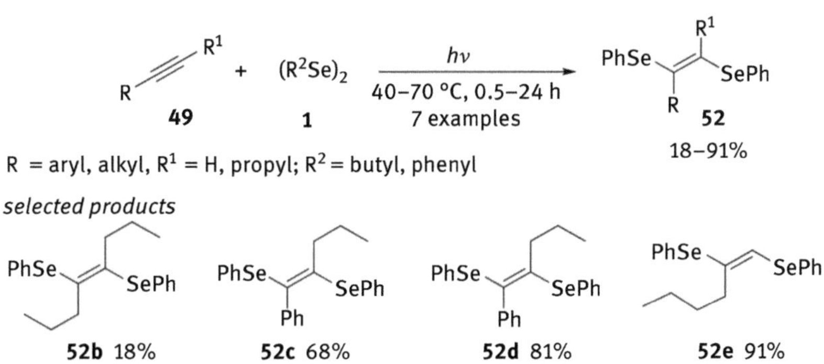

Scheme 6.64: Synthesis proposed by Ogawa and Sonoda [139].

The reaction mechanism involves the formation of the selenium-centered radical **II**, by reaction of PhSeH with oxygen (Scheme 6.65). Then, compound **I** was added slowly to a CCl_4 solution of diselenide **1** under oxygen atmosphere in the dark, generating species **IV** and product **52a**. Alternatively, it is predicted that the addition prompted the

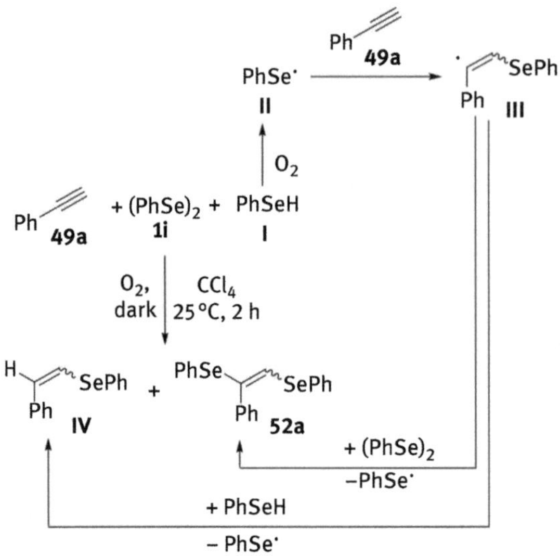

Scheme 6.65: Mechanism of the photoaddition [139].

formation of radical **III**, which undergoes a substitution with diselenide leading to the formation of product **52a** and species **IV**, with the regeneration of radical **II**.

Few years later, in 2003, Ananikov and Beletskaya [140, 141] developed a palladium-catalyzed procedure under solvent-free conditions to obtain bis-organoselanyl alkenes **52** in only 2 h at 100 °C (Scheme 6.66). The addition of diaryl diselenides **1** to terminal alkynes **49** was catalyzed by 1 mol% of $Pd(OAc)_2$, $PdCl_2$, or $Pd(PPh_3)_4$ complex in the presence of PPh_3 in excess (15 mol%). The products were obtained in excellent yields (95–99%) and high selectivity for the (*Z*)-isomer. They also performed the addition of disulfides **3** to alkynes **49**, and in all cases the products were obtained with a high selectivity (*Z*:*E* > 97:3).

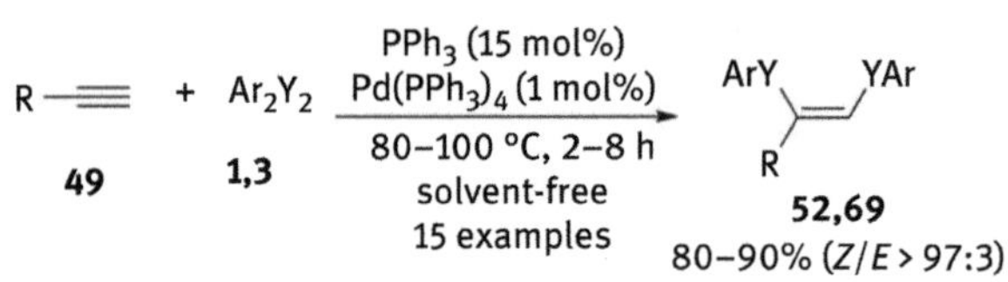

Y = Se, S; R = alkyl, alkynyl, propargyl amine and alcohols, homopropargyl alcohols
Selected products

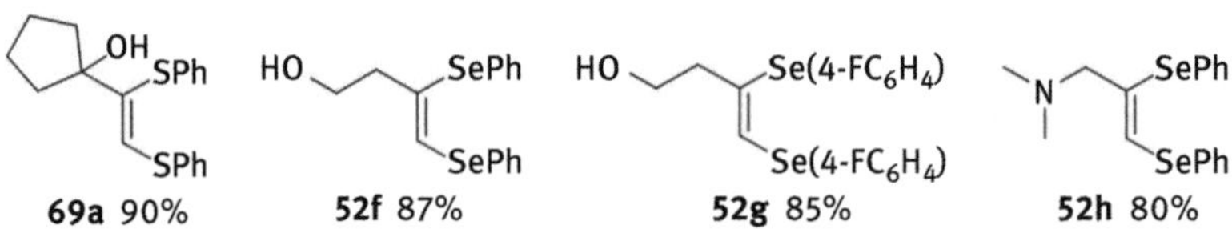

Scheme 6.66: Synthesis of Ananikov and Beletskaya [140, 141].

The mechanism of the solvent-free reaction using $Pd(PPh_3)_4$ **I** as the catalyst was studied by NMR. The formation of the dinuclear complexes **II** and **III** from the reaction of **I** with the dichalcogenide **1** or **3** is favored in the presence of an excess of PPh_3. Following this intermediate reacts with alkyne **49** to form the key adduct intermediate **V** that leads to products **52** and **69** after reaction with another molecule of **1** or **3**, respectively (Scheme 6.67).

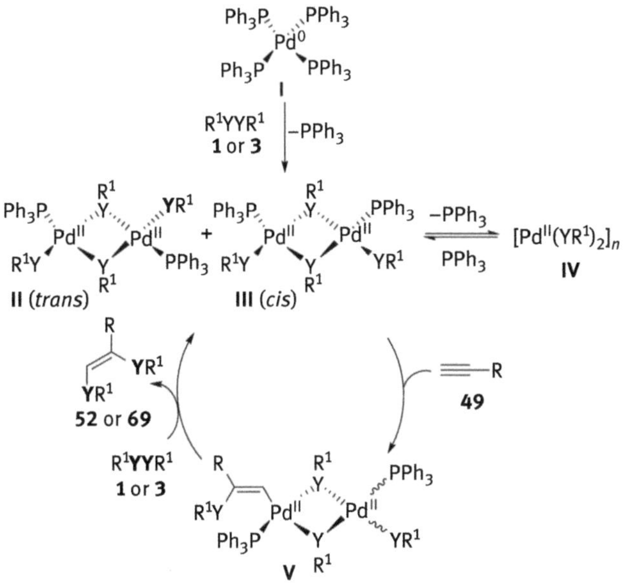

Scheme 6.67: Mechanism of the Pd-catalyzed reaction [140, 141].

In 2006, Perin et al. [122] described the solvent-free addition of chalcogenolate to propargylic alcohols and phenyl acetylene **49** to form the respective bis-organoselanyl alkenes **52** in good yields (Scheme 6.68). The reaction was performed using alumina-supported $NaBH_4$ as reducing agent to cleave the Se–Se bond in diphenyl diselenide **1**. (*E*)-Bis-organoselanyl alkenes **52** and **70** were preferentially obtained starting from phenyl acetylene (*E*:*Z* ratio = 71:29 to 88:12), while monosubstituted (*Z*)-vinyl chalcogenides **50** and **51** (*Z*:*E* ratio = 70:30 to 91:09) were obtained starting from propargylic alcohols. The reaction was conducted using three different conditions: at room temperature, heating at 60 °C (oil bath), and under microwave irradiation, allowing the formation of the products in only few minutes.

In 2007, Cai et al. [142] reported a new method to obtain (*Z*)-1,2-bis-chalcogen alkenes **52** and **69** using IL [bmim][PF_6] as the solvent (Scheme 6.69). The addition of diaryl diselenides **1** to aliphatic and aromatic terminal alkynes **49** was catalyzed by Pd (PPh_3) at 60 °C, and products **52** were obtained in excellent yields (94–98%) and high selectivity (*Z*:*E* ratio > 99:1). Vinyl sulfide **69** analogues were equally formed in

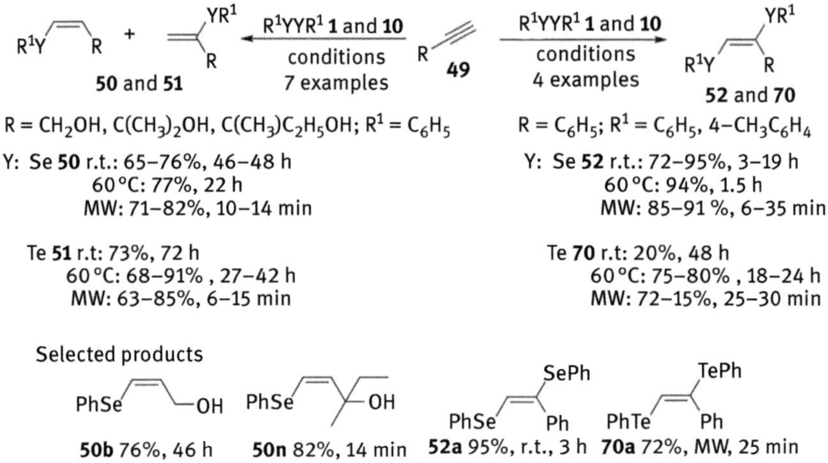

Scheme 6.68: Vinyl chalcogenides prepared by Perin et al. [122].

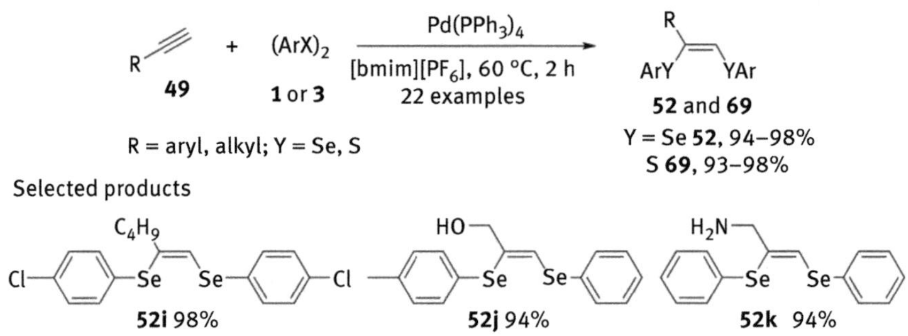

Scheme 6.69: Vinyl chalcogenides prepared by Cai et al. [142].

excellent yields (93–98%) and with the same selectivity than selenides **52**. Moreover, the solvent and the palladium catalyst were reused for four successive reactions and the product (*Z*)-1,2-bis(phenylselanyl)-hex-1-ene was obtained without decrease in yield.

In 2010, Perin and coworkers [143] reported the thiolation of phenylselenoalkynes **71** under solvent-free conditions using KF/alumina as base (Scheme 6.70). A mixture of (*Z*)- and (*E*)-isomers **72** was obtained in good yields (49–90%) after stirring at 60 °C for 2.5–3.0 h. Aromatic and aliphatic phenylselanyl alkynes **71** and thiols **31** were employed; however, the best yields were obtained using aromatic thiols. In all the studied examples, the presence of the organoselenium group in alkyne **71** directed the regiochemistry in favor of the (*Z*)-1,2-bis-chalcogeno alkenes (*Z*:*E* ratio = 62:38 to 75:25).

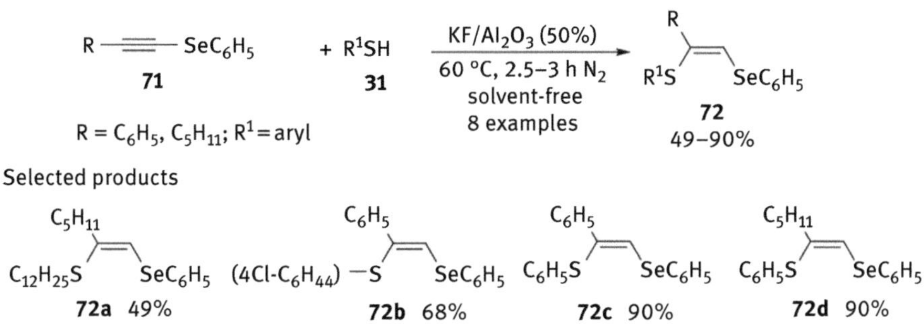

Scheme 6.70: Synthesis of mixed vinyl chalcogenides by Perin and coworkers [143].

The mechanism involves the reaction of the thiolate anion **I** with selanylalkyne **71**, forming the vinyl anion intermediates **II** and **III**. Subsequently, a protonation occurs giving the expected product **72** (Scheme 6.71).

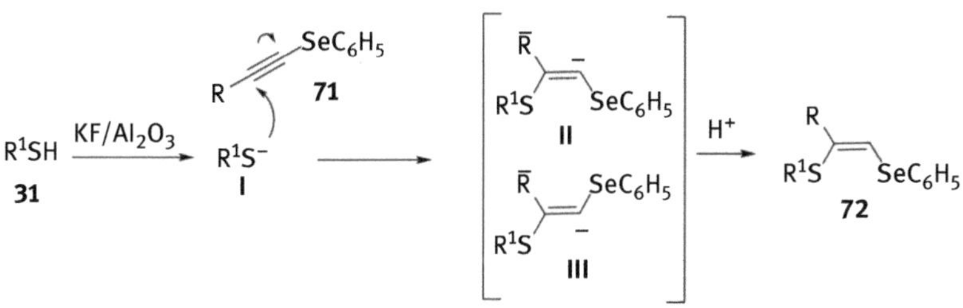

Scheme 6.71: Mechanism proposed by Lara et al. [143].

The same group in 2012 described [144] the use of KF/Al_2O_3 (50%)/PEG-400 as a recyclable system for the stereoselective synthesis of (*Z*)-1,2-bis-arylselanyl alkenes **52**, both under microwave irradiation and at 90 °C (oil bath). The formation of (*Z*)-alkenes **52** almost exclusively was found to be in 32–90% yields using the conventional heating and 22–98% yields under MW irradiation (Scheme 6.72). The reaction was extended to aromatic, aliphatic, and propargyl alkynes **49** and several diaryl diselenides **1,** and the products were obtained with high selectivity (*Z*:*E* ratio > 97:3). Interesting, when the sterically hindered dimesityl diselenide was used in the reaction with phenyl acetylene **49a**, 1-mesitylselanyl-2-phenylethyne was obtained in 72% yield in place of the expected vinyl selenide. Moreover, the reuse study of the KF/Al_2O_3 (50%)/PEG-400 system was made for the reaction between diphenyl diselenide and phenylacetylene. The results indicated the need for addition of more base to the reaction medium after each run. The expected product **52a** was isolated in 68% yield in the second reaction and in only 26% in the third one.

In 2009, Perin and coworkers [124] developed a new protocol to prepare 1,2-bis-phenylchalcogeno styrenes **52a** and **70a** in moderate yields using glycerol, at 60 °C in

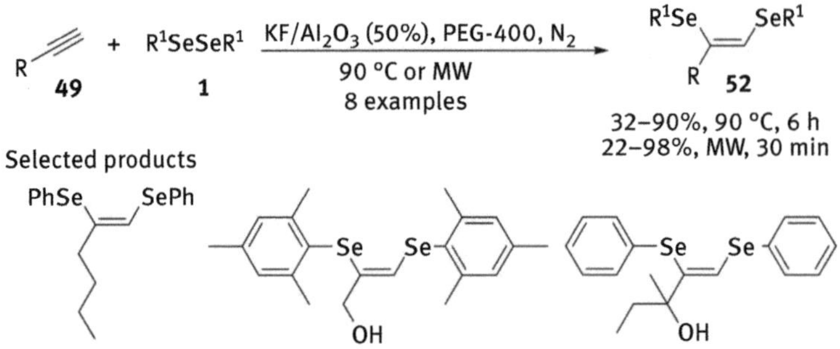

52l 22%, 90 °C 52m 90%, 90 °C 52m 98%, MW

Scheme 6.72: Synthesis of (*Z*)-1,2-bis-arylselanyl alkenes **52** [144].

the presence of $NaBH_4$ (Scheme 6.73). By the reaction between phenylacetylene **49a** and diphenyl diselenide **1i**, the products **52a** and **70a** were isolated in 75% and 45% yields and with a *Z*:*E* ratio of 77:23 to 82:18, respectively.

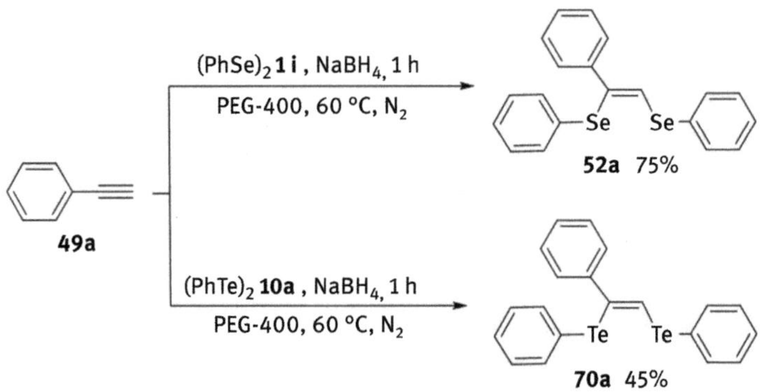

Scheme 6.73: Synthesis of 1,2-bis-phenylchalcogeno styrenes **52a** and **70a** [124].

The mechanism for the reaction using glycerol involves the cleavage of the chalcogen–chalcogen bond by $NaBH_4$ and the formation of the respective chalcogenolate anion **I.** Anion **I** reacts with phenylacetylene through *anti*-addition, possibly via the transition state **II,** leading to products **52a** and **70a** after abstraction of a proton from the reaction medium (Scheme 6.74).

The same reaction conditions were used by Perin et al. [145] to form (*Z*)-1,2-bis-chalcogeno alkenes (Se–Se, S–Se, and Te–Se) **52, 72,** and **73** through the addition

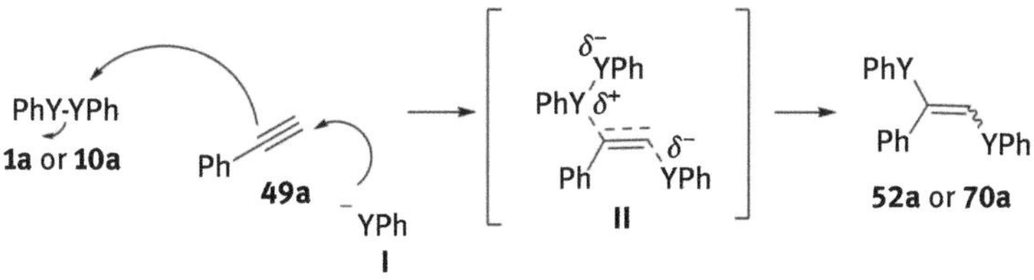

Scheme 6.74: Mechanism of the *anti*-addition [124].

of chalcogenolate anions to phenylselenoalkynes **71** (Scheme 6.75). The reactive species of chalcogen was generated in situ using $NaBH_4$ as reducing agent in PEG-400 as the solvent. Products **52, 72,** and **73** were obtained in short times and poor to very good yields (20–83% overall), with a high selectivity for the *Z* isomer (*Z*:*E* ratio = 97:3 to 100:0).

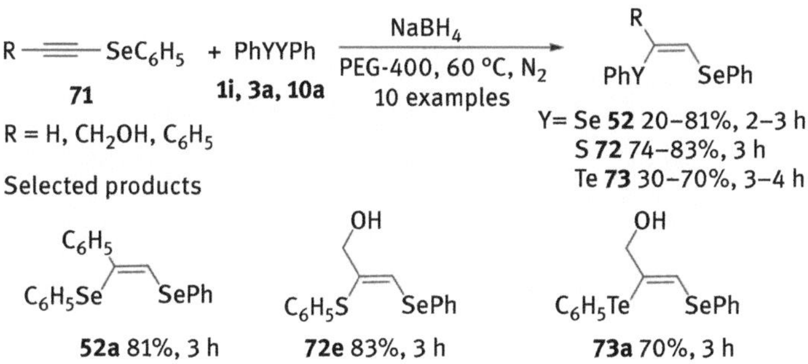

Scheme 6.75: Synthesis of (*Z*)-1,2-bis-chalcogeno alkenes using $NaBH_4$/PEG-400 [145].

In 2016, Perin, Alves and coworkers [146] reported an alternative method to obtain mainly (*E*)-bis-selanyl alkenes **52** in moderate to excellent yields (56–96%; *E*:*Z* ratio = 85:15 to 100:0) in aqueous H_3PO_2 as a reducing system. The protocol is based on the reaction of terminal alkynes **49** with organylselenols, generated in situ by the reaction of diorganyl diselenides **1** with H_3PO_2 at 90 °C under N_2 atmosphere. The reactions proceeded efficiently using a range of terminal alkynes and diorganyl diselenides, both containing electron-donating and electron-withdrawing groups. When the reaction was conducted in the absence of solvent, (*Z*)-vinyl selenides **50** were obtained in good to excellent yields (55–98%) with *Z*:*E* ratio of 70:30 to 100:0. Further, the H_3PO_2/H_2O system was recovered and directly reused for five times in the reaction between benzeneselenol **28** and phenylacetylene **49a** to prepare bis-selanyl alkene **52a** (Scheme 6.76).

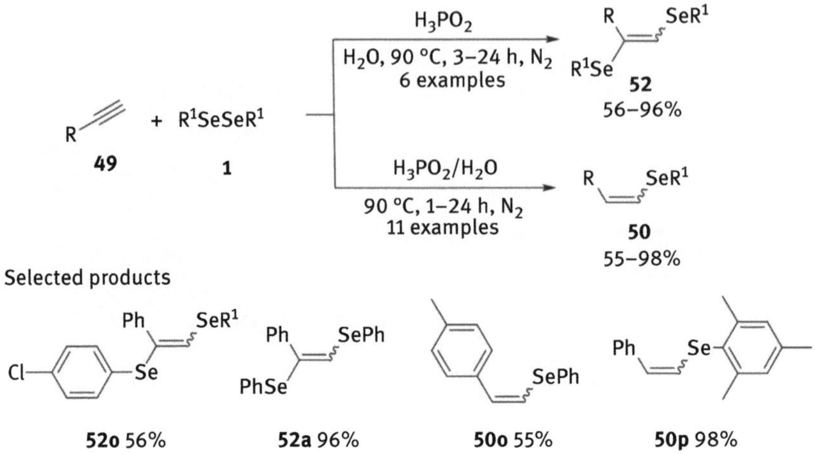

Scheme 6.76: Synthesis of vinyl selenides using H_3PO_2/H_2O [146].

6.10 Selenoalkynes

Selenoalkynes are used as versatile building blocks for a wide range of chemical transformations due to their stable triple bond and have been used as useful intermediates in organic synthesis [147]. These compounds have been reported as precursors to the bis-phenylchalcogen alkenes [145], 9-iodo-10-organochalcogen-phenanthrenes [148], 2-selanyl-benzo[*b*]furans [149], and 2-arylselenanylbenzo[*b*]selenophenes [150].

Several procedures for the synthesis of this class of compounds were described in the literature; however, only a few of them are eco-friendly. For example, an atom-economic protocol using $CuFe_2O_4$ NPs in PEG-400 was developed by Ranu and coworkers in 2013 to prepare different chalcogenoalkynes [65]. More specifically, selenoalkynes **71** and telluroalkynes **74** were prepared through the coupling reaction of alkynyl boronic acids **75** with diaryl dichalcogenides **1** and **10** (Se and Te) (Scheme 6.77). These

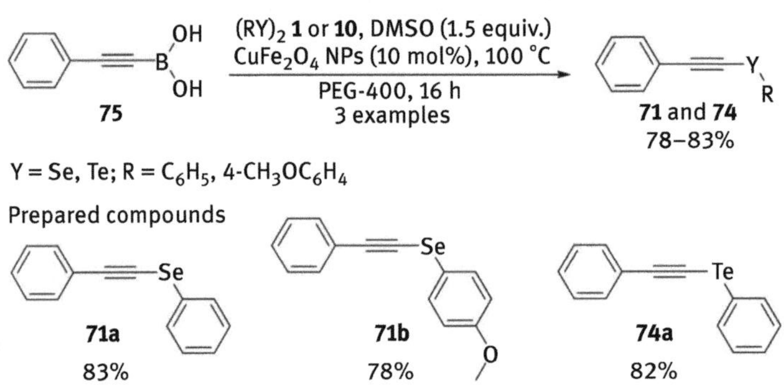

Scheme 6.77: Synthesis of chalcogenoalkynes by coupling reaction [65].

base-free reactions were carried out at 100 °C for 16 h, affording the chalcogenoalkynes **71** and **74** in 78–83% yields.

In 2018, Wu et al. [151] described a metal-free synthesis of alkynyl alkyl selenides **71** in aqueous medium by a three-component coupling reaction via a double C–Se bond formation process (Scheme 6.78). In this procedure, elemental selenium reacts with several epoxides **27** and a variety of functionalized terminal alkynes **49**, including aryl, alkyl, naphthyl, thienyl, and pyridyl. This efficient and straightforward route is highly regioselective for the preparation of selenoalkynes **71** in moderate to excellent yields at 45 °C with tolerance to a wide range of functional groups. In addition, the authors performed with efficience a gram-scale synthesis of compound **71c** (72% yield; 3.95 g). Further, this convenient pathway was applied to the selenation of pargyline (bioactive molecule) giving the product **71g** in acceptable yield.

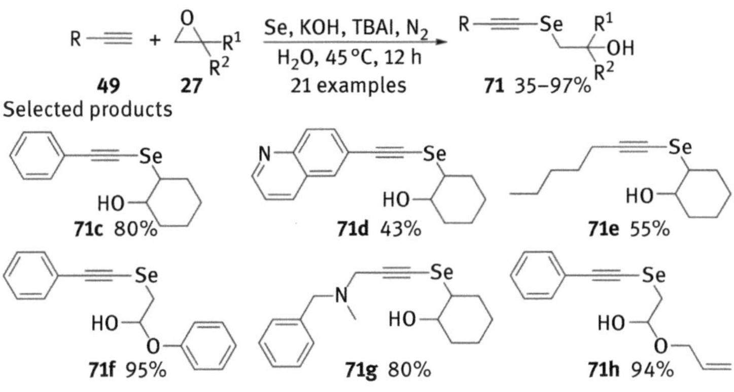

Scheme 6.78: Synthesis of alkynyl alkyl selenides [151].

The proposed mechanism for the double C–Se bond formation involves disproportionation of elemental selenium under basic conditions to generate a selenide anion. Subsequently, Se_2^{2-} attacks epoxide **27a** to form the ring-opened alkylselenide anion species **I**. This unstable species is quickly oxidized to the diselenide intermediate **II**. Finally, the terminal alkyne underwent alkylselenation in the presence of a base, to afford the alkynyl alkyl selenide **71** (Scheme 6.79).

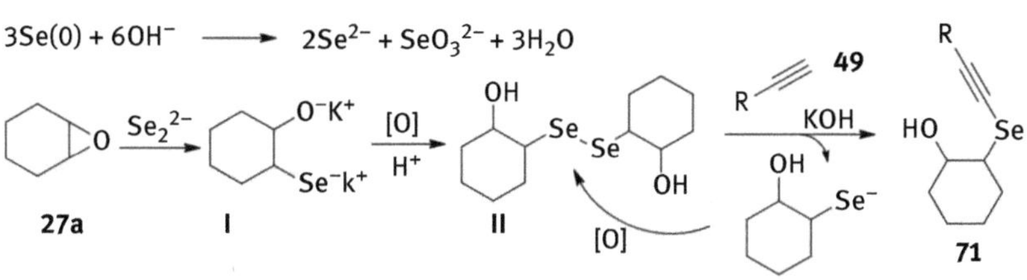

Scheme 6.79: Proposed mechanism to prepare alkynyl alkyl selenides [151].

6.11 Seleno-functionalized heterocycles

The heterocyclic motif is present in most of the natural occurring bioactive compounds [152]. For this reason, the development of methods to prepare new functionalized heterocycles is a hot topic in organic synthesis. The introduction of one or more organochalcogen substituents in the chemical structure is a strategy to increase the biological activities or to prepare molecules applied in asymmetric synthesis [153, 154].

In 2003, Fujita et al. [155] developed the first intramolecular oxyselenenylation, namely selenolactonization, and the subsequent deselenenylation in water using a polymer-supported organoselenium reagents, the selenyl bromide **76a** and the arylselenenyl bromides **76b** and **76c**. The use of a polymer support avoids the decomposition of the organoselenium reagents to form diselenide, due to the immobilization of the organoselenium on the polymer (Scheme 6.80). Compounds **76** were synthesized from aminomethyl-polystyrene and ArgoGel-NH_2, respectively. In the optimization tests, (*E*)-styrylacetic acid **77** was used as starting material in the reaction using three different resins, and the expected butenolide **79** was obtained in 29% yield with **76a**, 59% yield with **76b,** and 62% yield with **76c**. Compound **76c** was then used in various intramolecular oxyselenenylation and deselenenylation reactions to form the corresponding lactones **79** in moderate to good yields (41–83%). When unsaturated alcohol was used, however, the expected allylic ether was not formed, probably due to the interruption of the intramolecular nucleophile attack by water in the oxyselenenylation step.

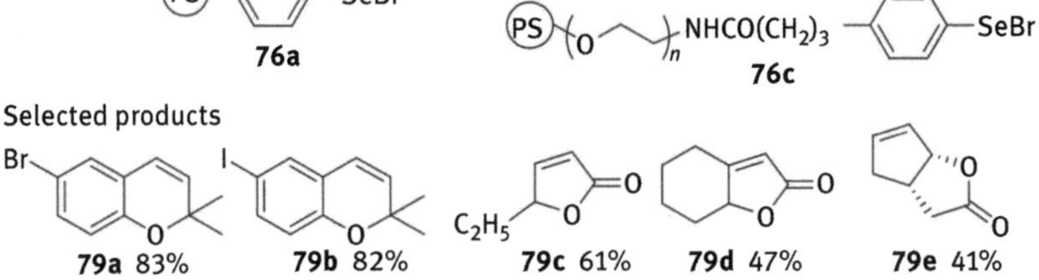

Scheme 6.80: Oxyselenenylation and deselenenylation in water [155].

In 2004, Ericsson and Engman [156] reported a microwave-assisted group-transfer cyclization of organoselenides to obtain tetrahydrofuran derivatives **80** (Scheme 6.81).

The cyclization of benzylic selenide **6af** was carried out forming the product **80** in 91% yield after 5 min, in a *cis/trans* ratio of 1/2.1.

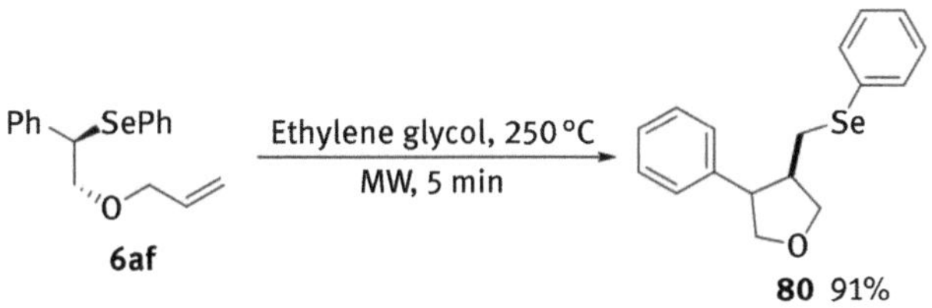

Scheme 6.81: Microwave-assisted group-transfer cyclization [156].

In 2012, Perin et al. [157] described the Zn-catalyzed preparation of new 5-arylchalcogenoalkyl-1H-tetrazoles **81** and **82** by the 1,3-dipolar cycloaddition of arylchalcogenoalkyl nitriles **83** with sodium azide in aqueous solution (Scheme 6.82). Products **81** and **82** were formed in moderate to good yields (61–86%) in 24 h. In general, electronic effect of the aromatic ring in the arylselenium species did not affect the reaction. Among the prepared compounds, **81a** showed good antifungal activity against *Trichosporon asahii* and *Candida lipolytica*.

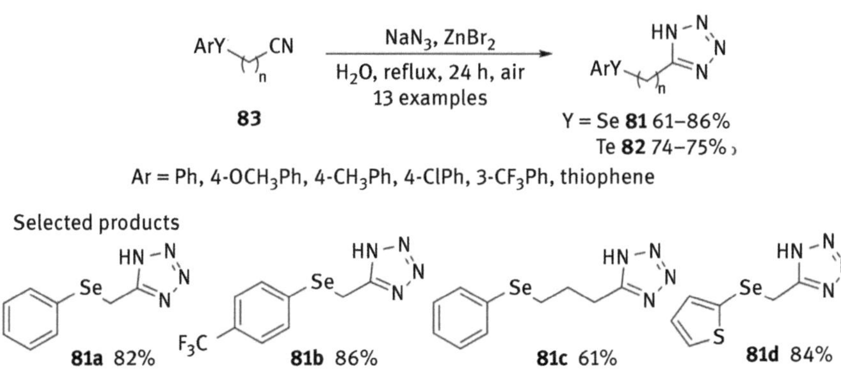

Scheme 6.82: Synthesis of new 5-arylchalcogenoalkyl-1H-tetrazole **81** and **82** [157].

Lenardão and coworkers in 2013 [158] published a method to prepare 3-arylselenylindoles **84** through the reaction between ArSeCl **11** and indole **85** at room temperature, under N_2 atmosphere using IL [bmim][$SeO_2(OCH_3)$] (Scheme 6.83). Products **84** were obtained in good yields (53–78%) in a relatively short reaction time (2–3 h). Moreover, it was observed that the presence of electron-donor or electron-withdrawing groups in the organoselenium species and in the indole did not affect the reaction. In addition, the IL could be reused up to four times with good results to prepare the 3-phenylselenylindole **84c**.

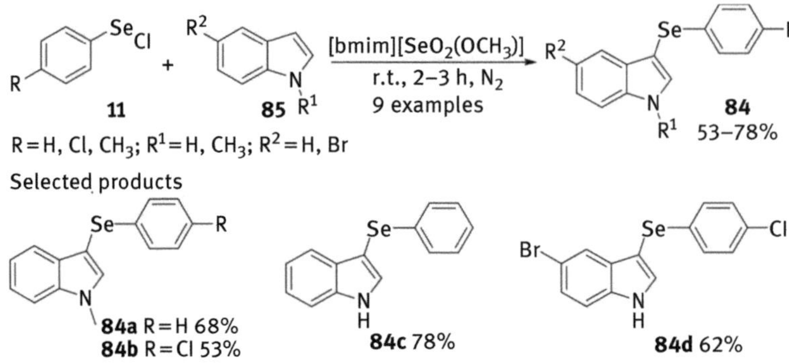

Scheme 6.83: Synthesis of 3-selenylindoles **13** in ionic liquid [158].

2-Organylselanyl pyridines **86** were prepared by the reaction of 2-chloropyridines **87** and organylselenols generated in situ from diorganyl diselenides **1** in glycerol, using hypophosphorous acid as reducing agent (Scheme 6.84) [159]. Products **86** were prepared in short reaction times (1.5–5.5 h) in 44–97% yields. The reaction was not sensitive to electronic effect in the aromatic ring of diaryl diselenides nor in the 2-chloropyridines. The reducing solvent–system glycerol/H_3PO_2 were reused for five times maintaining good yields (80–99 %) in the synthesis of 2-phenylselanyl pyridine. The authors described that glycerol also acts as a reducing agent that regenerates H_3PO_2 for new successive reactions with diselenide.

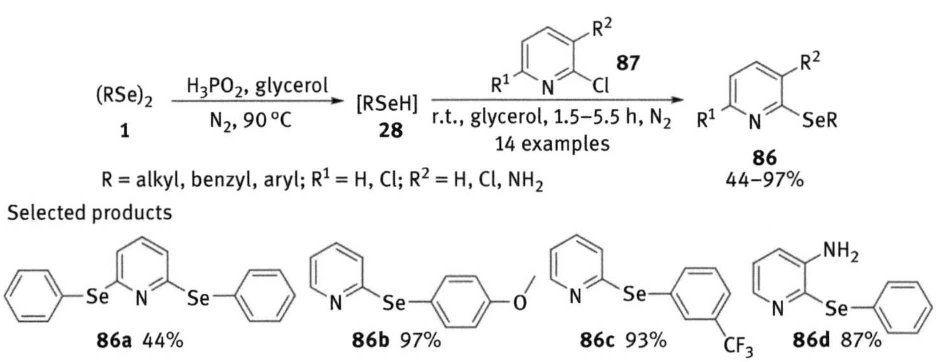

Scheme 6.84: Synthesis of 2-organylselanyl pyridines **86** [159].

In 2014, Potapov et al. [160] developed the green synthesis of two compounds: 1,5-bis[(3,5)-dimethylpyrazol-1-yl]-3-selena pentane **88** and 1,3-bis(1,2,3-benzotriazol-1-yl)-2-selena propane **89** (Scheme 6.85). The reaction was conducted using an inexpensive reagent: elemental selenium and sodium formaldehydesulfoxylate ($HOCH_2SO_2Na$, Rongalite®) in aqueous NaOH, to generate in situ the selenide ion. Products **88** and **89** were isolated in 70% and 90% yields, respectively. The

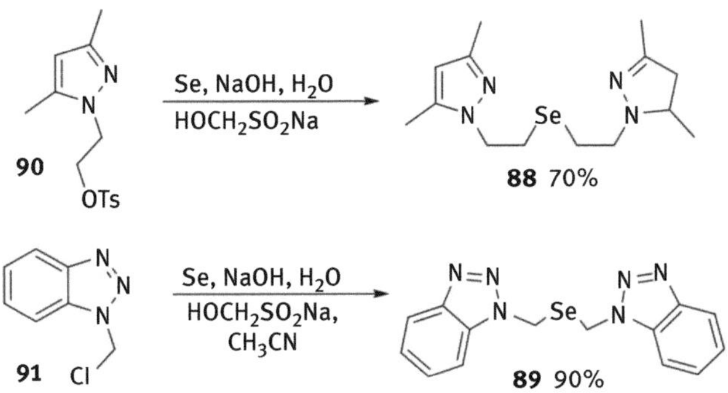

Scheme 6.85: Synthesis of azole-selenoethers **88** and **89** [160].

nitrogen atmosphere was unnecessary, probably due to the reductive atmosphere created by the formation of SO_2.

In 2016, Alves and coworkers [161] described the preparation of (arylselanyl)phenyl-1H-1,2,3-triazoles **92** using microwave irradiation as a nonclassic energy source (Scheme 6.86). Products **92** were obtained in good to excellent yields (85–97%) after short reaction time (30 min) through the azide–alkyne cycloaddition catalyzed by copper (CuAAC), in the presence of sodium ascorbate. A variety of terminal alkynes **49** with different substituents, aryl, alkyl, vinyl, ester, and alcohols, were used in the reaction with azidophenyl arylselenides **93**. Among the obtained products, compounds **92d** and **92e**, which could present liquid-crystalline properties, were successfully prepared.

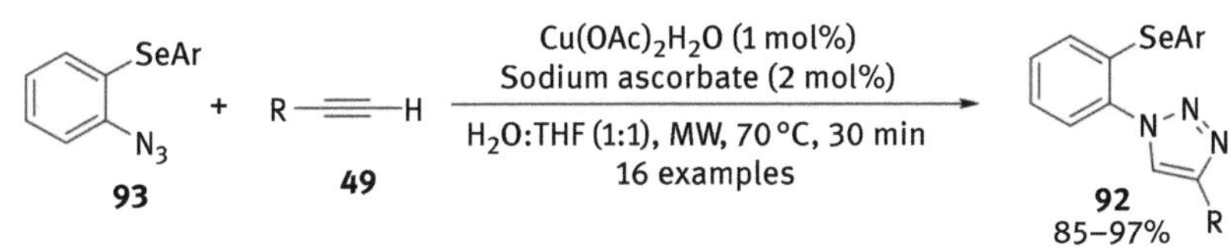

R = PH, 4-CH_3Ph, 4-ClPh, 4-OCH_3Ph, 2-OCH_3Ph, 2-OC_4H_9Ph, 4-OC_4O_9Ph, 2-OC_8H_{17}Ph, C_4H_9, CH_2CH_2OH, $COOCCH_3$, $CHCH_2CH_3$

Selected products

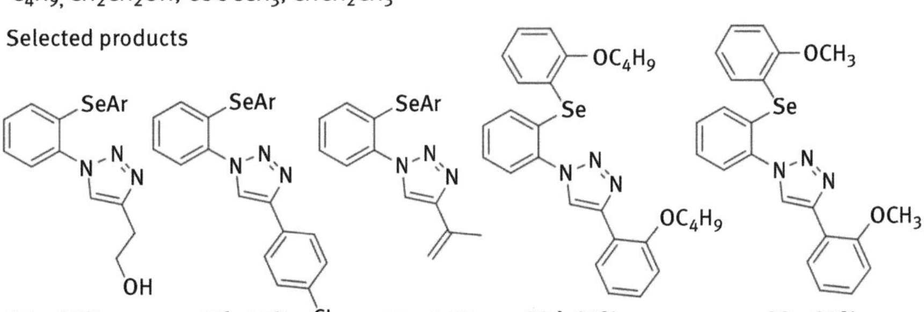

92a 85% **92b** 97% **92c** 94% **92d** 87% **92e** 92%

Scheme 6.86: Microwave-assisted synthesis of (arylselanyl)phenyl-1H-1,2,3-triazoles 92 [161].

In 2017, Lenardão and coworkers [162] described the copper-catalyzed ultrasound-promoted synthesis of new Se-containing derivatives of chrysin **94**, starting from diaryl diselenides **1** and chrysin **95** (Scheme 6.87). Products **94** were obtained in good to excellent yields (60–89%) after 3–8 h of reaction. Different diorganyl diselenides were used and the yield of the reaction was not affected by the presence of electron-donating or electron-withdrawing groups in the aromatic portion of the diselenide. When bis(2-aminophenyl)diselenide was used, no product was formed, probably due to the presence of the amino group, which could deactivate the copper catalyst. Further, the reaction doesn't work well with dibutyl diselenide, and the expected compound **94e** was obtained only in trace amounts.

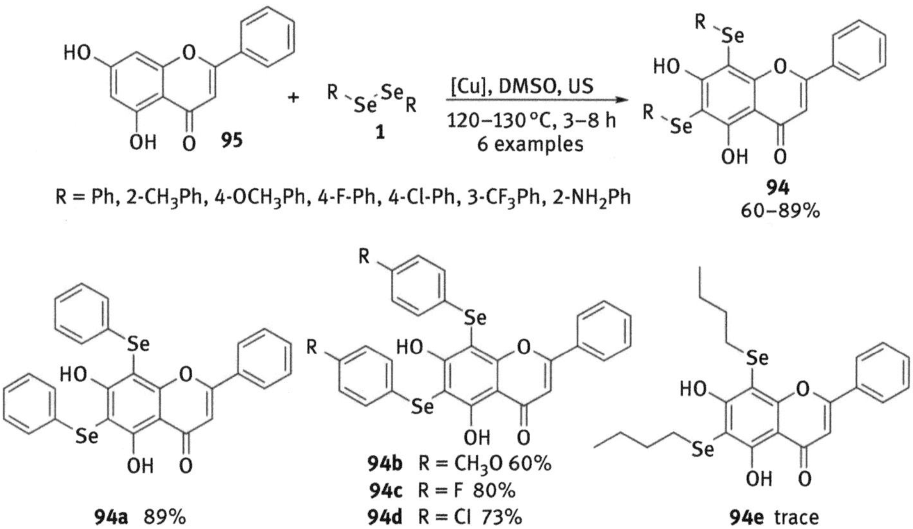

Scheme 6.87: Synthesis of new Se derivatives of chrysin **94** [162].

The mechanism of the selenation reaction involves the reaction of diphenyl diselenide with Cu, forming intermediate **I** (Scheme 6.88). Subsequently, **I** undergoes a nucleophilic attack from chrysin, generating species **II** and **III**. Then, species **II** donates a proton to **III**, affording the mono-Se-chrysin **IV**, PhSeH and regenerating CuI for a new reaction. A second selenation occurs from **IV** in the same way to give the expected product **94**.

In the same year, Lenardão and coworkers [163] published a method to obtain 3-selanyl-1H-indoles **84** and 3-selanyl-2-arylimidazo[1,2-α]pyridines **96** through a one-step reaction between 1H-indoles **85** or 2-arylimidazo[1,2-*α*]pyridines **97** and diorganyl diselenides **1** (Scheme 6.89). This copper-catalyzed reaction proceeded under ultrasound in few minutes (15–60 min), affording the expected products in good to excellent yields (57–96%). A variety of heterocycles and diselenides were satisfactorily used as substrates and no apparent influence of the substituents was observed. Six of the

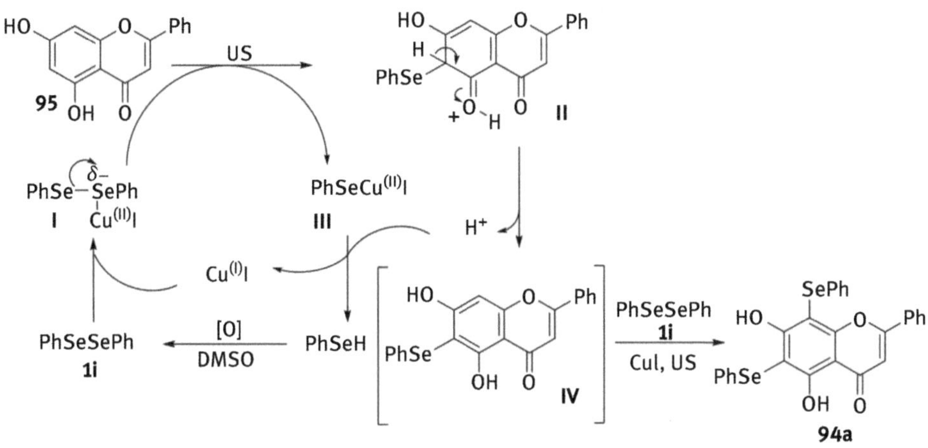

Scheme 6.88: Mechanism proposed for the synthesis of **94** [162].

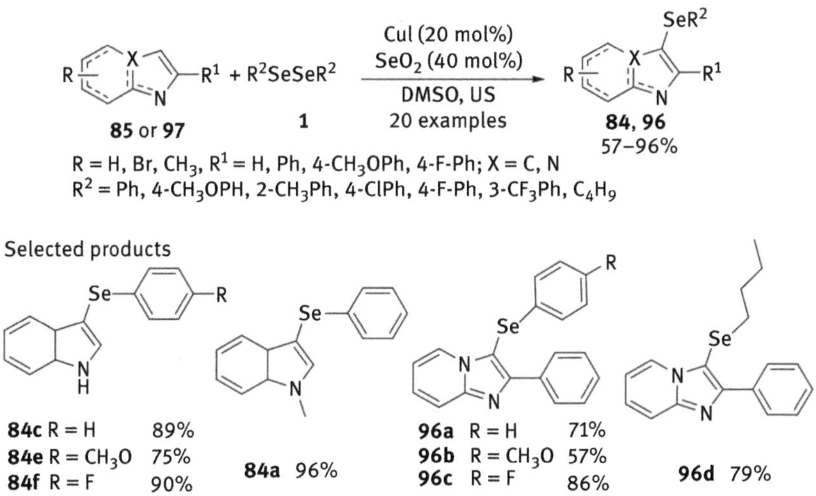

Scheme 6.89: Synthesis of 3-selanyl-1H-indoles **84** and 3-selanyl-2-arylimidazo[1,2-α]pyridines **96** [163].

prepared compounds (**84a–d** and **96a–d**) showed significant antioxidant activity in vitro, placing them as promising molecules for additional pharmacological studies.

In 2018, Sun and coworkers [164] reported a metal-free, iodide-catalyzed electrochemical C–H selenation of various indoles **85** and imidazo[1,2-α]pyridines **97** (Scheme 6.90). The method used an iodide salt both as electrolyte and catalyst, and the electrolysis was performed using a graphite plate anode and platinum plate cathode under galvanostatic node (constant current ~18 mA). The reaction has an excellent regioselectivity and the products **84** substituted in the C3-position

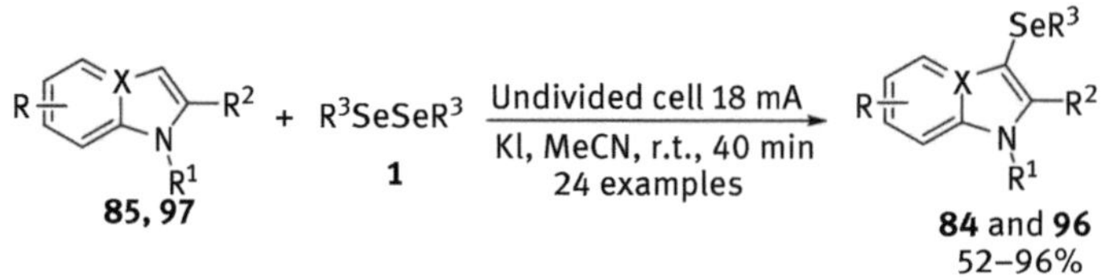

R = H, 4-CH_3, 5-CH_3, 6-CH_3, 5-F, 6-F, 5-Cl, 5-Br, 4-$COOCH_3$, 5-$COOCH_3$, 6-$COOCH_3$, 5-NO_2, 7-NO_2; R^1 = H, CH_3, Boc, Ts; R^2 = H, $COOCH_3$, CH_3, Ph; R^3 = Ph, Br, CH_3, X = C, N

Selected products

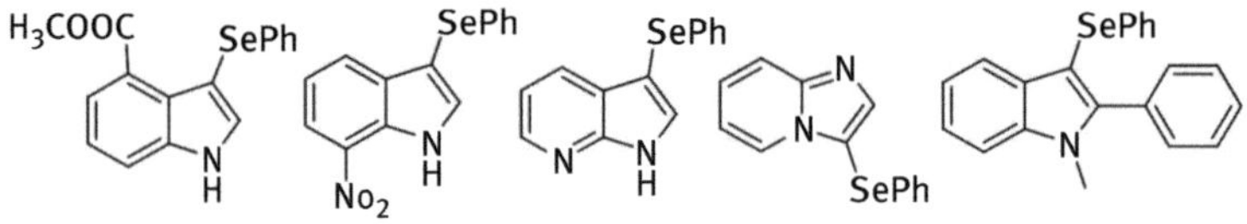

84g 52% **84h** 63% **84i** 84% **96e** 56% **84j** 86%

Scheme 6.90: Electrochemical C–H selenation [164].

were formed exclusively in moderate to excellent yields (52–96%). The protocol worked well with different indoles, unprotected indoles bearing electron-donor or electron-withdrawing groups and *N*-methyl indoles. The less reactive substrates were methyl-4-indolecarboxylate **84g** and 7-nitroindole **84h**, which afforded the respective products in 52% and 63%, respectively. This low reactivity was attributed to the steric hindrance and the strong electron-withdrawing effect, respectively. Moreover, the reaction didn't work with *t*-butyl-1H-indole-1-carboxylate and *N*-(4-toluenesulfonyl)indole, due to the strong electron-withdrawing effect of the substituents. Good results were obtained using dialkyl diselenides **1** as substrate, affording the respective products **84** in very good yields (83–87%). 7-Azaindole and imidazo [1,2-α]pyridine were suitable substrates for the reaction, and the respective products **84j** and **96e** were isolated in 84% and 56% yields, showing the versatility of the reaction.

Two possible mechanisms were proposed for the reaction (Scheme 6.91). In pathway A, oxidation of iodide at the anode occurs, forming iodonium, which reacts with the indole to give 3-iodo-indole **I**. This species is then captured by diselenide, forming the expected product **84** and molecular iodine. In parallel, iodine is reduced to iodide anion, which completes the catalytic cycle (Scheme 6.91A). The other possibility, pathway B, involves the formation of electrophile RSeI in situ, by the reaction of the pre-formed iodine with diselenide. This species undergoes heterolysis to form the electrophile RSe^+, which reacts with the indole, giving product **84**.

In the same year, Yang et al. [165] described a regioselective visible light promoted C–H selenation of 4-amino-substituted coumarin derivatives **98**, forming the selenylated products **99** in moderate to good yields (37–91%) after 24 h (Scheme 6.92). In the variation of the reaction scope, it was verified that electron-donating and

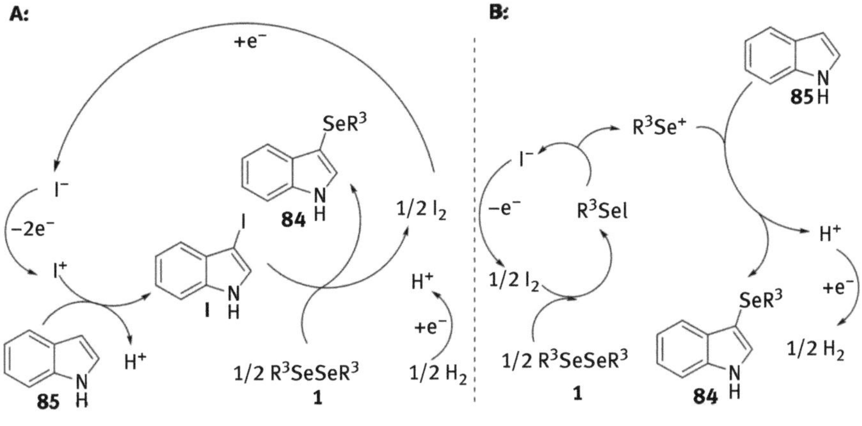

Scheme 6.91: Proposed mechanism pathways to prepare 3-selanylindole **84** [164].

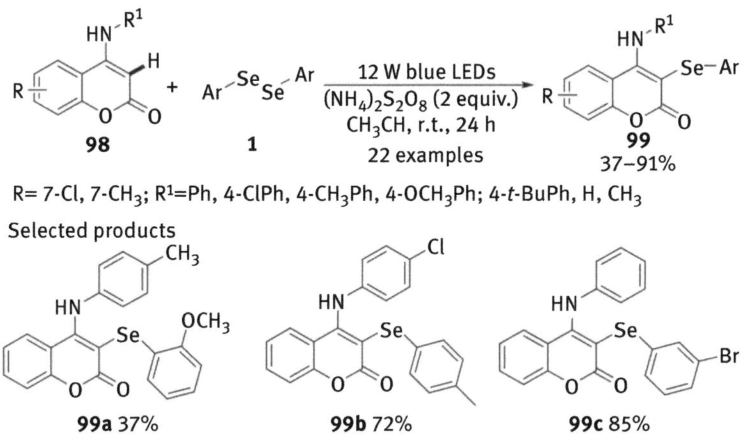

Scheme 6.92: Selenylation of 4-amino-substituted coumarin derivatives **98** [165].

electron-withdrawing groups in the coumarin moiety were well tolerated and did not affect the reactivity. Steric hindrance in the substituted diselenides in turn affected the reaction, and product **99a** was formed in only 37% yield. When *N*-substituted-4-(phenylamino)-2H-chromen-2-ones **98** were used, the di-selenylated products **100** were obtained in 69–80% yields. Substrates with substituent at the *para*-position of the 4-aminophenyl group afforded exclusively the C-3-monoselenylated product (Scheme 6.93).

Two plausible mechanisms could be involved in the reaction. In both cases, initially the radical anion $SO_4^{-\bullet}$ is formed from the persulfate anion under visible light. Subsequently, it reacts with the coumarin derivative to form the key radical cation intermediate **I**, which coexists with **II** (Scheme 6.94). In pathway A (R^1 = H in **98**), species **III** (a resonance structure of **II**) reacts with diselenide

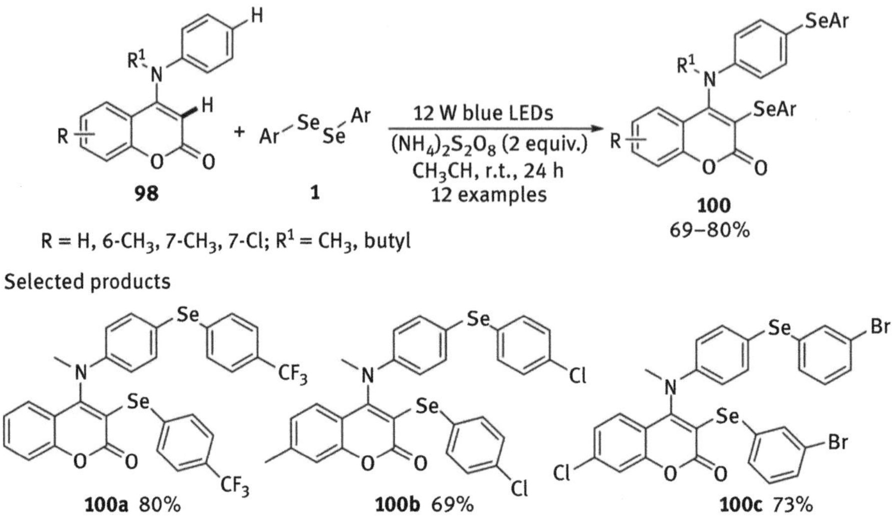

Scheme 6.93: Selenylation of *N*-substituted 4-amino coumarin derivatives **98** [165].

to afford the iminium intermediate **IV**. After deprotonation, **IV** leads to the expected product monoselenylated **99** (Scheme 6.94A). Instead, in pathway B ($R^1 = CH_3$ in **98**), species **III′** reacts with diselenide to form intermediate **IV′**, which leads to the monoselenylated intermediate species **99′**. Subsequently, **99′** reacts with the radical anion $SO_4^{-\bullet}$, affording the radical cation intermediate **V**, which exists with the resonance structure **VI**. Then, **VI** reacts with diselenide, generating intermediate **VII** that loses a proton to give the di-selenylated product **100** (Scheme 6.94-B).

6.12 Selenium-containing heterocycles

The molecular hybridization of two class of compounds is an interesting synthetic strategy to potentiate biological properties. In this sense, it has been an interest to develop new procedures to prepare heterocyclic compounds associated with chalcogen atoms (S, Se, or Te) [28].

As discussed in Section 6.11, compounds containing nitrogen and chalcogen atoms in their structure are an important class of molecules, having applications in asymmetric catalysis [33]; for example, Ebselen (Figure 6.3) is a selenium-containing heterocycle that has been used as a biological model capable of simulating catalytic functions of natural enzymes, acting as a peroxynitrite scavenger and a GPx mimic [33].

In 2008, Naik and coworkers [166] described the synthesis of selenopheno[2,3-*b*] quinoline derivatives **101** or **102** via the reaction between 2-seleno-3-formyl-quinolines **103** and phenacylbromide **104** or 2-chloroacetamide **105** under basic conditions and

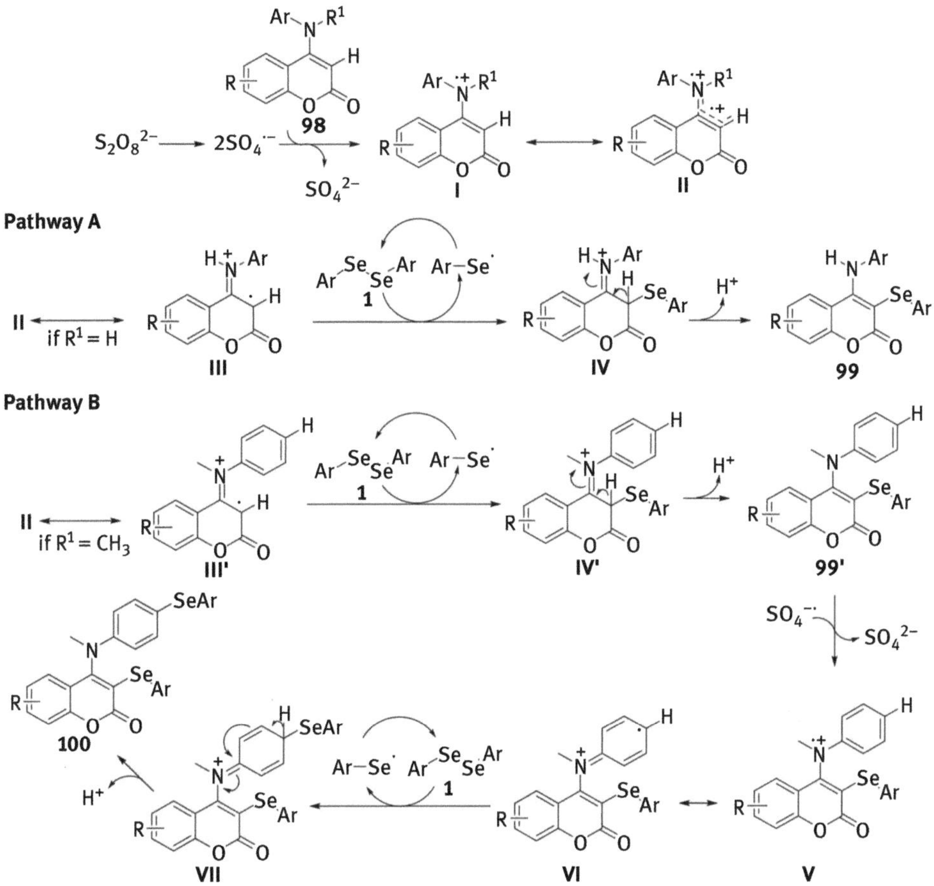

Scheme 6.94: Proposed mechanisms for the formation of **99** and **100** [165].

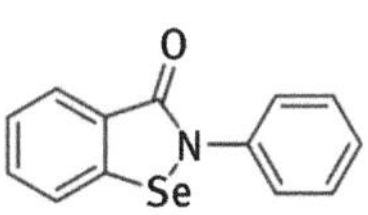

Figure 6.3: Chemical structure of Ebselen.

without solvent (Scheme 6.95). The preparation of five- and six-membered heterocyclic compounds containing one or two heteroatoms fused to quinoline ring is interesting from the chemical and biological points of view, due to the significant properties of the natural analogues. More specifically, the Se-containing quinoline core has received great attention due to the potential chemoprevention with low toxic effects. By this solvent-free procedure under microwave irradiation, a new class of quinolone derivatives **101** and **102** was prepared in 85–90% yields in only 7–8 min. It is necessary

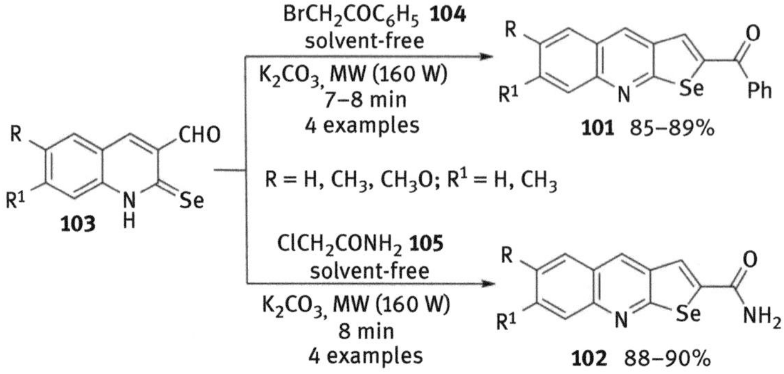

Scheme 6.95: Synthesis of selenopheno[2,3-*b*]quinoline derivatives **101** or **102** [166].

to point out that the isolation of selenopheno[2,3-*b*]quinoline derivatives **101** or **102** occurred just by adding water to the reaction mixture.

The synthesis and application of selenium-containing IL was proposed by Koketsu, Ishihara and Tanaka [167] in 2005. The IL 2-amino-4,5-dihydro-1,3-selenazol -4-iminium chloride was prepared by using *N,N*-unsubstituted selenoureas **106** and chloroacetonitrile **107** in a mixture of ethanol:water 99:1 as the solvent (Scheme 6.96). By this procedure, five different selenoureas **106** were used to prepare several examples of IL **108**. The reaction of IL **108a** with $NaBH_4$ (2 equiv.) in ethanol at room temperature for 4 h afforded 2-piperidino-1,3-selenazole **109** in 56% yield. The refluxing of 2-piperidino-4,5-dihydro-1,3-selenazol-4-iminium chloride **108a** (R= -$(CH_2)_5$-) for 24 h in water, in turn, gave 2-piperidino-4,5-dihydro-1,3-selenazol-4-one **110** in 99% yield. The use of these ILs shows as an alternative to prepare Se-containing five-membered heterocycle, specifically 1,3-selenazoles are of special interest as synthetic tools and in medicinal chemistry, due to their antibiotic and cancerostatic activities [168].

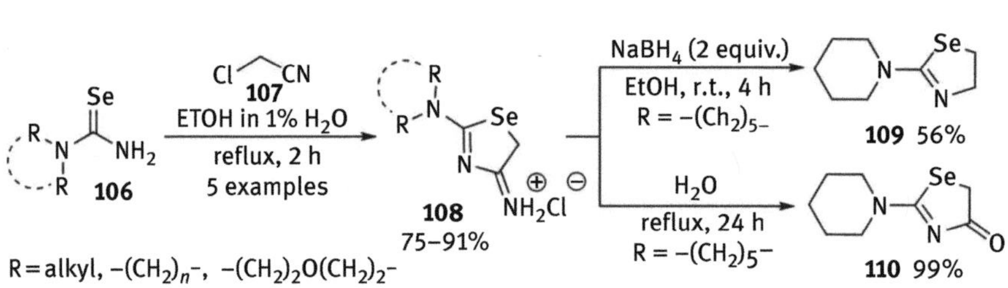

Scheme 6.96: Synthesis of ionic liquid 2-amino-4,5-dihydro-1,3-selenazol-4-iminium chloride **108** and their derivatives **109–110** [167].

Based on experimental evidences, a reaction mechanism for the formation of ILs **108** was proposed, which involves the participation of water (Scheme 6.97). First, a nucleophilic attack of the selenium atom of selenourea **106** to the methylene carbon of nitrile **107** occurs, forming salt **I**. Next, a water-assisted annulation occurs to give

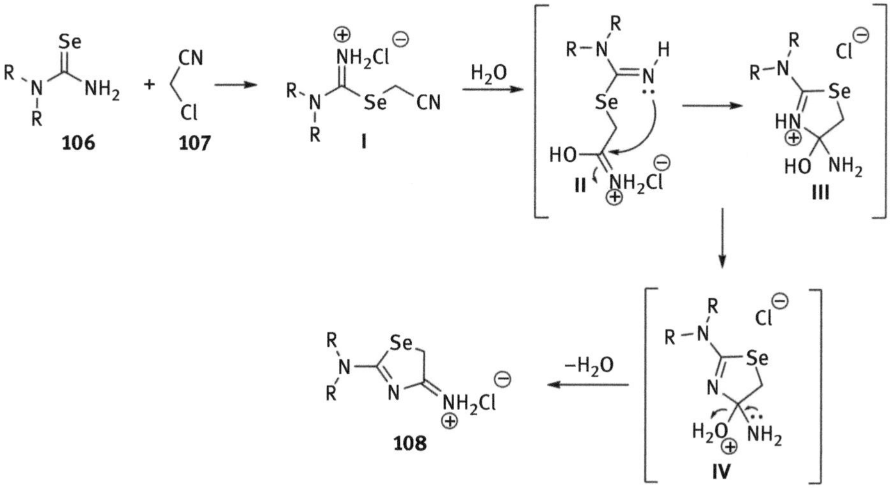

Scheme 6.97: Mechanism of the formation of five-membered heterocycles **108** [167].

the ammonium chloride intermediates **III** and **IV**. Finally, **108** is formed by releasing water to the reaction medium.

Also interested in selenium-containing heterocycles, Srinivasan and coworkers [169] described the synthesis of 2-amino-1,3-selenazoles **111** by the condensation of selenourea **106** with phenacyl bromides **112** under mild conditions (Scheme 6.98). In this chapter, selenazoles were isolated in excellent yields after few minutes of reaction, using only IL and water (1:1) as a solvent system. Regarding the use of water to facilitate the solubility of the selenourea, the IL shows a good miscibility with water, generating a homogeneous system. After completion of the reaction, the mixture was poured into water and the product **111a** was extracted using ethyl acetate. To reuse the IL [Hbim][BF_4], the aqueous layer containing IL was dried under reduced pressure (80 °C, at 10 mmHg). After water removal, [Hbim][BF_4]

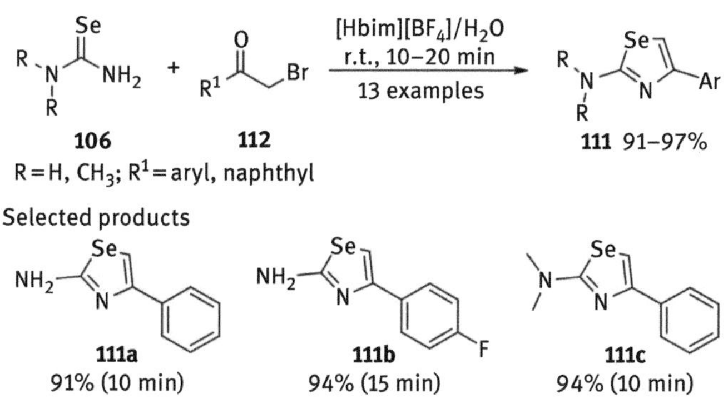

Scheme 6.98: Synthesis of 2-amino-1,3-selenazoles **111** [169].

could be easily recovered and reused three times, affording good yields of product **111a** in all the successive reactions.

Considering the chemical and biological importance of the selenazole unit, Alves and coworkers described in 2017 [170] the domino oxidative cyclization of methyl ketones with bis(2-aminophenyl) diselenide to prepare 2-acyl-benzo[1,3-*d*] selenazoles (Scheme 6.99). This direct one-pot strategy afforded 2-acyl-benzo[1,3-*d*] selenazoles **113** from good to excellent yields through in situ generation of 2-arylethane-1,2-diones **114** from commercially available arylmethyl ketones **35** using DMSO as solvent and the nontoxic reducing agent $Na_2S_2O_5$. This intramolecular cyclocondensation was more efficient using microwave irradiation compared to the conventional heating, with the benzo[1,3-*d*]selenazoles **113** being obtained in good to excellent yields (60–94%) after 2.3 h under MW versus 48 h under conventional heating (100 °C). The proposed mechanism involves three steps. First, the formation of ethane-1,2-dione derivative **114** by the Kornblum oxidation occurs. In the second step, the amino group of bis(2-aminophenyl) diselenide **1k** reacts with ethane-1,2-dione **114**, forming the imine diselenide intermediate **I**, followed by the Se–Se bond cleavage, leading to intermediates **II** and **III**. Then, radical **III** undergoes an intramolecular cyclization, affording the aminyl radical **IV**. In the final step, intermediate **IV** is oxidized to give the desired selenazole **113** (Scheme 6.99).

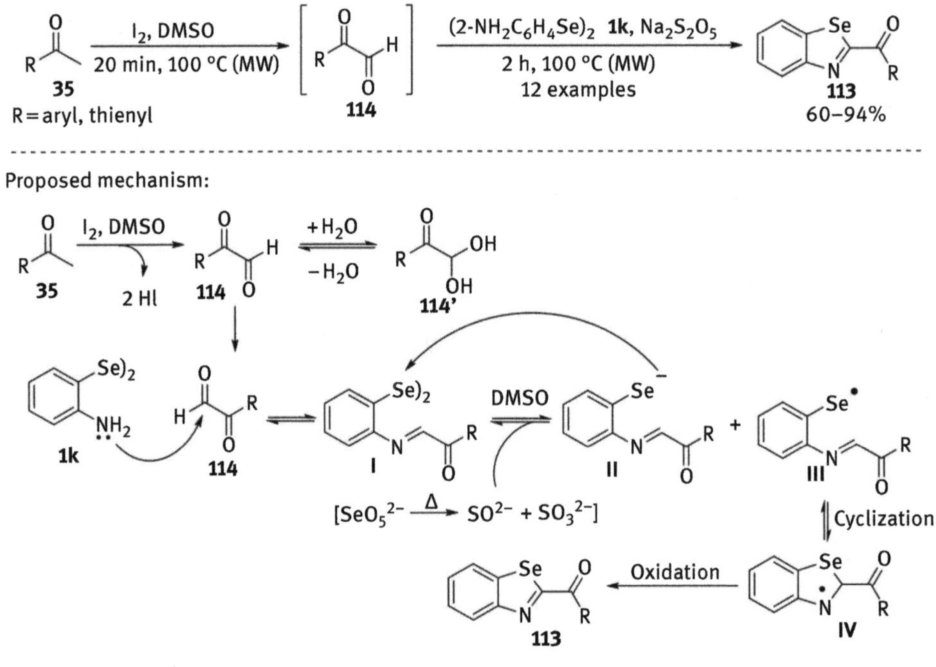

Scheme 6.99: Synthesis and proposed mechanism of 2-acyl-benzo[1,3-*d*]selenazoles **113** [170].

Besides selenoureas, other starting materials for the synthesis of selenium-containing heterocycles have been explored. For instance, isoselenocyanates **115** can be a useful reagent for cyclization reactions. Sashida et al. [171] reported the intramolecular cyclization reaction of a selenol intermediate, which is generated in situ from the reaction between the isoselenocyanate **115** and 2-ethynylanilines **116**, to prepare (*Z*)-4-methylene-3-selenaquinoline derivatives **117** (Scheme 6.100). The desired compounds were selectively obtained in short reaction time using microwave irradiation instead of conventional heating (oil bath). The reaction proved quite sensitive to steric factors, and no reaction occurred using secondary amines, since they were unable to attack the C-*sp* of the starting material **116**.

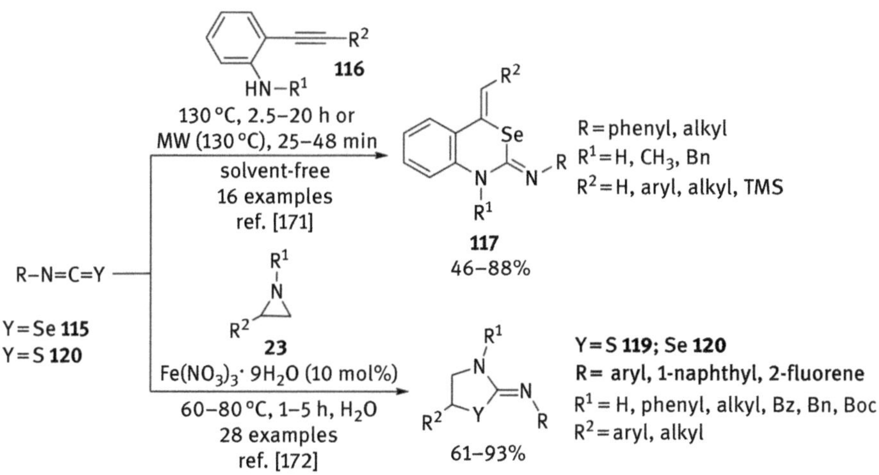

Scheme 6.100: Synthesis of heterocycles **117** [171], **119**, and **120** [172].

Both isothio- **120** and isoselenocyanates **115** were used to prepare S- and Se-containing heterocycles **119** or **120** "on water" conditions, as described by Sengoden and Punniyamurthy [172] (Scheme 6.100). The general method involves an iron-catalyzed [3 + 2] cycloaddition reaction of aziridines **23** with heterocumulenes (the isochalcogenocyanates **115** or **120**) using $Fe(NO_3)_3{\cdot}9H_2O$ (10 mol%) as the catalyst. Several aziridines **23** successfully reacted with isoselenocyanates **115**, giving the respective products **120** in good to excellent yields (61–93%). Moreover, the reaction was scaled up for the reaction between 1-isoselenocyanato-2-methoxybenzene (7.0 mmol) and 1-isopropyl-2-phenylaziridine (7.7 mmol), affording the respective iminoazoselenolidine in 91% yield. In this work, aziridines reacted with other heterocumulenes, including isocyanates, isothiocyanates, and carbodiimides, affording the respective compounds from moderate to good yields (65–79%).

In 2012, Kumar et al. [173] described the use of an IL as a novel soluble support to prepare 1,2,3-selenadiazoles **121** and 1,2,3-thiadiazoles **122** (Scheme 6.101). The title

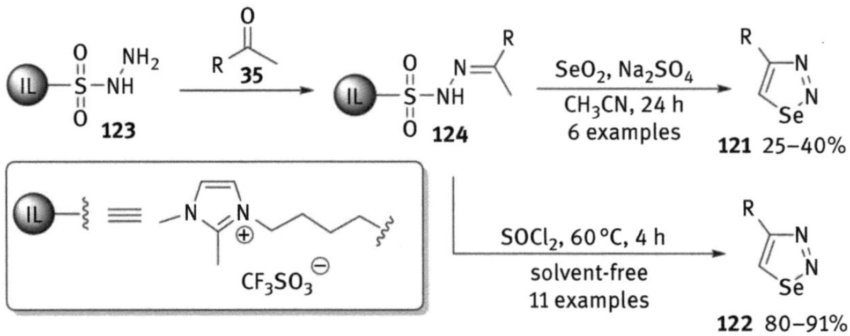

R = aryl, 1-naphthyl, 2-naphthyl, heteroaryl

Scheme 6.101: Synthesis of ionic liquid-supported sulfonyl hydrazones **124** followed by synthesis of 1,2,3-chalcogenodiazoles **121** and **122** [173].

compounds are important intermediates in the synthesis of several organic compounds, such as 2-thioindoles and 2-alkoxybenzo[*b*]thiophenes [174]. Besides, these compounds possess valuable medicinal properties, including anti-HIV activity [175]. In the first step of the synthesis, the previously synthesized IL-supported hydrazine **123** reacts with an aromatic ketone **35** containing both electron-withdrawing and electron-releasing groups in the aromatic ring, forming the IL-supported sulfonyl hydrazone **124**. Next, by using the IL-supported key intermediate **124**, the authors prepared 1,2,3-selenadiazoles **121** in acetonitrile medium and 1,2,3-thiadiazoles **122** under solvent-free conditions.

A new heterogeneous catalyst based on cesium loaded on silica was used by Lavanya and coworkers [176] to promote the one-pot condensation of dicyanomethane **125**, various substituted ketones **35**, and elemental selenium (Scheme 6.102). By this simple and efficient catalytic protocol, 2-amino-5-substituted selenophene-3-

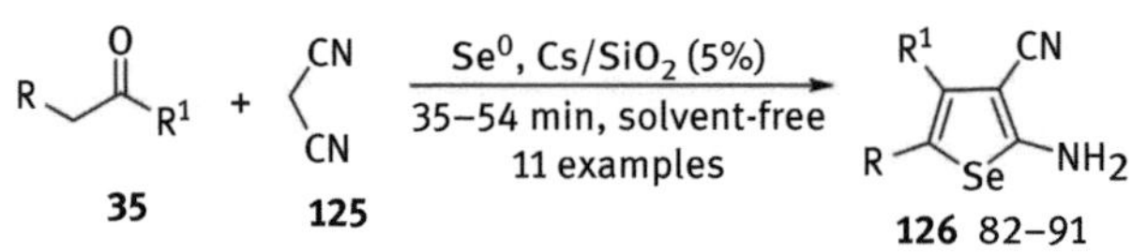

R = H, EtCOO, CH_3, C_6H_5, Br; R^1 = H, CH_3, C_6H_5, 4-$CH_3C_5H_4$, 4-BrC_5H_4, 4-$CH_3C_5H_4$, 4-$C_2H_5C_5H_4$, 2,4-$(CH_3O)_2C_5H_4$

Selected products

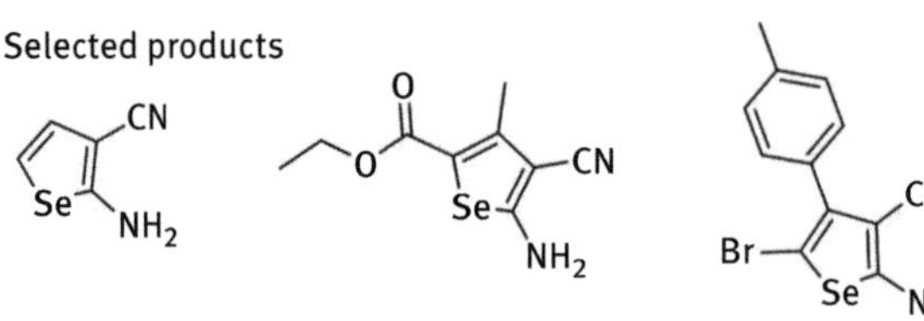

126a 87% **126b** 90% **126c** 82%

Scheme 6.102: Synthesis of heterocycles **126** [176].

carbonitrile derivatives **126** were prepared in 82–91% yields in short reaction times (35–54 min). The authors showed that in Cs/SiO_2 cesium moieties are covalently connected to the silica, and these groups provide active catalytic sites on the surface of silica. In addition, after the reaction the catalyst was recovered by simple filtration. The recovered Cs/SiO_2 catalyst was reused without significant loss in the product yield for three times.

Liang et al. [177] described the microwave-assisted syntheses of benzimidazole-containing selenadiazole derivatives **127**, targeting new selenium-containing organic compounds with anticancer potential (Scheme 6.103). In this reaction, a mixture of selenadiazole **128**, HBTU, and *N,N*-diisopropylethylamine were stirred at room temperature for 2 h, followed by the addition of *o*-phenylenediamine **129**. The mixture was stirred for additional 12 h at room temperature and, in the sequence, at 150 °C for 30 min under microwave irradiation. By the combination of a peptide coupling reagent (HBTU) and microwave irradiation, four benzimidazole-containing selenadiazole derivatives **127** were obtained in good yields via an intramolecular dehydration promoted by microwave irradiation. The synthetic heterocyclic compounds **127** were identified as potent antiproliferative agents against the human breast cancer cell lines MDA-MB-231 and MCF-7.

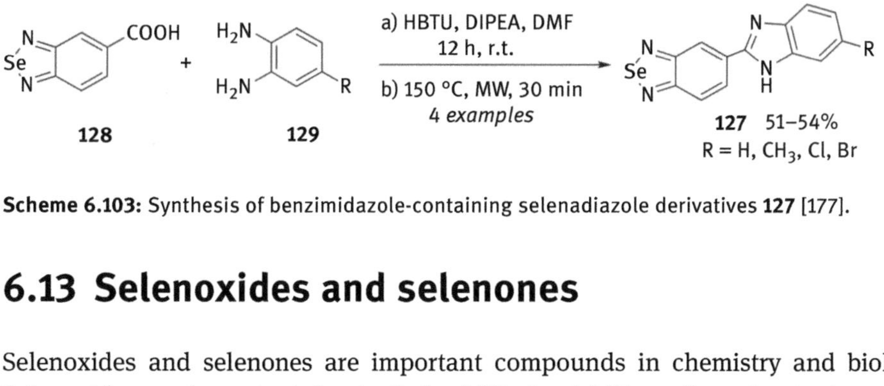

Scheme 6.103: Synthesis of benzimidazole-containing selenadiazole derivatives **127** [177].

6.13 Selenoxides and selenones

Selenoxides and selenones are important compounds in chemistry and biology. Selenoxides are important due to their ability to stabilize adjacent anionic centers [178] and their thermal stability [179–182]. In the selenone molecules, the selenoyl moiety as an efficient leaving group plays a role as a strong electron-withdrawing substituent [183]. Several methods to prepare these classes of compounds were reported in literature, by the oxidation of the corresponding selenides. The greener ones are described here.

In 2010, Khurana and Nand [184] developed the microwave-promoted solvent-free, chemoselective oxidation of selenides **6** to selenoxides **130** and selenones **131** using solid-supported sodium hypochlorite in aqueous medium (Scheme 6.104). According to the authors, selenides **6** were converted to the respective selenoxides **130** in 84–89% yields with the aid of neutral alumina, after 7–10 min. Selenones

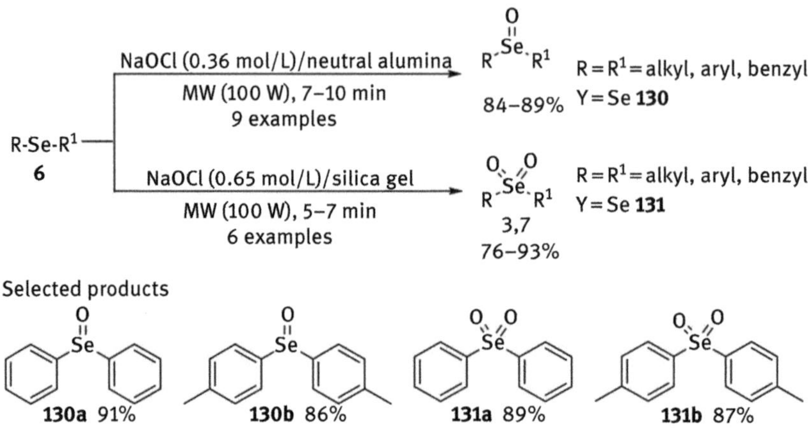

Scheme 6.104: Synthesis of selenoxides **130** and selenones **131** under MW [184].

131, instead, were prepared using silica gel as a solid support in 76–93% yields. Several substituted aromatic and aliphatic selenides were used for both the transformations. Sodium hypochlorite was added in two portions and the irradiation of MW was conducted in two times with a break of 1 min to cool the reaction mixture and to add the reagent.

In 2018, Marini and coworkers [185] reported an Oxone®-mediated oxidation of vinyl selenides **50** in water to obtain selenones **132** without organic cosolvents or additional catalysts (Scheme 6.105). In this work, phenyl vinyl selenones **132** were prepared using Oxone® (2.2 equiv.) as oxidant in 3-24 h of reaction at 60 °C under "on water" conditions. The reaction was highly selective, and only the desired products **132** were obtained in 45–89% yields, without any parallel reactions of epoxidation or addition of water on the C=C double bond. Moreover, the results were compared to those obtained using an excess of *m*-CPBA, a conventional procedure, showing that this new method is a greener and convenient alternative.

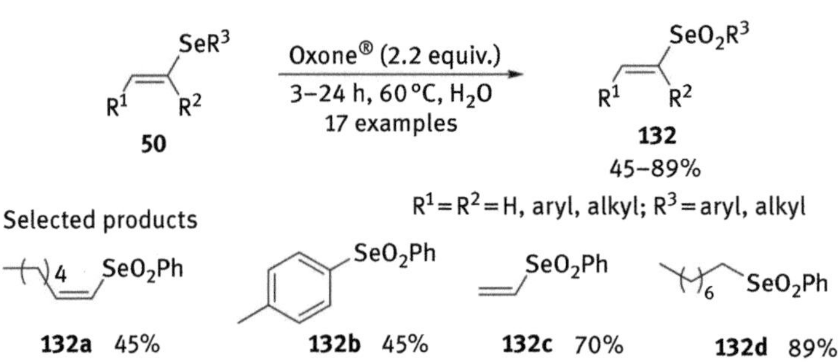

Scheme 6.105: Synthesis of selenones **132** promoted by Oxone® [185].

Selenoxides **133** were selectively obtained when the amount of Oxone® was reduced by half (0.6 equiv.) and the reaction was conducted at room temperature. A plausible mechanism for the oxidation of vinyl selenides **50a** indicates that water is not only the solvent but also a modulator of the Oxone® activity, which can facilitate the oxygen transfer process (Scheme 6.106) [185].

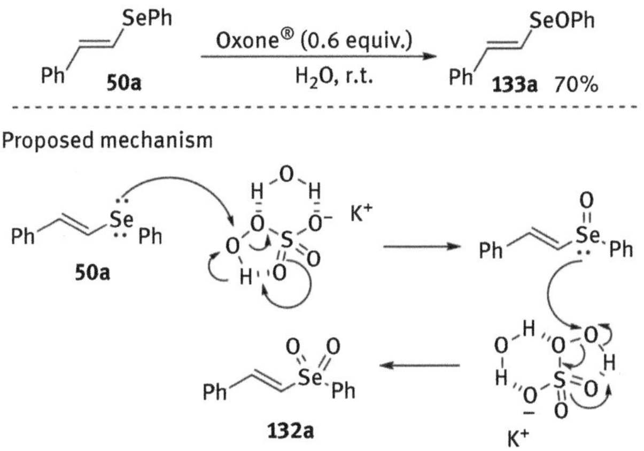

Scheme 6.106: Synthesis of selenoxide **133a** and the proposed mechanism [185].

6.14 Organophosphorus selenides

Organophosphorus compounds have a wide range of applications in different areas, including medicinal chemistry, industrial, and agricultural due to their unique biological and physical properties. Phosphorus-functionalized organic molecules offer fascinating possibilities for synthetic, structural, and mechanistic studies. The phosphorus-heteroatom bond formation thus persists as a valid and active field in chemical research, resulting in new organophosphorus compounds with potential multifaceted interest [186]. As an example, phosphine selenides were used as a catalyst in *aza*-Heck reactions [187].

Among the various green synthetic approaches to prepare phosphorus-containing organoselenium compounds, those that avoid the use of volatile organic solvents are advantageous over the conventional, solution phase synthesis. The minimization of the generation of toxic and nontoxic wastes and the problem regarding the disposal of the solvents are features of the solvent-free protocols. In this line, solvent-free mechanochemical protocols are a greener method, alternative to the traditional ones [188].

In 2018, Kumar and coworkers [189] described the eco-friendly synthesis of phosphine selenides **134** and sulfides **135** in moderate to excellent yields (almost quantitative), applying the solventless mechanochemical technique of ball milling and a simple workup procedure, without using chromatographic purification methods (Scheme 6.107). By this approach, a range of phosphines **136**, including tertiary ones, aminophosphines, and a variety of bisphosphines, reacted with stoichiometric amounts of elemental sulfur or selenium in a jar in the rotary ball mill containing a set of different sized ceramic balls (1 × 15 mm, 2 × 12 mm, 7 × 10 mm, 12 × 8 mm, and 40 × 5 mm) for 4 h at a frequency of 450 rpm. The solid-state reactions were monitored by $^{31}P\{^1H\}$ NMR spectroscopy and a good performance was observed even in a scale-up reaction.

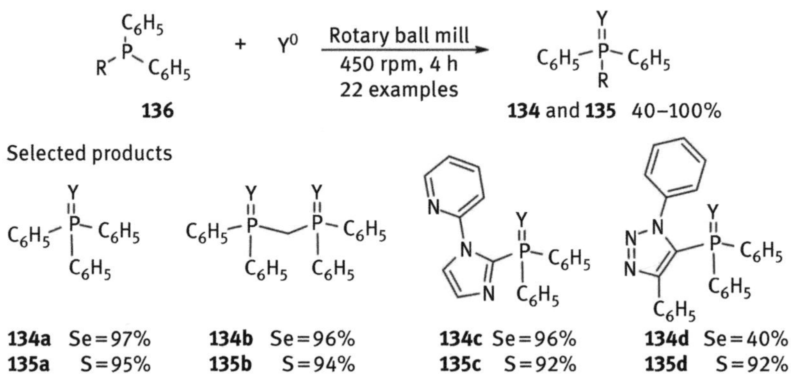

Scheme 6.107: Synthesis of phosphine selenides **134** and sulfides **135** [189].

In the same year, Sim et al. [190] developed the orthogonal "one-step one-pot" mechanochemical reaction to prepare cyclophosphazanes containing selenium **137** or sulfur **138** bonded to phosphorus atom (Scheme 6.108). In the optimal procedure, the phosphorous reagent **139**, an excess of elemental chalcogen, and an appropriate organic compound (like alcohol **34** or a primary amine **140**) were added into a 10 mL stainless steel milling jar containing a 10 mm ball, in the presence of Et_3N and the mixture was milled from 6 to 10 h at 30 Hz. The organophosphorus

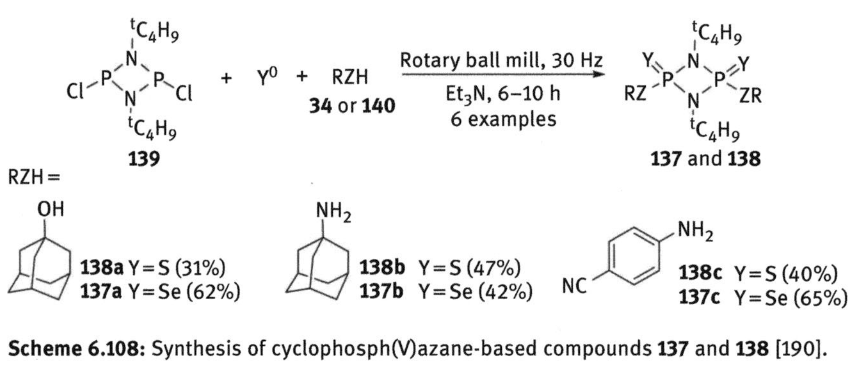

Scheme 6.108: Synthesis of cyclophosph(V)azane-based compounds **137** and **138** [190].

compound derivatives of the bulky adamantyl alcohol **34** or amine **140** adopt a *cis*-conformation when synthesized, as confirmed by NMR and XRD analysis. After completion of the reaction, the products were purified by recrystallization and submitted for hydrolytic- and air-stability studies. These studies, conducted for over 1 and 12 months, showed that the prepared compounds are bench-stable, without signs of decomposition.

6.15 Diselenocarbamates

The class of diselenocarbamates is used in organic synthesis as intermediates to prepare new complex molecules [191–194]. Despite its usefulness, there are few procedures to obtain this class of compounds [191, 195–197].

In 2012, Pan et al. [198] developed a one-pot catalyst- and solvent-free reaction between CSe_2, amines **140**, and alkyl halides **2** to prepare diselenocarbamates **141**. A wide range of products was prepared in 10–30 min at –10 °C in moderate to high yields (72–95%, Scheme 6.109). The reaction works well with a variety of alkyl halides and secondary amines.

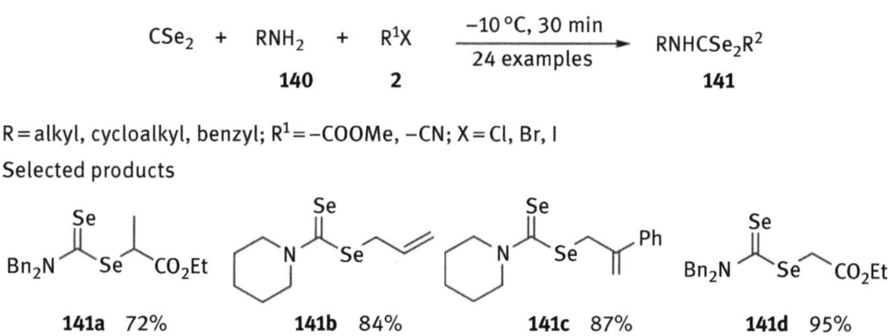

Scheme 6.109: Synthesis of diselenocarbamates **141** under solvent-free conditions [198].

In the same work [198], the authors presented another methodology to obtain products **141**, through a Michael-type addition of alkenes **32** with amines **140** and CSe_2 at room temperature for 30 min. The best results were obtained when the reaction was conducted over silica gel, which could absorb amine **140**, helping in the nucleophilic addition due to its Lewis acid activity. In this work, secondary amines and methyl acrylate or acrylonitrile were used as electron-deficient alkenes. By this alternative protocol, the expected diselenocarbamates **141** were obtained in good to excellent yields (83–91%) (Scheme 6.110).

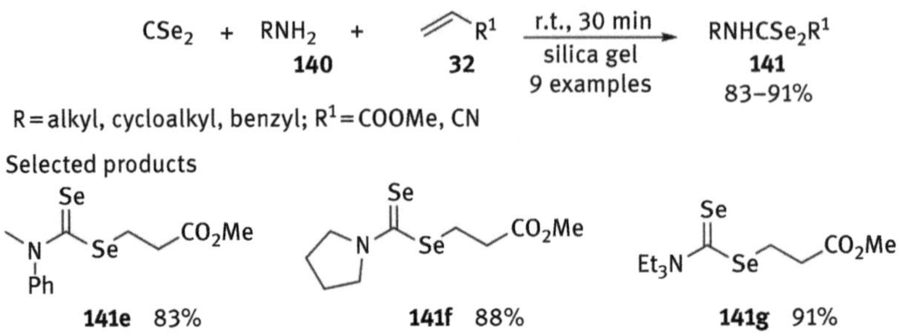

Scheme 6.110: Synthesis of diselenocarbamates **3** via Michael addition [198].

6.16 Conclusions and outlook

This chapter presented the panorama regarding the preparation of various classes of organoselenium compounds. The strategy of introducing a chalcogen atom in different structures is a way to access versatile compounds, which can be important tools in organic synthesis and medicinal chemistry. Moreover, they are by themselves interesting molecules, exhibiting several pharmacological activities.

The methods discussed in this chapter take into account principles of green chemistry like the noninvolvement of hazardous metals or solvents, high temperature, or long reaction times. The syntheses have been conducted in green solvents, such as glycerol, PEG-400, or ILs, or using alternative energy sources, such as ultrasound, microwave, or mechanochemistry.

The extensive collection of works summarized here demonstrates the synthesis of organoselenium compounds, by making it as greener as possible. There are many opportunities still to be explored in this field and we hope that this chapter can put some light in the need for cleaner and efficient protocols to prepare such class of compounds.

Acknowledgments: This study was financed in part by the Coordenação de Aperfeiçoamento de Pessoal de Nível Superior – Brasil (CAPES) – Finance Code 001. FAPERGS CNPq and FINEP are also acknowledged for their financial support.

References

[1] Constable, DJC., Jimenez-Gonzalez, C., and Henderson, RK. Perspective on solvent use in the pharmaceutical industry. Org Process Res Dev 2007, 11, 133–137.

[2] Kerton, FM., and Marriott, R. Alternative Solvents for Green Chemistry, 2nd ed, RSC Green Chemistry Book Series, RSC Publishing, Cambridge, 2013.

[3] Knochel, P. Modern Solvents in Organic Synthesis; Topics in Current Chemistry 206, Springer-Verlag, Heidelberg, 1999.

[4] Reichardt, C., and Welton, T. Solvents and Solvent Effects in Organic Chemistry, 4th ed, Wiley-VCH Verlag, Weinheim, 2011.
[5] Li, C-J., and Chan, T-H. Comprehensive Organic Reactions in Aqueous Media, 2nd ed, John Wiley & Sons, Hoboken, New Jersey, 2007.
[6] Grieco, PA. in Organic Synthesis in Water, Thomson Science, Glasgow, 1998.
[7] Tanaka, K. Solvent-free Organic Synthesis, WILEY-VCH Verlag, Weinheim, 2009.
[8] Sharma, SK., and Mudhoo, A. Green Chemistry for Environmental Sustainability, Taylor & Francis Group, Boca Raton, 2011.
[9] Tundo, P., Perosa, A., and Zecchini, F. Methods and Reagents for Green Chemistry, John Wiley & Sons, New Jersey, 2007.
[10] Lindström, UM. Stereoselective organic reactions in water. Chem Rev 2002, 102, 2751–2772.
[11] Li, C-J. Organic reactions in aqueous media-with a focus on carbon-carbon bond formation. Chem Rev 1993, 93, 2023–2035.
[12] Ranke, J., Stolte, S., Störmann, R., Arning, J., and Jastorff, B. Design of sustainable chemical products: The example of ionic liquids. Chem. Rev. 2007, 107, 2183–2206.
[13] Martins, MAP., Frizzo, CP., Moreira, DN., and Zanatta, N. Ionic liquids in heterocyclic synthesis. Chem Rev 2008, 108, 2015–2050.
[14] Hallett, JP., and Welton, T. Room-temperature ionic liquids: Solvents for synthesis and catalysis. 2. Chem Rev 2011, 111, 3508–3576.
[15] Welton, T. Room-temperature ionic liquids. solvents for synthesis and catalysis. Chem Rev 1999, 99, 2071–2083.
[16] Tanaka, K., and Toda, F. Solvent-free organic synthesis. Chem Rev 2000, 100, 1025–1074.
[17] Walsh, PJ., Li, H., and Parrodi, CA. A green chemistry approach to asymmetric catalysis: Solvent-free and highly concentrated reactions. Chem Rev 2007, 107, 2503–2545.
[18] Dallinger, D., and Kappe, CO. Microwave-Assisted Synthesis in Water as Solvent. Chem Rev 2007, 107, 2563–2591.
[19] Martins, MAP., Frizzo, CP., Moreira, DN., Buriol, L., and Machado, P. Solvent-free heterocyclic synthesis. Chem Rev 2009, 109, 4140–4182.
[20] Ivanković, A., Dronjić, A., Bevanda, AM., and Talić, S. Review of 12 principles of green chemistry in practice. IJSGE 2017, 6, 39–48.
[21] Perin, G., Alves, D., Jacob, RG., Barcellos, AM., Soares, LK., and Lenardão, EJ. Synthesis of organochalcogen compounds using non-conventional reaction media. ChemistrySelect 2016, 2, 205–258.
[22] Santoro, S., Azeredo, JB., Nascimento, V., Sancineto, L., Braga, AL., and Santi, C. The green side of the moon: Ecofriendly aspects of organoselenium chemistry. RSC Adv 2014, 4, 31521–31535.
[23] Beletskaya, IP., and Ananikov, VP. Transition-metal-catalyzed C-S, C-Se, and C-Te bond formation via cross-coupling and atom-economic addition reactions. Chem Rev 2011, 111, 1596–1636.
[24] Beletskaya, I. Element-element additions to unsaturated carbon-carbon bonds catalyzed by transition metal complexes. Chem Rev 2006, 106, 2320–2354.
[25] Zeni, G., Lüdtke, DS., Panatieri, RB., and Braga, AL. Vinylic Tellurides: From preparation to their applicability in organic synthesis. Chem Rev 2006, 106, 1032–1076.
[26] Jagodzinski, TS. Thioamides as useful synthons in the synthesis of heterocycles. Chem Rev 2003, 103, 197–227.
[27] Barcellos, AM., Abenante, AL., Sarro, MT. et al. New prospective for redox modulation mediated by organoselenium and organotellurium compounds. Curr Org Chem 2017, 21, 2044–2061.

[28] Mugesh, G., Du Mont, W-W., and Sies, H. Chemistry of biologically important synthetic organoselenium compounds. Chem Rev 2001, 101, 2125–2179.
[29] Comasseto, JV., and Gariani, RA. Biotransformations on organic selenides and tellurides: Synthetic applications. Tetrahedron 2009, 65, 8447–8459.
[30] Nogueira, CW., and Rocha, JBT. Toxicology and pharmacology of selenium: Emphasis on synthetic organoselenium compounds. Arch Toxicol 2011, 85, 1313–1359.
[31] Mukherjee, AJ., Zade, SS., Singh, HB., and Sunoj, RB. Organoselenium chemistry: Role of intramolecular interactions. Chem Rev 2010, 110, 4357–4416.
[32] Perin, G., Lenardão, EJ., Jacob, RG., and Panatieri, RB. Synthesis of Vinyl Selenides. Chem Rev 2009, 109, 1277–1301.
[33] Ninomiya, M., Garud, DR., and Koketsu, M. Biologically significant selenium-containing heterocycles. Coord Chem Rev 2011, 255, 2968–2990.
[34] Whith, T. Organoselenium chemistry in stereoselective reactions. Angew Chem Int Ed 2000, 39, 3740–3749.
[35] Nogueira, CW., Zeni, G., and Rocha, JBT. Organoselenium and organotellurium compounds: Toxicology and pharmacology. Chem Rev 2004, 104, 6255–6285.
[36] Nogueira, CW., and Rocha, JBT. Organoselenium and organotellurium compounds: Toxicology and pharmacology In PATAI's Chemistry of Functional Groups, John Wiley & Sons, New York, 2nd edition, 2012.
[37] Devilanova, FA. Handbook of Chalcogen Chemistry, RSC Publishing, Cambridge, UK, 2007.
[38] Wirth, T. Organoselenium Chemistry – Modern Developments in Organic Synthesis; Topics in Current Chemistry 208, Springer-Verlag, Heidelberg, 2000.
[39] Shahzad, SA. Novel Selenium-Mediated Rearrangements and Cyclisations, Springer, Cardiff, UK, 2013.
[40] Woollins, JD., and Laitinen, R. Selenium and Tellurium Chemistry, Springer, New York, 2011.
[41] Menezes, PH., and Zeni, G. Vinyl Selenides In Patai's Chemistry of Functional Groups, John Wiley & Sons, New York, 2011.
[42] Chivers, T. A Guide to Chalcogen-Nitrogen Chemistry, World Scientific Publishing, Singapore, 2005.
[43] Braga, AL., Schwab, RS., and Rodrigues, OED. Eco-Friendly Access and Application of Organoselenium Reagents: Advances toward Green Chemistry In, C. Santi, Ed., Organoselenium Chemistry: Between Synthesis and Biochemistry, Bentham Science, Sharjah, e-book. 2014. Doi: 10.2174/97816080583891140101,.
[44] Santi C. Ed., Organoselenium Chemistry: Between Synthesis and Biochemistry, Bentham Science, Sharjah, e-book. 2014. Doi: 10.2174/97816080583891140101,.
[45] Santi, C., and Tidei, C. Electrophilic Se/Te reagents: reactivity and their contribution to "Green Chemistry" In, Z. Rappoport, Ed., The Chemistry of Organic Selenium and Tellurium Compounds, Vol. 4, John Wiley & Sons, Chichester, UK, 2013, 569–655.
[46] Lenardão, EJ., Santi, C., and Sancineto, L. New Frontiers in Organoselenium Compound, Springer International Publishing, 2018. Doi: 10.1007/978-3-319-92405-2,.
[47] Tan, K-S., Arnold, AP., and Rabenstein, DL. Selenium-77 nuclear magnetic resonance studies of selenols, diselenides, and selenenyl sulfides. Can J Chem 1988, 66, 54–60.
[48] Wirth, T. Organoselenium Chemistry. Synthesis and Reactions, Wiley-VCH Verlag & Co. KGaA, Weinheim, 2012, 462.
[49] Freudendahl, DM., Shahzad, SA., and Wirth, T. Recent advances in organoselenium chemistry. Eur J Org Chem 2009, 1649–1664.
[50] Maiti, SN., Spevak, P., Singh, MP., Micetich, RG., and Reddy, AVN. Reductive cleavage of symmetrical disulfides with hydrazines. Synth Commun 1988, 18, 575–581.

[51] Hu, X., Tian, Z., Lu, X., and Chen, Y. A synthetic method for diselenides under phase-transfer conditions. Synth Commun 1997, 27, 553–557.
[52] Li, Z., Ke, F., Deng, H., Xu, H., Xiang, H., and Zhou, X. Synthesis of disulfides and diselenides by copper-catalyzed coupling reactions in water. *Org Biomol Chem* 2013, 11, 2943–2946.
[53] Soleiman-Beigi, M., Yavari, I., and Sadeghizadeh, F. A rapid, efficient and green procedure for transformation of alkyl halides/ tosylates to organochalcogens in water. Phosphorus Sulfur Silicon 2018, 193, 41–44.
[54] Kholshin, SV., Yagunov, SE., Kandalintseva, NV., and Prosenko, AE. Synthesis of selenium-containing derivatives of para-bromopropyl-substituted phenols. *Russ Chem Bull Int. Ed.* 2018, 67, 852–857.
[55] Braga, AL., Lüdtke, DS., and Vargas, F. Enantioselective synthesis mediated by catalytic chiral organoselenium compounds. Curr Org Chem 2006, 10, 1921–1938.
[56] Princival, JL., Santos, AA., and Comasseto, JV. Reactive organometallics from organotellurides: Application in organic synthesis. J Braz Chem Soc 2010, 21, 2042–2054.
[57] Singh, D., Alberto, EE., Rodrigues, OED., and Braga, AL. Eco-friendly cross-coupling of diaryl diselenides with aryl and alkyl bromides catalyzed by CuO nanopowder in ionic liquid. Green Chem 2009, 11, 1521–1524.
[58] Saha, A., Saha, D., and Ranu, BC. Copper nano-catalyst: Sustainable phenyl-selenylation of aryl iodides and vinyl bromides in water under ligand free conditions. Org Biomol Chem 2009, 7, 1652–1657.
[59] Narayanaperumal, S., Alberto, EE., Gul, K., Rodrigues, OED., and Braga, AL. Synthesis of diorganyl selenides mediated by zinc in ionic liquid. J Org Chem 2010, 75, 3886–3889.
[60] Narayanaperumal, S., Alberto, EE., Gul, K., Kawasoko, CY., Dornelles, L., Rodrigues, OED., and Braga, AL. Zn in ionic liquid: An efficient reaction media for the synthesis of diorganyl chalcogenides and chalcogenoesters. Tetrahedron 2011, 67, 4723–4730.
[61] Narayanaperumal, S., Alberto, EE., Andrade, FM., Lenardão, EJ., Taube, PS., and Braga, AL. Ionic liquid: An efficient and recyclable medium for synthesis of unsymmetrical diorganyl selenides promoted by InI. Org Biomol Chem 2009, 7, 4647–4650.
[62] Gul, K., Narayanaperumal, S., Dornelles, L., Rodrigues, OED., and Braga, AL. Bimetallic system for the synthesis of diorganyl selenides and sulfides, chiral β-seleno amines, and seleno- and thioesters. Tetrahedron Lett 2011, 52, 3592–3596.
[63] Ricordi, VG., Freitas, CS., Perin, G. et al. Glycerol as a recyclable solvent for copper-catalyzed cross-coupling reactions of diaryl diselenides with aryl boronic acids. Green Chem 2012, 14, 1030–1034.
[64] Reddy, KHV., Satish, G., Ramesh, K., Karnakar, K., and Nageswar, YVD. Magnetically separable $CuFe_2O_4$ nanoparticle catalyzed C-Se cross coupling in reusable PEG medium. Chem Lett 2012, 41, 585–587.
[65] Kundu, D., Mukherjee, N., and Ranu, BC. A general and green procedure for the synthesis of organochalcogenides by $CuFe_2O_4$ nanoparticle catalysed coupling of organoboronic acids and dichalcogenides in PEG-400. RSC Adv 2013, 3, 117–125.
[66] Xu, H., Wang, Y., Wang, Z., Liu, J., Mario, S., and Wim, D. Dendritic tellurides acting as Antioxidants. Chin Sci Bull 2006, 51, 2315–2321.
[67] Kumar, AV., Reddy, VP., Reddy, CS., and Rao, KR. Potassium selenocyanate as an efficient selenium source in C–Se cross-coupling catalyzed by copper iodide in water. Tetrahedron Lett 2011, 52, 3978–3981.

[68] Zhao, H., Hao, W., Xi, Z., and Cai, M. Palladium-catalyzed cross-coupling of $PhSeSnBu_3$ with aryl and alkyl halides in ionic liquids: A practical synthetic method of diorganyl selenides. New J Chem 2011, 35, 2661–2665.
[69] Freitas, CS., Barcellos, AM., Ricordi, V. et al. Synthesis of diaryl selenides using electrophilic selenium species and nucleophilic boron reagents in ionic liquid. Green Chem 2011, 12, 2931–2938.
[70] Banerjee, S., Adak, L., and Ranu, BC. Ionic liquid/PPh_3 promoted cleavage of diphenyl disulphide and diselenide: A straight-forward metal-free one-pot route to the synthesis of unsymmetrical sulfides and selenides. Tetrahedron Lett 2012, 53, 2149–2152.
[71] Mukherjee, N., Chatterjee, T., and Ranu, BC. Reaction under ball-milling: Solvent-, ligand-, and metal-free synthesis of unsymmetrical diaryl chalcogenides. J Org Chem 2013, 78, 11110–11114.
[72] Kumar, RU., Reddy, KHV., Satish, G., Swapna, K., and Nageswar, YVD. Metal free synthesis of diaryl selenides using SeO_2 as a selenium source. Tetrahedron Lett 2016, 57, 3138–3141.
[73] Neto, PBR., Santana, SO., Levitre, G. et al. Electrochemical synthesis of organochalcogenides in aqueous medium. Green Chem 2016, 18, 657–661.
[74] Rodrigues, J., Saba, S., Joussef, A C., Rafique, J., and Braga, L A. KIO_3-catalyzed C(sp^2)-H bond selenylation/sulfenylation of (hetero)arenes: Synthesis of chalcogenated (hetero) arenes and their evaluation for anti-Alzheimer activity. Asian J Org Chem 2018, 7, 1819–1824.
[75] Silveira, CC., Perin, G., Braga, LA., Dabdoub, MJ., and Jacob, RG. Phenyltelluroacrylonitriles and phenylselenoacrylonitriles as precursors of (*Z*)-α-phenylseleno-α,β-unsaturated aldehydes, β-amino-α-phenylselenonitriles and Diels-Alder adducts. Tetrahedron 2001, 57, 5953–5959.
[76] Jacob, R G., Mesquita, K D., Borges, E L. et al. Síntese de sulfetos e selenetos derivados do Óleo de mamona. Rev Virtual Quim 2014, 6, 60–72.
[77] Perin, G., Araujo, DR., Nobre, PC. et al. Ultrasound-promoted synthesis of 2-organoselanyl-naphthalene using Oxone® in aqueous medium as an oxidizing agent. PeerJ 2018. Doi: 10.7717/peerj.4706.
[78] Perin, G., Goulart, H A., Soares, L K. et al. Synthesis and pharmacological evaluation of novel selenoethers glycerol derivatives for the treatment of pain and inflammation: Involvement of nitrergic and glutamatergic systems. Appl Biochem Biotechnol 2018. Doi: 10.1007/s12010-018-2887-z.
[79] Peglow, TJ., Schumacher, RF., Cargnelutti, R. et al. Preparation of bis (2-pyridyl) diselenide derivatives: Synthesis of selenazolo-[5, 4-*b*]-pyridines and unsymmetrical diorganyl selenides, and evaluation of antioxidant and anticholinesterasic activities. Tetrahedron Lett 2017, 58, 3734–3738.
[80] Soares, LC., Alberto, EE., Schwab, RS. et al. Ephedrine-based diselenide: A promiscuous catalyst suitable to mimic the enzyme glutathione peroxidase (GPx) and to promote enantioselective C–C coupling reactions. Org Biomol Chem 2012, 10, 6595–6599.
[81] Braga, AL., Milani, P., Vargas, F., Paixão, MW., and Sehnem, JA. Modular chiral thiazolidine catalysts in asymmetric aryl transfer reactions. Tetrahedron: Asymmetry 2006, 17, 2793–2797.
[82] Orian, L., and Toppo, S. Organochalcogen peroxidase mimetics as potential drugs: Along story of a promise still unfulfilled. Free Radical Bio Med 2014, 66, 65–74.
[83] Braga, AL., Schneider, PH., Paixão, MW., Deobald, AM., Peppe, C., and Bottega, DP. Chiral seleno-amines from indium selenolates. A straightforward synthesis of selenocysteine derivatives. J Org Chem 2006, 71, 4305–4307.
[84] Braga, AL., Schwab, RS., Alberto, EE., Salman, SM., Vargas, J., and Azeredo, JB. Ring opening of unprotected aziridines by zinc selenolates in a biphasic system. Tetrahedron Lett 2009, 50, 2309–2311.

[85] Hu, XE. Nucleophilic ring opening of aziridines. Tetrahedron 2004, 60, 2701–2743.
[86] Wang, J-Y., Hu, Y., Wang, D-X., Pan, J., Huang, Z-T., and Wang, M-X. Unprecedented carbon–Carbon bond cleavage in nucleophilic aziridine ring opening reaction, efficient ring transformation of aziridines to imidazolidin-4-ones. Chem Commun 2009, 422–424.
[87] Salman, SM., Schwab, RS., Alberto, EE. et al. Efficient ring opening of protected and unprotected aziridines promoted by stable zinc selenolate in ionic liquid. Synlett 2011, 69–72.
[88] Salman, SM., Narayanaperumal, S., Schwab, RS., Bender, CR., Rodrigues, OED., and Dornelles, L. CuO nano particles and [bmim]BF_4: An application towards the synthesis of chiral β-seleno amino derivatives via ring opening reaction of aziridines with diorganyl diselenides. RSC Adv 2012, 2, 8478–8482.
[89] Vargas, J., Narayanaperumal, S., Gul, K. et al. Synthesis of chiral β-chalcogen amine derivatives and Gram-positive bacteria activity. Tetrahedron 2012, 68, 10444–10448.
[90] Narayanaperumal, S., Gul, K., Kawasoko, CY. et al. Transition Metal Oxide Nanopowder and Ionic Liquid: An Efficient System for the Synthesis of Diorganyl Selenides, Selenocysteine and Derivatives. J Braz Chem Soc 2010, 21, 2079–2087.
[91] Azuma, H., Tamagaki, S., and Ogino, K. Stereospecific total syntheses of sphingosine and its analogues from L-serine. J Org Chem 2000, 65, 3538–3541.
[92] Knapp, S., and Zhao, D. Synthesis of the sialidase inhibitor siastatin B. Org Lett 2000, 2, 4037–4040.
[93] Yang, M-H., Yan, G-B., and Zheng, Y-F. Regioselective ring-opening reactions of 1,2-epoxides with thiols and arylselenols directly promoted by [Bmim]BF_4. Tetrahedron Lett 2008, 49, 6471–6474.
[94] Vieira, AA., Azeredo, JB., Godoi, M., Santi, C., da Silva, JEN., and Braga, AL. Catalytic chalcogenylation under greener conditions: A solvent-free sulfur- and seleno-functionalization of olefins via I_2/DMSO oxidant system. J Org Chem 2015, 80, 2120–2127.
[95] Liu, M., Li, Y., Yu, L., Xu, Q., and Jiang, X. Visible light-promoted, iodine-catalyzed selenoalkoxylation of olefins with diselenides and alcohols in the presence of hydrogen peroxide/air oxidant: An efficient access to α-alkoxyl selenides. Sci China Chem 2018, 61, 294–299.
[96] Gerzson, MFB., Victoria, FN., Radatz, CS. et al. In vitro antioxidant activity and in vivo antidepressant-like effect of α-(phenylselanyl) acetophenone in mice. Pharmacol Biochem Behav 2012, 102, 21–29.
[97] Martinez, DM., Barcellos, AM., Casaril, AM. et al. Twice acting antioxidants: Synthesis and antioxidant properties of selenium and sulfur-containing zingerone derivatives. Tetrahedron Lett 2015, 56, 2243–2246.
[98] Victoria, FN., Anversa, R., Penteado, F., Castro, M., Lenardão, EJ., and Savegnago, L. Antioxidant and antidepressant-like activities of semi-synthetic α-phenylseleno citronellal. Eur J Pharmacol 2014, 742, 131–138.
[99] Victoria, FN., Radatz, CS., Sachini, M. et al. KF/Al_2O_3 and PEG-400 as a recyclable medium for the selective α-selenation of aldehydes and ketones. Preparation of potential antimicrobial agents. Tetrahedron Lett 2009, 50, 6761–6763.
[100] Chu, C-M., Gao, S., Sastry, MNV., Kuo, C-W., Lu, C., Liu, J-T., and Yao, C-F. Ceric ammonium nitrate (CAN) as a green and highly efficient promoter for the 1,4-addition of thiols and benzeneselenol to α,β-unsaturated ketones. Tetrahedron 2007, 63, 1863–1871.
[101] Battistelli, B., Lorenzo, T., Tiecco, M., and Santi, C. "On-Water" Michael-type addition reactions promoted by PhSeZnCl. Eur J Org Chem 2011, 1848–1851.
[102] Perin, G., Borges, EL., Rosa, PC., Carvalho, PN., and Lenardão, EJ. Simple cleavage of diorganyl diselenides with $NaBH_4$/PEG-400 and direct Michael addition to electron-deficient alkenes. Tetrahedron Lett 2013, 54, 1718–1721.

[103] Heppke, G., Martens, J., Praefcke, K., and Simon, H. Selenol esters-a novel class of liquid-crystal compounds. Angew Chem Int Ed Engl 1977, 16, 318–319.

[104] Victoria, FN., Martinez, DM., Castro, M. et al. Antioxidant properties of (*R*)-Se-aryl thiazolidine-4-carboselenoate. Chem Bio Inter 2013, 205, 100–107.

[105] Domínguez-Álvarez, E., Plano, D., Font, M. et al. Synthesis and antiproliferative activity of novel selenoester derivatives. Eur J Med Chem 2014, 73, 153–166.

[106] Sanmartín, C., Plano, D., Domínguez, E. et al. Synthesis and pharmacological screening of several aroyl and heteroaroyl selenylacetic acid derivatives as cytotoxic and antiproliferative agents. Molecules 2009, 14, 3313–3338.

[107] Subramanyam, C., Naguchu, M., and Weinreb, SM. An approach to amphimedine and related marine alkaloids utilizing an intramolecular Kondrat'eva pyridine synthesis. J Org Chem 1989, 54, 5580–5585.

[108] Bennasar, ML., Zulaica, E., Solé, D., Roca, T., García-Díaz, D., and Alonso, S. Total synthesis of the bridged indole alkaloid apparicine. J Org Chem 2009, 74, 8359–8368.

[109] Martin, SF., Chen, KX., and Eary, CT. An enantioselective total synthesis of (+)-geissoschizine. Org Lett 1999, 1, 79–81.

[110] Makriyannis, A., Günther, WHH., and Mautner, HG. Selenol esters as specific reagents for the acylation of thiol groups. J Amer Chem. Soc 1973, 8403–8406.

[111] Singh, D., Narayanaperumal, S., Gul, K., Godoi, M., Rodrigues, OED., and Braga, AL. Efficient synthesis of selenoesters from acyl chlorides mediated by CuO nanopowder in ionic liquid. Green Chem 2010, 12, 957–960.

[112] Tabaralli, G., Alberto, EE., Deobalda, AM. et al. Ionic liquid: An effcient and reusable media for seleno- and thioester synthesis promoted by indium. Tetrahedron Lett 2010, 51, 5728–5731.

[113] Godoi, M., Ricardo, EW., Bottesselle, GV., Galetto, FZ., Azeredo, JB., and Braga, AL. Synthesis of selenol esters from diorganyl diselenides and acyl chlorides under solvent-free conditions and microwave irradiation. Green Chem 2012, 14, 456–460.

[114] Santi, C., Battistelli, B., Testaferri, L., and Tiecco, M. On water preparation of phenylselenoesters. Green Chem 2012, 14, 1277–1280.

[115] Sancineto, L., Vargas, JP., Monti, B. et al. Atom effcient preparation of zinc selenates for the synthesis of selenol esters under "On Water" conditions. Molecules 2017, 22, 953–966.

[116] Perin, G., Silveira, MB., Barcellos, AM., Jacob, RG., and Alves, D. Polyethylene glycol-400/ H_3PO_2: An eco-friendly reductive system for the synthesis of selanylesters. Org Chem Front 2015, 2, 1531–1535.

[117] Perin, G., Silveira, MB., Barcellos, AM. et al. Rongalite®/PEG-400 as reducing system in the synthesis of new glycerol-derived selenol esters using anhydrides and bis-(2,2-dimethyl-1,3-dioxolanylmethyl)diselenide as substrates. Arkivoc 2017, (ii), 138-48.

[118] Kuligowski, C., Bezzenine-Lafollée, S., Chaume, G. et al. Approach toward the total synthesis of griseoviridin: Formation of thioethynyl and thiovinyl ether-containing nine-membered lactones through a thioalkynylation-macrolactonization-hydrostannylation sequence. J Org Chem 2002, 67, 4565–4568.

[119] Alonso, F., Beletskaya, IP., and Yus, M. Transition-metal-catalyzed addition of heteroatom-hydrogen bonds to alkynes. Chem Rev 2004, 104, 3079–3159.

[120] Ananikov, VP., Orlov, N., and Beletskaya, IP. Efficient and convenient synthesis of â-vinyl sulfides in nickel-catalyzed regioselective addition of thiols to terminal alkynes under solvent-free conditions. Organometallics 2006, 25, 1970–1977.

[121] Lenardão, EJ., Dutra, LG., Saraiva, MT., Jacob, RG., and Perin, G. Hydroselenation of alkynes using $NaBH_4$/$BMIMBF_4$: Easy access to vinyl selenides. Tetrahedron Lett 2007, 48, 8011–8801.

[122] Perin, G., Jacob, RG., Dutra, LG., Azambuja, F., Santos, GFF., and Lenardão, EJ. Addition of chalcogenolate anions to terminal alkynes using microwave and solvent-free conditions: Easy access to bis-organochalcogen alkenes. Tetrahedron Lett 2006, 47, 935–938.
[123] Lenardão, EJ., Gonçalves, LCC., Mendes, SR. et al. $NaBH_4$/[bmim]BF_4: A new reducing system to access vinyl selenides and tellurides. J Braz Chem Soc 2010, 21, 2093–2099.
[124] Lenardão, EJ., Silva, MS., Sachini, M., Lara, RG., Jacob, RG., and Perin, G. Synthesis of alkenyl selenides and tellurides using PEG-400. ARKIVOC 2009, xi, 221–227.
[125] Perin, G., Jacob, RG., Azambuja, F. et al. The first synthesis of b-phenylchalcogeno-a,b-unsaturated esters via hydrochalcogenation of acetylenes using microwave and solvent-free conditions. Tetrahedron Lett 2005, 46, 1679–1682.
[126] Perin, G., Mendes, SR., Silva, MS., and Lenardão, EJ. Synthesis of β-phenylchalcogeno α,β-unsaturated ketones via hydrochalcogenation of acetylenes using microwave and solvent-free conditions. Synth Commun 2006, 36, 2587–2595.
[127] Lenardão, EJ., Silva, MS., Mendes, SR. et al. Synthesis of β-phenylchalcogeno-α,β-unsaturated esters, ketones and nitriles using microwave and solvent-free conditions. J Braz Chem Soc 2007, 18, 943–950.
[128] Kabalka, GW., and Venkataiah, B. Preparation of (Z)- and (E)-vinyl selenides utilizing vinylboronic acids and vinylboronic esters in ionic liquids. Tetrahedron Lett 2002, 43, 3703–3705.
[129] Wang, Z., Mo, H., and Bao, W. Mild, efficient and highly stereoselective synthesis of (Z)-vinyl chalcogenides from vinyl bromides catalyzed by copper(I) in ionic liquids based on amino acids. Synlett 2007, 91–94.
[130] Santoro, S., Battistelli, B., Testaferri, L., Tiecco, M., and Santi, C. Vinylic substitutions promoted by PhSeZnCl: Synthetic and theoretical investigations. Eur J Org Chem 2009, 4921–4925.
[131] Gonçalves, LC., Fiss, GF., Perin, G., Alves, D., Jacob, RG., and Lenardão, EJ. Glycerol as a promoting medium for cross-coupling reactions of diaryl diselenides with vinyl bromides. Tetrahedron Lett 2010, 51, 6772–6775.
[132] Gonçalves, LCC., Lima, DB., Borba, PMY. et al. Glycerol/CuI/Zn as a recyclable catalytic system for synthesis of vinyl sulfides and tellurides. Tetrahedron Lett 2013, 54, 3475–3480.
[133] Webber, R., Peglow, TJ., Nobre, PC. et al. 1,1-Dibromoalkenes as versatile reagents to a transition metal-free and stereoselective synthesis of (*E*)-1-bromo-1-selenoalkenes and ketene selenoacetals. Tetrahedron Lett 2016, 57, 4128–4132.
[134] Perin, G., Jacob, RG., Botteselle, GV. et al. Clean and Atom-Economic Synthesis of α-Phenylselenoacrylonitriles and α-Phenylseleno-α,β-Unsaturated Esters by Knoevenagel Reaction under Solvent-Free Conditions. J Braz Chem Soc 2005, 16, 857–862.
[135] Lara, RG., Soares, LK., Jacob, RG., Schumacher, RF., and Perin, G. Selective synthesis of (*Z*)-chalcogenoenynes and (*Z, Z*)-1,4-bis-chalcogenbuta-1,3-dienes using PEG-400. J Braz Chem Soc 2016, 27, 2046–2054.
[136] Perin, G., Barcellos, AM., Luz, EQ. et al. Green hydroselenation of aryl alkynes: Divinyl selenides as a precursor of resveratrol. Molecules 2017, 22, 327–336.
[137] Martynov, AV., Potapov, VA., Amosova, SV., Makhaeva, NA., Beletskaya, IP., and Hevesi, L. Cross-coupling of (Z)-1,2-bis(ethylseleno)ethene with the Grignard reagents. J Organomet Chem 2003, 674, 101–103.
[138] Comasseto, JV., Ling, LW., Petragnani, N., and Stefani, HA. Vinylic selenides and tellurides – Preparation, reactivity and synthetic applications. Synthesis 1997, 373–403.
[139] Ogawa, A., Yokoyama, H., Yokoyama, K., Masawaki, T., Kambe, N., and Sonoda, N. Photoinitiated addition of diphenyl diselenide to acetylenes. J Org Chem 1991, 56, 5721–5723.

[140] Ananikov, VP., and Beletskaya, IP. Palladium-catalyzed addition of disulfides and diselenides to alkynes under solvent free conditions. Org Biomol Chem 2004, 4, 284–287.
[141] Ananikov, VP., and Beletskaya, IP. Solvent-free palladium-catalyzed addition of diaryl dichalcogenides to alkynes. Russ Chem Bull Int Ed 2004, 53, 561–565.
[142] Cai, M., Wang, Y., and Hao, W. Palladium-catalyzed addition of diaryl disulfides and diselenides to terminal alkynes in room temperature ionic liquids. Green Chem 2007, 9, 1180–1184.
[143] Lara, RG., Borges, EL., Lenardão, EJ., Alves, D., Jacob, RG., and Perin, G. Addition of thiols to phenylselenoalkynes using KF/alumina under solvent-free conditions. J Braz Chem Soc 2010, 21, 2125–2129.
[144] Lara, RG., Rosa, PC., Soares, LK., Silva, MS., Jacob, RG., and Perin, G. A simple and stereoselective synthesis of (Z)-1,2-bis-arylselanyl alkenes from alkynes using KF/Al_2O_3. Tetrahedron 2012, 68, 10414–10418.
[145] Perin, G., Borges, EL., and Alves, D. Highly stereoselective method to prepare bis-phenylchalcogen alkenes via addition of chalcogenate to phenylseleno alkynes. Tetrahedron Lett 2012, 53, 2066–2069.
[146] Luz, EQ., Lopes, EF., Ricordi, VG. et al. Water-dependent selective synthesis of mono-or Bis-selanyl alkenes from terminal alkynes. ChemistrySelect 2016, 1, 4289–4294.
[147] Movassagh, B., and Mohammadi, E. Green trends in synthesis of alkenyl and alkynyl chalcogenides. Curr Green Chem 2016, 3, 18–35.
[148] Grimaldi, TB., Lutz, G., Back, DF., and Zeni, G. (Biphenyl-2-alkyne) derivatives as common precursors for the synthesis of 9-iodo-10-organochalcogen-phenanthrenes and 9-organochalcogen-phenanthrenes. Org Biomol Chem 2016, 14, 10415–10426.
[149] Roehrs, JA., Gay, RM., Branda, R., Menezes, PH., Nogueira, CW., and Zeni, G. Electrophilic cyclization of 2-chalcogenealkynylanisoles: Versatile access to 2-chalcogen-benzo[*b*]furans. J Org Chem 2009, 74, 2153–2162.
[150] Perin, G., Roehrs, JA., Hellwig, PS. et al. Synthesis of 2-organylchalcogenyl–benzo[b] selenophenes: Precursors to access alkynes susceptible to cyclization. ChemistrySelect 2017, 2, 4561–4566.
[151] Wu, G., Min, L., Li, H. et al. Metal-free synthesis of alkynyl alkyl selenides via three-component coupling of terminal alkynes, Se, and epoxides. Green Chem 2018, 20, 1560–1563.
[152] Seus, N., Goldani, B., Lenardão, EJ., Savegnago, L., Paixão, MW., and Alves, D. Organocatalytic synthesis of (arylselanyl)phenyl-1*H*-1,2,3-triazole-4-caroxamides by cycloaddition between azidophenyl arylselenides and β-oxo-amides. Eur J Org Chem 2014, 1059–5.
[153] Tiecco, M., Testaferri, L., Santi, C. et al. Asymmetric azidoselenenylation of alkenes: A key for the synthesis of enantiomerically enriched nitrogen-containing compounds. Angew Chem 2003, 115, 3239–3241.
[154] Santi, C., Santoro, S., and Battistelli, B. Organoselenium compounds as catalysis in nature and laboratory. Curr Org Chem 2010, 14, 2442–2462.
[155] Fujita, K-I., Hashimoto, S., Kanakubo, M., Oishia, A., and Taguchia, Y. Solid-phase intramolecular oxyselenenylation and deselenenylation reactions using polymer-supported selenoreagents and their application to aqueous media organic synthesis. Green Chem 2003, 5, 549–553.
[156] Ericsson, C., and Engman, L. Microwave-assisted group-transfer cyclization of organotellurium compounds. J Org Chem 2004, 69, 5143–5146.
[157] Libero, FM., Xaviera, MCD., Victoria, FN. et al. Synthesis of novel selenium and tellurium-containing tetrazoles: A class of chalcogen compounds with antifungal activity. Tetrahedron Lett 2012, 53, 3091–3094.
[158] Zimmerman, EG., Thurow, S., Freitas, CS. et al. A selenium-based ionic liquid as a recyclable solvent for the catalyst-free synthesis of 3-selenylindoles. Molecules 2013, 18, 4081–4090.

[159] Thruow, S., Webber, R., Perin, G., Lenardão, EJ., and Alves, D. Glycerol/hypophosphorous acid: An efficient system solvent-reducing agent for the synthesis of 2-organylselanyl pyridines. Tetrahedron Lett 2013, 54, 3215–3218.
[160] Potapov, AS., Chernova, NP., Ogorodnikow, VD., Petrenko, TV., and Khlebnikov, A. Facile synthesis of pyrazole- and benzotriazole-containing selenoethers. Sci World J 2014, 1–5.
[161] Xavier, MCDF., Xavier, DM., Seus, N., Lenardão, EJ., Perin, G., and Alves, D. Microwave assisted rapid synthesis of (arylselanyl)phenyl-1*H*-1,2,3-triazoles by copper catalyzed 1,3-dipolar cycloaddition. Curr. Microwave Chem 2016, 3, 14–23.
[162] Fonseca, SF., Padilha, NB., and Thurow, et al. Ultrasound-promoted copper-catalyzed synthesis of bis-arylselanyl chrysin derivatives with boosted antioxidant and anticancer activities. Ultrason Sonochem 2017, 39, 827–836.
[163] Vieira, BM., Thurow, S., da Costa, M. et al. Ultrasound-assisted synthesis and antioxidant activity of 3-selanyl-1*H*-indole and 3-selanylimidazo[1,2-*a*]pyridine derivatives. Asian J Org Chem 2017, 6, 1635–1646.
[164] Zhang, X., Wang, C., Jiang, H., and Sun, L. Convenient synthesis of selenyl-indoles *via* iodide ion-catalyzed electrochemical C-H selenation. Chem Commun 2018, 54, 8781–8784.
[165] Yang, D., Li, G., Xing, C., Cui, W., Li, K., and Wei, W. Metal- and photocatalyst-free visible-light-promoted regioselective selenylation of coumarin derivatives *via* oxidation-induced C-H functionalization. Org Chem Front 2018, 5, 2974–2979.
[166] Raghavendra, M., Naik, HSB., and Sherigara, BS. A facile one-pot microwave-induced synthesis of some novel selenolo[2,3-*b*]quinoline derivatives under solvent-free conditions. Phosphorus, Sulfur Silicon Relat Elem 2008, 183, 1501–1509.
[167] Koketsu, M., Tanaka, H., and Ishihara, H. A facile preparation of 2-amino-1,3-selenazoles by reactions of *N, N*-unsubstituted selenoureas with chloroacetonitrile. Chem Lett 2005, 34, 1260–1261.
[168] Koketsu, M., and Ishihara, H. Synthesis of 1,3-selenazine and 1,3-selenazole and their biological activities. Curr Org Chem 2003, 7, 175–185.
[169] Potewar, TM., Ingale, SA., and Srinivasan, KV. An efficient and eco-friendly synthesis of 2-amino-1,3-selenazoles in an ionic liquid/water system under ambient conditions. ARKIVOC 2008, xii, 117–125.
[170] Balaguez, RA., Betin, ES., Barcellos, T., Lenardão, EJ., Alves, D., and Schumacher, RF. Synthesis of 2-acyl-benzo[1,3-*d*]selenazoles via domino oxidative cyclization of methyl ketones with bis(2-aminophenyl) diselenide. New J Chem 2017, 41, 1483–1487.
[171] Sashida, H., Pan, C., Kaname, M., and Minoura, M. A facile and practical solvent-free one-pot synthesis of (*Z*)-4-methylene-3-selenaquinoline derivatives from *o*-ethynylanilines and isoselenocyanates. Synthesis 2010, 3091–3096.
[172] Sengoden, M., and Punniyamurthy, T. "On Water": Efficient iron-catalyzed cycloaddition of aziridines with heterocumulenes. Angew Chem Int Ed 2013, 52, 572–575.
[173] Kumar, A., Muthyala, MK., Choudhary, S., Tiwari, RK., and Parang, K. Ionic liquid as soluble support for synthesis of 1,2,3-thiadiazoles and 1,2,3-selenadiazoles. J Org Chem 2012, 77, 9391–9396.
[174] Androsov, DA., and Neckers, DC. Synthesis and reactivity of 4-(2-Chloro-5-nitrophenyl)-1,2,3-thiadiazole. A Novel One-pot Synthesis of *N*-Substituted Indole-2-thiols. J Org Chem 2007, 72, 5368–5373.
[175] Tian, Y., Rai, D., Zhan, P. et al. Design, synthesis, and biological evaluation of novel 3,5-disubstituted-1,2,6-thiadiazine-1,1-dione derivatives as HIV-1 NNRTIs. Chem. Biol. Drug. Des. 2013, 82, 384–393.

[176] Maddila, S., Lavanya, P., and Jonnalagadda, SB. Cesium loaded on silica as an efficient and recyclable catalyst for the novel synthesis of selenophenes. Arabian J Chem 2016, 9, 891–897.
[177] Liang, Y., Zhou, Y., Deng, S., and Chen, T. Microwave-assisted syntheses of benzimidazole-containing selenadiazole derivatives that induce cell- cycle arrest and apoptosis in human breast cancer cells by activation of the ROS/AKT pathway. ChemMedChem 2016, 11, 2339–2346.
[178] Reich, HJ., and Shah, SK. Organoselenium chemistry, α-lithio selenoxides and selenides. preparation and further transformation to olefins, dienes, and allylic alcohols. J Am Chem Soc 1975, 97, 3250–3252.
[179] Sharpless, KB., Lauer, RF., and Teranishi, AY. Electrophilic and nucleophilic organoselenium reagents. new routes to α,β-unsaturated carbonyl compounds. J Am Chem Soc 1973, 95, 6137–6139.
[180] Reich, HJ., Renga, GM., and Reich, IL. Organoselenium chemistry. conversion of ketones to enones by selenoxide syn elimination. J Am Chem Soc 1975, 97, 5434–5447.
[181] Shimizu, M., Ando, R., and Kuwajima, I. Reactions of 3-hydroxyvinyl selenones with alkoxides. oxetane formation and fragmentation reactions. J Org Chem 1984, 49, 1230–1238.
[182] Uemura, S., and Fukuzawa, S. New aspects of the telluroxide elimination. a facile elimination of sec-alkyl phenyl telluroxide leading to olefins, allylic alcohols, and allylic ethers. J Am Chem Soc 1983, 105, 2748–2752.
[183] Buyck, T., Wang, Q., and Zhu, J. Triple role of phenylselenonyl group enable a one-pot synthesis of 1,3-oxazinan-2-ones from α-isocyanoacetates, phenyl vinyl selenones, and water. J Am Chem Soc 2014, 136, 11524–11528.
[184] Khurana, JM., and Nand, B. Aqueous sodium hypochlorite mediated chemoselective oxidation of chalcogenides to monoxides and dioxides by microwave exposure. Can J Chem 2010, 88, 906–909.
[185] Palomba, M., Trappetti, F., Bagnoli, L., Santi, C., and Marini, F. Oxone-mediated oxidation of vinyl selenides in water. Eur J Org Chem 2018, 3914–3919.
[186] Allen, D., Loakes, D., and Tebby, JC. Organophosphorus Chemistry. Volume 45, The Royal Society of Chemistry, Cambridge, UK, 2016.
[187] Zheng, T., Tabor, JR., Stein, ZL., and Michael, FE. Regioselective metal-free Aza-Heck reactions of terminal alkenes catalyzed by phosphine selenides. Org Lett 2018, 20, 6975–6978.
[188] James, SL., Adams, CJ., Bolm, C. et al. Mechanochemistry: Opportunities for new and cleaner synthesis. Chem Soc. Rev. 2012, 41, 413–447.
[189] Kumar, R., Kumar, S., Pandey, MK., Kashid, VS., Radhakrishna, L., and Balakrishna, MS. Synthesis of phosphine chalcogenides under solvent-free conditions using a rotary ball mill. Eur J Inorg Chem 2018, 1028–1037.
[190] Sim, Y., Tan, D., Ganguly, R., Li, Y., and García, F. Orthogonality in main group compounds: A direct one-step synthesis of air- and moisture-stable cyclophosphazanes by mechanochemistry. Chem Commun 2018, 54, 6800–6803.
[191] Kobetsu, M., Fukuta, Y., and Ishihara, H. Reaction of *N, N*-dimethylselenocarbamoyl chloride with nucleophiles. preparation of diselenocarbamates, selenothiocarbamates, and selenoureas. J Org Chem 2002, 67, 1008–1011.
[192] Li, Q-H., Ding, Y., and Huang, N-W. Synthesis and biological activities of dithiocarbamates containing 1,2,3-triazoles group. Chin Chem Lett 2014, 25, 1469–1472.
[193] Morf, P., Raimondi, F., Nothofer, H-G. et al. Dithiocarbamates: Functional and Versatile Linkers for the Formation of Self-Assembled Monolayers. Langmuir 2006, 22, 658–663.

[194] Zhang, D., Chen, J., Liang, Y., and Zhou, H. Facile synthesis of novel ionic liquids containing dithiocarbamate. Synth Commun 2005, 35, 521–526.
[195] Garud, D R., Tanahashi, N., Ninomiya, M., and Koketsu, M. Synthesis of 2-selenoxoperhydro -1,3-selenazin-4-ones via diselenocarbamate intermediates. Tetrahedron 2009, 65, 4775–4778.
[196] Koketsu, M., Yamamura, Y., and Ishihara, H. Synthesis of 2-Selenoxoperhydro-1,3-selenazolidin-4-ones via Diselenocarbamate Intermediates, Synthesis, 2006, 2738–2742.
[197] Ishihara, H., Koketsu, M., Fukuta, Y., and Nada, F. Reaction of lithium aluminum hydride with elemental selenium: Its application as a selenating reagent into organic molecules. J Am Chem Soc 2001, 123, 8408–8409.
[198] Pan, X., Zhu, J., Zou, J. et al. Straightforward and highly efficient synthesis of diselenocarbamates. Org Lett 2012, 14, 6170–6173.

Guilherme M. Martins and Samuel R. Mendes

7 Synthesis and Biological Activity of Five- and Six-Membered Se-Containing Heterocycles

7.1 Introduction

In the last decades, there has been tremendous scientific progress in the development of chemotherapeutic drugs, which are applied in several treatments, for example, for autoimmune diseases such as multiple sclerosis and rheumatoid arthritis, in addition to being used to suppress the rejections to several transplants. These drugs usually have fewer side effects and lower costs. In addition, selenium-containing heterocycles are considered to be privileged structures of extreme interest to the scientific community, not only due to their synthetic structural reactivity, but also due to their diverse medicinal applications, such as antibacterial, antiviral, antitumor, antioxidant, antidepressant, cytotoxic, among others [1]. Five- and six-membered heterocycles include the structural units of various biologically active compounds, being found in several medicines[2] and comprising a range of natural products [3]. Considering this wide range of uses as well as biological activities, selenium-containing heterocycles continue to motivate researchers to develop new compounds as well as to perform biological tests.

7.2 Five-membered Se-containing heterocycles

In this chapter we will discuss the importance of heterocyclic derivatives containing selenium, demonstrating its importance not only as a synthetic tool, but also in medicinal chemistry. The synthesis and biological activity of five-membered Se-containing heterocycles published in recent years will be described below.

7.2.1 Selenophene derivatives

Selenophene nucleus belongs to a heterocyclic-aromatic class, being present in a wide variety of organic compounds with promising pharmacological properties [4]. It has several biological activities, such as antioxidant [5], anti-inflammatory [6],

Guilherme M. Martins, Departamento de Química, Universidade Federal de Santa Maria, RS, Brazil
Samuel R. Mendes, Laboratório de Síntese e Catálise – SINCA – Departamento de Química, Universidade do Estado de Santa Catarina, SC, Brazil

https://doi.org/10.1515/9783110625110-007

antihypertensive [7], anticonvulsant [8, 9], antibacterial [10], antitumor [11] and cytostatic [12, 13], being incorporated in natural product derivatives [14, 15].

Zhou et al. [16] conducted a study to further explore antibreast cancer drugs, suggesting new treatment possibilities, with fewer side effects than tamoxifen drugs or other therapeutic agents. The objective is to evaluate the importance of the selenophene derivatives as estrogen receptors (ER). The authors explored the structural diversity as well as the activities of these selenophenes in two ERs, ERα and ERβ, in which ERα is found mainly in the female reproductive system, and ERβ is found mainly in the prostate, colon, central nervous system, and cardiovascular system [17]. A wide variety of heterocyclic cores has been explored in the development of SERMs (selective estrogen receptor modulators) [18, 19]. Their antiproliferative activities in cancer and normal cell lines have also been investigated, revealing what may be a general strategy for obtaining superagonists to other members of the nuclear receptor superfamily. Four series of novel selenophene core compounds are shown in Schemes 7.1 and 7.2.

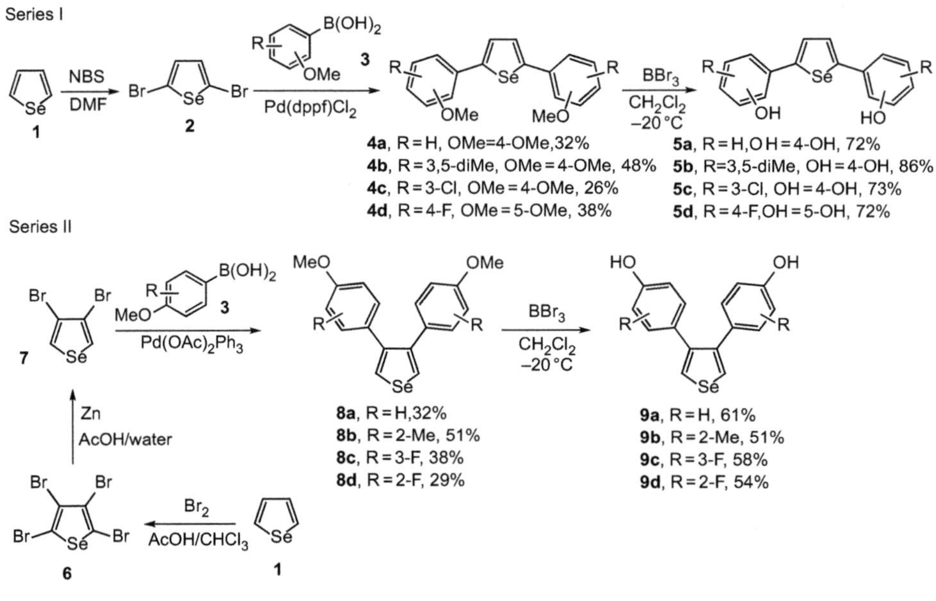

Scheme 7.1: Representative examples for dissubstituted selenophene synthesis.

Dibrominated selenophene precursor **2** was prepared by bromination of selenophene **1** using *N*-bromosuccinimide in dimethylformamide (DMF; Scheme 7.1 – Series I). After, Suzuki cross-coupling was performed, providing products **4a–d** in modest yields (26–48%). Finally, the compounds **4a–d** were treated with boron tribromide, providing the final 2,5-disubstituted diphenolic selenophenes **5a–d** in

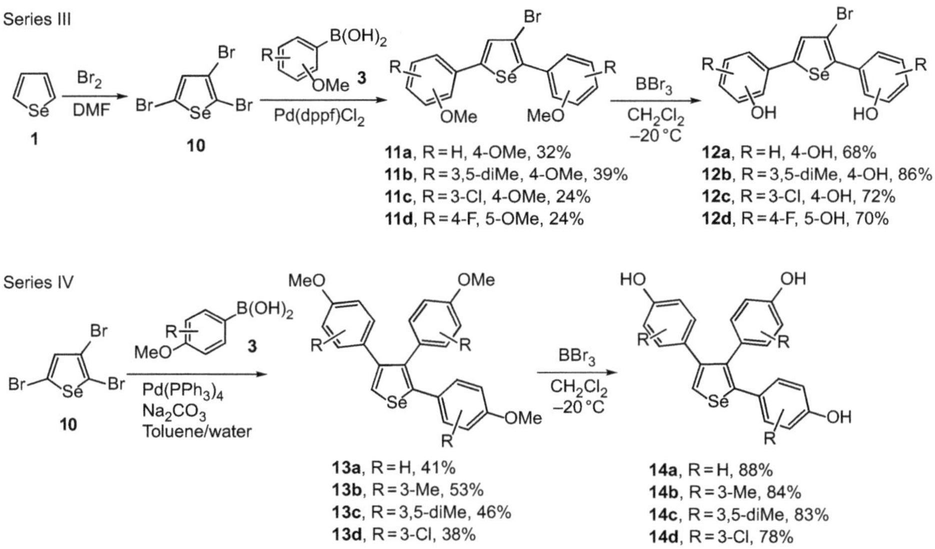

Scheme 7.2: Representative examples for trisubstituted selenophene synthesis.

good yields (72–86%). Regarding compounds **9a–d** (Scheme 7.1 – Series II), the 3,4-dibromoselenophen **7** intermediate cannot be obtained by direct bromination of selenophene **1**. Thus, a debromination of 2,3,4,5-tetrabromoselenophen **6** was performed with Zn^0 and AcOH in water. The respective **6** was achieved through bromination of selenophene **1** with Br_2, using $CHCl_3$ and AcOH as solvent. Thereafter, the treatment of 3,4-dibromoselenophene **7** with aryl boronic acid **3** using $Pd(OAc)_2/PPh_3$ as a catalyst afforded the products **8a–d** in moderate yields (29–51%). Then, with cleavage of the methoxy groups of **8a–d** by BBr_3 in CH_2Cl_2, the desired products **9a–d** was obtained in good yields (51–61%).

To provide compounds **12a–d, 14a–d** (Scheme 7.2 – Series III and IV), the tribrominated selenophene precursor **10** was prepared by bromination of selenophene **1** with Br_2 in DMF. For the series of compounds **11a–d**, 2,3,5-tribromoselenophene **10** was treated with aryl boronic acid **3** using $Pd(dppf)Cl_2$ as catalyst to give **11a–d,** followed by cleavage of the ether group, affording the products **12a–d** in good yields (68–86%). To achieve **14a–d** compounds, Suzuki cross-coupling was applied using **10** and aryl boronic acid **3**, followed by cleavage of the ether group, producing good yields (78–88%).

According to the results of ER binding affinity, it has been observed that selenophene core ligands are largely ERβ selective, and the position of the phenolic group has a marked effect on their binding affinity. This result is similar to that already reported, with selenophene, thiophene, and furan derivatives [19, 20]. In transcription assays, most selenophenes exhibited partial or complete ERβ agonist activity, while in ERα a wide range of activities were shown, covering antagonists and agonists, with some being outlined as superagonists. In comparison to the drug tamoxifen, compound **14c** showed significant antiproliferative potency in MCF-7 breast cancer cell lines (75.3% inhibition).

Considering that resveratrol core is a natural phenol found in plants with the function of protecting, Chovanec et al. [21] proposed that benzo[*b*]selenophene derivatives inspired by resveratrol could act as an antioxidant in yeast. The study was conducted in lower eukaryotes *Saccharomyces cerevisiae*, evaluating the toxicity, DNA damage, and reactive oxygen species (ROS), as well as their ability to act as active antimicrobials redox agents. Additionally, Arsenyan et al. [22] developed an aryl/alkyl cyclization pathway via selenobromination, followed by induced displacement of 3,2-aryl group as a synthetic route for the preparation of resveratrol-inspired polyhydroxy-2- and -3-arylbenzo[*b*]selenophenes (Scheme 7.3). The redox properties, free radical scavenging ability, and cytotoxicity against malignant cell lines (MCF-7, MDA-MB-231, HepG2, and 4T1) were explored.

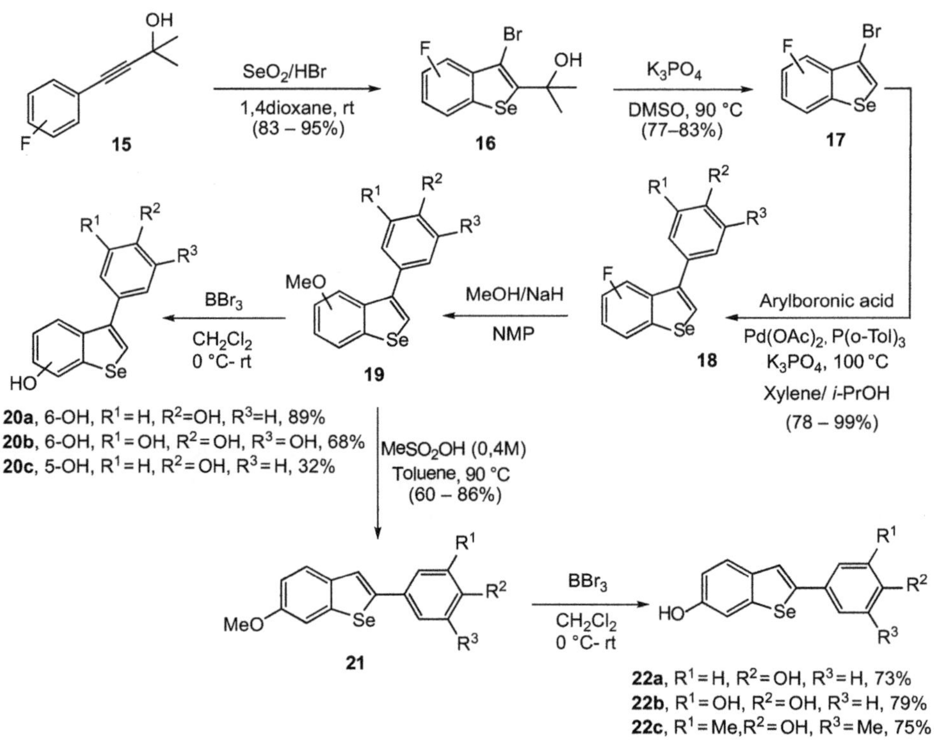

Scheme 7.3: Representative examples for hydroxy substituted benzo[*b*]selenophenes synthesis.

In order to obtain the respective hydroxy substituted benzo[*b*]selenophene derivatives, Chovanec et al. performed a five-step procedure to achieve the 3-aryl derivatives; however, for the derivatives with aryl at the 2-position, an additional step was required (Scheme 7.3). Initially, the corresponding propargyl alcohol **15**, under selenobromination conditions, produced benzo[*b*]selenophenes **16** in excellent yields (83–95%). Thereafter, the 2-unsubstituted derivatives **17** were obtained via deacetonation (77–83%). Suzuki cross-coupling of **17** with the appropriate boronic acid followed by methoxylation via substitution of the fluorine atom provided methoxylated precursors **19**. Then, **19** was heated at 90 °C for 4 h using 0.4 M methanesulfonic acid solution in toluene, promoting the rearrangement of the aryl group from position 3 to position 2 of the selenophenic ring, providing good yields of products **21** (60–86%). Finally, the respective benzo[*b*]selenophenes **19** and **21** were demethylated to provide the desired products **20a–c** (32–89%) and **22a–c** (73–79%).

The biological activities of this benzo[*b*]selenophene derivative were examined in terms of toxic effects and the ability to induce damage in DNA and ROS, and the data collected provided promising results. Some benzo[*b*]selenophenes exhibited toxic effects against yeast cells and could be used as antifungal agents. Finally, most benzo[*b*]selenophenes demonstrated antioxidant activity, suggesting their use as antioxidant supplements in the human diet in those without toxic effects.

In 2019, Martins et al. [23] reported an easy and convenient pathway for the synthesis of 2-aryl-selenopheno[2,3-*b*]indole derivatives (**24**). By electrophilic cyclization of the respective 3(arylalkynyl)indoles **23**, using $SeCl_2$ as the electrophilic source of selenium, the authors obtained 17 new compounds with yields up to 82%. Electronic and steric properties of the substituent attached to the nitrogen atom influenced the reaction yield. The halogen atoms attached to the alkyne group and to the indole ring proved an average 20% reduction in yield (Scheme 7.4).

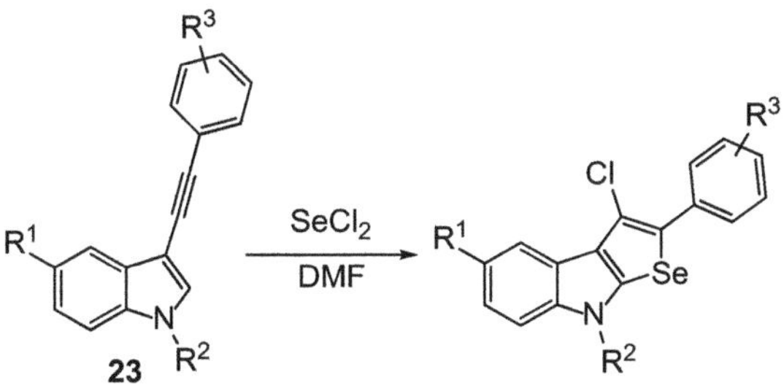

23

24a, R^1 = H, R^2 = H, R^3 = H, 80%
24b, R^1 = CN, R^2 = H, R^3 = H, 72%
24c, R^1 = H, R^2 = H, R^3 = 3-Me, 82%
24d, R^1 = H, R^2 = Me, R^3 = 4-Br, 51%
24e, R^1 = H, R^2 = Ph, R^3 = 3-Me, 22%
24f, R^1 = Br, R^2 = H, R^3 = Ph, 70%

Scheme 7.4: Representative examples for 2-aryl-selenopheno[2,3-*b*]indoles synthesis.

The antifungal activity of selected derivatives was evaluated against several strains of *Cryptococcus* and *Candida*, as well as *Saccharomyces cerevisiae*. The values of minimum inhibitory concentrations (MIC) and minimum lethal concentrations (MLC) were obtained from nystatin and fluconazole comparators. *Cryptococcus neoformans*, *Cryptococcus gatti*, and *Candida krusei* were most susceptible to this class of compound; **24c** showed optimal MIC and MLC values (MIC = 36.3; MLC = 72.5 μM), (MIC = 36.3; MLC = 145.1 μM), and (MIC = 4.5; MLC = 290 μM), respectively, showing magnitude below or equivalent to its comparator.

7.2.2 Selenazole derivatives

Selenazoles are an important class of heterocycles, showing significant biological effects and considerable pharmacological relevance [24,25,26]. However, to date, there are no organoselium compounds in clinical use as anticancer agents, even with the promising results already demonstrated. In 2017, Zhang et al. [27] proposed the synthesis of multiheterocyclic molecules from 1,3-selenazole as a template, modified by pyrazole, 1,2,4-triazole, tetrazole, 1,2,4-triazine, and tetrazole. Inhibitory activities against the cell division cycle 25B phosphatase (Cdc25B) were tested, and more than 10 compounds showed activities. Additionally, Supuran et al. [28] synthesized a series of disubstituted selenazole derivatives and evaluated their activity as inhibitors of carbonic anhydrase (CA, EC 4.2.1.1) against human (h) isoforms hCA I, II, IV, AV, VB, and IX, which are involved in several diseases, such as glaucoma, epilepsy, retinitis pigmentosa, arthritis, and tumors.

Initially, 4-cyanobenzenesulfonamide **26** was obtained by reaction of sulfonyl chloride derivative **25** with aqueous ammonium hydroxide solution. Therefore, 4-sulfamoylbenzoselenoamide **27** was prepared by reacting **26** with Na_2Se as a selenating reagent under reflux in ethanol. Finally, **27** was treated with different α-haloketones, incorporating aromatic **30a–f** or aliphatic **28a–c** moieties by refluxing in ethanol, obtaining several 2,5-disubstituted 1,3-selenazoles (**31a–f** and **29a–c**) in good yields (55–83%, Scheme 7.5). Additionally, the authors reported the synthesis of functionalized 2,5-selenazoles (**33**, **35**, **37**, and **39**), starting from **29c**, as shown in Scheme 7.6.

To obtain 1,3-selenazole **33**, compound **29c** was treated with thiophenol (**32**) and Et_3N in acetonitrile, yielding the product in 80%. Tellurite and selenide derivatives were also assessed (**35a–c** and **37a–c**) to provide the desired products in good yields (74–78%). Finally, the synthesis of 1,3-selenazole **39** was performed, which has great pharmacological interest.

In general, the compounds showed potent inhibition against the tumor-associated IX hCA transmembrane, with K_i's in nanomolar to subnanomolar inhibition range. They were evaluated for their effects on cell viability against human prostate (PC3) and breast (MDA-MB-231) cancer cells lines, in which it showed excellent values for antitumor activity. The authors emphasize that these selenazole derivatives are interesting for developing new Carbonic anhydrases IX inhibitors [29].

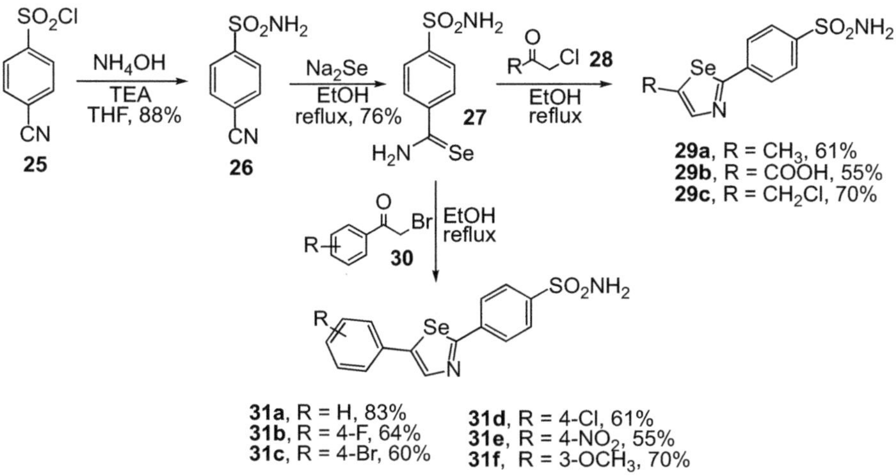

Scheme 7.5: Examples for the synthesis of selenazole derivatives.

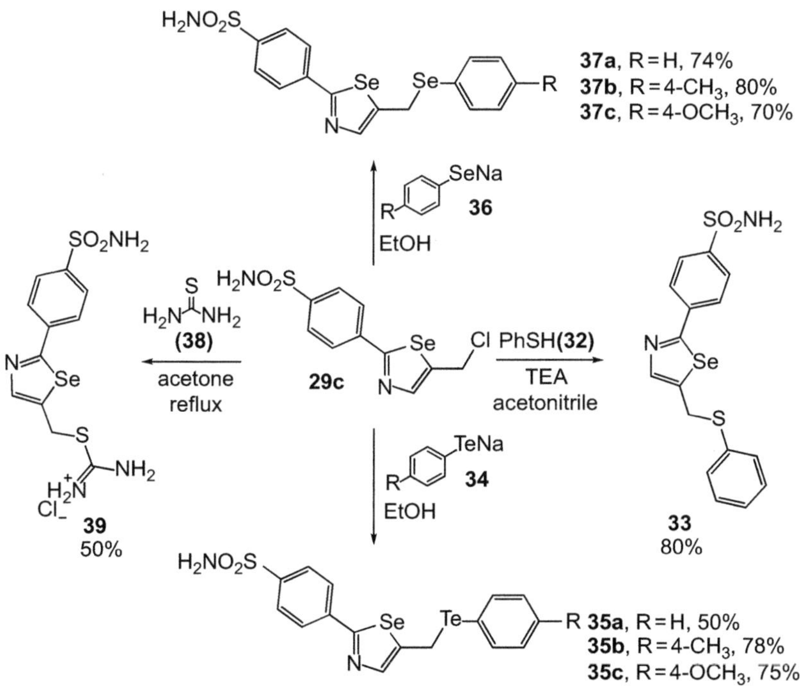

Scheme 7.6: Synthesis of substituted 2,5-selenazoles **33, 35, 37**, and **39**.

7.2.3 Selenadiazoles derivatives

Sanmartín et al. [30] developed a series of benzo[*c*][1,2,5]selenadiazole-5-carboxylic acids (BSCA) (Scheme 7.7, **43a–k** and **44**), to evaluate their antiproliferative effect against five human tumor cell lines, including prostate (PC-3), colon (HT-29), leukemia (CCRF-CEM), lung (HTB-54), and breast (MCF-7), by standard MTT assay and antioxidant activity using the DPPH test. The authors noted that other 1,2,5-selenadiazole derivatives have already been identified as potent agents with antiproliterative effect against human cancer cells, proving to be less toxic to nontumor cells [31, 32].

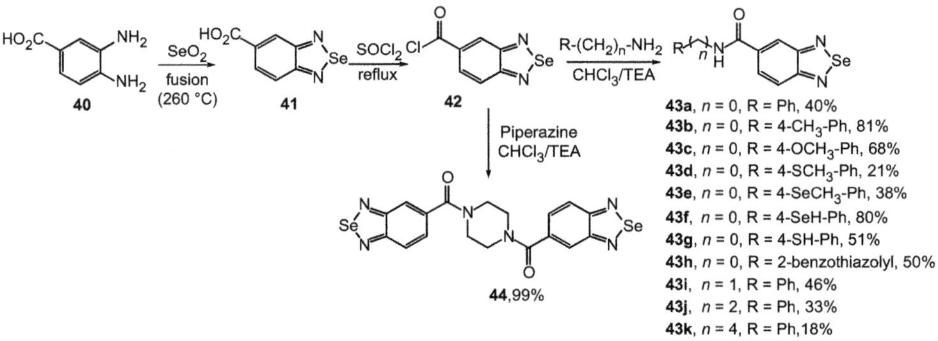

Scheme 7.7: Representative examples for the synthesis of selenadiazole derivatives.

Derivatives **41** were obtained from the appropriate orthoaromatic diamine **40** and selenium dioxide in the 1:1 molar ratio in the absence of solvent, heating to the melting point. However, derivatives **43** and **44** were obtained in two different steps starting from the corresponding acyl chloride **42**, which was treated without prior purification with the appropriate amine or diamine in chloroform in the presence of triethylamine at room temperature providing products in moderate-to-excellent yields (18–99%). The authors highlight the compounds **43c–e** and **43h**, which showed potent inhibitory activity with GI_{50} values (concentration at 50% of maximal inhibition of cell proliferation) below 10 µM in cancer cell lines, being analyzed in non-malignant cell lines from breast (184B5) and lung (BEAS-2B). In addition, compound **43e** showed promising antiproliferative activity in breast cancer cells (MCF-7). Induction of cell death by compound **43e** was independent of the apoptotic process, not affecting cell cycle progression. Likewise, the radical removal properties of the novel selenadiazole derivatives were confirmed by testing their ability to eliminate DPPH radicals. Compound **43e** has been identified as a cell growth inhibitory agent and selectively toxic to cancer cells; however, compound **43g** has proven to be the most potent antioxidant among the synthesized selenadioneol derivatives. Considering the high radical scavenging activity and low toxicity shown by most compounds, the authors emphasize the importance of this class of compound to serve as an excellent support for reaching novel synthetic antioxidant derivatives useful for the treatment of various diseases, such as cancer, neurodegenerative, cardiac, and leishmaniasis diseases [33]. Likewise, Chen et al. [34] reported that benzimidazole-containing selenadiazole derivatives may induce cell-cycle arrest and apoptosis in human breast cancer cell lines by activation of the ROS/AKT (inhibition of protein kinase B) pathway.

Additionally, substituted 1,2,3-selenadiazoles derivatives have demonstrated excellent antifungal activity against *Cryptococcus neoformans* [35], as well as excellent antitumor activity in the growth of human melanoma cells (A375) [36], and anti-HIV-1 activity against HIV-I in MT-4 cells [37]. Khanna et al. [38] reported the synthesis of novel 4,5-disubstituted 1,2,3-selenadiazole derivatives, which were proved to be active against various bacteria and fungi. The authors proposed a solvent-free path for the preparation of 1,2,3-selenadiazoles **46** from the respective semicarbazones **45** (Scheme 7.8). Semicarbazone derivatives **45** can be readily prepared from the respective ketones and semicarbazide hydrochloride, under heating in MeOH.

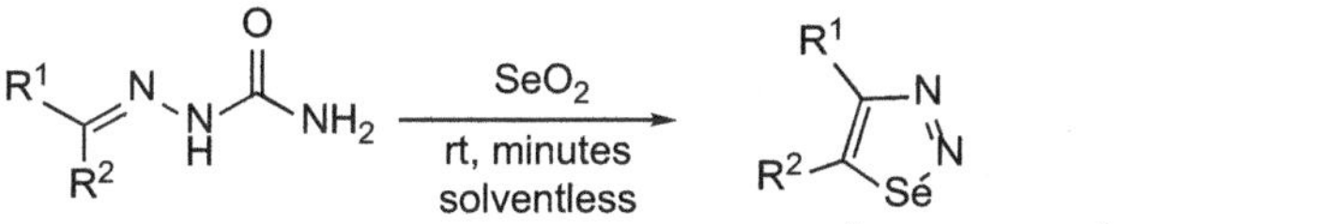

45

46a, $R^1 = CH_2CH_3$, $R^2 = H$, 50%
46b, $R^1 = CH_2CH_3$, $R^2 = CH_3$, 51%
46c, $R^1 = Ph$, $R^2 = H$, 45%
46d, $R^1 = Ph$, $R^2 = CH_3$, 45%
46e, $R^1 = CH_2CH_3$, $R^2 = H$, 50%

Scheme 7.8: Representative examples for the synthesis of 1,2,3-selenadiazole derivatives.

The respective 1,2,3-selenadiazoles (**46a–e**) showed activity against bacterial and fungal stains using agar disc diffusion, as well as agar well diffusion, showing better antibacterial properties compared to established antibiotics, such as tetracycline. The 4-ethyl-5-methyl-1,2,3-selenadiazole **46b** demonstrated better antimicrobial activity among the compounds tested. These compounds were also evaluated for antifungal activity against *Aspergillus niger* and *Penicillium notatum*, showing to be highly active even in extremely low concentrations. The compound **46a** showed good inhibition against resistant *Pseudomonas aeruginosa*.

7.2.4 Cyclic selenamides derivatives

Selenoenzyme glutathione peroxidase (GPx) was discovered in 1973 and has since been studied because of its important role in mechanisms of endogenous response to exposure to ROS [39, 40]. Several molecules are studied for the purpose of obtaining GPx enzyme activity [41, 42]. In addition, Ebselen derivatives have demonstrated GPx mimicking characteristics [43, 44, 45, 46], as well as antibacterial activity [47], showing several advantages. There are several routes to reach Ebselen derivatives, which were first prepared by Weiss in 1924 [48]. The general procedure for obtaining cyclic selenamides (Ebselen) and their analogues are described in Scheme 7.9.

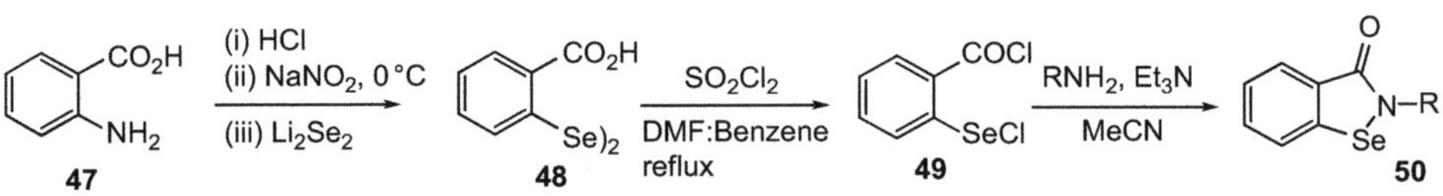

Scheme 7.9: General procedure for obtaining the cyclic selenamides.

From the anthranilic acid **47**, diazotization was carried out with sodium nitrite in aqueous HCl solution, followed by selenenylation with dilithium diselenide, with the elimination of nitrogen gas to provide 2,2′-diselenodibenzoic acid **48**. The respective diselenide with excess of thionyl chloride in the presence of DMF produces 2-(chloroseleno)benzoyl chloride **49**. Finally, acylation with the appropriate amine provides the respective Ebselen **50** (Scheme 7.9).

Another possible path is selenium-nitrogen coupling reaction catalyzed by copper (Scheme 7.10). Lars Engman [49] described the preparation of Ebselen derivatives from the respective N-substituted benzamides via ortho-lithiation with *n*-BuLi, with elemental selenium and copper bromide as the oxidant. Additionally, Kumar et al. [50, 51] demonstrated that 2-chloro-, 2-bromo-, and 2-iodo-arylamides substrates can be applied in the selenium and nitrogen coupling reaction using CuI and 1,10 phenanthroline as ligand with potassium carbonate and DMF as solvent.

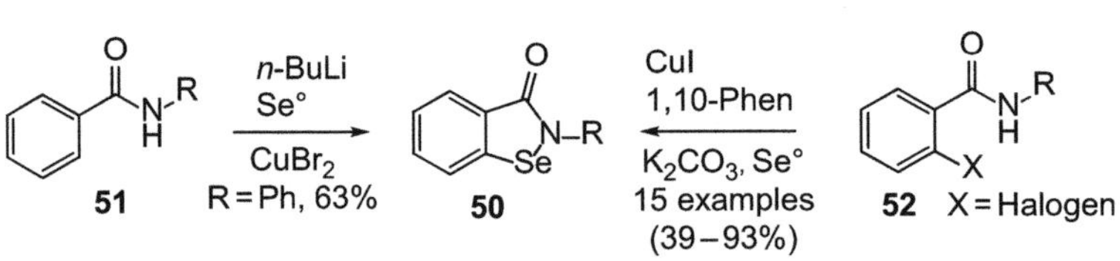

Scheme 7.10: Selenium-nitrogen coupling reaction catalyzed by copper.

Lopez-Ribot and coworkers [52] demonstrated a broad spectrum of antifungal action for Ebselen derivatives against a variety of important fungi. The authors emphasize that *Candida auris* is resistant to Fluconazole and Amphotericin B, and new possibilities need to be obtained to treat this emerging pathogen. The authors have identified that Ebselen **50a** (Figure 7.1) shows 100% inhibition of *C. auris* growth, as well as the ability to inhibit *C. auris* biofilm formation. According to the results, Ebselen showed a broad spectrum of antifungal actions against a wide variety of fungi, including yeasts and molds. Previous studies have shown that Ebselen can restore the efficacy of Meropenem against a laboratory strain that produces NDM-1 (an enzyme that makes bacteria resistant to a variety of β-lactam antibiotics) [53].

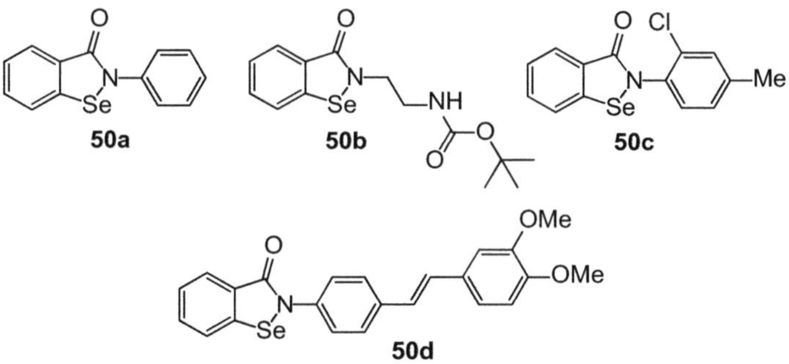

Figure 7.1: Ebselen derivatives.

Chan et al. [54] reported trials of 46 analogues of Ebselen, analyzing the structural relation of these derivatives with their biological potential to increase the antimicrobial efficacy of Meropenem against Carbapenems resistant NDM-1 producing Enterobacteriaceae. In general, the compound **50b** (Figure 7.1) showed strong synergistic antimicrobial activity with carbapenems having low cytotoxicity; it was observed that the application of these compounds with carbapenem adjuvants needs to be considered. Mucha and coworkers [55] developed a series of 25 analogues of Ebselen, providing a novel approach for the inhibition of human methionine aminopeptidase 2 (MetAP2) activity. Inhibition of human methionine aminopeptidase 2 was identified as the major route to inhibit angiogenesis during growth and metastasis of solid tumors. The authors evaluated their inhibitory activity against three neutral aminopeptidases (MetAP2, alanine, and leucine aminopeptidases), demonstrating that these Ebselen derivatives are selective inhibitors of MetAP2 slow binding. Most of Ebselen analogues exhibited moderate potency (IC_{50} = 1–12 μM), being the most promising obtained with analog **50c** (Figure 7.1), which showed IC_{50} = 0.121 ± 0.066 μM. In addition, Li et al.[56] have identified that hybrid derivatives of Ebselen and resveratrol may be related to cancer treatment (Figure 7.1 – **50d**). The results indicated that four human cancer cell lines showed TrxR inhibitory activities, being able to cause apoptosis induced by oxidative stress in cancer cells.

7.3 Six-membered Se-containing heterocycles

The development of six-membered Se-containing heterocycles has increased over the past 20 years [57]. Like the five-membered showed before, six-membered heterocycles have attracted attention due to their biological activities [58]. Thus, the synthesis and biological activity of these compounds published in recent years will be described here.

7.3.1 Selenazine derivatives

Alzheimer's disease (AD) is a complex and most prevalent neurodegenerative disorder with multiple dysfunctional pathways. This disease is the most common form of dementia [59]. In 2015, a series of phenoselenazines (PSZ, **58**) were synthesized and evaluated as multitargeting ligands aimed at the cholinergic, amyloid, and oxidative stress pathways of AD [60]. Rao et al. prepared the PSZ as shown in Scheme 7.11.

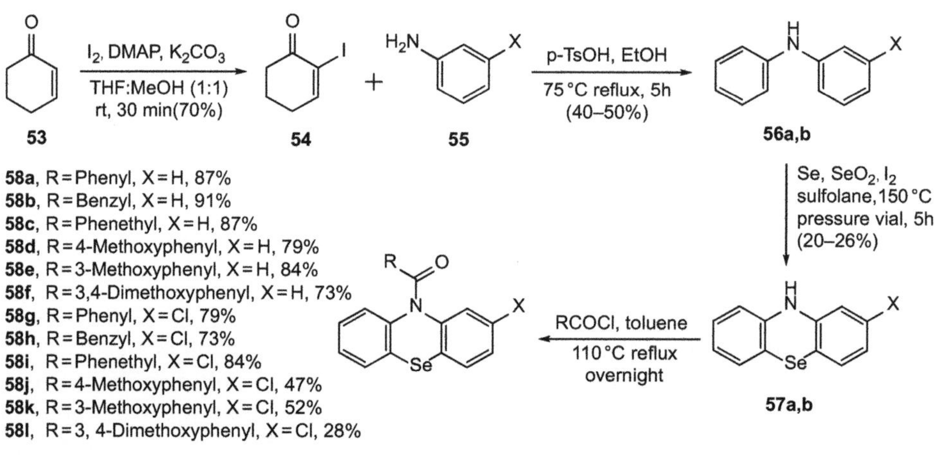

Scheme 7.11: Synthesis of tricyclic phenoselenazines **58**.

Initially, iodinated 2-cyclohex-2-enone **54** was prepared from the reaction between cyclohex-2-enone and iodine in the presence of dimethylaminopyridine and potassium carbonate. The starting precursor diphenylamine **56** was synthesized by coupling α-iodinated 2-cyclohex-2-enone (**54**) with substituted anilines **55** in a metal-free approach by refluxing overnight in presence of trace amounts of *p*-TSOH to produce **56** in 40-50% yields. The ring closing of **56** to obtain the PSZ tricyclic ring **57** was achieved by heating them at 150 °C in a pressure vial, in the presence of selenium,

selenium dioxide (SeO_2), and iodine, using sulfolane as the solvent. The final yield of PSZ derivatives **57a** and **57b** were 20–26%. In the final step, PSZ derivatives **57a** and **57b** were subjected to nucleophilic addition/elimination reaction with acid chlorides to produce PSZ derivatives **58a–l** in moderate-to-good yields (28–91%, Scheme 7.11).

In the phenoselenazine series, **58j** (2-chloro-10*H*-phenoselenazin-10-yl-(4-methoxyphenyl)methanone) showed good nonselective cholinesterase inhibition (AChE IC_{50} = 5.8 ± 0.4 μM; BuChE IC_{50} = 4.9 ± 0.5 μM). Interestingly, *N*-10 unsubstituted phenoselenazine **57a** (AChE IC_{50} = 5.6 ± 0.4 μM; BuChE IC_{50} = 3.0 ± 0.5 μM; Aβ1-42 aggregation inhibition = 45.6%; DPPH scavenging = 84.4%) was able to show multitargeting ability by demonstrating cholinesterase inhibition, β-amyloid aggregation, and antioxidant properties. These results show that fused tricyclic ring systems based on either phenoselenazine templates can be useful to develop hybrid small molecules to target multiple pathological routes associated with Alzheimer's disease.

Similarly, in 2016, Viglianisi et al. synthesized different benzo[*b*][1,4]selenazines and conducted preliminary investigation of the GPx-like activity. Selenazines **60** were prepared from 2-sulfonylaminoaryl diselenides substituted with electron-withdrawing or -donating groups (Scheme 7.12) [61].

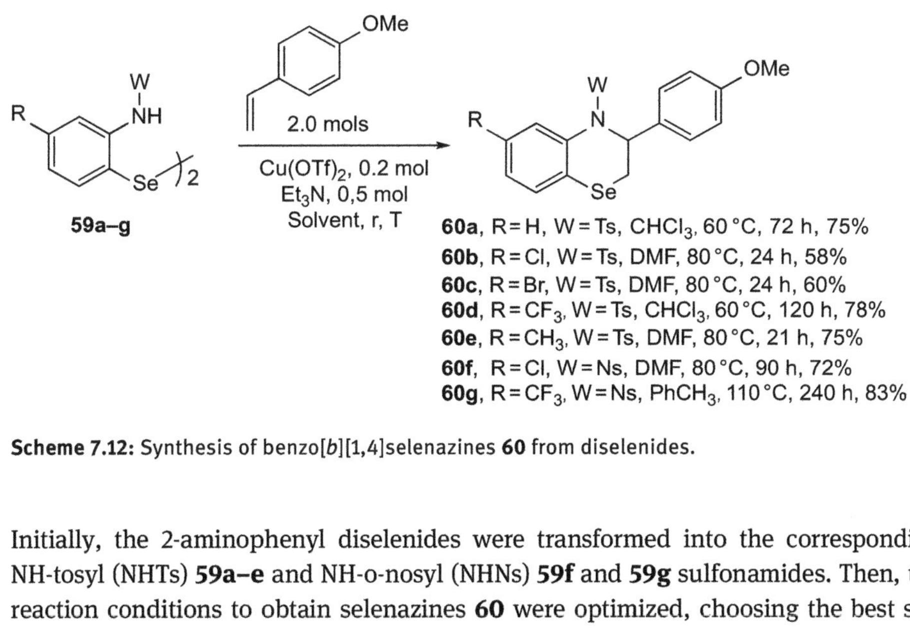

Scheme 7.12: Synthesis of benzo[*b*][1,4]selenazines **60** from diselenides.

Initially, the 2-aminophenyl diselenides were transformed into the corresponding NH-tosyl (NHTs) **59a–e** and NH-o-nosyl (NHNs) **59f** and **59g** sulfonamides. Then, the reaction conditions to obtain selenazines **60** were optimized, choosing the best solvent and reaction time for each of the different diselenides, as shown in Scheme 7.12. The procedure takes place in one pot using a substoichiometric amount of $Cu(OTf)_2$ and a weak base (Et_3N). The respective products **60a-g** were isolated in moderate to good yields (58–83%).

A preliminary investigation of the GPx-like activity was realized with selenazine **60b** and the corresponding *N*-unsubstituted derivative **61** (Figure 7.2) [62, 63]. As shown in Figure 7.2, when *N*-unsubstituted benzo[*b*]selenazine **61** was used as catalyst, 50% of oxidized dithiothreitol (DTT^{OX}) was formed within 150 min. These results showed that the catalytic function of the heterocycles is strongly influenced by the nucleophilic character of the selenium atom, and by the presence of free amine groups. Diphenyl diselenide was employed to compare the activity of **61** with a selenium derivative commonly used as a standard, and, as expected, diselenide was appreciably more active.

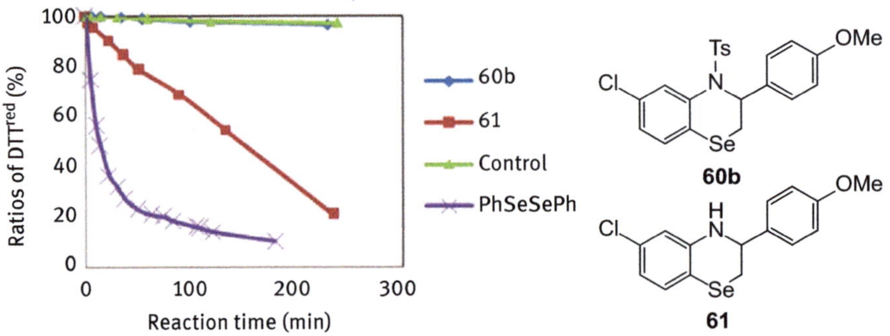

Figure 7.2: Adapted from Viglianisi et al. [61]. GPx-like activity of compounds **60b, 61,** and diphenyl diselenide in the formation of DTT^{OX} from DTT^{RED} with H_2O_2 in CD_3OD. The oxidation of the substrate was monitored by ^1H NMR spectroscopy. A control experiment was carried out in the absence of the catalyst.

7.3.2 Flavanone derivatives

Flavones are well known for their antioxidant, anti-inflammatory, and anticancer activities [64]. The corresponding selenoflavones were also the subject of research that revealed their potential as a neuroprotective agent [65]. Recently, Jeong et al. synthesized selenoflavanones **64** and evaluated their neuroprotective effects. The synthetic pathway is shown in Scheme 7.13. Bromobenzene **62** was acylated with cinnamoyl chloride using $AlCl_3$ under Friedel–Crafts conditions. In the next step,

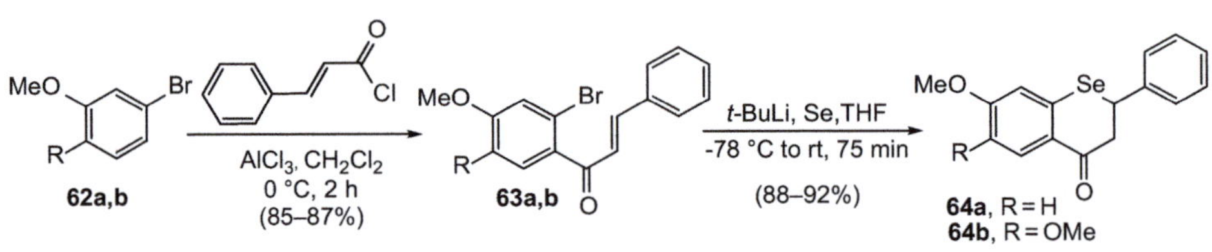

Scheme 7.13: Synthesis of Selenoflavanones **64**.

selenium was introduced to the heterocycle by reaction of **63** with *t*-BuLi, followed by elemental selenium. The respective selenoflavanones **64** were produced in good overall yields (88–92%).

The authors compared the biological activity and physicochemical properties with the respective flavones and reported that selenoflavanones showed lower polarity and higher lipophilicity than the corresponding flavanones, which suggests that they would be able to more easily penetrate the blood–brain barrier. The antioxidant activity was confirmed by in vitro assay. Moreover, hydrogen peroxide-induced cell death decreased with selenoflavanone treatment. Total infarction volumes in the transient ischemia mouse model were significantly reduced by the selenoflavanone treatment. Furthermore, selenoflavanones led to more potent neuroprotective activity than flavanones. Based on these observations, and because selenoflavanones did not cause cytotoxicity at low concentrations, the authors concluded that selenoflavanones **64** could be an effective approach for developing a neuroprotective agent.

7.3.3 Thienopyrimidine derivatives

On the other hand, Sharga et al. prepared fused thienopyrimidine derivatives of phenylselenyl tribromide (Scheme 7.14). First, the compounds **66a–f** were synthesized by reaction of the corresponding thieno[2,3-*d*]-pyrimidines **65a–f** with propargyl bromide

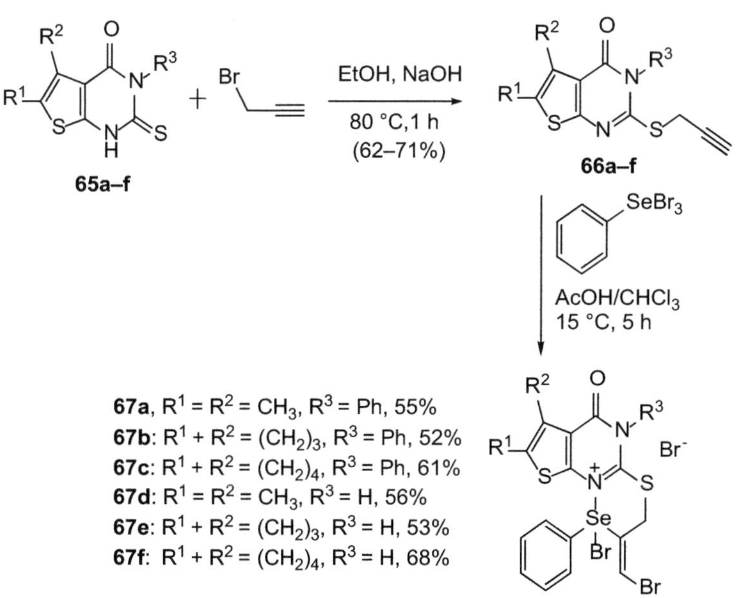

Scheme 7.14: Synthesis of fused thienopyrimidines **67** derivatives of phenylselenyl tribromide.

in ethanol at equimolar presence of NaOH. Thereafter, thiaselenazinium salts **67a–f** were obtained by cyclization reaction between **66a–f** and phenylselenyl tribromide.

The compounds **67a–f** were evaluated for toxicity (method of Poroikov et al. [66]) and antimicrobial activity. The results revealed that the lowest toxicity level was estimated for the derivative **67c**. Then, **67c** and phenylselenyl tribromide were chosen for the antimicrobial activity tests. The studied compounds were most active against yeasts and presented poor to moderate activity against bacteria.

7.3.4 Selenorhodamine derivatives

Six-membered Se-containing heterocycles have also been applied as photosensitizers in photodynamic therapy (PDT) [67], which employs these compounds and light irradiation to induce cell death in the target region [68]. Photosensitizers are compounds that produce ROS such as singlet oxygen ($^1O_2$) [69], but only when they are light irradiated [68, 70]. The mechanism of action of PDT is best explained in the review *Photodynamic therapeutics: basic principles and clinical applications* [71]. This therapy is employed as treatment for numerous cancers, for example, prostate, bladder, and actinic keratosis [67]. In this context, Detty et al. prepared selenorhodamines with an angular **73** or linear **74** fused benzo group (Scheme 7.15) [72]. These compounds were evaluated for their potential as photosensitizers for PDT in Colo-26 cells. Initially, selenides **69** and **70** were obtained by addition of bis-3-dimethylaminophenyl diselenide to 1-lithio-6-dimethylamino-2-naphthamide or 3-lithio- 6-dimethylamino-2-naphthamide, respectively. Xanthones **71** and **72** were prepared by cyclization of diaryl selenides **69** and **70**, respectively. Later, **71** and **72** were separated by chromatography on SiO_2. Finally, the addition of phenylmagnesium bromide to a stirred suspension of **71** or **72** in THF followed by work up with 10% aqueous HPF_6 produced dyes **73** and **74**.

Compounds **73** and **74** were examined for their photophysical properties (absorption, fluorescence, and ability to generate singlet oxygen), for their dark and phototoxicity toward Colo-26 cells, and for their colocalization with mitochondrial-specific agents in Colo-26 and HUT-78 cells. Dyes **73** and **74** provide the first rhodamine photosensitizers with $\lambda_{max} > 640$ nm. Compound **73** was proved to be an effective photosensitizer in vitro toward Colo-26 cells with values of EC_{50} of 6.4×10^{-8} M, with only 1.0 Jcm^{-2} of laser light delivered at $\lambda_{max} \pm 2$ nm. Longer wavelengths of absorption gave **73** greater potential for use in vivo. Extended selenorhodamine dye **74** was proved to be a photosensitizer in vitro toward Colo-26 cells using broad-band light with values of EC_{50} of 1.8×10^{-7} M, with 10 J cm^{-2} of broad-band light.

More recently, Hanaoka and coworkers developed a photosensitizer (**75**) that is activated under hypoxic conditions (Figure 7.3) [73]. The azo-based photosensitizer was synthetized as shown in Scheme 7.16. Selenide **78** was obtained by

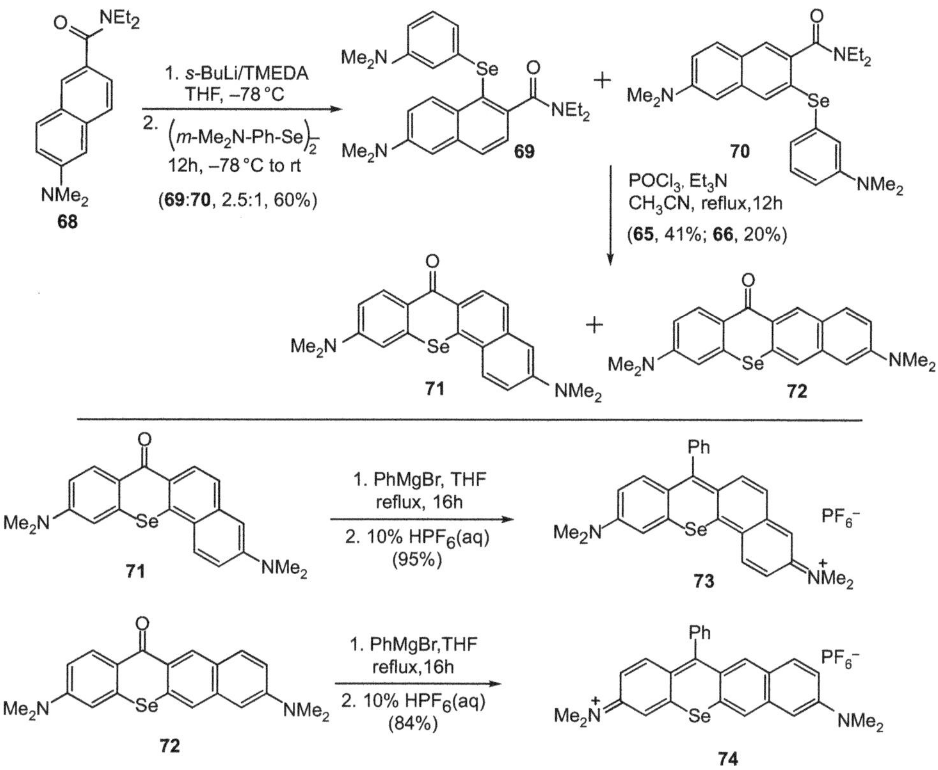

Scheme 7.15: Synthesis of extended rhodamine dyes **73** and **74**.

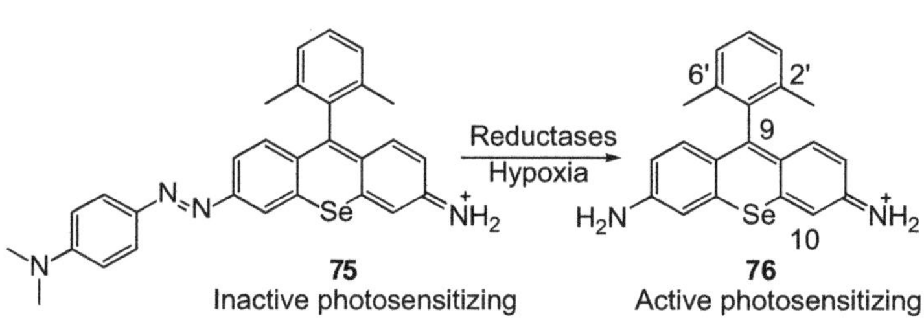

Figure 7.3: Design strategy and chemistry structure of an activatable photosensitizer [73].

addition of bis-4-diallylaminophenyl diselenide to 2-lithio-4-(diallylamino)-*N,N*-diethylbenzamide. Compound **79** was prepared by the cyclization of diaryl selenide **78**. Later, the addition of 2,6-dimethylphenylmagnesium bromide to a stirred suspension of **79** in THF followed by reflux produced compound **80**, which was treated with 1,3-dimethylbarbituric acid and $Pd(PPh_3)_4$ in CH_2Cl_2 for the removal

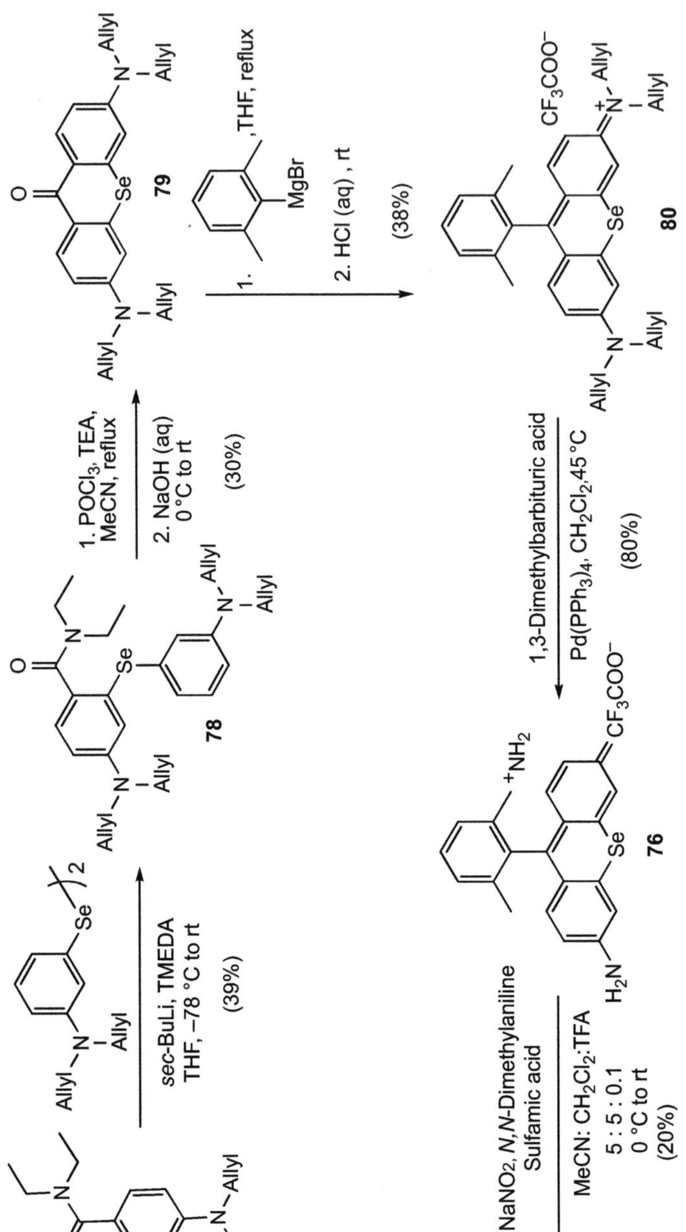

Scheme 7.16: Synthesis of photosensitizer **75**.

of the allyl groups. **75** was obtained by formation of the diazonium salt from **76** and subsequent reaction with *N,N*-dimethylaniline.

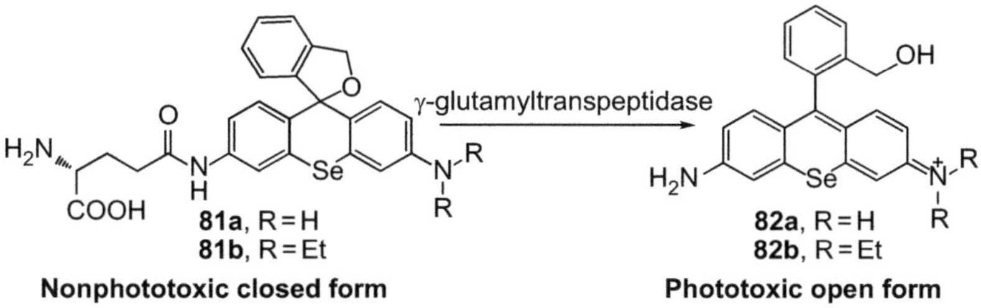

Figure 7.4: Chemical structures of photosensitizers [76].

The novel photosensitizer **75** is selective and reductively activated specifically in cells under mild hypoxic conditions, allowing the production of $^1O_2$ (Figure 7.3). The hypoxic condition (around 5% oxygen concentration) is common in solid tumors [74, 75].

Urano et al. also designed γ-glutamil hydroxymethyl selenorhodamine green **81** as a photoinactive compound that can be activated by a tumor-specific peptidase after topical administration (Figure 7.4) [76].

The synthesis of selenides **83a–b** is similar to the one shown in Scheme 7.16. Compounds **85** were obtained in three steps from selenides **83** as shown in Scheme 7.17. Finally, the reaction between **85a–b** and Boc-Glu-O*t*Bu mediated by 1-[bis(dimethylamino)methylene]-1*H*-1,2,3-triazolo[4,5-*b*]pyridinium 3-oxid hexafluorophosphate, followed deprotection reaction provided γ-glutamil hydroxymethyl selenorhodamine **81**.

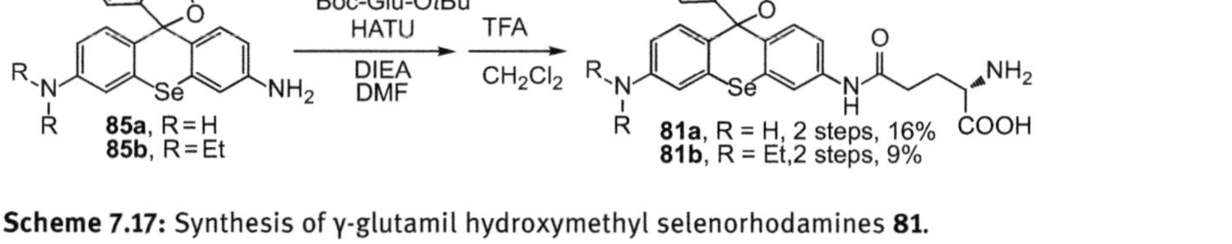

Scheme 7.17: Synthesis of γ-glutamil hydroxymethyl selenorhodamines **81**.

After preliminary studies, the authors decided to focus on **81a** for evaluation as a candidate activatable photosensitizer. This occurred since **81b** is in the its phototoxic form at physiological pH (2.5% of **81b** exists in its open form at pH 7.4), which can mediate nonspecific phototoxicity. In contrast, in the case of **81a**, only 0.1% exists in the open form at pH 7.4; therefore, background phototoxicity is strongly suppressed. Photosensitizer **81a** is activated by aminopeptidase, which is very interesting, because various peptidases are overexpressed in different types of tumors. The authors consider that **81a** is converted into **82a** on the surface of the cells, and **82a** is then internalized into the cells due to its greater hydrophobicity, accumulating mainly in lysosomes. They also confirmed that high-GGT-expressing cells were specifically killed by PDT with **81a**. Furthermore, in a tumor-bearing CAM model, tumors were selectively ablated by PDT with **81a**, without damage to adjacent healthy tissues.

Mclver et al. prepared new selenorhodamines photosensitizers for extracorporeal photopheresis (ECP; Figure 7.5) [77]. ECP is a combination of leukapheresis and photodynamic therapy in which blood is treated with photoactivable drugs, which are then activated with ultraviolet light and re-infused to the patient. ECP has been used in the treatment of erythrodermic cutaneous T-cell lymphoma and other T-cell-mediated disorders [78].

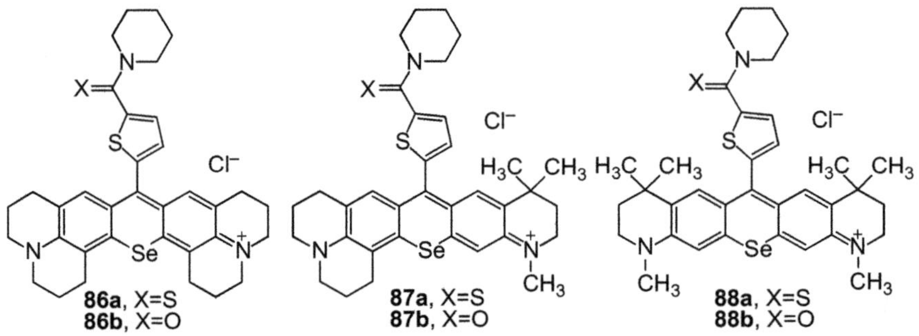

Figure 7.5: Structures of selenorhodamines **86-88**.[77].

Selenorhodamines **86a** and **86b** were synthesized as described by Kryman et al. [79]. Selenorhodamines **87–88** were synthesized from selenoxanthones **89–90** (Scheme 7.18), which were prepared following the procedure previously reported [80]. First, piperidin-1-yl(thiophen-2-yl)methanethione was deprotonated with LDA, and the respective 2-lithiothiophene **91** was then added to THF solutions of **89** and **90** at −78 °C. Workup with aqueous HPF_6 gave **87a** and **88a** in 88% and 79% isolated yields, respectively, as the PF_6 salts (Scheme 7.18). Amides **87b** and **88b** were

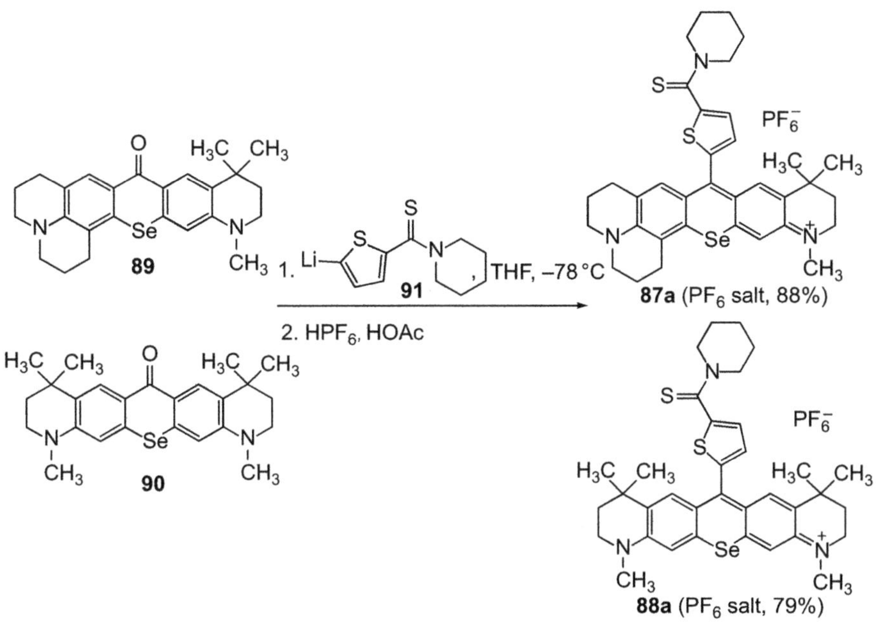

Scheme 7.18: Synthesis of selenorhodamines **87a-88a** as the PF_6 salts.

obtained by reaction of respective thioamides with trifluoracetic anhydride in 56% and 55% yields, respectively. All PF_6 salts of the rhodamines were converted to chloride salts **87a–b** and **88a–b** with a chloride ion-exchange resin.

7.3.5 Selenomorpholine derivatives

Hypochlorous acid (HClO), as one of the highly ROS [69], plays a main role in immune defense against microorganisms and also in inflammation [81]. Thus, the detection of HClO in biological samples is of significant interest [82]. In 2018, Guo et al. developed a reversible and mitochondria targetable fluorescent probe **92** for real-time detection of HClO/ClO^- based on the mechanism of inhibition of photoinduced electron transfer (Figure 7.6) [83]. Probe **92** showed desirable features, including good water solubility, high sensitivity, fast response time, and good selectivity toward HClO/ClO^- over other species. The bioimaging experiments showed that the probe could be used to sense the exogenous and endogenous HClO in RAW264.7 cells. During the tests, the strongly fluorescent compound **93** was detected by ESI-MS analysis after incubation of the probe with NaClO, proving that it was the fluorescent species.

Probe **92** was prepared via integration of selenomorpholine unit on the naphthalimide fluorophore (Scheme 7.19). Initially, compound **94** was converted to

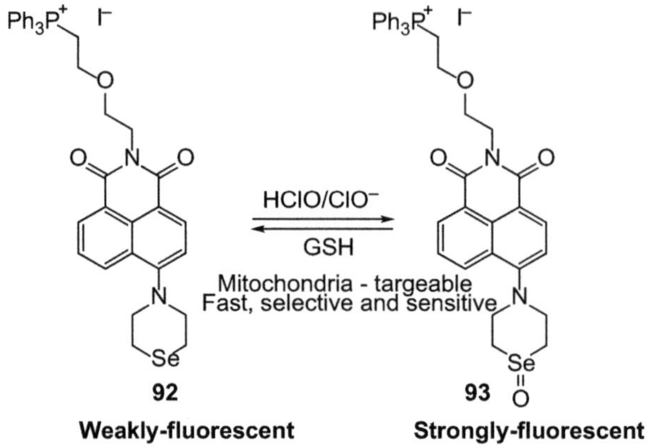

Figure 7.6: The structure and sensing mechanism of probe **92** for HClO/ClOCSF]?>/CSF]?> [83].

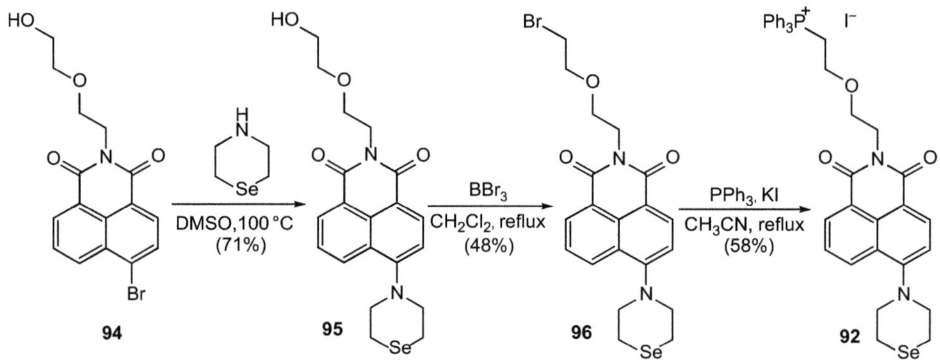

Scheme 7.19: Synthesis of probe **92**.

selenomorpholine substituted derivative **95** and then **96** was obtained by treating **95** with phosphorus tribromide (BBr_3) in dichloromethane. Finally, **96** was reacted with triphenylphosphine in the presence of potassium iodide under reflux conditions in acetonitrile to produce **92**.

7.4 Conclusions

As seen in this chapter, Se-containing heterocycles may be considered relatively new chemical compounds. However, the promising results that cover the pharmacological area motivate the scientific community to investigate the action of these derivatives. Selenium-containing five- and six-membered heterocyclics have an enormous biological and pharmacological potential, exploited in several countries. Consequently, this encourages chemists to evaluate their reactivity by considering various combinations between the respective heteroatoms and substituents, showing that organoselenium chemistry encompasses not only synthetic chemistry but also explores several applications of medicinal interest.

References

[1] Jain, VK. An Overview of Organoselenium Chemistry: From Fundamentals to Synthesis. In: Organoselenium Compounds in Biology and Medicine, Cambridge, Royal Society of Chemistry, 2018, 1–33.

[2] Taylor, AP., Robinson, RP., Fobian, YM., Blakemore, DC., Jones, LH., and Fadeyi, O. Modern advances in heterocyclic chemistry in drug discovery. Org Biomol Chem 2016, 14, 6611–6637.

[3] Majumdar KC, Chattopadhyay SK editors. Heterocycles in Natural Product Synthesis, Weinheim, Germany, Wiley-VCH Verlag GmbH & Co. KGaA, 2011.

[4] Abdelwahab Mahmoud, AB., Kirsch, G., and Peagle, E. Biologically active selenophenes and benzo[*b*]selenophenes. Curr Org Synth 2017, 14, 1091–1101.

[5] Tavadyan, L., Manukyan, Z., Harutyunyan, L., Musayelyan, M., Sahakyan, A., and Tonikyan, H. Antioxidant properties of selenophene, thiophene and their aminocarbonitrile derivatives. Antioxidants 2017, 6, 22.

[6] Donato, F., Pavin, NF., Goes, ATR. et al. Antinociceptive and anti-hyperalgesic effects of bis(4-methylbenzoyl) diselenide in mice: Evidence for the mechanism of action. Pharm Biol 2015, 53, 395–403.

[7] Grange, RL., Ziogas, J., North, AJ., Angus, JA., and Schiesser, CH. Selenosartans: Novel selenophene analogues of milfasartan and eprosartan. Bioorg Med Chem Lett 2008, 18, 1241–1244.

[8] Wilhelm, EA., Jesse, CR., Bortolatto, CF., Nogueira, CW., and Savegnago, L. Anticonvulsant and antioxidant effects of 3-alkynyl selenophene in 21-day-old rats on pilocarpine model of seizures. Brain Res Bull 2009, 79, 281–287.

[9] Wilhelm, EA., Gai, BM., Souza, ACG., Bortolatto, CF., Roehrs, JA., and Nogueira, CW. Involvement of GABAergic and glutamatergic systems in the anticonvulsant activity of 3-alkynyl selenophene in 21 day-old rats. Mol Cell Biochem 2012, 365, 175–180.

[10] Wiles, JA., Phadke, AS., Bradbury, BJ., Pucci, MJ., Thanassi, JA., and Deshpande, M. Selenophene-containing inhibitors of type IIA bacterial topoisomerases. J Med Chem 2011, 54, 3418–3425.

[11] Khalifa, ME., Abdel-Hafez, SH., Gobouri, AA., and Kobeasy, MI. Synthesis and biological activity of novel arylazothiazole disperse dyes containing selenium for dyeing polyester fibers. Phosphorus Sulfur Silicon Relat Elem 2015, 190, 461–476.

[12] Mhetre, AB., Lee, H., Yang, H., Lee, K., Nam, D-H., and Lim, D. Synthesis and anticancer activity of benzoselenophene and heteroaromatic derivatives of 1, 2, 9, 9a-tetrahydrocyclopropa[*c*]benzo[*e*]indol-4-one (CBI). Org Biomol Chem 2017, 15, 1198–1208.

[13] Wilhelm, EA., Jesse, CR., Prigol, M., Alves, D., Schumacher, RF., and Nogueira, CW. 3-Alkynyl selenophene protects against carbon-tetrachloride-induced and 2-nitropropane-induced hepatic damage in rats. Cell Biol Toxicol 2010, 26, 569–577.

[14] Arsenyan, P., Paegle, E., Domracheva, I., Gulbe, A., Kanepe-Lapsa, I., and Shestakova, I. Selenium analogues of raloxifene as promising antiproliferative agents in treatment of breast cancer. Eur J Med Chem 2014, 87, 471–483.
[15] Vasiljeva, J., Domracheva, I., and Arsenyan, P. Selenium analogues of (S)-clopidogrel: Preparation method and properties. Tetrahedron Lett 2016, 57, 196–198.
[16] Zhang, S., Wang, Z., Hu, Z. et al. Selenophenes: introducing a new element into the core of non-steroidal estrogen receptor ligands. ChemMedChem 2017, 12, 235–249.
[17] Swedenborg, E., Power, KA., Cai, W., Pongratz, I., and Rüegg, J. Regulation of estrogen receptor beta activity and implications in health and disease. Cell Mol Life Sci 2009, 66, 3873–3894.
[18] Zhou, H-B., Carlson, KE., Stossi, F., Katzenellenbogen, BS., and Katzenellenbogen, JA. Analogs of methyl-piperidinopyrazole (MPP): Antiestrogens with estrogen receptor α selective activity. Bioorg Med Chem Lett 2009, 19, 108–110.
[19] Min, J., Wang, P., Srinivasan, S. et al. Thiophene-core estrogen receptor ligands having superagonist activity. J Med Chem 2013, 56, 3346–3366.
[20] Luo, J., Hu, Z., Xiao, Y. et al. Rational design and optimization of selenophenes with basic side chains as novel potent selective estrogen receptor modulators (SERMs) for breast cancer therapy. Medchemcomm 2017, 8, 1485–1497.
[21] Mániková, D., Šetáková, Z., Rendeková, J. et al. Resveratrol-inspired benzo[*b*]selenophenes act as anti-oxidants in yeast. Molecules 2018, 23, 1–13.
[22] Paegle, E., Domracheva, I., Turovska, B. et al. Natural-antioxidant-inspired benzo[*b*]selenophenes: synthesis, redox properties, and antiproliferative activity. Chem – An Asian J 2016, 11, 1929–1938.
[23] Martins, GM., Do Carmo, G., Morel, AF., Kaufman, TS., and Silveira, CC. A convenient and atom-economic one-pot selenium-chloride-mediated synthesis of 2-arylselenopheno[2,3-*b*]indoles and their antifungal activity. Asian J Org Chem 2019, 8, 369–375.
[24] Below, H., Pfeiffer, W-D., Geisler, K., Lalk, M., and Langer, P. 1,3-Selenazole. European J Org Chem 2005, 2005, 3637–3639.
[25] Facchinetti, V., Nery, A., Avellar, M., Gomes, C., de Souza, M., and Vasconcelos, T. Highlights on the synthesis and biological activity of 1,3-selenazoles. Curr Org Synth 2015, 12, 140–149.
[26] Khalifa, ME., and Gobouri, AA. Biological screening and assessment of certain substituted monoazo heterocycles containing sulphur and / or nitrogen and their seleno like moieties. Polish J Chem Technol 2017, 19, 28–35.
[27] Zhang, C., Sun, X., Pu, Y. et al. Design, synthesis and activities of multiheterocyclic modified novel molecules using 1,3-selenazole as template. Chinese J Org Chem 2017, 37, 440–454.
[28] Angeli, A., Trallori, E., Ferraroni, M. Di Cesare Mannelli L, Ghelardini, C., and Supuran, CT. Discovery of new 2, 5-disubstituted 1,3-selenazoles as selective human carbonic anhydrase IX inhibitors with potent anti-tumor activity. Eur J Med Chem 2018, 157, 1214–1222.
[29] Supuran, CT., and Winum, J-Y. Carbonic anhydrase IX inhibitors in cancer therapy: An update. Future Med Chem 2015, 7, 1407–1414.
[30] Ruberte, AC., Plano, D., Encío, I., Aydillo, C., Sharma, AK., and Sanmartín, C. Novel selenadiazole derivatives as selective antitumor and radical scavenging agents. Eur J Med Chem 2018, 157, 14–27.
[31] Yang, Y., Deng, S., Zeng, Q., Hu, W., and Chen, T. Highly stable selenadiazole derivatives induce bladder cancer cell apoptosis and inhibit cell migration and invasion through the activation of ROS-mediated signaling pathways. Dalt Trans 2016, 45, 18465–18475.
[32] He, L., Ji, S., Lai, H., and Chen, T. Selenadiazole derivatives as theranostic agents for simultaneous cancer chemo-/radiotherapy by targeting thioredoxin reductase. J Mater Chem B 2015, 3, 8383–8393.

[33] Plano, D., Moreno, E., Font, M., Encío, I., Palop, JA., and Sanmartín, C. Synthesis and in vitro anticancer activities of some selenadiazole derivatives. Arch Pharm (Weinheim) 2010, 343, 680–691.
[34] Liang, Y., Zhou, Y., Deng, S., and Chen, T. Microwave-assisted syntheses of benzimidazole-containing selenadiazole derivatives that induce cell-cycle arrest and apoptosis in human breast cancer cells by activation of the ROS/AKT pathway. ChemMedChem 2016, 11, 2339–2346.
[35] Jalilian, A., Sattari, S., Bineshmarvasti, M., Daneshtalab, M., and Shafiee, A. Synthesis and in vitro antifungal and cytotoxicity evaluation of substituted 4,5-dihydronaphtho[1,2-*d*][1,2,3] thia(or selena)diazoles. Farm 2003, 58, 63–68.
[36] El-Desoky, SI., Badria, FA., Abozeid, MA., Kandeel, EA., and Abdel-Rahman, AH. Synthesis and antitumor studies of novel benzopyrano-1,2,3-selenadiazole and spiro[benzopyrano]-1,3,4-thiadiazoline derivatives. Med Chem Res 2013, 22, 2105–2114.
[37] Zhan, P., Liu, X., Fang, Z., Pannecouque, C., and Clercq, E De. 1,2,3-selenadiazole thioacetanilides: synthesis and anti-HIV activity evaluation. Bioorg Med Chem 2009, 17, 6374–6379.
[38] Jadhav, AA., Dhanwe, VP., Joshi, PG., and Khanna, PK. Solventless synthesis of new 4,5-disubstituted 1,2,3-selenadiazole derivatives and their antimicrobial studies. Cogent Chem 2016, 2, 1–12.
[39] Flohe, L., Günzler, WA., and Schock, HH. Glutathione peroxidase: A selenoenzyme. FEBS Lett 1973, 32, 132–134.
[40] Jiang, J-H., Zhou, H., Li, H-J. et al. Comprehensive modeling of the antioxidant mechanism of ebselen. J Theor Comput Chem 2015, 14, 1550053.
[41] Jiao, Y., Wang, Y., Guo, S., and Wang, G. Glutathione peroxidases as oncotargets. Oncotarget 2017, 8, 80093–80102.
[42] Singh, VP., Poon, J., Yan, J. et al. Nitro-, Azo-, and amino derivatives of ebselen: synthesis, structure, and cytoprotective effects. J Org Chem 2017, 82, 313–321.
[43] Cheng, B., Zhong, J., Wu, F. et al. Ebselen protects rat hearts against myocardial ischemia reperfusion injury. Exp Ther Med 2019, 17, 1412–1419.
[44] Slusarczyk, W., Olakowska, E., Larysz-Brysz, M. et al. Use of ebselen as a neuroprotective agent in rat spinal cord subjected to traumatic injury. Neural Regen Res 2019, 14, 1255–1261.
[45] Martini, F., Rosa, SG., Klann, IP. et al. A multifunctional compound ebselen reverses memory impairment, apoptosis and oxidative stress in a mouse model of sporadic Alzheimer's disease. J Psychiatr Res 2019, 109, 107–117.
[46] Xie, Y., Tan, Y., Zheng, Y., Du, X., and Liu, Q. Ebselen ameliorates β-amyloid pathology, tau pathology, and cognitive impairment in triple-transgenic Alzheimer's disease mice. JBIC J Biol Inorg Chem 2017, 22, 851–865.
[47] Facchinetti, V., De Souza, MVN., Nery, ACS. et al. Synthetic aspects and first-time assessment of 2-amino-1,3-selenazoles against mycobacterium tuberculosis. Lett Drug Des Discov 2018, 15, 1224–1229.
[48] Lesser, R., and Weiß, R. Über selenhaltige aromatische Verbindungen (VI). Berichte der Dtsch Chem Gesellschaft (A B) Ser 1924, 57, 1077–1082.
[49] Engman, L., and Hallberg, A. Expedient synthesis of ebselen and related compounds. J Org Chem 1989, 54, 2964–2966.
[50] Balkrishna, SJ., Bhakuni, BS., and Kumar, S. Copper catalyzed/mediated synthetic methodology for ebselen and related isoselenazolones. Tetrahedron 2011, 67, 9565–9575.
[51] Balkrishna, SJ., Kumar, S., Azad, GK. et al. An ebselen like catalyst with enhanced GPx activity via a selenol intermediate. Org Biomol Chem 2014, 12, 1215–1219.

[52] Wall, G., Chaturvedi, AK., Wormley, FL. et al. Screening a repurposing library for inhibitors of multidrug-resistant candida auris identifies ebselen as a repositionable candidate for antifungal drug development. Antimicrob Agents Chemother 2018, 62, 1–13.
[53] Chiou, J., Wan, S., Chan, KF. et al. Ebselen as a potent covalent inhibitor of New Delhi metallo-β-lactamase (NDM-1). Chem Commun 2015, 51, 9543–9546.
[54] Jin, W Bin., Xu, C., Cheng, Q. et al. Investigation of synergistic antimicrobial effects of the drug combinations of meropenem and 1,2-benzisoselenazol-3(2H)-one derivatives on carbapenem-resistant Enterobacteriaceae producing NDM-1. Eur J Med Chem 2018, 155, 285–302.
[55] Węglarz-Tomczak, E., Burda-Grabowska, M., Giurg, M., and Mucha, A. Identification of methionine aminopeptidase 2 as a molecular target of the organoselenium drug ebselen and its derivatives/analogues: Synthesis, inhibitory activity and molecular modeling study. Bioorg Med Chem Lett 2016, 26, 5254–5259.
[56] Yan, J., Guo, Y., Wang, Y., Mao, F., Huang, L., and Li, X. Design, synthesis, and biological evaluation of benzoselenazole-stilbene hybrids as multi-target-directed anti-cancer agents. Eur J Med Chem 2015, 95, 220–229.
[57] Jain VK, Priyadarsini KI editors. Organoselenium Compounds in Biology and Medicine, Cambridge, Royal Society of Chemistry, 2018.
[58] Martynov, A V., and Amosova, S V. Synthetic approaches to bis(chloromethylidene) substituted six-membered rings with two heteroatoms including S, Se, N in various combinations. Mini Rev Org Chem 2017, 15, 433–442.
[59] Elfakhri, KH., Abdallah, IM., Brannen, AD., and Kaddoumi, A. Multi-faceted therapeutic strategy for treatment of Alzheimer's disease by concurrent administration of etodolac and α-tocopherol. Neurobiol Dis 2019, 125, 123–134.
[60] Tin, G., Mohamed, T., Gondora, N., Beazely, MA., and Rao, PPN. Tricyclic phenothiazine and phenoselenazine derivatives as potential multi-targeting agents to treat Alzheimer's disease. Medchemcomm 2015, 6, 1930–1941.
[61] Menichetti, S., Capperucci, A., Tanini, D., Braga, AL., Botteselle, G V., and Viglianisi, C. A One-Pot Access to Benzo[*b*][1,4]selenazines from 2-Aminoaryl Diselenides. European J Org Chem 2016, 2016, 3097–3102.
[62] Kumakura, F., Mishra, B., Priyadarsini, KI., and Iwaoka, M. A water-soluble cyclic selenide with enhanced glutathione peroxidase-like catalytic activities. European J Org Chem 2010, 440–445.
[63] Iwaoka, M., and Kumakura, F. Applications of water-soluble selenides and selenoxides to protein chemistry. Phosphorus, Sulfur Silicon Relat Elem 2008, 183, 1009–1017.
[64] Verma, AK., and Pratap, R. The biological potential of flavones. Nat Prod Rep 2010, 27, 1571–1593.
[65] Choi, YS., Kim, YJ., Lee, JY., Lee, J., and Jeong, JH. Synthesis and evaluation of selenoflavones that have potential neuroprotective effects. Heterocycles 2014, 89, 2794–2805.
[66] Poroikov, V V., Filimonov, DA., Borodina, Y V. et al. Quantitative relationships between structure and delayed neurotoxicity of chemicals studied by the self-consistent regression method using the pass program. Pharm Chem J 2004, 38, 188–190.
[67] Detty, MR., Gibson, SL., and Wagner, SJ. Current clinical and preclinical photosensitizers for use in photodynamic therapy. J Med Chem 2004, 47, 3897–3915.
[68] Dougherty, TJ., Gomer, CJ., Henderson, BW. et al. Photodynamic therapy. JNCI J Natl Cancer Inst 1998, 90, 889–905.
[69] Flors, E. by SN and C. Contents. In: Comprehensive Series in Photochemistry and Photobiology. 2016. p. P007–P025.

[70] Lovell, JF., Liu, TWB., Chen, J., and Zheng, G. Activatable photosensitizers for imaging and therapy. Chem Rev 2010, 110, 2839–2857.
[71] Sharman, WM., Allen, CM., and Van Lier, JE. Photodynamic therapeutics: basic principles and clinical applications. Drug Discov Today 1999, 4, 507–517.
[72] Davies, KS., Linder, MK., Kryman, MW., and Detty, MR. Extended rhodamine photosensitizers for photodynamic therapy of cancer cells. Bioorganic Med Chem 2016, 24, 3908–3917.
[73] Piao, W., Hanaoka, K., Fujisawa, T. et al. Development of an azo-based photosensitizer activated under mild hypoxia for photodynamic therapy. J Am Chem Soc 2017, 139, 13713–13719.
[74] Petrova, V., Annicchiarico-Petruzzelli, M., Melino, G., and Amelio, I. The hypoxic tumour microenvironment. Oncogenesis 2018, 7, 10.
[75] Muz, B., de la Puente, P., Azab, F., and Azab, AK. The role of hypoxia in cancer progression, angiogenesis, metastasis, and resistance to therapy. Hypoxia 2015, 3, 83–92.
[76] Chiba, M., Ichikawa, Y., Kamiya, M. et al. An activatable photosensitizer targeted to γ-glutamyltranspeptidase. Angew Chemie – Int Ed 2017, 56, 10418–10422.
[77] McIver, ZA., Kryman, MW., Choi, Y. et al. Selective photodepletion of malignant T cells in extracorporeal photopheresis with selenorhodamine photosensitizers. Bioorganic Med Chem 2016, 24, 3918–3931.
[78] Worel, N., and Leitner, G. Clinical results of extracorporeal photopheresis. Transfus Med Hemotherapy 2012, 39, 254–262.
[79] Kryman, MW., Davies, KS., Linder, MK., Ohulchanskyy, TY., and Detty, MR. Selenorhodamine photosensitizers with the Texas-red core for photodynamic therapy of cancer cells. Bioorg Med Chem 2015, 23, 4501–4507.
[80] Kryman, MW., Schamerhorn, GA., Hill, JE. et al. Synthesis and properties of heavy chalcogen analogues of the texas reds and related rhodamines. Organometallics 2014, 33, 2628–2640.
[81] Lapenna, D., and Cuccurullo, F. Hypochlorous acid and its pharmacological antagonism: an update picture. Gen Pharmacol Vasc Syst 1996, 27, 1145–1147.
[82] Cheng, G., Fan, J., Sun, W. et al. A highly specific BODIPY-based probe localized in mitochondria for HClO imaging. Analyst 2013, 138, 6091–6096.
[83] Lv, X., Yuan, X., Wang, Y., and Guo, W. A naphthalimide based fast and selective fluorescent probe for hypochlorous acid/hypochlorite and its application for bioimaging. New J Chem 2018, 42, 15105–15110.

Pablo A. Nogara, Cláudia S. Oliveira and João Batista T. Rocha

8 Chemistry and pharmacology of synthetic organoselenium compounds

8.1 Introduction

Active support for the importance of selenium (Se) in the cell physiology of mammals was provided in 1973. Two groups of investigators identified Se as part of the enzyme glutathione peroxidase (GPx) [1, 2]. Up till now, 25 different selenoproteins have been identified in the human genome [3–5].

After the clear demonstration of Se essentiality to vertebrates, the search for low-molecular-mass organoselenium compounds that can mimic the selenoproteins has increased considerably [6]. In this context, the organic chemists have been investing their efforts in search of synthetic organoselenium compounds, which could mimic the activity of selenoenzymes (e.g., GPx), as well as to find new Se-containing molecules with pharmacological potential, not necessarily linked to the imitation of selenoproteins [6–8]. Some organoselenium compounds were synthesized a few years after Se discovery, but they were little explored from the biological point of view [9]. Nowadays, there are a large number of structurally different organoselenium compounds, such as diselenides, selenophenes, quinoline derivatives, bis-selenides, selenazoles, and selenides [6]. Here, we highlight the major synthetic pathways, as well as the pharmacological properties of some important organoselenium compounds (Figure 8.1).

8.2 History of selenium: emphasis on synthetic organoselenium compounds

Selenium was discovered accidentally in 1817 by Jöns Jacob Berzelius (1779–1848) while examining a foul-smelling red mud found in the lead chambers of their sulfuric acid factory (Figure 8.2). With rudimentary technology, Berzelius was also able to characterize with an exactness some of the chemical properties of Se, as well as its similarities with tellurium (Te) and sulfur (S) [10, 11]. In terms of toxicology, Se can be seen as a classic example of Paracelsus adage ("the dose makes the poison"),

Pablo A. Nogara and João Batista T. Rocha, Departamento de Bioquímica e Biologia Molecular, Centro de Ciências Naturais e Exatas, Universidade Federal de Santa Maria, RS, Brazil
Cláudia S. Oliveira, Instituto de Pesquisa Pelé Pequeno Príncipe, PR, Brazil; Faculdades Pequeno Príncipe, PR, Brazil

https://doi.org/10.1515/9783110625110-008

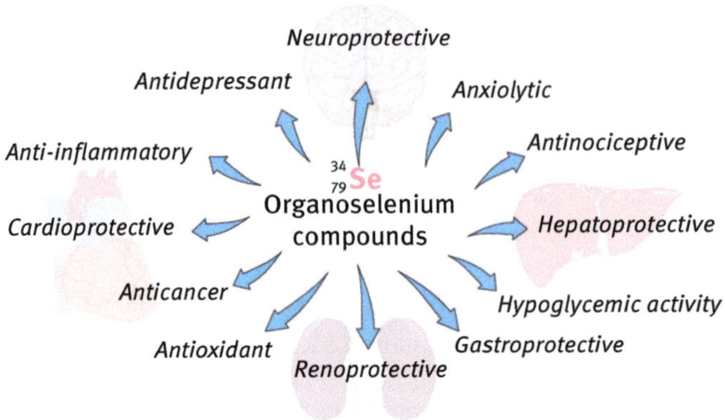

Figure 8.1: Examples of the pharmacological effects presented by synthetic organoselenium compounds.

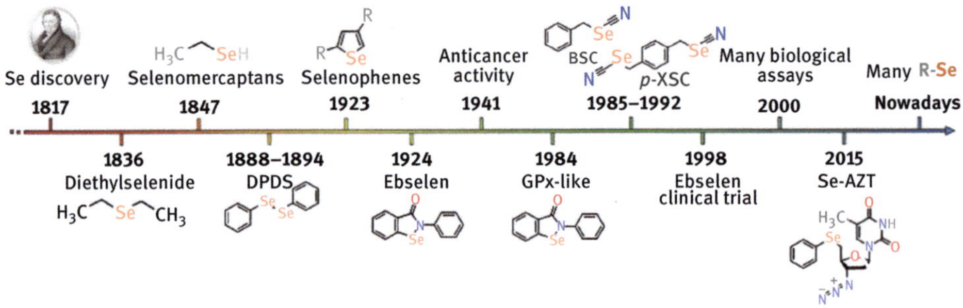

Figure 8.2: From selenium discovery to synthetic organoselenium molecules (R-Se).

i.e., when consumed in excess, this essential element is toxic [9, 12–15]. Selenosis was first identified in farm animals fed with Se accumulator plants. The animals are affected with loss of hair, and hooves were cracking and sloughing [16]. Moreover, it was reported that a Chinese population living in a seleniferous area underwent through symptoms of Se intoxication, such as loss of hair and nails, and skin lesions [17]. Currently, there is great concern about the dietary overexposure to Se, which may facilitate the development of neuropathological diseases (amyotrophic lateral sclerosis and Parkinson's disease), diabetes type 2, and hypertension [18–20].

In contrast, about five decades ago, Schwarz and Foltz described, for the first time, the importance of Se to vertebrates. Briefly, they observed that the simultaneous absence of vitamin E and factor 3 (lately identified as an organoselenium compound), produced by the American yeasts G and K, caused necrotic liver degeneration in rats [21, 22]. The molecular physiological role of Se was proved when it was identified as a component of 25 mammalian proteins. The majority of the selenoproteins of the mammalian selenoproteome are classified as oxidoreductases, which regulate the cell redox state, neutralizing the reactive species generated during the cell metabolism [3, 5, 23, 24].

Nineteen years after the discovery of Se, the first organoselenium compound, diethylselenide, was synthesized by C. J. Löwig [9, 25–27]. In 1847, Siemens isolated a selenomercaptan molecule, bis(ethylselanyl)mercury, possibly after reacting ethylselenol with $HgCl_2$, which in a letter to Berzelius was called "child of selenium" [10, 28]. However, the first synthetic organoselenium compounds tested as pharmacological agents were diselenide diacetic acid and diselenide dibutyric acid, where they had no anticancer effects in tumor-bearing mice [9]. Nowadays, the number of synthetic organoselenium compounds that are being tested against tumorigenic cells is enormous [15, 29–34]. Here it is worth mentioning that the organoselenocyanates (BSC and *p*-XSC) and the selenozidovudine (derivatives of the AZT molecule) have been demonstrating promising results [35–40].

Some organoselenium compounds that have extensively been studied in recent times, for example, diphenyl diselenide ($(PhSe)_2$ or DPDS) and ebselen, were synthesized at the end of the nineteenth century or beginning of the twentieth century. However, their pharmacological effects were recognized long time after their synthesis. The pieces of evidence suggested that $(PhSe)_2$ was first synthesized between 1888 and 1894 by M. C. Chabrie, who believed that he had synthesized chlorobenzene, diphenyl selenide, and phenylselenol from selenium tetrachloride ($SeCl_4$) and benzene in the presence of anhydrous aluminum chloride. However, Krafft and colleagues, after repeating the Chabrie's experiment, observed that phenylselenol was actually $(PhSe)_2$ [41–43].

The first use of diphenyl diselenide as a therapeutic agent was given to a woman with cancer (acute stem cell leukemia) in 1956. However, ironically, this study remained unnoticed, and it has been barely cited in recent literature. Specifically, diphenyl diselenide was given to a patient with leukemia for 7 days to determine whether it could or not have some beneficial effects, without causing the side-effects caused by selenocystine. Selenocystine was tested in four patients with severe toxic effects (hair loss, severe nausea, vomiting, among others). In contrast to selenocystine, diphenyl diselenide did not cause any side-toxic effect nor therapeutic effects against leukemia [44].

To our knowledge, the first investigation of the chemical properties of diphenyl diselenide and analogs with potential pharmacological interest was reported by Wilson et al. [45]. The authors compared the GPx-like activity of ebselen, diphenyl diselenide, and dozens of its analogs. Notably, they observed that diphenyl diselenide was about 1.6 times more effective than ebselen as GPx-mimic. In the last decades, the number of studies about the toxicological, biochemical, and pharmacological properties of diphenyl diselenide and its derivatives has increased dramatically [46–62].

Ebselen (2-phenyl-1,2-benzoselenazol-3-one, also called PZ 51) was synthesized in 1924 by R. Lesser and R. Weiß, together with a series of Se-containing aromatic compounds [9, 63]. The first studies about the biochemical properties of ebselen showed its GPx-like activity [64]. Since then, the toxicological and pharmacological properties of ebselen have been well studied in several experimental models [65–73]. Tests with human have been carried out with ebselen in cases of acute ischemic stroke [74], aneurysmal subarachnoid hemorrhages [75], complete occlusion of the middle cerebral artery [76], as potential lithium mimetic [77–79], cardiovascular and oxidative stress in diabetes patients [80], and noise-induced hearing loss [81]. However, the majority of the results were not satisfactory, which did not justify its therapeutic use in the tested pathologies.

8.3 Synthesis of organoselenium compounds

The first organoselenium compounds reported in the nineteenth century were elementary molecules, such as diethyl selenide ($CH_3CH_2SeCH_2CH_3$) and ethylselenol (CH_3CH_2SeH) [9]. Nowadays, several synthetic methodologies are available to obtain a set of different types of organoselenium molecules; however, here we will highlight only a few of them, including the classical approaches to produce ebselen, diphenyl diselenide, organoselenocyanates, and of the most recent selenozidovudine derivatives.

8.3.1 Ebselen

2-Phenyl-1,2-benzoselenazol-3-one (ebselen) can be synthesized mainly by three routes (Figure 8.3). The diazotization of anthranilic acid followed by the selenenylation of the diazonium salt with disodium diselenide leads to the 2,2′-diselanediyldibenzoic acid, which is treated with thionyl chloride, giving the 2-(chloroseleno)benzoyl chloride, which reacted with aniline forming the ebselen (**Route A**) [82]. The one-pot preparation from *N*-phenylbenzamide, using ortholithiation, selenium insertion, and oxidative cyclization, is also used (**Route B**) [83]. In addition, a Cu-catalyzed method can be applied, using 2-halo-*N*-phenylbenzamide and Se powder (**Route C**) [84].

8.3.2 Diphenyl diselenide

Diphenyl diselenide is generally prepared by air oxidation of selenol or selenolate intermediates from bromobenzene (Figure 8.4, **Route A**) [85–87] or diazonium salts (**Route C**) [63, 88], or can be obtained from Cu-catalyzed coupling reaction between aryl halides and elemental selenium (**Route B**) [89]. Also, these synthetic pathways can be used to produce diaryl diselenide derivatives.

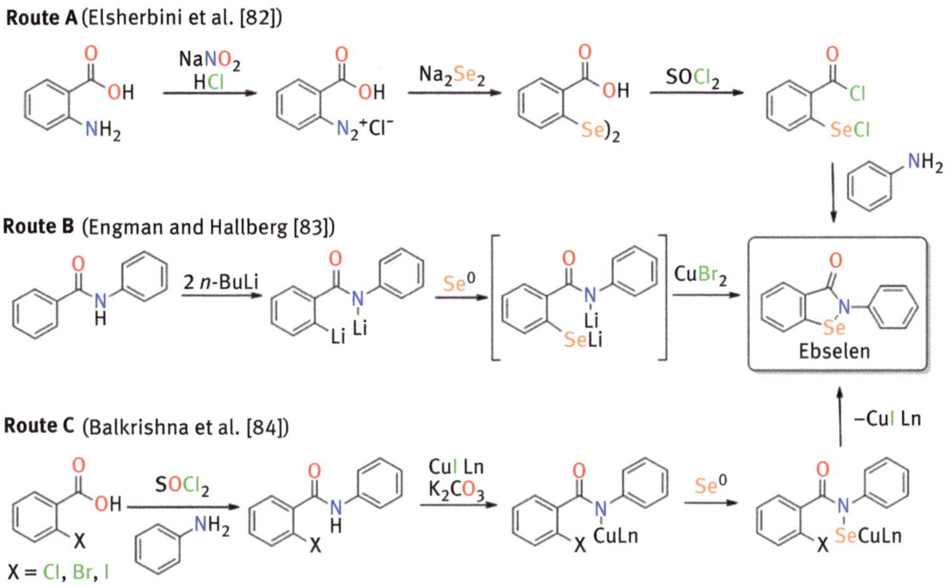

Figure 8.3: Ebselen's synthesis.

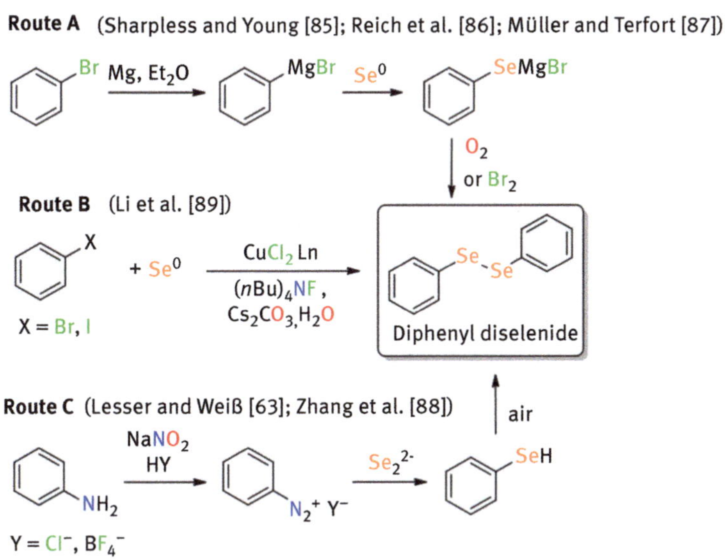

Figure 8.4: Diphenyl diselenide's synthesis.

8.3.3 Selenides

The diaryl diselenide derivatives can be used to produce monoselenides (Figure 8.5), by the reduction of the respective diselenide by sodium borohydride ($NaBH_4$) or potassium hydroxide (KOH), and the reaction of the generated selenol/selenolate to a electrophilic center, such as tosyl (Ts), mesyl (Ms), or aryl halide (chlorobenzene moiety) [35, 36, 90–92]. In this way, compounds such as pyridylselenide glycerol derivatives (Figure 8.5A), alkylseleno-carbohydrates (Figure 8.5B), selenozidovudine derivatives (AZT-Se) (Figure 8.5C), and 4-arylselenyl-7-chloroquinolines (Figure 8.5D) can be obtained with good yields. The approach used by the group of Dr. O. E. D. Rodrigues [35, 36], which uses AZT as the pharmacophore for the synthesis of S-, Se-, or Te-containing organochalogens, seems to be much more rational than empirically searching for pharmacological targets for organoselenides [6, 7]. Furthermore, the AZT derivatives have low toxicity to mice and healthy human cells [93–95].

The selenium-containing *N*-heterocycle molecules can be obtained from CuI/SeO_2-catalyzed reactions under ultrasound irradiation (Figure 8.6) [96]. Starting from 1H-indoles (Figure 8.6A) or imidazo[1,2-a]pyridines (Figure 8.6B) and diphenyl diselenide derivatives, the corresponding products 3-(organylselanyl)-1H-indole and 3-(organylselanyl)imidazo[1,2-a]pyridine are obtained. About selenophenes, these molecules can be prepared from $FeCl_3$–diorganyl diselenide-mediated intramolecular cyclization of (*Z*)-selenoenynes, with good yields (Figure 8.6C) [97].

8.3.4 Organoselenocyanates

The production of organoselenocyanates is effortless [98–100]. Starting from an electrophile molecule, such as benzyl bromide or *p*-xylylene dibromide (1,4-bis(bromomethyl)benzene), in the presence of potassium selenocyanate (KSeCN), benzyl selenocyanate – BSC (Figure 8.7A) or *p*-xyleneselenocyanate – *p*-XSC (Figure 8.7B) are formed. The anticancer properties of these compounds have been revised recently by Rocha et al. (2018) [15] and will be discussed briefly in the following sections of this chapter.

8.4 Pharmacological properties of organoselenium compounds

The significant advances in the synthesis of organoselenium compounds lead to a big library of molecules, the majority of them without known applicability. However, many compounds have been empirically tested and some therapeutic properties, such as anti-inflammatory, anticancer, antidepressant, antinociceptive, anxiolytic,

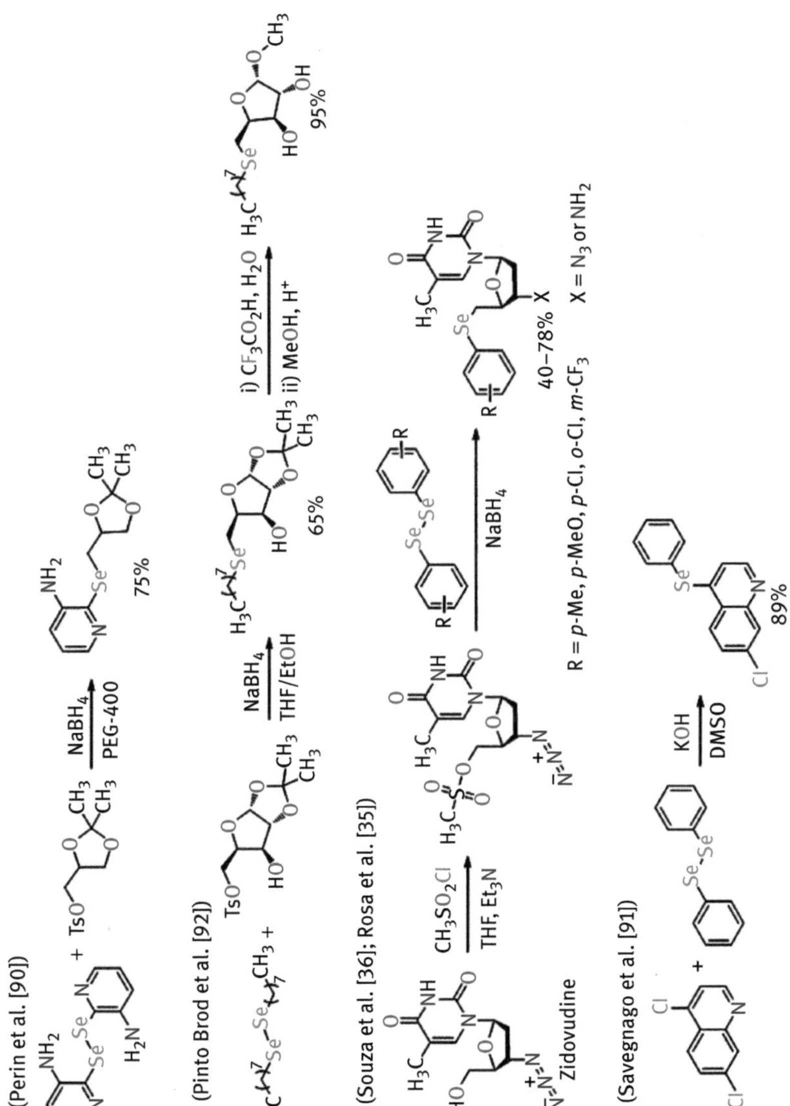

Figure 8.5: Synthesis of selenides derivatives.

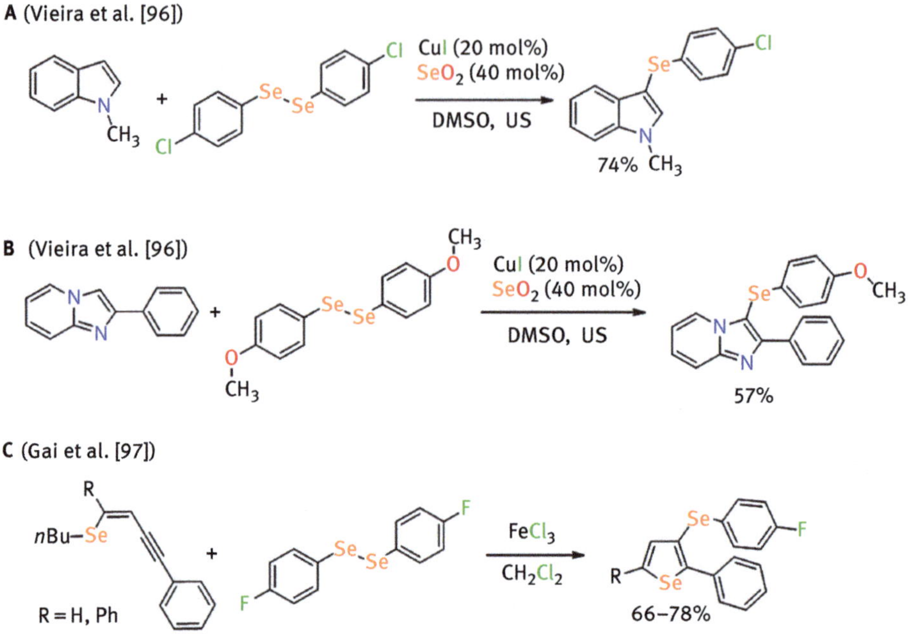

Figure 8.6: Synthesis of selenium-containing *N*-heterocycles (A, B) and selenophenes (C).

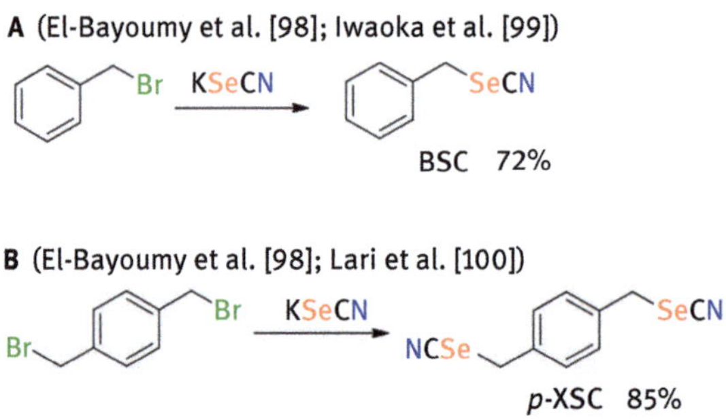

Figure 8.7: Synthetic approach to producing organoselenocyanates. Benzyl selenocyanate (BSC) and 1,4-phenylene-bis(methylene)-selenocyanate (*p*-XSC) (B).

cardioprotective, hepatoprotective, gastroprotective, neuroprotective, and renoprotective, have been observed. Here, we will highlight some organoselenium compounds that showed promising results (Table 8.1). Nonetheless, we have to emphasize that a good part of the promising molecules have such a generic, and sometimes a wide range of biological effects, making it difficult to identify their molecular targets. As mentioned earlier, the use of a specific pharmacophore in the organoselenium compounds can be considered a winning strategy [35–37, 93–95, 101, 102].

8.4.1 Antinociceptive activity

The treatment of pain requires analgesics and anti-inflammatory drugs, which frequently have undesirable side effects [103, 104]. A myriad of studies have demonstrated the antinociceptive activity of organoselenium compounds, such as ebselen, diphenyl diselenides, monoselenides, and bis selenide derivatives, in different rodent models of pain (formalin-, acetic acid-induced abdominal writhing, tail-flick test, capsaicin, and thermal model) [105–109].

The diphenyl diselenide antinociceptive activity appears to involve the interaction with the glutamatergic system, L-arginine-nitric oxide pathway, activation of opioid, dopaminergic, and muscarinic cholinergic receptors [106, 110, 111]. In the case of *m*-trifluoromethyldiphenyl diselenide, the antinociceptive effect involves the modulation of the μ-opioid and δ-opioid receptors, and the serotonergic system [112, 113]. For *p*-methoxy-diphenyl diselenide, the antinociceptive effects seem to be mediated via modulation of the glutamatergic and GABAergic systems and protein kinase A pathway [114–116]. However, there is no conclusive evidence for a direct interaction of organoselenium compounds with the systems or pathways described earlier.

The pyridine derivative of diphenyl diselenide, the 2,2′-dipyridyl diselenide, exerted nociceptive actions associated with anti-inflammatory effects, and the beneficial effects were associated with changes in the inducible nitric oxide synthase (iNOS), nuclear factor-kappa B (NF-κB) and c-Jun N-terminal kinase (JNK) phosphorylation levels in the spinal cord of mice [117]. The antinociception of bis (4-methylbenzoyl) diselenide was reported to be mediated by modulation of the nitrergic system [118].

The 3-(4-chlorophenylselanyl)-1-methyl-1H-indole has been reported to have antinociceptive and anti-inflammatory effects in mice via modulation of monoaminergic, opioidergic, and adenosinergic systems [107]. For the bis selenide alkene derivatives, their nociceptive effects seemed to involve the modulation of the serotoninergic system, nitric oxide, cyclic GMP and ATP-sensitive, and voltage-gated K^+ channels. The authors also indicated the potential interaction of 3-(4-chlorophenylselanyl)-1-methyl-1H-indole with kainate and trans-ACPD receptors and with pro-inflammatory cytokines [109, 119, 120].

8.4.2 Anxiolytic activity

Diphenyl diselenide, *p*-methoxyl- and *m*-trifluoromethyl-diphenyl diselenide have been proposed to have anxiolytic-like effects in different animal models (mice, chicken, and zebrafish) [121–123]. The diselenides presented anxiolytic action in the elevated plus-maze and light–dark box tests [121, 124–126]. The anxiolytic-like effect of diphenyl diselenide can be associated with modulation of the 5-HT and $GABA_A$ receptors [125], while for *m*-trifluoromethyl-diphenyl diselenide, the serotonergic system can be involved [121]. The compound 4-phenylselenyl-7-chloroquinoline exhibited anxiolytic activity, and the glutamatergic system seems to be involved [127]. However, as commented earlier, a direct interaction of organoselenium compounds with specific molecular components of the neurotransmitter systems cited earlier has not yet been demonstrated.

8.4.3 Antidepressant-like activity

The organoselenium compounds have been evaluated in animal models of depression. The mechanisms involved in their activity appear to affect the modulation of multiple sites. For instance, the antidepressive-like effect of diphenyl diselenide has been associated with modulation of the serotonergic, noradrenergic, and dopaminergic systems [128, 129], while ebselen showed interactions only with the noradrenergic and dopaminergic systems [130]. Of particular therapeutic significance, ebselen is under clinical trials as a potential lithium substitute. The studies of ebselen as lithium mimetic have indicated that it can inhibit the enzyme inositol monophosphatase (IMPase), which is considered as the molecular target of Li^{+}. In contrast with ebselen, diphenyl diselenide did not inhibit the IMPase [77, 79].

The *m*-trifluoromethyl-diphenyl diselenide and *p*-chloro-diphenyl diselenide [$(p\text{-ClPhSe})_2$] had an antidepressant-like activity that was associated with modulation of the serotonergic system [131, 132]. The triazole derivatives (4-phenyl-1-(phenylselanylmethyl)-1,2,3-triazole and phenylselanyl-1H-1,2,3-triazole-4-carbonitrile), and the selenophene derivatives, which have quite distinct chemical structures from diselenides, have also been reported to modulate the serotoninergic system [97, 133, 134].

The antidepressant-like activity of methylphenyl selenide and α-(phenylselanyl) acetophenone was associated with modulation of the dopaminergic system [135, 136], whereas the compound bis-selenide [(*Z*)-2,3-bis(4-chlorophenylselanyl) prop-2en-1-ol] modulated the L-arginine–nitric oxide–cyclic guanosine monophosphate pathway, and the serotonergic $5\text{-HT}_{2A/C}$ and 5-HT_3 receptors [137, 138].

8.4.4 Hepatoprotective activity

The organoselenium compounds have been showing hepatoprotective action [7]. More recently, diphenyl diselenide was demonstrated to blunt the hepatic oxidative stress induced by bisphenol A in male mice apparently via upregulation of the Nrf2/ Keap-1 signaling pathway [49]. In silver catfish, diphenyl diselenide was reported to attenuate the hepatic oxidative damage induced by clomazone [139]. In the mice thioacetamide-induced hepatic toxicity model, diphenyl diselenide blunted the thioacetamide hepatotoxicity. In contrast, the diselenide analogs (*o*-methoxyl- and *p*-methyl-diphenyl diselenides), and β-selenoamines did not modulate the thioacetamide-induced hepatic toxicity [140, 141].

In a rat model of sodium glutamate-induced hepatotoxicity, $(p\text{-ClPhSe})_2$ blunted glutamate toxicity, which was associated with preservation of the mitochondrial functionality and modulation of the poly(ADP-ribose) polymerase, iNOS, and p38 proteins [142].

Freitas et al. (2012) demonstrated that the addition of chlorine moiety in the para position of the aromatic ring of diaryl diselenide reduced the liver and kidney toxicity in mice exposed to $HgCl_2$ [58]. The binaphthyl diselenide showed a protective effect on 2-nitropropane-induced hepatotoxicity in rats [143]. Also, the bis(4-methylbenzoyl) diselenide had hepatoprotective effects in the tetrachloride (CCl_4)-induced oxidative damage in mice, possibly by modulating the antioxidant status [144].

8.4.5 Cardioprotective activity

Organoselenium compounds have been studied in different models of cardiovascular damage. For instance, ebselen was effective against daunorubicin-induced cardiomyopathy in rats. Ebselen normalized the serum levels of cardiac enzymes creatine kinase and lactate dehydrogenase, as well as serum GPx [145]. Mordente et al. (2015) demonstrated the beneficial properties of ebselen for the treatment of anthracycline-induced cardiotoxicity. The authors demonstrated that ebselen and ebselen diselenide were capable of inhibiting human myocardial cytosolic reductases, which are responsible for the formation of anthracycline cardiotoxic metabolites, probably by forming covalent adducts with cysteine residues of the reductases (carbonyl reductase and aldehyde reductase). The pharmacological cardioprotective effects of ebselen could be mediated by its capacity to inhibit g-butyrobetaine hydroxylase (BBOX). The BBOX inhibition was reported to aid in the recovery after cardiac dysfunction induced by ischemia/reperfusion [146, 147]. Ebselen (IC_{50} = 0.7 μM) has been reported to be a more potent inhibitor of BBOX than other organoselenium compounds (IC_{50} = 5–55 μM). The inhibitory mechanism of BBOX by ebselen involves the displacement of structural zinc after the interactions of ebselen with Cys residues from the Cys_3-His motif, which are involved in the coordination with Zn^{2+} [148].

The chronic iron overload can cause cardiac failure, and in this sense, ebselen was used in a model of chronic iron overload in mice. Ebselen exhibited cardioprotective effects against oxygen free radical damage, decreasing the levels of cytotoxic aldehydes (hexanal, 4-hydroxyl-2-nonenal, and malondialdehyde) and iron in heart tissue. Moreover, mice treated with ebselen showed an increase in GPx activity [149].

Considering that the lipoprotein oxidation is involved in the development of atherosclerosis, and it affects the vascular wall and leads to coronary artery diseases, the diphenyl diselenide was tested against copper- and peroxyl radical-induced human low-density lipoprotein (LDL) oxidation in vitro. The diphenyl diselenide inhibited lipid peroxidation and prevented the oxidation of protein moieties of human LDL, which could be mediated by its thiol-peroxidase-like activity [150]. Hypercholesterolemia and oxidative stress are well-known risk factors in coronary artery diseases, and diphenyl diselenide was reported to reduce the serum levels of total cholesterol and the oxidative tissue stress in cholesterol-fed rabbits [151]. However, de Bem et al. [151] did not evaluate the possible participation of inorganic selenium (derived from the metabolism of diphenyl diselenide) in the hypocholesterolemic effects of diselenide in rabbits.

8.4.6 Hypoglycemic activity

The diphenyl diselenide was reported to prevent diabetic complications possibly via the reduction of blood glucose levels (and consequently, glycated proteins and oxidative stress) in streptozotocin-induced diabetes in rats [7]. Similarly, diphenyl diselenide reduced biochemical alterations associated with oxidative stress in rats fed with fructose [152] and modulate glucose metabolism [153]. However, as discussed earlier for the modulation of cholesterol levels, the participation of inorganic selenium in the protective effects of diphenyl diselenide was not investigated in the cited studies.

$(p\text{-ClPhSe})_2$ reversed the hyperglycemia induced by fructose administration via modulation of liver enzymes involved in glucose metabolism [154]. The treatment with $(p\text{-ClPhSe})_2$ was also reported to restore most of the metabolic parameters altered by monosodium glutamate administration in rats, including obesity [155]. As indicated earlier, the involvement of inorganic selenium in the observed effects is highly plausible, because organoselenides are well known to release selenium to the inorganic pool, which modulates the synthesis of selenoproteins [9].

8.4.7 Gastroprotective activity

Gastrointestinal ulcer is a significant disease that affects a considerable number of people in the world. Chronic alcohol intake, chronic usage of nonsteroidal anti-inflammatory drugs, smoking, excessive stress, and *Helicobacter pylori* infection are some of the factors that contribute to the development of gastric ulcers [156, 157]. In this context, ebselen prevented ulceration induced by diclofenac [70], ethanol [158], aspirin, and water-immersion restraint stress [159]. Moreover, ebselen inhibited acid secretion in the guinea-pig parietal cells by interaction with thiol groups of the H^+/K^+-ATPase (a gastric proton pump) [160]. In the same way, dihydroxy-1-selenolane, diphenyl diselenide, and binaphthyl diselenide also presented protective activities against indomethacin- and ethanol-induced gastric lesions in rats, via inhibition of gastric H^+/K^+-ATPase and due to its antioxidant activities [61, 143, 161, 162]. The proton pump inhibitors are used for the reduction of acid secretion from the stomach, where the H^+/K^+-ATPase enzyme is an important target [163]. Thus, the organoselenium compounds could inhibit the HCl secretion [7].

8.4.8 Renoprotective activity

The antinephrotoxic effects of organoselenium compounds have been presented in the literature. Ebselen reduced the nephrotoxicity in cisplatin- and gentamicin-induced renal damage in rats [164–166]. This organoselenium also protected against acute renal ischemia by improving renal function, associated with the reduction of the lipid peroxidation and oxidative DNA damage, due to suppression of peroxynitrite production or its scavenging activity [167]. Also, ebselen significantly reduced the expression of proteins implicated in kidney fibrosis and inflammation [168].

p-Methoxyl-diphenyl diselenide and (*E*)-2-benzylidene-4-phenyl-1,3-diselenole reduced renal injury induced by cisplatin in rats, probably because of their antioxidant activity [169, 170]. In addition, diphenyl diselenide and binaphthyl diselenide are demonstrated to be effective against acute renal damage induced by glycerol in rats [171, 172].

8.4.9 Anticancer activity

As discussed in Section 8.2, synthetic organoselenium have been tested as chemopreventive agents. The mechanism of the chemopreventive activity of Se is still not well elucidated, but, probably, its pro-oxidant action is involved [15, 173].

Ebselen, diphenyl diselenide and derivatives, and selenocyanates have been tested in different types of tumorigenic cells [31], presenting promising results. These molecules can be reduced forming the very reactive moiety, selenol, which can react with the intracellular reduced thiol or even trigger the expression of apoptotic genes, leading to cell death [15].

Ethaselen, an ebselen derivative, demonstrated anticancer activity in many types of tumors (MCF-7, H1666, A549, and LoVo cell lines). The mechanism of action of this compound is related to the inhibition of the thioredoxin reductase enzyme (TrxR) that is overexpressed in many cancer types. Ethaselen binds in the C-terminal redox center of TrxR, forming two covalent bonds with Cys497 and Sec498 residues. As a consequence, the increase in oxidized thioredoxin levels enhances the levels of cellular reactive oxygen species (ROS), which could be involved in the anticancer activity [174, 175].

The new class of organoselenium compounds, the AZT-Se derivatives, has been tested in tumorigenic cell lines. The results have been promising, but little is known about the mechanism involved. It seems that the AZT-Se compounds could induce the expression of proapoptotic genes [35–37]. More studies are necessary to study the effects of these molecules in healthy and cancerous cells.

8.4.10 Neuroprotective activity

Neuroprotective activity is one of the described pharmacological properties of the synthetic organoselenium compounds [176]. The neuroprotective effects of ebselen and diphenyl diselenide were widely tested *in vitro* and *in vivo* in different experimental models [53, 177–180]. Moreover, ebselen was tested in human clinical trials in cases of acute ischemic stroke [74], aneurysmal subarachnoid hemorrhages [75], and complete occlusion of the middle cerebral artery [76].

The neuroprotective activity of the organoselenium compounds (particularly diselenides) can be due to the ability to feed the Se pool to the selenoprotein synthesis. The selenoproteins are fundamental to the homeostasis of the central nervous system [181, 182] once, in cases of Se deprivation, the brain is one of the last organs to show a reduction in the Se levels [183]. Moreover, the synthetic organoselenium compounds demonstrate an antioxidant and anti-inflammatory activity, which could be directly involved in their neuropharmacological effects.

8.5 Possible general mechanisms of action of organoselenides

As briefly discussed in Section 8.4 and Table 8.1, the organoselenium compounds are found to have significant pharmacological properties. The compounds have been tested *in vitro* and *in vivo* in various animal models. Despite the vast number of studies showing the pharmacological properties of the organoselenium compounds, the exact molecular mechanism of their action still needs to be clarified.

The putative pharmacological/nutritional mechanism of action of organoselenium compounds is depicted in Figure 8.8. Organoselenium compounds can feed the inorganic selenium pool and can enhance the synthesis of physiologically relevant selenoproteins [23, 184, 185]. Selenium forms weaker sigma-bonds than sulfur analogs, and the cleavage of such bonds is fast and proceeds under mild reaction conditions [186]. Physiologically, a variety of organoselenium compounds can use Se to selenoprotein synthesis, as indirectly observed in the classical study of Schwarz and Foltz, where they demonstrated the protective effects of a myriad of structurally distinct organoselenium compounds against hepatic

Table 8.1: Examples of organoselenium compounds and their pharmacological activities.

Compound	Activity	Experimental model	Dose	Mode of action	Ref.
Ebselen 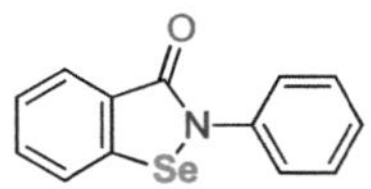	Antinociceptive and anti-inflammatory	Carrageenin-induced paw edema, tail-flick, formalin, acetic acid-induced abdominal writhing, and capsaicin models of pain in mice	10–100 mg/kg (i.p. or s.c.)	–	[105]
	Antidepressant-like	OFT, FST, and TST in mice	3–30 mg/kg (s.c.)	Interaction with the noradrenergic and dopaminergic systems, but not with the serotonergic system	[130]
	Gastroprotective	Diclofenac-induced gastric lesions in rats	31–316 mg/kg (p.o.)	–	[70]
		Aspirin-induced gastric ulceration in rats	30–300 mg/kg (p.o.)	Inhibition of gastric acid secretions	[159]
		Ethanol-induced gastric lesions in rats	30–300 mg/kg (p.o.)	Inhibitory action on lipid peroxides	[158]

(continued)

Table 8.1 (continued)

Compound	Activity	Experimental model	Dose	Mode of action	Ref.
Diphenyl diselenide 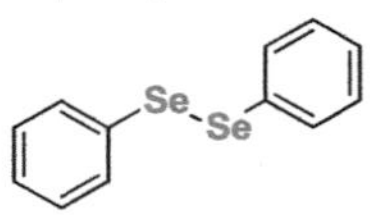	Antinociceptive	Acetic acid-, capsaicin-, glutamate-, bradykinin- and phorbol myristate acetate-induced pain in mice	1–100 mg/kg (p.o.)	Interaction with the glutamatergic system and L-arginine–nitric oxide pathway	[110]
		Formalin-induced nociception in mice	10 and 20 nmol/paw (s.c.)	Activation of opioid, dopaminergic D2, muscarinic cholinergic receptors or the interaction with α1 and α2 adrenergic receptors	[106]
	Antidepressant-like and anxiolytic-like	FST, TST, elevated plus-maze, light–dark box, and OFT in mice	0.1–100 mg/kg (p.o.)	Interaction with L-arginine–nitric oxide–cyclic guanosine monophosphate pathway	[126]
	Antidepressant-like	Depressive-like behavior induced by monosodium glutamate in rats	10 mg/kg (i.g.)	Improvement of neuronal plasticity re-established by cortical 5-HT uptake and Na^+, K^+-ATPase activity	[53]
		Depressive-like behavior triggered by methimazole in rats	5 ppm on the diet	Restoration of the thyroid hormones and ROS levels, and MAO-B activity	[190]
		Post-menopausal depression-like behavior in ovariectomized female mice	10 mg/kg/day for 7 days (p.o.)	Involvement of $5\text{-HT}_{2A/2C}$ and 5-HT_3 serotonergic receptors	[129]
		TST and OFT in mice	1–5 mg/kg (p.o.)	Modulation of K^+ channels and PPARγ receptors	[191]
		TST and OFT in rats treated with malathion	50 mg/kg (p.o.)	Involvement of Na^+, K^+-ATPase activity	[192]

		Serotonin uptake in rat platelets and synaptosomes	0.01–100 µM	Inhibition of serotonin uptake into platelets and synaptosomes	[193]
		FST in rats	0.1–30 mg/kg (p.o.)	Involvement of the monoaminergic system	[128]
	Anxiolytic	Anxiety induced by glutamate in rats	1 mg/kg/day/ 7 days (i.g.)	Modulation of hippocampal BDNF-Akt pathway and GABA and 5-HT uptake	[194]
		Spatiotemporal analysis of behavior, vertical exploration, and homebase formation in zebrafish	0.1–1 µM	–	[123]
		Chick social separation-stress behavior	1–50 mg/kg (p. o.)	–	[122]
		Elevated plus maze task in rats	5–50 µmol/kg (i.p.)	Involvement of the $GABA_A$ and 5-HT receptors	[125]
	Gastroprotective	Ethanol- and indomethacin-induced gastric lesions in rats	0.01–50 mg/ kg (i.p.)	Inhibition of gastric K^+-ATPase activity	[161]
		Ethanol-induced gastric lesions in rats	0.01–10 mg/ kg (p.o.)	Antioxidant activity	[61]
	Neuroprotective	Model of Parkinson's disease induced by 6-OHDA in rats	1 mg/kg/day/ 30 days (i.g.)	Involvement of the BDNF/TrkB pathway	[177]

(continued)

Table 8.1 (continued)

Compound	Activity	Experimental model	Dose	Mode of action	Ref.
m-Trifluoromethyl-diphenyl diselenide 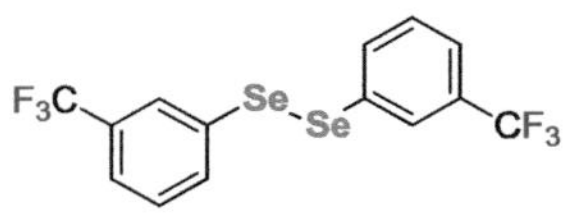	Antinociceptive	Capsaicin- and acetic acid-induced nociception in mice, and in tail-immersion and hot-plate tests	10–100 mg/kg (p.o.)	Interaction with μ-opioid and δ-opioid receptors	[113]
		Glutamate-induced licking behavior model in mice	1–50 mg/kg (p.o.)	Involvement of serotonergic system	[112]
	Anxiolytic-like	Light/dark box and elevated plus-maze tests in mice	0.1 – 100 mg/kg (p.o.)	Involvement of the serotonergic system	[121]
	Antinociceptive and antidepressant-like	Comorbidity of chronic pain and depression induced by partial sciatic nerve ligation in mice	0.1–10 mg/kg (i.g.)	Reduction of the levels of pro-inflammatory cytokines, p38MAPK activation, and reduction of glutamate release and 5-HT uptake	[54]
	Antidepressant-like	Forced swimming test in mice	1–100 mg/kg (p.o.)	Interaction with serotonergic and opioid systems	[131]
		FST and modified tail suspension test in mice	50 mg/kg (i.g.)	μ and δ-opioid receptor activation and the κ-opioid receptor blockade	[195]

p-Methoxyl-diphenyl diselenide 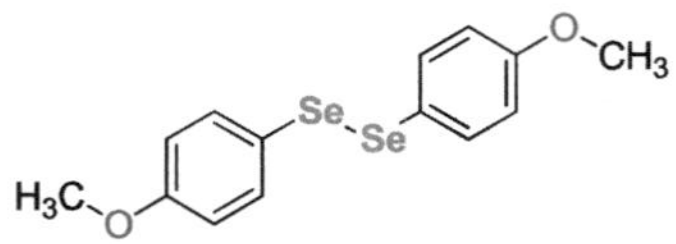	Antinociceptive	Acetic acid-induced nociception in mice	1–50 mg/kg (p.o.)	Interaction with glutamatergic and GABAergic systems, and inhibition of protein kinase A pathway	[115]
	Antinociceptive and anti-inflammatory	Partial sciatic nerve ligation surgery and CFA-induced inflammatory pain model in mice	25 mg/kg/day/7 days (i.g.)	Modulation of inflammatory protein contents	[196]
	Antinociceptive and antidepressant-like	Pain-depression dyad induced by reserpine in rats	10 mg/kg/day/ 30 days (p.o.)	Antioxidant activity	[116]
	Anxiolytic and antidepressant-like	Locomotor activity monitor, FST, TST, elevated plus maze test and light–dark test in mice	0.1–5 mg/kg (i.g.)	–	[124]
p-Chloro-diphenyl diselenide 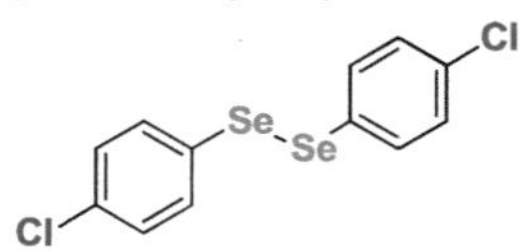	Neuroprotective	Memory impairment related to stress caused by corticosterone in mice	1 or 5 mg/kg/day/7 days (i.g.)	Modulation of hippocampal glutamate uptake	[178]
	Antidepressant-like	Cognitive impairment caused by aging in male rats	10 or 25 mg/kg/day/ 7 days (p.o.)	Antioxidant action and modulation of serotonin receptors	[132]
2,2′-Dipyridyl diselenide 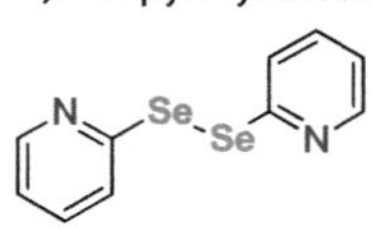	Antinociceptive and anti-inflammatory	Formalin and tail immersion tests, paw and ear edema in mice	0.01–50 mg/kg (i.g.)	Reduction of iNOS, Nf-κB, and c-Jun N-terminal kinase phosphorylation levels	[117]

(continued)

Table 8.1 (continued)

Compound	Activity	Experimental model	Dose	Mode of action	Ref.
Binaphthyl diselenide 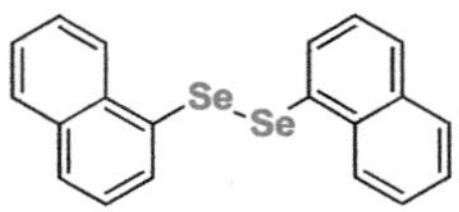	Gastroprotective	Ethanol-induced gastric lesions in rats	5–10 mg/kg (p.o.)	Antioxidant and free radical scavenging activity	[143]
Methylphenyl selenide 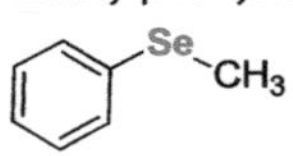	Antidepressant-like	FST and TST in mice	10–50 mg/kg (i.g.)	Interaction with the dopaminergic system	[135]
Bis(4-methylbenzoyl) diselenide 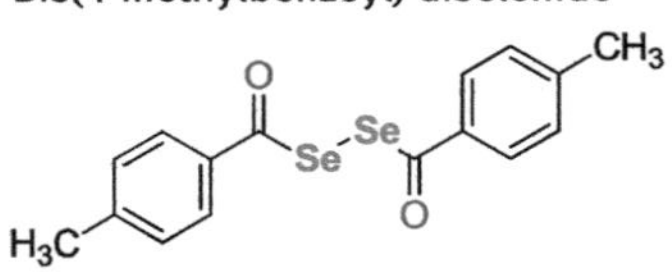	Antinociceptive and antihyperalgesic	Formalin, capsaicin, bradykinin, and glutamate-induced nociception; and hyperalgesia induced by carrageenan or CFA in mice	0.1–50 mg/kg (p.o.)	Nitrergic system	[118]
	Hepatoprotective	CCl_4–induced oxidative damage in mice	25 mg/kg (p.o.)	Modulation of the antioxidant status	[144]

α-(Phenylselanyl) acetophenone 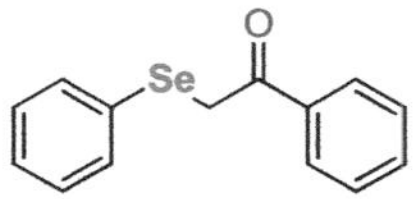	Anxiolytic, antinociceptive, antidepressant-like, antihyperalgesic, and antioxidant	Formalin and glutamate nociception-induced tests and acute stress restriction induced in mice	1–50 mg/kg (i.g.)	Dopaminergic and adrenergic systems	[197, 198]
	Antidepressant-like	OPT, FST, and TST in mice	0.1–10 mg/kg (p.o)	Interaction with the serotonergic system	[136]
3-(4-Chlorophenyl selanyl)-1-methyl-1H-indole 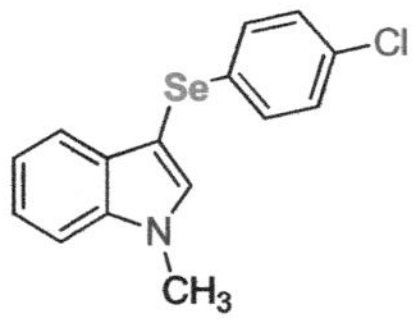	Antinociceptive and anti-inflammatory	Formalin and glutamate tests in mice	0.01–10 mg/kg (i.g.)	Modulation of the dopaminergic, adenosinergic, opioid, serotonergic, and noradrenergic systems	[107]
		Paw edema induced by formalin and ear edema induced by croton oil in mice	0.01–10 mg/kg (i.g.)		
	Antidepressant-like	Lipopolysaccharide-induced depressive-like behavior, neuroinflammation, and oxidative stress in mice	20–50 mg/kg (i.g.)	Modulation of the inflammatory and oxidative pathways	[179]
Selanylimidazo pyridine 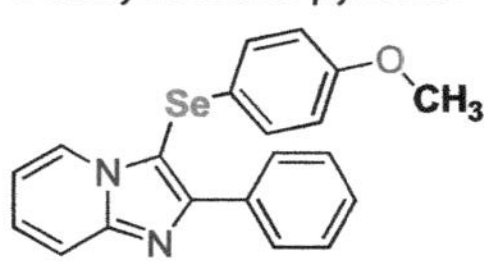	Antidepressant-like and anti-inflammatory	Acute lipopolysaccharide-induced depressive-like behavior, neuroinflammation, and oxidative stress in mice	20–50 mg/kg (i.g.)	Modulation of the inflammatory, antioxidant, and neurotrophic systems	[180]
	Anxiolytic	Acute restraint stress in mice	1–50 mg/kg (i.g.)	Modulation of the oxidative stress	[199]

(continued)

Table 8.1 (continued)

Compound	Activity	Experimental model	Dose	Mode of action	Ref.
4-Phenylselenyl-7-chloroquinoline 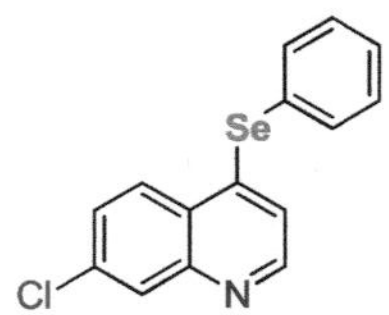	Antinociceptive and anti-inflammatory	Acetic acid, formalin, hot-plate and OFT, ear edema induced by croton oil, and carrageenan-induced pleurisy in mice	0.1–25 mg/kg (i.g.) and 0.01–25 mg/kg (p.o.)	Modulation of serotonergic, nitrergic, and glutamatergic systems, and MPO inhibition	[108, 200]
	Anxiolytic	EPM, light–dark and OFT; kainate-induced anxiety-related behavior in mice	5–50 mg/kg (p.o.)	Involvement of the glutamatergic system	[127]
4-Arylselanyl-7-chloroquinolines 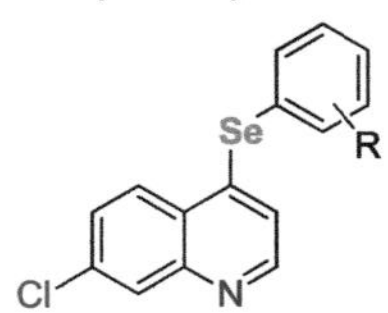	Cognitive enhancement	Step-down inhibitory avoidance task in mice	10 mg/kg (i.g.)	Acetylcholinesterase inhibition (in vitro) and cognitive enhancement	[201]
4-Phenyl-1-(phenylselanylmethyl)-1,2,3-triazole 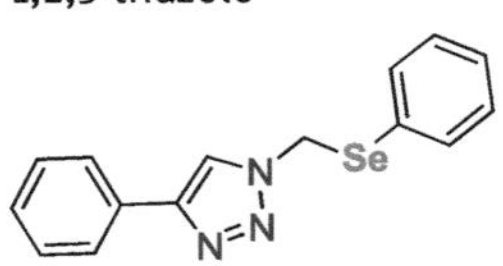	Antidepressant-like	OFT and TST in mice	1–50 mg/kg (p.o.)	Modulation of the dopaminergic and serotoninergic systems	[202]

Phenylselanyl-1H-1,2,3-triazole-4-carbonitrile 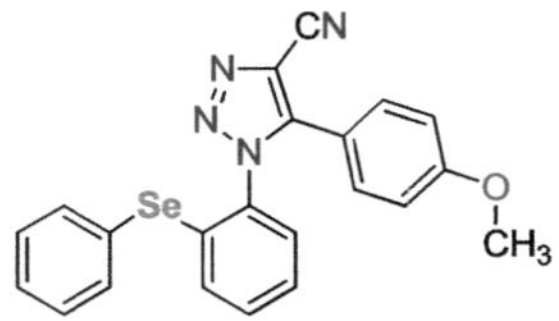	Antidepressant-like	OFT and FST in mice	0.01–20 mg/kg (i.g.)	Modulation of the serotonergic pathway	[203]
1,2-Bis-(4-methoxyphenylselanyl) styrene 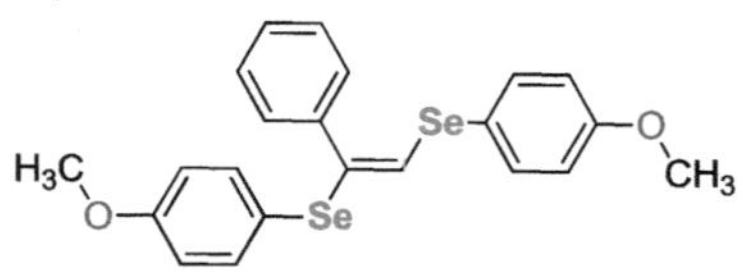	Antinociceptive and anti-inflammatory	Formalin and glutamate tests in mice	0.1–50 mg/kg (i.g.)	Serotoninergic system	[120]
(*Z*)-2, 3-Bis(4-chlorophenylselanyl) prop-2-en-1-ol 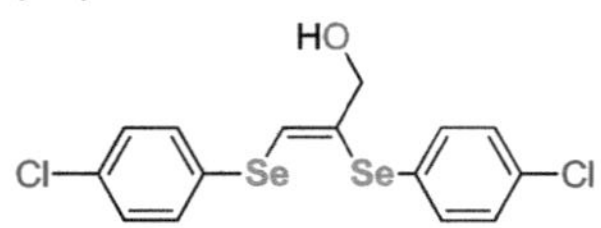	Antinociceptive	Hot plate test in mice	5–50 mg/kg (p.o.)	Activation of ATP-sensitive and voltage-gated K^+ channels; interaction with kainate and *trans*-ACPD receptors and vanilloid and pro-inflammatory cytokines	[109]
	Antidepressant-like	Chronic constriction injury and FST in mice	1–5 mg/kg (p.o.)	–	[137]
		FST and TST in mice	0.5–5 mg/kg (p.o.)	Involvement of the serotoninergic system	[138]

(continued)

Table 8.1 (continued)

Compound	Activity	Experimental model	Dose	Mode of action	Ref.
Selenoethers Glycerol derivatives 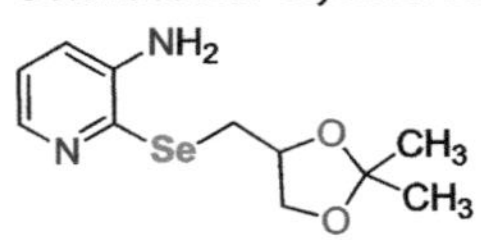	Antinociceptive and anti-inflammatory	Formalin, glutamate, and test hot-plate test in mice	50 mg/kg (p.o.)	Glutamatergic and/or nitrergic pathways	[90]
Bis(phenylimidazoselenazolyl) diselenide 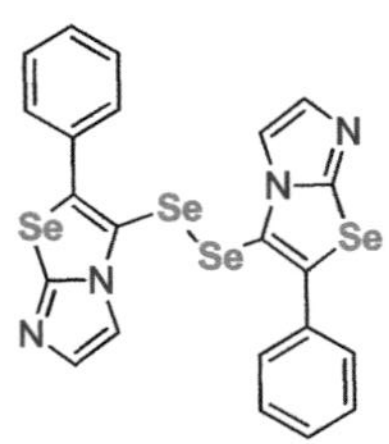	Antinociceptive and anti-inflammatory	CFA-induced inflammatory pain model and collagen-induced arthritis model in mice	0.1–1 mg/kg (p.o.)	Interaction with the L-arginine–nitric oxide pathway	[204]
Dihydroxy-1-selenolane 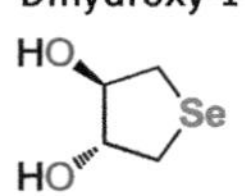	Gastroprotective	Indomethacin-induced stomach ulceration in mice	0.5–5.0 mg/kg (p.o.)	Antioxidant activity	[162]

3-(4-Fluoro phenylselenyl)-2,5-diphenylselenophene 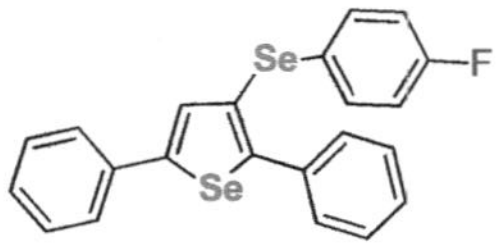	Antiallodynic and anxiolytic	PSNL, FST, TST, and light/dark test in mice	0.1 mg/kg (i.g.)	Interaction with 5-HT receptors	[205]
	Antidepressant-like	FST and TST in mice	50–100 mg/kg (i.g.)	5-HT reuptake inhibition	[97]
		FST and TST in mice	50 mg/kg (i.g.)	ERK signaling activation	[134]
	Antidepressant- and anxiolytic-like	Anxiety/depressant-like behavior induced by corticosterone	0.1 mg/kg/day/7 days (i.g.)	Modulation of serotoninergic and glutamatergic systems	[206]
3-(4-Fluorophenyl selanyl)-2-phenyl selenophene 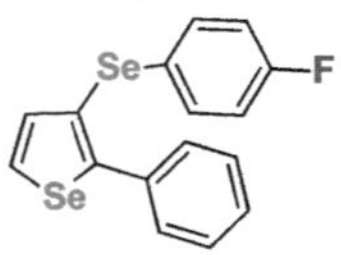	Antidepressant-like	FST in mice	50 mg/kg (i.g.)	–	[97]
2-Benzoyl 4-iodoselenophene 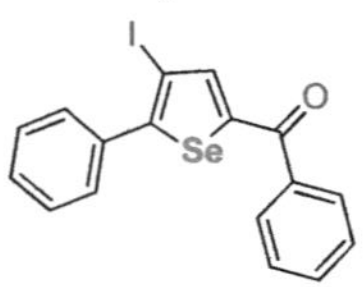	Antidepressant-like	TST and FST in mice	5–50 mg/kg (i.g.)	Modulation of the serotoninergic system (MAO-A inhibition)	[133]

(continued)

Table 8.1 (continued)

Compound	Activity	Experimental model	Dose	Mode of action	Ref.
α-Phenylseleno citronellal 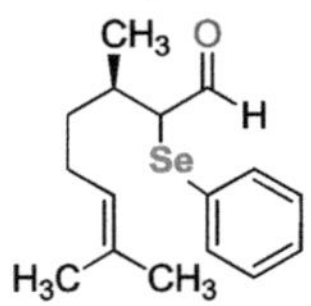	Antidepressant-like	OFT, TST, and FST in mice	1–100 mg/kg (p.o.)	–	[207]
(Octylseleno)-xylofuranoside 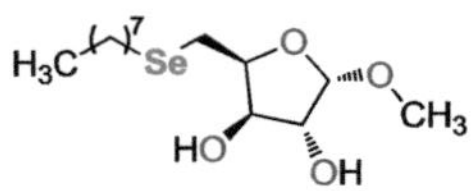	Antidepressant-like	TST in mice	0.001–10 mg/kg (p.o.)	Interaction with the serotonergic, noradrenergic, and dopaminergic systems	[92]
		TST and OFT in mice	0.01 mg/kg (p.o.)	Activation of protein kinases PKA, PKC, CAMKII, and ERK 1/2	[208]
Ethaselen 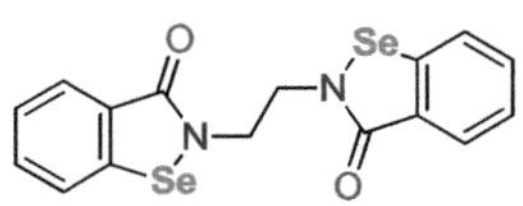	Anticancer	MCF-7, A549, H1666, and LoVo cell lines	0.1–20 μM	TrxR inhibition	[174, 175]

Abbreviations: 5-HT, 5-hydroxytryptamine or serotonina; 6-OHDA, 6-hydroxydopamine; ACPD, 1-aminocyclopentane-*trans*-1,3-dicarboxylic acid; BDNF, brain-derived neurotrophic factor; CAMKII, Ca^{2+}/calmodulin-dependent protein kinase II; CFA, complete Freund's adjuvant; CCl_4, carbon tetrachloride; COX-2, cyclooxygenase-2; EPM, elevated plus-maze; ERK 1/2, extracellular-regulated protein kinase ½; FST, forced swimming test; GABA, gamma-aminobutyric acid; i.g., intragastrically; iNOS, inducible nitric oxide synthase; IL-1β, interleukin 1β; MAO, monoamine oxidase; MPO, myeloperoxidase; Nf-κB, nuclear factor-κB p65; OFT, open field tests; OLT, object location test; ORT, object recognition test; p.o., orally; PKA, protein kinase A; PKC, protein kinase C; PPARγ, peroxisome proliferator-activated receptor γ; PSNL, partial sciatic nerve ligation; ROS, reactive oxygen species; s.c., subcutaneous injection; SDPA, step-down passive avoidance test; TNF-α, tumor necrosis factor alpha; TrkB, tropomyosin receptor kinase B; TrxR, thioredoxin reductase; TST, tail suspension test.

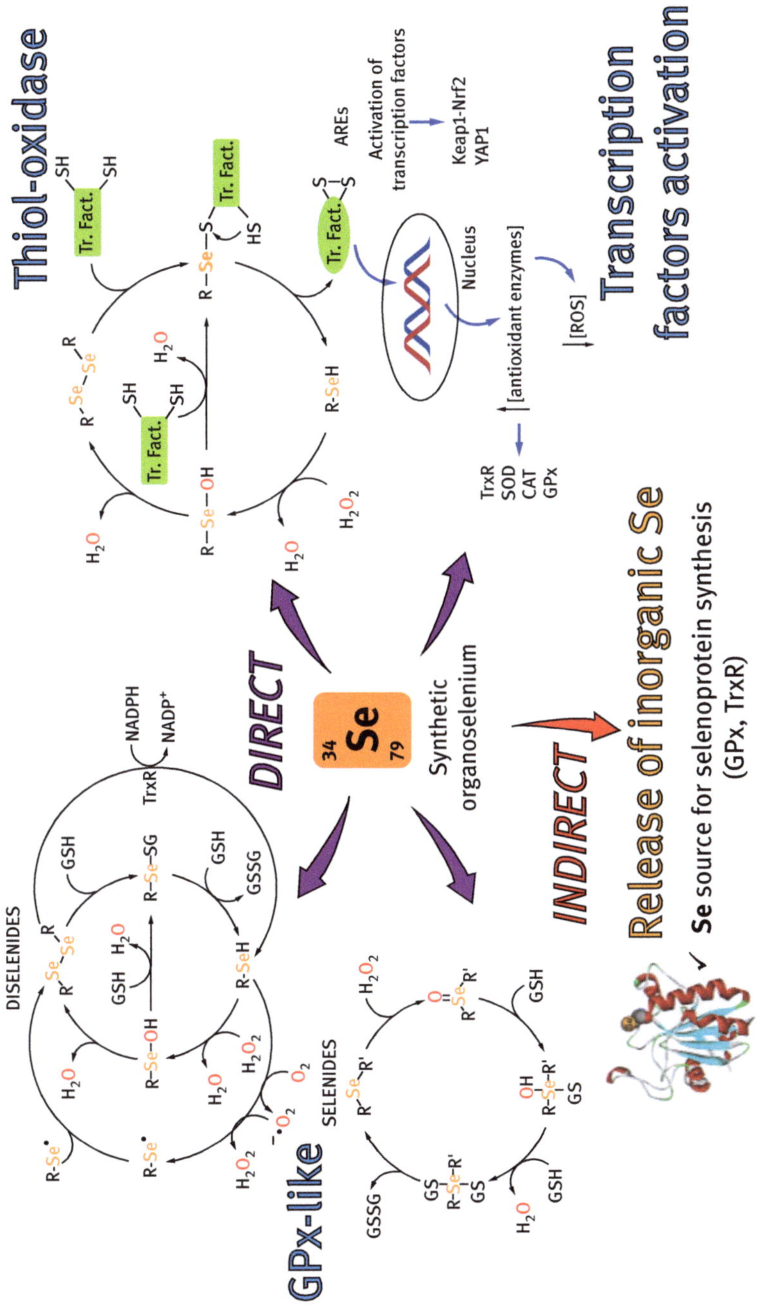

Figure 8.8: The proposed mechanism of action of organoselenium compounds.

necrosis induced by vitamin E and Se- deficient diet in rodents [187]. Except for ebselen, which was reported not to release inorganic selenium, there are only a few studies demonstrating that diselenides release inorganic selenium from the organic moiety [6, 184]. Thus, the reported effects of organoselenium compounds can be indirect via the release of Se to the synthesis of selenoproteins.

Organoselenium compounds can also weakly imitate the activity of selenoproteins [6]. For example, as commented in Section 8.2, diselenides and ebselen have a GPx-like activity [6, 45, 188], that is, they can reduce the H_2O_2 or ROOH to H_2O or ROH. However, organoselenium compounds can have a thiol-oxidase or thiol-modifying activity, which can activate transcription factors, such as the Nrf2. The activation of electrophilic responsive elements or antioxidant response elements increases the transcription and transduction of antioxidant enzymes, which decrease the ROS levels in the cellular milieu (Figure 8.8). On the other hand, the anticancer activity of organic compounds of selenium could be related to its pro-oxidant capacity, due to the generation of superoxide anion ($O_2^{\bullet-}$), H_2O_2, and selenyl radicals (R-Se$^{\bullet}$) (Figure 8.8) [15, 173].

8.6 Conclusion

The organoselenium compounds have been demonstrated to modulate the cellular redox status by different mechanisms (Figure 8.8). However, the use of organoselenium compounds as potential therapeutic agents still needs much more detailed *in vitro* and *in silico* studies. The empirical approaches used until now are obsolete and should be abandoned. As discussed in the elegant critical review published by Orian and Troppo [189], organoselenium compounds have no specific molecular targets and can modulate any protein having reactive thiol groups. Accordingly, the thiol-modifying properties of organoselenium compounds seem to be the significant determinant of their pharmacological effects [6]. The quite similar and nonspecific pharmacological effects (e.g., antinociceptive and antidepressant) of structurally unrelated organoselenium compounds possibly can be related to the release of inorganic selenium. In this way, the search of organoselenium compounds with higher selectivity for specific protein targets still need to be developed. For this purpose, the design of new molecules with pharmacophore groups (that have much high probability of interacting with specific molecular targets) can be considered more appropriate and rational than synthesizing molecules by chance and testing them *in vivo* using only the obsolete trial-and-error approaches.

Acknowledgments: The authors thank the financial support from Coordination for Improvement of Higher Education Personnel CAPES/PROEX (no. 23038.005848/2018-31; no. 0737/2018; no. 88882.182123/2018-01), the National Council for Scientific and Technological Development (CNPq), the Rio Grande do Sul Foundation for Research Support (FAPERGS – Brazil), and the Instituto de Pesquisa Pelé Pequeno Príncipe.

Conflicts of Interest: The authors declare no conflict of interest.

References

[1] Flohé, L. The labour pains of biochemical selenology: The history of selenoprotein biosynthesis. Biochim Biophys Acta – Gen Subj 2009, 1790, 1389–1403.
[2] Flohé, L., Günzler, W.A., and Schock, H.H. Glutathione peroxidase: a selenoenzyme. FEBS Lett 1973, 32, 132–134.
[3] Lobanov, A. V., Hatfield, D.L., and Gladyshev, V.N. Eukaryotic selenoproteins and selenoproteomes. Biochim Biophys Acta 2009, 1790, 1424–1428.
[4] Hatfield, D.L., Tsuji, P.A., Carlson, B.A., and Gladyshev, V.N. Selenium and selenocysteine: Roles in cancer, health, and development. Trends Biochem Sci 2014, 39, 112–120.
[5] Labunskyy, V.M., Hatfield, D.L., and Gladyshev, V.N. Selenoproteins: molecular pathways and physiological roles. Physiol Rev 2014, 94, 739–777.
[6] Barbosa, N. V., Nogueira, C.W., Nogara, P.A., De Bem, A.F., Aschner, M., and Rocha, J.B.T. Organoselenium compounds as mimics of selenoproteins and thiol modifier agents. Metallomics 2017, 9, 1703–1734.
[7] Nogueira, C.W., and Rocha, J.B.T. Toxicology and pharmacology of selenium: Emphasis on synthetic organoselenium compounds. Arch Toxicol 2011, 85, 1313–1359.
[8] Singh, F. V., and Wirth, T. Selenium reagents as catalysts. Catal Sci Technol 2019, 9, 1073–1091.
[9] Rocha, J.B.T., Piccoli, B.C., and Oliveira, C.S. Biological and chemical interest in selenium: A brief historical account. Arkivoc 2017, part ii, 457–491.
[10] Fredja, A. Organic selenium chemistry. Ann New York Acad Sci 1972, 1, 1–9.
[11] Arnér, E.S.J. History of Selenium Research, In Selenium: Its Molecular Biology and Role in Human Health, 2012, 1–19.
[12] Frost, D. V., and Olson, O.E. The two faces of selenium – can selenophobia be cured? Crit. Rev Toxicol 1972, 1, 467–514.
[13] Koller, L.D., and Exon, J.H. The two faces of selenium – deficiency and toxicity – are similar in animals and man. Can J Vet Res 1986, 50, 297–306.
[14] Tinggi, U. Essentiality and toxicity of selenium and its status in Australia: A review. Toxicol Lett 2003, 137, 103–110.
[15] Rocha, J.B.T., Oliveira, C.S., and Nogara, P.A. Toxicology and Anticancer Activity of Synthetic Organoselenium Compounds, In Organoselenium Compounds in Biology and Medicine: Synthesis, Biological and Therapeutic Treatments, 2018, 342–376.
[16] Moxon, A.L. Alkali disease or selenium poisoning. Bulletins 1937, 5, 311.
[17] Yang, G.Q., Wang, S.Z., Zhou, R.H., and Sun, S.Z. Endemic selenium intoxication of humans in China. Am J Clin Nutr 1983, 37, 872–881.

[18] Vinceti, M., Filippini, T., and Wise, L.A. Environmental selenium and human health: an update. Curr Environ Heal reports 2018, 5, 464–485.
[19] Vinceti, M., Filippini, T., and Rothman, K.J. Selenium exposure and the risk of type 2 diabetes: A systematic review and meta-analysis. Eur J Epidemiol 2018, 33, 789–810.
[20] Vinceti, M., Chawla, R., Filippini, T., Dutt, C., Cilloni, S., Loomba, R., Bargellini, A., Orsini, N., Dhillon, K.S., and Whelton, P. Blood pressure levels and hypertension prevalence in a high selenium environment: Results from a cross-sectional study. Nutr Metab Cardiovasc Dis 2019, 29, 398–408.
[21] Schwarz, K. A hitherto unrecognized factor against dietary necrotic liver degeneration in American yeast (factor 3). Exp Biol Med 1951, 78, 852–856.
[22] Schwarz, K., and Foltz, C.M. Selenium as an integral part of factor 3 against dietary necrotic liver degeneration. J Am Chem Soc 1957, 79, 3292–3293.
[23] Papp, L.V., Lu, J., Holmgren, A., and Khanna, K.K. From selenium to selenoproteins: synthesis, identity, and their role in human health. Antioxid Redox Signal 2007, 9, 775–806.
[24] Reeves, M.A., and Hoffmann, P.R. The human selenoproteome: recent insights into functions and regulation. Cell Mol Life Sci 2009, 66, 2457–2478.
[25] Mlochowski, J., Kloc, K., Lisiak, R., Potaczek, P., and Wójtowicz, H. Developments in the chemistry of selenaheterocyclic compounds of practical importance in synthesis and medicinal biology. ARKIVOC 2007, vi, 14–46.
[26] Mugesh, G., Du Mont, W.W., and Sies, H. Chemistry of biologically important synthetic organoselenium compounds. Chem Rev 2001, 101, 2125–2179.
[27] Löwig, C. Ueber Schwefelwasserstoff- und Selenwasserstoffäther. Ann Der Phys Und Chemie 1836, 113, 550–553.
[28] Siemens, C. Ueber das Selenmercaptan. Ann Der Chemie Und Pharm 1847, 61, 360–362.
[29] Spengler, G., Gajdács, M., Marć, M.A., Domínguez-Álvarez, E., and Sanmartín, C. Organoselenium compounds as novel adjuvants of chemotherapy drugs – a promising approach to fight cancer drug resistance. Molecules 2019, 24, 1–15.
[30] Frieben, E.E., Amin, S., and Sharma, A.K. Development of isoselenocyanate compounds' syntheses and biological applications. J Med Chem 2019, 62, 5261–5275.
[31] Gandin, V., Khalkar, P., Braude, J., and Fernandes, A.P. Organic selenium compounds as potential chemotherapeutic agents for improved cancer treatment. Free Radic Biol Med 2018, 127, 80–97.
[32] Bhattacherjee, D., Basu, C., Bhardwaj, Q., Mal, S., Sahu, S., Sur, R., and Bhabak, K.P. Design, synthesis and anti-cancer activities of benzyl analogues of garlic-derived diallyl disulfide (DADS) and the corresponding diselenides. ChemistrySelect 2017, 2, 7399–7406.
[33] Mhetre, A.B., Lee, H., Yang, H., Lee, K., Nam, D.H., and Lim, D. Synthesis and anticancer activity of benzoselenophene and heteroaromatic derivatives of 1,2,9,9a-tetrahydrocyclopropa[c]benzo[e]indol-4-one (CBI). Org Biomol Chem 2017, 15, 1198–1208.
[34] Jantová, S., Mrvová, N., Hudec, R., Sedlák, J., Pánik, M., and Milata, V. Pro-apoptotic effect of new quinolone 7- ethyl 9-ethyl-6-oxo-6,9-dihydro[1,2,5]selenadiazolo [3,4-h]quinoline-7-carboxylate on cervical cancer cell line HeLa alone/with UVA irradiation. Toxicol Vitr 2016, 33, 35–44.
[35] Da Rosa, R.M., Piccoli, B.C., Da Silva, F.D.A., Dornelles, L., Rocha, J.B.T., Sonego, M.S., Begnini, K.R., Collares, T., Seixas, F.K., and Rodrigues, O.E.D. Synthesis, antioxidant and antitumoral activities of 5′-arylchalcogeno-3-aminothymidine (ACAT) derivatives. MedChemComm 2017, 8, 408–414.
[36] De Souza, D., Mariano, D.O.C., Nedel, F., Schultze, E., Campos, V.F., Seixas, F., Da Silva, R.S., Munchen, T.S., Ilha, V., Dornelles, L. et al. New organochalcogen multitarget drug: Synthesis

and antioxidant and antitumoral activities of chalcogenozidovudine derivatives. J Med Chem 2015, 58, 3329–3339.
[37] Wagner, M.S., Schultze, E., Oliveira, T.L., de Leon, P.M.M., Thurow, H.S., Campos, V.F., Oliveira, I., Souza, D. de., Rodrigues, O.E.D., Collares, T. et al. Revitalizing the AZT through of the selenium: an approach in human triple negative breast cancer cell line. Front Oncol 2018, 8, 1–10.
[38] Facompre, N.D., El-Bayoumy, K., Sun, Y.W., Pinto, J.T., and Sinha, R. 1,4-Phenylenebis (methylene)selenocyanate, but not selenomethionine, inhibits androgen receptor and Akt signaling in human prostate cancer cells. Cancer Prev Res 2010, 3, 975–984.
[39] Sugie, S., Reddy, B.S., El-Bayoumy, K., and Tanaka, T. Inhibition by dietary benzylselenocyanate of hepatocarcinogenesis induced by azoxymethane in Fischer 344 rats. Japanese J Cancer Res 1989, 80, 952–957.
[40] Ronai, Z., Tillotson, J.K., Traganos, F., Darzynkiewicz, Z., Conaway, C.C., Upadhyaya, P., and El-Bayoumy, K. Effects of organic and inorganic selenium compounds on rat mammary tumor cells. Int J Cancer 1995, 63, 428–434.
[41] Bradt, W.E., and Green, J.F. The interaction of selenium tetrachloride and benzene in the presence of anhydrous aluminum chloride. J Org Chem 1937, 1, 540–543.
[42] Krafft, F., and Vorster, W. Ueber umwandlung des diphenylsulfons in diphenylsulfid und diphenylselenid. Berichte Der Dtsch Chem Gesellschaft 1893, 26, 2813–2822.
[43] Krafft, F., and Kaschau, A. Ueber thianthren und selenanthren, C12H8Se2. Berichte Der Dtsch Chem Gesellschaft 1896, 29, 443–445.
[44] Weisberger, A.S., and Suhrland, L.G. Studies on analogues of L-Cysteine and L-cystine III. The effect of selenium cystine on leukemia. Blood 1956, 11, 19–30.
[45] Wilson, S.R., Zucker, P.A., Huang, R.R.C., and Spector, A. Development of synthetic compounds with glutathione peroxidase activity. J Am Chem Soc 1989, 111, 5936–5939.
[46] Poester, V.R., Mattei, A.S., Mendes, J.F., Klafke, G.B., Ramis, I.B., Sanchotene, K.O., and Xavier, M.O. Antifungal activity of diphenyl diselenide alone and in combination with itraconazole against Sporothrix brasiliensis. Med Mycol 2019, 57, 328–331.
[47] Quispe, R.L., Jaramillo, M.L., Galant, L.S., Engel, D., Dafre, A.L., Teixeira da Rocha, J.B., Radi, R., Farina, M., and de Bem, A.F. Diphenyl diselenide protects neuronal cells against oxidative stress and mitochondrial dysfunction: Involvement of the glutathione-dependent antioxidant system. Redox Biol 2019, 20, 118–129.
[48] Adedara, I.A., Owoeye, O., Awogbindin, I.O., Ajayi, B.O., Rocha, J.B.T., and Farombi, E.O. Diphenyl diselenide abrogates brain oxidative injury and neurobehavioural deficits associated with pesticide chlorpyrifos exposure in rats. Chem Biol Interact 2018, 296, 105–116.
[49] Müller, S.G., Jardim, N.S., Quines, C.B., and Nogueira, C.W. Diphenyl diselenide regulates Nrf2/Keap-1 signaling pathway and counteracts hepatic oxidative stress induced by bisphenol A in male mice. Environ Res 2018, 164, 280–287.
[50] Meinerz, D.F., Branco, V., Aschner, M., Carvalho, C., and Rocha, J.B.T. Diphenyl diselenide protects against methylmercury-induced inhibition of thioredoxin reductase and glutathione peroxidase in human neuroblastoma cells: A comparison with ebselen. J Appl Toxicol 2017, 37, 1073–1081.
[51] Lovato, F.L., Teixeira Da Rocha, J.B., and Dalla Corte, C.L. Diphenyl diselenide protects against methylmercury-induced toxicity in saccharomyces cerevisiae via the Yap1 transcription factor. Chem Res Toxicol 2017, 30, 1134–1144.

[52] Leite, M.R., Cechella, J.L., Pinton, S., Nogueira, C.W., and Zeni, G. A diphenyl diselenide-supplemented diet and swimming exercise promote neuroprotection, reduced cell apoptosis and glial cell activation in the hypothalamus of old rats. Exp Gerontol 2016, 82, 1–7.
[53] Quines, C.B., Rosa, S.G., Velasquez, D., Da Rocha, J.T., Neto, J.S.S., and Nogueira, C.W. Diphenyl diselenide elicits antidepressant-like activity in rats exposed to monosodium glutamate: A contribution of serotonin uptake and Na+, K+-ATPase activity. Behav Brain Res 2016, 301, 161–167.
[54] Brüning, C.A., Martini, F., Soares, S.M., Sampaio, T.B., Gai, B.M., Duarte, M.M.M.F., and Nogueira, C.W. m-Trifluoromethyl-diphenyl diselenide, a multi-target selenium compound, prevented mechanical allodynia and depressive-like behavior in a mouse comorbid pain and depression model. Prog Neuro-Psychopharmacology Biol Psychiatry 2015, 63, 35–46.
[55] Bortolatto, C.F., Heck, S.O., Zborowski, V.A., Gai, B.M., Neto, J.S.S., and Nogueira, C.W. Evidence for the contribution of multiple mechanisms in the feeding pattern of rats exposed to p-chloro-diphenyl diselenide-supplemented diets. Physiol Behav 2015, 151, 298–307.
[56] Fiuza, T. da L., Oliveira, C.S., da Costa, M., Oliveira, V.A., Zeni, G., and Pereira, M.E. Effectiveness of (PhSe)2 in protect against the HgCl2 toxicity. J Trace Elem Med Biol 2015, 29, 255–262.
[57] Pinton, S., Souza, A.C., Sari, M.H.M., Ramalho, R.M., Rodrigues, C.M.P., and Nogueira, C.W. P,p'-Methoxyl-diphenyl diselenide protects against amyloid-β induced cytotoxicity in vitro and improves memory deficits in vivo. Behav Brain Res 2013, 247, 241–247.
[58] De Freitas, M.L., Andreia, R.H., Roman, S.S., and Brandão, R. Effects of 4, 4-dichloro-diphenyl diselenide (ClPhSe)2 on toxicity induced by mercuric chloride in mice : A comparative study with diphenyl diselenide (PhSe)2. Environ Toxicol Pharmacol 2012, 34, 985–994.
[59] Nogueira, C.W., and Rocha, J.B.T. Diphenyl diselenide a janus-faced molecule. J Braz Chem Soc 2010, 21, 2055–2071.
[60] Jesse, C.R., Rocha, J.B.T., Nogueira, C.W., and Savegnago, L. Further analysis of the antinociceptive action caused by p-methoxyl-diphenyl diselenide in mice. Pharmacol Biochem Behav 2009, 91, 573–580.
[61] Ineu, R.P., Pereira, M.E., Aschner, M., Nogueira, C.W., Zeni, G., and Rocha, J.B.T. Diphenyl diselenide reverses gastric lesions in rats: Involvement of oxidative stress. Food Chem Toxicol 2008, 46, 3023–3029.
[62] Barbosa, N.B.V., Rocha, J.B.T., Soares, J.C.M., Wondracek, D.C., Gonçalves, J.F., Schetinger, M.R.C., and Nogueira, C.W. Dietary diphenyl diselenide reduces the STZ-induced toxicity. Food Chem Toxicol 2008, 46, 186–194.
[63] Lesser, R., and Weiß, R. Über selenhaltige aromatische Verbindungen (VI). Berichte Der Dtsch Chem Gesellschaft 1924, 57, 1077–1082.
[64] Müller, A., Cadenas, E., Graf, P., and Sies, H. A novel biologically active seleno-organic compound-1. Glutathione peroxidase-like activity in vitro and antioxidant capacity of PZ 51 (Ebselen) Biochem Pharmacol 1984, 33, 3235–3239.
[65] Slusarczyk, W., Olakowska, E., Larysz-Brysz, M., Woszczycka-Korczyńska, I., De Carrillo, D. G., Węglarz, W.P., Lewin-Kowalik, J., and Marcol, W. Use of ebselen as a neuroprotective agent in rat spinal cord subjected to traumatic injury. Neural Regen Res 2019, 14, 1255–1261.
[66] Martini, F., Rosa, S.G., Klann, I.P., Fulco, B.C.W., Carvalho, F.B., Rahmeier, F.L., Fernandes, M.C., and Nogueira, C.W. A multifunctional compound ebselen reverses memory impairment, apoptosis and oxidative stress in a mouse model of sporadic Alzheimer's disease. J Psychiatr Res 2019, 109, 107–117.

[67] Capper, M.J., Wright, G.S.A., Barbieri, L., Luchinat, E., Mercatelli, E., McAlary, L., Yerbury, J.J., O'Neill, P.M., Antonyuk, S. V., Banci, L. et al. The cysteine-reactive small molecule ebselen facilitates effective SOD1 maturation. Nat Commun 2018, 9, 1–9.
[68] Antoniadou, I., Kouskou, M., Arsiwala, T., Singh, N., Vasudevan, S.R., Fowler, T., Cadirci, E., Churchill, G.C., and Sharp, T. Ebselen has lithium-like effects on central 5-HT 2A receptor function. Br J Pharmacol 2018, 175, 2599–2610.
[69] Santofimia-Castaño, P., Izquierdo-Alvarez, A., de la Casa-Resino, I., Martinez-Ruiz, A., Perez-Lopez, M., Portilla, J.C., Salido, G.M., and Gonzalez, A. Ebselen alters cellular oxidative status and induces endoplasmic reticulum stress in rat hippocampal astrocytes. Toxicology 2016, 357–358, 74–84.
[70] Leyck, S., and Parnham, M.J. Acute antiinflammatory and gastric effects of the seleno-organic compound ebselen. Agents Actions 1990, 30, 426–431.
[71] Husain, K., Morris, C., Whitworth, C., Trammell, G.L., Rybak, L.P., and Somani, S.M. Protection by ebselen against cisplatin-induced nephrotoxicity: antioxidant system. Mol Cell Biochem 1998, 178, 127–133.
[72] Yang, C.F., Shen, H.M., and Ong, C.N. Ebselen induces apoptosis in HepG2 cells through rapid depletion of intracellular thiols. Arch Biochem Biophys 2000, 374, 142–152.
[73] Namura, S., Nagata, I., Takami, S., Masayasu, H., and Kikuchi, H. Ebselen reduces cytochrome c release from mitochondria and subsequent DNA fragmentation after transient focal cerebral ischemia in mice. Stroke 2001, 32, 1906–1911.
[74] Yamaguchi, T., Sano, K., Takakura, K., Saito, I., Shinohara, Y., Asano, T., and Yasuhara, H. Ebselen in acute ischemic stroke. Stroke 1998, 29, 12–17.
[75] Saito, I., Asano, T., Sano, K., Takakura, K., Abe, H., Yoshimoto, T., Kikuchi, H., Ohta, T., and Ishibashi, S. Neuroprotective effect of an antioxidant, ebselen, in patients with delayed neurological deficits after aneurysmal subarachnoid hemorrhage. Neurosurgery 1998, 42, 269–278.
[76] Ogawa, A., Yoshimoto, T., Kikuchi, H., Sano, K., Saito, I., Yamaguchi, T., and Yasuhara, H. Ebselen in acute middle cerebral artery occlusion: A placebo-controlled, double-blind clinical trial. Cerebrovasc Dis 1999, 9, 112–118.
[77] Singh, N., Sharpley, A.L., Emir, U.E., Masaki, C., Herzallah, M.M., Gluck, M.A., Sharp, T., Harmer, C.J., Vasudevan, S.R., Cowen, P.J. et al. Effect of the putative lithium mimetic ebselen on brain myo-inositol, sleep, and emotional processing in humans. Neuropsychopharmacology 2016, 41, 1768–1778.
[78] Masaki, C., Sharpley, A.L., Godlewska, B.R., Berrington, A., Hashimoto, T., Singh, N., Vasudevan, S.R., Emir, U.E., Churchill, G.C., and Cowen, P.J. Effects of the potential lithium-mimetic, ebselen, on brain neurochemistry: A magnetic resonance spectroscopy study at 7 tesla. Psychopharmacology (Berl) 2016, 233, 1097–1104.
[79] Singh, N., Halliday, A.C., Thomas, J.M., Kuznetsova, O., Baldwin, R., Woon, E.C.Y., Aley, P.K., Antoniadou, I., Sharp, T., Vasudevan, S.R. et al. A safe lithium mimetic for bipolar disorder. Nat Commun 2013, 4, 1332–1337.
[80] Beckman, J.A., Goldfine, A.B., Leopold, J.A., and Creager, M.A. Ebselen does not improve oxidative stress and vascular function in patients with diabetes: A randomized, crossover trial. Am J Physiol Circ Physiol 2016, 311, H1431–H1436.
[81] Kil, J., Lobarinas, E., Spankovich, C., Griffiths, S.K., Antonelli, P.J., Lynch, E.D., and Le Prell, C.G. Safety and efficacy of ebselen for the prevention of noise-induced hearing loss: A randomised, double-blind, placebo-controlled, phase 2 trial. Lancet 2017, 390, 969–979.
[82] Elsherbini, M., Hamama, W., Zoorob, H., Bhowmick, D., Mugesh, G., and Wirth, T. Synthesis and antioxidant activities of novel chiral ebselen analogues. Heteroat Chem 2014, 25, 320–325.

[83] Engman, L., and Hallberg, A. Expedient synthesis of ebselen and related compounds. J Org Chem 1989, 54, 2964–2966.

[84] Balkrishna, S.J., Bhakuni, B.S., Chopra, D., and Kumar, S. Cu-catalyzed efficient synthetic methodology for ebselen and related Se-N heterocycles. Acc Chem Res 2010, 12, 5394–5397.

[85] Sharpless, K.B., and Young, M.W. Olefin synthesis. Rate enhancement of the elimination of alkyl aryl selenoxides by electron-withdrawing substituents. J Org Chem 1975, 40, 947–949.

[86] Reich, H.J., Cohen, M.L., and Clark, P.S. Reagents for synthesis of organoselenium compounds: diphenyl diselenide and benzeneselenenyl chloride. Org Synth 1988, 6, 533.

[87] Müller, J., and Terfort, A. Synthesis of pure aromatic, aliphatic, and araliphatic diselenides. Inorgan Chim Acta 2006, 359, 4821–4827.

[88] Zhang, Y., Jia, X., and Zhou, X. Samarium diiodide-induced reduction of amorphous selenium: a facile synthesis of diaryl diselenides. Synth Commun 1994, 24, 1247–1252.

[89] Li, Z., Ke, F., Deng, H., Xu, H., Xiang, H., and Zhou, X. Synthesis of disulfides and diselenides by copper-catalyzed coupling reactions in water. Org Biomol Chem 2013, 11, 2943–2946.

[90] Perin, G., Goulart, H.A., Soares, L.K., Peglow, T.J., Schumacher, R.F., Pinz, M.P., Reis, A.S., Luchese, C., and Wilhelm, E.A. Synthesis and pharmacological evaluation of novel selenoethers glycerol derivatives for the treatment of pain and inflammation: involvement of nitrergic and glutamatergic systems. Appl Biochem Biotechnol 2018, 187, 1398–1423.

[91] Savegnago, L., Vieira, A.I., Seus, N., Goldani, B.S., Castro, M.R., Lenardão, E.J., and Alves, D. Synthesis and antioxidant properties of novel quinoline-chalcogenium compounds. Tetrahedron Lett 2013, 54, 40–44.

[92] Pinto Brod, L.M., Fronza, M.G., Vargas, J.P., Lüdtke, D.S., Luchese, C., Wilhelm, E.A., and Savegnago, L. Involvement of monoaminergic system in the antidepressant-like effect of (octylseleno)-xylofuranoside in the mouse tail suspension test. Prog Neuro-Psychopharmacology Biol Psychiatry 2016, 65, 201–207.

[93] Ecker, A., da Silva, R.S., Dos Santos, M.M., Ardisson-Araújo, D., Rodrigues, O.E.D., da Rocha, J.B.T., and Barbosa, N.V. Safety profile of AZT derivatives: Organoselenium moieties confer different cytotoxic responses in fresh human erythrocytes during in vitro exposures. J Trace Elem Med Biol 2018, 50, 240–248.

[94] Ecker, A., Ledur, P.C., da Silva, R.S., Leal, D.B.R., Rodrigues, O.E.D., Ardisson-Araújo, D., Waczuk, E.P., da Rocha, J.B.T., and Barbosa, N.V. Chalcogenozidovudine derivatives with antitumor activity: Comparative toxicities in cultured human mononuclear cells. Toxicol Sci 2017, 160, 30–46.

[95] Mariano, D.O.C., De Souza, D., Meinerz, D.F., Allebrandt, J., De Bem, A.F., Hassan, W., Rodrigues, O.E.D., and Da Rocha, J.B.T. The potential toxicological insights about the anti-HIV drug azidothymidine-derived monoselenides in human leukocytes: Toxicological insights of new selenium-azidothymidine analogs. Hum Exp Toxicol 2017, 36, 910–918.

[96] Vieira, B.M., Thurow, S., da Costa, M., Casaril, A.M., Domingues, M., Schumacher, R.F., Perin, G., Alves, D., Savegnago, L., and Lenardão, E.J. Ultrasound-assisted synthesis and antioxidant activity of 3-selanyl-1 H-indole and 3-selanylimidazo[1,2-a]pyridine derivatives. Asian J Org Chem 2017, 6, 1635–1646.

[97] Gai, B.M., Stein, A.L., Roehrs, J.A., Bilheri, F.N., Nogueira, C.W., and Zeni, G. Synthesis and antidepressant-like activity of selenophenes obtained via iron(iii)-PhSeSePh-mediated cyclization of Z-selenoenynes. Org Biomol Chem 2012, 10, 798–807.

[98] El-bayoumy, K., Chae, Y., Upadhyaya, P., and Reddy, B.S. Inhibition of 7,12-dimethylbenz(a) anthracene-induced tumors and DNA adduct formation in the mammary glands of female Sprague-Dawley rats by the synthetic organoselenium compound, 1,4-phenylenebis (methylene)selenocyanate. Cancer Res 1992, 52, 2402–2407.

[99] Iwaoka, M., Katsuda, T., Komatsu, H., and Tomoda, S. Experimental and theoretical studies on the nature of weak nonbonded interactions between divalent selenium and halogen atoms. J Org Chem 2005, 70, 321–327.

[100] Lari, A., Gleiter, R., and Rominger, F. Supramolecular organization based on van der Waals forces: Syntheses and solid state structures of isomeric [6.6] cyclophanes with 2,5-diselenahex-3-yne bridges. Eur J Org Chem 2009, 2267–2274.

[101] Thomé, G.R., Oliveira, V.A., Chitolina Schetinger, M.R., Saraiva, R.A., Souza, D., Dorneles Rodrigues, O.E., Teixeira Rocha, J.B., Ineu, R.P., and Pereira, M.E. Selenothymidine protects against biochemical and behavioral alterations induced by ICV-STZ model of dementia in mice. Chem Biol Interact 2018, 294, 135–143.

[102] Munchen, T.S., Sonego, M.S., de Souza, D., Dornelles, L., Seixas, F.K., Collares, T., Piccoli, B. C., da Silva, F.D.Á., da Rocha, J.B.T., Quoos, N. et al. New 3′-triazolyl-5′-aryl-chalcogenothymidine: synthesis and anti-oxidant and antiproliferative bladder carcinoma (5637) activity. ChemistrySelect 2018, 3, 3479–3486.

[103] Woolf, C.J. What is this thing called pain? J. Clin Invest 2010, 120, 3742–3744.

[104] Pathan, S.A., Mitra, B., and Cameron, P.A. A systematic review and meta-analysis comparing the efficacy of nonsteroidal anti-inflammatory drugs, opioids, and paracetamol in the treatment of acute renal colic. Eur Urol 2018, 73, 583–595.

[105] Nogueira, C.W., Quinhones, E.B., Jung, E.A.C., Zeni, G., and Rocha, J.B.T. Anti-inflammatory and antinociceptive activity of diphenyl diselenide. Inflamm Res 2003, 52, 56–63.

[106] Zasso, F.B., Goncales, C.E.P., Jung, E.A.C., Araldi, D., Zeni, G., Rocha, J.B.T., and Nogueira, C. W. On the mechanisms involved in antinociception induced by diphenyl diselenide. Environ Toxicol Pharmacol 2004, 19, 283–289.

[107] Birmann, P.T., Sousa, F.S.S., De Oliveira, D.H., and Domingues, M. 3-(4-Chlorophenylselanyl) -1-methyl-1 H -indole, a new selenium compound elicits an antinociceptive and anti-inflammatory effect in mice. Eur J Pharmacol 2018, 827, 71–79.

[108] Pinz, M., Reis, A.S., Duarte, V., Márcia, J., Goldani, B.S., Alves, D., Savegnago, L., Luchese, C., and Wilhelm, E.A. 4-Phenylselenyl-7-chloroquinoline, a new quinoline derivative containing selenium, has potential antinociceptive and anti-inflammatory actions. Eur J Pharmacol 2016, 780, 122–128.

[109] Jesse, C.R., Savegnago, L., and Nogueira, C.W. Spinal mechanisms of antinociceptive effect caused by oral administration of bis-selenide in mice. Brain Res 2008, 31.

[110] Savegnago, L., Pinto, L.G., Jesse, C.R., Alves, D., Rocha, J.B.T., Nogueira, C.W., and Zeni, G. Antinociceptive properties of diphenyl diselenide: Evidences for the mechanism of action. Eur J Pharmacol 2007, 555, 129–138.

[111] Savegnago, L., Santos, A.R.S., Jesse, C.R., Nogueira, C.W., and Rocha, J.B.T. Mechanisms involved in the antinociceptive effect caused by diphenyl diselenide in the formalin test. J Pharm Pharmacol 2008, 60, 1679–1686.

[112] Brüning, C.A., Gai, B.M., Soares, S.M., Martini, F., and Nogueira, C.W. Serotonergic systems are implicated in antinociceptive effect of m-trifluoromethyl diphenyl diselenide in the mouse glutamate test. Pharmacol Biochem Behav 2014, 125, 15–20.

[113] Bruning, C.A., Prigol, M., Roehrs, J.A., Zeni, G., and Nogueira, C.W. Evidence for the involvement of mu-opioid and delta-opioid receptors in the antinociceptive effect caused by oral administration of m-trifluoromethyl-diphenyl diselenide in mice. Behav Pharmacol 2010, 21, 621–626.

[114] Sari, M.H.M., Zborowski, V.A., Ferreira, L.M., Jardim, N.S., Araujo, P.C.O., Brüning, C.A., Cruz, L., and Nogueira, C.W. Enhanced pharmacological actions of p, p ' -methoxyl-diphenyl diselenide- loaded polymeric nanocapsules in a mouse model of neuropathic pain : Behavioral and molecular insights. J Trace Elem Med Biol 2018, 46, 17–25.

[115] Pinto, L.G., Jesse, C.R., Nogueira, C.W., and Savegnago, L. Evidence for the involvement of glutamatergic and GABAergic systems and protein kinase A pathway in the antinociceptive effect caused by p-methoxy-diphenyl diselenide in mice. Pharmacol Biochem Behav 2008, 88, 487–496.
[116] Oliveira, C.E.S., Marcondes Sari, M.H.M., Zborowski, V.A., Prado, V.C., Nogueira, C.W., and Zeni, G. Pain-depression dyad induced by reserpine is relieved by p,p'-methoxyl-diphenyl diselenide in rats. Eur J Pharmacol 2016, 791, 794–802.
[117] Rosa, S.G., Brüning, C.A., Pesarico, A.P., Souza, A.C.G., De, Souza, G., and Nogueira, C.W. Anti-inflammatory and antinociceptive effects of 2,2′-dipyridyl diselenide through reduction of inducible nitric oxide synthase, nuclear factor-kappa B and c-Jun N-terminal kinase phosphorylation. J Trace Elem Med Biol 2018, 48, 38–45.
[118] Donato, F., Pavin, N.F., Tiago, A., Goes, R., Souza, L.C., Soares, L.C., Dorneles, O.E., Jesse, C.R., Savegnago, L., Jesse, R. et al. Antinociceptive and anti-hyperalgesic effects of bis (4-methylbenzoyl) diselenide in mice: Evidence for the mechanism of action. Pharm Biol 2015, 53, 395–403.
[119] Jesse, C.R., Savegnago, L., and Nogueira, C.W. Role of nitric oxide/cyclic GMP/K+ channel pathways in the antinociceptive effect caused by 2,3-bis(mesitylseleno)propenol. Life Sci 2007, 81, 1694–1702.
[120] Anversa, R.G., Sousa, F.S.S., Birmann, P.T., Lima, D.B., Lenardão, E.J., Bruning, C.A., and Savegnago, L. Antinociceptive and anti-inflammatory effects of 1,2-bis-(4-methoxyphenylselanyl) styrene in mice: Involvement of the serotonergic system. J Pharm Pharmacol 2018, 70, 901–909.
[121] Brüning, C.A., Prigol, M., Roehrs, J.A., Nogueira, C.W., and Zeni, G. Involvement of the serotonergic system in the anxiolytic-like effect caused by m-trifluoromethyl-diphenyl diselenide in mice. Behav Brain Res 2009, 205, 511–517.
[122] Prigol, M., Luchese, C., Pinton, S., Ferreira, M., Santos, J.P.A., Karkow, A.K., Mortari, A.C., and Nogueira, C.W. Diphenyl diselenide induces anxiolytic-like and sedative effects on the chick social separation-stress behavior. Neurosci Lett 2011, 495, 140–143.
[123] Ibrahim, M., Mussulini, B.H.M., Moro, L., de Assis, A.M., Rosemberg, D.B., de Oliveira, D.L., Rocha, J.B.T., Schwab, R.S., Schneider, P.H., Souza, D.O. et al. Anxiolytic effects of diphenyl diselenide on adult zebrafish in a novelty paradigm. Prog Neuro-Psychopharmacol Biol Psychiatry 2014, 54, 187–194.
[124] Oliveira, C.E.S., Sari, M.H.M., Zborowski, V.A., Araujo, P.C.O., Nogueira, C.W., and Zeni, G. p, p′-Methoxyl-diphenyl diselenide elicits an antidepressant-like effect in mice without discontinuation anxiety phenotype. Pharmacol Biochem Behav 2017, 154, 31–38.
[125] Ghisleni, G., Kazlauckas, V., Both, F.L., Pagnussat, N., Mioranzza, S., Batista, J., Rocha, T., Souza, D.O., and Porciúncula, L.O. Diphenyl diselenide exerts anxiolytic-like effect in Wistar rats: Putative roles of GABAA and 5HT receptors. Prog Neuropsychopharmacol Biol Psychiatry 2008, 32, 1508–1515.
[126] Savegnago, L., Jesse, C.R., Pinto, L.G., Rocha, J.B., Barancelli, D.A., Nogueira, C.W., and Zeni, G. Diphenyl diselenide exerts antidepressant-like and anxiolytic-like effects in mice: Involvement of l-arginine-nitric oxide-soluble guanylate cyclase pathway in its antidepressant-like action. Pharmacol Biochem Behav 2007, 88, 418–426.
[127] Reis, A.S., Pinz, M., Duarte, L.F.B., Roehrs, J.A., Alves, D., Luchese, C., and Wilhelm, E.A. 4-phenylselenyl-7-chloroquinoline, a novel multitarget compound with anxiolytic activity: Contribution of the glutamatergic system. J Psychiatr Res 2017, 84, 191–199.
[128] Savegnago, L., Jesse, C.R., Pinto, L.G., Rocha, J.B.T., Nogueira, C.W., and Zeni, G. Monoaminergic agents modulate antidepressant-like effect caused by diphenyl diselenide in rats. Prog Neuro-Psychopharmacol Biol Psychiatry 2007, 31, 1261–1269.

[129] Rocha, J.T., Gai, B.M., Pinton, S., Sampaio, T.B., Nogueira, C.W., and Zeni, G. Effects of diphenyl diselenide on depressive-like behavior in ovariectomized mice submitted to subchronic stress: Involvement of the serotonergic system. Psychopharmacology (Berl) 2012, 222, 709–719.
[130] Posser, T., Kaster, M.P., Baraúna, S.C., Rocha, J.B.T., Rodrigues, A.L.S., and Leal, R.B. Antidepressant-like effect of the organoselenium compound ebselen in mice: Evidence for the involvement of the monoaminergic system. Eur J Pharmacol 2009, 602, 85–91.
[131] Brüning, C.A., Souza, A.C.G., Gai, B.M., Zeni, G., and Nogueira, C.W. Antidepressant-like effect of m-trifluoromethyl-diphenyl diselenide in the mouse forced swimming test involves opioid and serotonergic systems. Eur J Pharmacol 2011, 658, 145–149.
[132] Bortolatto, C.F., Wilhelm, E.A., Chagas, P.M., and Nogueira, C.W. P-chloro-diphenyl diselenide, an organoselenium compound, with antidepressant-like and memory enhancer actions in aging male rats. Biogerontology 2012, 13, 237–249.
[133] Velasquez, D., Quines, C., Pistóia, R., Zeni, G., and Nogueira, C.W. Selective inhibition of MAO-A activity results in an antidepressant-like action of 2-benzoyl 4-iodoselenophene in mice. Physiol Behav 2017, 170, 100–105.
[134] Gai, B.M., Sanna, M.D., Stein, A.L., Zeni, G., Galeotti, N., and Nogueira, C.W. ERK1/2 phosphorylation is involved in the antidepressant-like action of 2,5-diphenyl-3-(4-fluorophenylseleno)-selenophene in mice. Eur J Pharmacol 2014, 736, 44–54.
[135] Oliveira, C.E.S., Gai, B.M., Godoi, B., Zeni, G., and Nogueira, C.W. The antidepressant-like action of a simple selenium-containing molecule, methyl phenyl selenide, in mice. Eur J Pharmacol 2012, 690, 119–123.
[136] Gerzson, M.F.B., Victoria, F.N., Radatz, C.S., Gomes, M.G., Boeira, S.P., Jacob, R.G., Alves, D., Jesse, C.R., and Savegnago, L. In vitro antioxidant activity and in vivo antidepressant-like effect of α-(phenylselanyl) acetophenone in mice. Pharmacol Biochem Behav 2012, 102, 21–29.
[137] Jesse, C.R., Wilhelm, E.A., and Nogueira, C.W. Depression-like behavior and mechanical allodynia are reduced by bis selenide treatment in mice with chronic constriction injury: A comparison with fluoxetine, amitriptyline, and bupropion. Psychopharmacology (Berl) 2010, 212, 513–522.
[138] Jesse, C.R., Wilhelm, E.A., Bortolatto, C.F., and Nogueira, C.W. Evidence for the involvement of the serotonergic 5-HT2A/C and 5-HT3 receptors in the antidepressant-like effect caused by oral administration of bis selenide in mice. Prog Neuro-Psychopharmacology Biol Psychiatry 2010, 34, 294–302.
[139] Menezes, C., Leitemperger, J., Toni, C., Santi, A., Lópes, T., Berenice, N., Barbosa, V., Radünz, J., and Lucia, V. Comparative study on effects of dietary with diphenyl diselenide on oxidative stress in carp (Cyprinus carpio) and silver catfish (Rhamdia sp.) exposed to herbicide clomazone. Environ Toxicol Pharmacol 2013, 36, 706–714.
[140] Stefanello, S.T., Hartmann, D.D., Amaral, G.P., Courtes, A.A., Leite, M.T.B., da Silva, T.C., Gonçalves, D.F., Souza, M.B., da Rosa, P.C., Dornelles, L. et al. Antioxidant protection by β-selenoamines against thioacetamide-induced oxidative stress and hepatotoxicity in mice. J Biochem Mol Toxicol 2017, 31, 1–7.
[141] Stefanello, S.T., Rosa, E.J.F., Amaral, G.P., Carvalho, N.R., Luz, C.A., Bender, C.R., Schwab, R.S., Dornelles, L., and Soares, F.A.A. Effect of diselenide administration in thioacetamide-induced acute neurological and hepatic failure in mice. Toxicol Res (Camb) 2015, 4, 707–717.
[142] Quines, C.B., Chagas, P.M., Hartmann, D., Carvalho, N., Soares, F., and Nogueira, C.W. (p-ClPhSe)2 reduces hepatotoxicity induced by monosodium glutamate by improving mitochondrial function in rats. J Cell Biochem 2017, 118, 2877–2886.

[143] Ibrahim, M., Ibrahim, M., Muhammad, N., Shah, M.I.A., de Oliveira Leite, G., and Rocha, J.B. T. Pharmacological mechanisms underlying gastroprotective activities of binapthyl diselenide in Wistar rats. Inflammopharmacology 2018, 26, 1117–1123.
[144] Filho, C.B., Fabbro, L. Del., Boeira, S.P., Furian, A.F., Savegnago, L., Soares, C., Braga, A. L., and Jesse, C.R. Hepatoprotective effect of bis (4-methylbenzoyl) diselenide against CCl4 -induced oxidative damage in mice. Cell Biochem Funct 2013, 31, 152–158.
[145] Saad, S.Y., Najjar, T.A., and Arafah, M.M. Cardioprotective effects of subcutaneous ebselen against daunorubicin-induced cardiomyopathy in rats. Basic Clin Pharmacol Toxicol 2006, 99, 412–417.
[146] Dambrova, M., Liepinsh, E., and Kalvinsh, I. Mildronate: cardioprotective action through carnitine-lowering effect. Trends Cardiovasc Med 2002, 12, 275–279.
[147] Hayashi, Y., Tajima, K., Kirimoto, T., Miyake, H., and Matsuura, N. Cardioprotective effects of MET-88, a gamma-butyrobetaine hydroxylase inhibitor, on cardiac dysfunction induced by ischemia/reperfusion in isolated rat hearts. Pharmacology 2000, 61, 238–243.
[148] Rydzik, A.M., Brem, J., Struwe, W.B., Kochan, G.T., Benesch, J.L.P., and Schofield, C.J. Ejection of structural zinc leads to inhibition of c -butyrobetaine hydroxylase. Bioorg Med Chem Lett 2014, 24, 4954–4957.
[149] Davis, M.T., and Bartfay, W.J. Production and Iron Concentrations in the Hearts of Chronically Iron-Overloaded Mice. Biol Res Nurs 2004, 6, 37–45.
[150] Bem, A.F., Farina, M., Lima, R. De., Wayne, C., and Dinis, T.C.P. Diphenyl diselenide, a simple glutathione peroxidase mimetic, inhibits human LDL oxidation in vitro. Atherosclerosis 2008, 201, 92–100.
[151] De Bem, A.F., Portella, R.D.L., Colpo, E., Maria, M., Frescura, M., Frediane, A., Taube, P.S., Nogueira, C.W., Farina, M., Luiz, E. et al. Diphenyl diselenide decreases serum levels of total cholesterol and tissue oxidative stress in cholesterol-fed rabbits. Basic Clin Pharmacol Toxicol 2009, 17–23.
[152] Ribeiro, M.C.P., Ávila, D.S., Schiar, V.P.P., Dos Santos, D.B., Meinerz, D.F., Duarte, M.M.F., Monteiro, R., Puntel, R., De Bem, A.F., Hassan, W. et al. Diphenyl diselenide supplementation reduces biochemical alterations associated with oxidative stress in rats fed with fructose and hydrochlorothiazide. Chem Biol Interact 2013, 204, 191–199.
[153] Nogueira, C.W., Rocha, J.T., Segatto, M., Pallottini, V., Zeni, G., Trapani, L., and Rosa, P. Molecular effects of diphenyl diselenide on cholesterol and glucose cell metabolism. Curr Med Chem 2013, 20, 4426–4434.
[154] Quines, C.B., Rosa, S.G., Chagas, P.M., Velasquez, D., Prado, V.C., and Nogueira, C.W. (p-ClPhSe)2 stimulates carbohydrate metabolism and reverses the metabolic alterations induced by high fructose load in rats. Food Chem Toxicol 2017, 107, 122–128.
[155] Quines, C.B., Rosa, S.G., Velasquez, D., Prado, V.C., Neto, J.S.S., and Nogueira, C.W. (p-ClPhSe)2 stabilizes metabolic function in a rat model of neuroendocrine obesity induced by monosodium glutamate. Food Chem Toxicol 2018, 118, 168–180.
[156] Roberts-Thomson, I.C. Rise and fall of peptic ulceration: A disease of civilization?. J Gastroenterol Hepatol 2018, 33, 1321–1326.
[157] Yegen, B.C. Lifestyle and peptic ulcer disease. Curr Pharm Des 2018, 24, 2034–2040.
[158] Tabuchi, Y., Sugiyama, N., Horiuchi, T., Furusawa, M., and Furuhama, K. Ebselen, a seleno-organic compound, protects against ethanol-induced murine gastric mucosal injury in both in vivo and in vitro systems. Eur J Pharmacol 1995, 272, 195–201.
[159] Tabuchi, Y., and Kurebayashi, Y. Antisecretory and antiulcer effects of ebselen, a seleno-organic compound, in rats. Jpn J Pharmacol 1993, 61, 255–257.

[160] Beil, W., Staar, U., and Sewing, K.F. Interaction of the anti-inflammatory seleno-organic compound ebselen with acid secretion in isolated parietal cells and gastric H+/K+-ATPase. Biochem Pharmacol 1990, 40, 1997–2003.
[161] Savegnago, L., Trevisan, M., Alves, D., Rocha, J.B.T., Nogueira, C.W., and Zeni, G. Antisecretory and antiulcer effects of diphenyl diselenide. Environ Toxicol Pharmacol 2006, 21, 86–92.
[162] Chakraborty, S., Yadav, S.K., Subramanian, M., Priyadarsini, K.I., Iwaoka, M., and Chattopadhyay, S. DL- trans -3,4-Dihydroxy-1-selenolane (DHS red) accelerates healing of indomethacin-induced stomach ulceration in mice. Free Radic Res 2012, 46, 1378–1386.
[163] Shin, J.M., Munson, K., Vagin, O., and Sachs, G. The gastric HK-ATPase: Structure, function, and inhibition. Pflugers Arch Eur J Physiol 2009, 457, 609–622.
[164] Yoshida, M., Iizuka, K., Terada, A., Hara, M., Nishijima, H., Shimada, A., Nakada, K., Satoh, Y., and Akama, Y. Prevention of nephrotoxicity of cisplatin by repeated oral administration of ebselen in rats. Tohoku j Exp Med 2000, 191, 209–220.
[165] Lynch, E.D., Gu, R., Pierce, C., and Kil, J. Reduction of acute cisplatin ototoxicity and nephrotoxicity in rats by oral administration of allopurinol and ebselen. Hear Res 2005, 201, 81–89.
[166] Dhanarajan, R., Abraham, P., and Isaac, B. Protective effect of ebselen, a selenoorganic drug, against gentamicin-induced renal damage in rats. Basic Clin Pharmacol Toxicol 2006, 99, 267–272.
[167] Noiri, E., Nakao, A., Uchida, K., Tsukahara, H., Ohno, M., Fujita, T., Brodsky, S., and Goligorsky, M.S. Oxidative and nitrosative stress in acute renal ischemia. Am J Physiol Physiol 2017, 281, F948–F957.
[168] Chew, P., Yuen, D.Y.C., Stefanovic, N., Pete, J., Coughlan, M.T., Jandeleit-Dahm, K.A., Thomas, M.C., Rosenfeldt, F., Cooper, M.E., and De Haan, J.B. Antiatherosclerotic and renoprotective effects of ebselen in the diabetic apolipoprotein E/GPx1-double knockout mouse. Diabetes 2010, 59, 3198–3207.
[169] Bortolatto, C.F., Wilhelm, E.A., Roman, S.S., and Nogueira, C.W. (E)-2-Benzylidene-4-phenyl-1,3-diselenole ameliorates signals of renal injury induced by cisplatin in rats. J Appl Toxicol 2014, 34, 87–94.
[170] Wilhelm, E.A., Bortolatto, C.F., and Nogueira, C.W. p-Methoxyl-diphenyl diselenide protects against cisplatin-induced renal toxicity in mice. Food Chem Toxicol 2012, 50, 1187–1193.
[171] Brandão, R., Acker, C.I., Leite, M.R., Barbosa, N.B.V., and Nogueira, C.W. Diphenyl diselenide protects against glycerol-induced renal damage in rats. J Appl Toxicol 2009, 29, 612–618.
[172] Ibrahim, M., Luchese, C., Pinton, S., Roman, S.S., Hassan, W., Nogueira, C.W., and Rocha, J. B.T. Involvement of catalase in the protective effect of binaphthyl diselenide against renal damage induced by glycerol. Exp Toxicol Pathol 2011, 63, 331–335.
[173] Tan, H.W., Mo, H.Y., Lau, A.T.Y., and Xu, Y.M. Selenium species: Current status and potentials in cancer prevention and therapy. Int J Mol Sci 2019, 20.
[174] Wang, L., Yang, Z., Fu, J., Yin, H., Xiong, K., Tan, Q., Jin, H., Li, J., Wang, T., Tang, W. et al. Ethaselen: A potent mammalian thioredoxin reductase 1 inhibitor and novel organoselenium anticancer agent. Free Radic Biol Med 2012, 52, 898–908.
[175] Zheng, X., Chen, Y., Bai, M., Liu, Y., Xu, B., Sun, R., and Zeng, H. The antimetastatic effect and underlying mechanisms of thioredoxin reductase inhibitor ethaselen. Free Radic Biol Med 2019, 131, 7–17.
[176] Hassan, W., S. Oliveira, C., Noreen, H., P. Kamdem, J., W. Nogueira, C., and J. Rocha, B.T. Organoselenium compounds as potential neuroprotective therapeutic agents. Curr Org Chem 2016, 20, 218–231.

[177] Sampaio, T.B., Pinton, S., da Rocha, J.T., Gai, B.M., and Nogueira, C.W. Involvement of BDNF/TrkB signaling in the effect of diphenyl diselenide on motor function in a Parkinson's disease rat model. Eur J Pharmacol 2017, 795, 28–35.
[178] Zborowski, V.A., Sari, M.H.M., Heck, S.O., Stangherlin, E.C., Neto, J.S.S., Nogueira, C.W., and Zeni, G. p-Chloro-diphenyl diselenide reverses memory impairment-related to stress caused by corticosterone and modulates hippocampal [3H]glutamate uptake in mice. Physiol Behav 2016, 164, 25–33.
[179] Casaril, A.M., Domingues, M., Fronza, M., Vieira, B., Begnini, K., Lenardão, E.J., Seixas, F. K., Collares, T., Nogueira, C.W., and Savegnago, L. Antidepressant-like effect of a new selenium-containing compound is accompanied by a reduction of neuroinflammation and oxidative stress in lipopolysaccharide-challenged mice. J Psychopharmacol 2017, 31, 1263–1273.
[180] Domingues, M., Casaril, A.M., Birmann, P.T., Lourenço, D. de A., Vieira, B., Begnini, K., Lenardão, E.J., Collares, T., Seixas, F.K., and Savegnago, L. Selanylimidazopyridine prevents lipopolysaccharide-induced depressive-like behavior in mice by targeting neurotrophins and inflammatory/oxidative mediators. Front Neurosci 2018, 12, 1–10.
[181] Steinbrenner, H., and Sies, H. Selenium homeostasis and antioxidant selenoproteins in brain: Implications for disorders in the central nervous system. Arch Biochem Biophys 2013, 536, 152–157.
[182] Solovyev, N.D. Importance of selenium and selenoprotein for brain function: From antioxidant protection to neuronal signalling. J Inorg Biochem 2015, 153, 1–12.
[183] Kühbacher, M., Bartel, J., Hoppe, B., Alber, D., Bukalis, G., Bräuer, A.U., Behne, D., and Kyriakopoulos, A. The brain selenoproteome: Priorities in the hierarchy and different levels of selenium homeostasis in the brain of selenium-deficient rats. J Neurochem 2009, 110, 133–142.
[184] Adams, W.J., Kocsis, J.J., and Snyder, R. Acute toxicity and urinary excretion of diphenyldiselenide. Toxicol Lett 1989, 48, 301–310.
[185] Thompson, H.J. Role of low molecular weight, selenium-containing compounds in human health, In Selenium, Hatfield, D.L, Ed., Springer, Boston, MA, 2001, 283–297.
[186] Wirth, T. Selenium, In Comprehensive Organometallic Chemistry III, 2007, 457–499.
[187] Schwarz, K., and Foltz, C.M. Factor 3 activity of selenium compounds. J Biol Chem 1958, 233, 245–251.
[188] Müller, A., Gabriel, H., and Sies, H. A novel biologically active seleno-organic compound. II. Activity of PZ 51 in relation to glutathione peroxidase. Biochem Pharmacol 1984, 33, 3241–3245.
[189] Orian, L., and Toppo, S. Organochalcogen peroxidase mimetics as potential drugs: A long story of a promise still unfulfilled. Free Radic Biol Med 2014, 66, 65–74.
[190] Dias, G.R.M., de Almeida, T.M., Sudati, J.H., Dobrachinski, F., Pavin, S., Soares, F.A.A., Nogueira, C.W., and Barbosa, N.B.V. Diphenyl diselenide supplemented diet reduces depressive-like behavior in hypothyroid female rats. Physiol Behav 2014, 124, 116–122.
[191] Wilhelm, E.A., Jesse, C.R., Bortolatto, C.F., Barbosa, N.B.V., and Nogueira, C.W. Evidence of the involvement of K+ channels and PPARγ receptors in the antidepressant-like activity of diphenyl diselenide in mice. J Pharm Pharmacol 2010, 62, 1121–1127.
[192] Acker, C.I., Luchese, C., Prigol, M., and Nogueira, C.W. Antidepressant-like effect of diphenyl diselenide on rats exposed to malathion: Involvement of Na+K+ ATPase activity. Neurosci Lett 2009, 455, 168–172.
[193] Borges, V.C., Savegnago, L., Dadalt, G., and Nogueira, C.W. Disubstituted diaryl diselenides inhibit [3H]-serotonin uptake in rats. Neurotox Res 2009, 15, 57–61.

[194] Rosa, S.G., Quines, C.B., Stangherlin, E.C., and Nogueira, C.W. Diphenyl diselenide ameliorates monosodium glutamate induced anxiety-like behavior in rats by modulating hippocampal BDNF-Akt pathway and uptake of GABA and serotonin neurotransmitters. Physiol Behav 2016, 155, 1–8.
[195] Rosa, S.G., Pesarico, A.P., Tagliapietra, C.F., Da Luz, S.C.A., and Nogueira, C.W. Opioid system contribution to the antidepressant-like action of m -trifluoromethyl-diphenyl diselenide in mice: A compound devoid of tolerance and withdrawal syndrome. J Psychopharmacol 2017, 31, 1250–1262.
[196] Sari, M.H.M., Zborowski, V.A., Ferreira, L.M., Jardim, N.S., Barbieri, A.V., Cruz, L., and Nogueira, C.W. p,p′-Methoxyl-diphenyl diselenide-loaded polymeric nanocapsules as a novel approach to inflammatory pain treatment: Behavioral, biochemistry and molecular evidence. Eur J Pharm Sci 2018, 111, 38–45.
[197] Sousa, F.S.S., Anversa, R.G., Birmann, P.T., De Souza, M.N., Balaguez, R., Alves, D., Luchese, C., Wilhelm, E.A., and Savegnago, L. Contribution of dopaminergic and noradrenergic systems in the antinociceptive effect of a-(phenylalanyl) acetophenone. Pharmacol Reports 2017, 69, 871–877.
[198] Sousa, F.S.S., Birmann, P.T., Balaguez, R., Alves, D., Brüning, C.A., and Savegnago, L. A-(phenylselanyl) acetophenone abolishes acute restraint stress induced-comorbid pain, depression and anxiety-related behaviors in mice. Neurochem Int 2018, 120, 112–120.
[199] Domingues, M., Casaril, A.M., Birmann, P.T., Bampi, S.R., Lourenço, D. de A., Vieira, B.M., Dapper, L.H., Lenardão, E.J., Sonego, M., Collares, T. et al. Effects of a selanylimidazopyridine on the acute restraint stress-induced depressive- and anxiety-like behaviors and biological changes in mice. Behav Brain Res 2019, 366, 96–107.
[200] Silva, V.D.G., Reis, A.S., Pinz, M.P., da Fonseca, C.A.R., Duarte, L.F.B., Roehrs, J.A., Alves, D., Luchese, C., and Wilhelm, E.A. Further analysis of acute antinociceptive and anti-inflammatory actions of 4-phenylselenyl-7-chloroquinoline in mice. Fundam Clin Pharmacol 2017, 31, 513–525.
[201] Duarte, L.F.B., Barbosa, E.S., Oliveira, R.L., Pinz, M.P., Godoi, B., Schumacher, R.F., Luchese, C., Wilhelm, E.A., and Alves, D. A simple method for the synthesis of 4-arylselanyl-7-chloroquinolines used as in vitro acetylcholinesterase inhibitors and in vivo memory improvement. Tetrahedron Lett 2017, 58, 3319–3322.
[202] Donato, F., de Gomes, M.G., Goes, A.T.R., Seus, N., Alves, D., Jesse, C.R., and Savegnago, L. Involvement of the dopaminergic and serotonergic systems in the antidepressant-like effect caused by 4-phenyl-1-(phenylselanylmethyl)-1,2,3-triazole. Life Sci 2013, 93, 393–400.
[203] Fronza, M.G., Brod, L.M.P., Casaril, A.M., Sacramento, M., Alves, D., and Savegnago, L. Computational and biological evidences on the serotonergic involvement of SeTACN antidepressant-like effect in mice. PLoS One 2017, 12, e0187445.
[204] Chagas, P.M., Fulco, B.C.W., Sari, M.H.M., Roehrs, J.A., and Nogueira, C.W. Bis (phenylimidazoselenazolyl) diselenide elicits antinociceptive effect by modulating myeloperoxidase activity, NOx and NFkB levels in the collagen-induced arthritis mouse model. J Pharm Pharmacol 2017, 69, 1022–1032.
[205] Gai, B.M., Bortolatto, C.F., Brüning, C.A., Zborowski, V.A., Stein, A.L., Zeni, G., and Nogueira, C.W. Depression-related behavior and mechanical allodynia are blocked by 3-(4-fluorophenylselenyl)-2,5-diphenylselenophene in a mouse model of neuropathic pain induced by partial sciatic nerve ligation. Neuropharmacology 2014, 79, 580–589.
[206] Gai, B.M., Bortolatto, C.F., Heck, S.O., Stein, A.L., Duarte, M.M.M.F., Zeni, G., and Nogueira, C.W. An organoselenium compound improves behavioral, endocrinal and neurochemical changes induced by corticosterone in mice. Psychopharmacology 2014, 231, 2119–2130.

[207] Victoria, F.N., Anversa, R., Penteado, F., Castro, M., Lenardão, E.J., and Savegnago, L. Antioxidant and antidepressant-like activities of semi-synthetic α-phenylseleno citronellal. Eur J Pharmacol 2014, 742, 131–138.
[208] Pinto Brod, L.M., Fronza, M.G., Vargas, J.P., Lüdtke, D.S., Brüning, C.A., and Savegnago, L. Modulation of PKA, PKC, CAMKII, ERK 1/2 pathways is involved in the acute antidepressant-like effect of (octylseleno)-xylofuranoside (OSX) in mice. Psychopharmacology 2017, 234, 717–725.

Bubun Banerjee and Brindaban Chandra Ranu

9 Selenoamides, selenazadienes, and selenocarbonyls in organic synthesis

9.1 Introduction

During the first discovery of selenium in 1817 [1], this red amorphous powder was treated as a dangerous component responsible for livestock poisoning [2]. Later on, in the 1950s, it was established that selenium is an essential nutrient available in many selenoproteins [3, 4]. Selenium is a homologous element with oxygen and sulfur in the periodic table but it shows different specific features and reactivity in many occasions [5]. Various selenium containing molecules were found to possess significant biological efficacies such as anticancer [6–9], anti-inflammatory [10, 11], antimelanogenesis [12, 13], and neuroprotective [14] activities. Organoselenium scaffolds are common in commercially available drug molecules [15–21]. Figure 9.1 represents a glimpse of selenium containing bioactive scaffolds [22]. As a result, during the last decade, synthesis of organoselenium compounds dragged considerable attention and thus a large number of methods are available for the synthesis of selenium containing scaffolds under various reaction conditions [23].

In many occasions, selenoamides, selenazadienes, and selenocarbonyls are being used as key starting materials for the synthesis of diverse organoselenium scaffolds under various reaction conditions [24]. This chapter summarizes the synthesis of various organoselenium scaffolds, where either selenoamide or selenazadiene or selenocarbonyl plays the key role.

9.2 Synthesis of organoselenium scaffolds involving selenamide

9.2.1 Synthesis of 4-benzyl-2-phenyl-1,3-selenazole derivatives

A facile and simple method was developed for the synthesis of 4-benzyl-2-phenyl-1,3-selenazole derivatives (**4**) via cycloaddition reactions of 3-selanylpropargyl alcohols (**1**) and selenamide (**2**) using scandium(III) triflate ($Sc(OTf)_3$) as catalyst in the

Bubun Banerjee, Department of Chemistry, Indus International University, Himachal Pradesh, India
Brindaban Chandra Ranu, School of Chemical Sciences, Indian Association for the Cultivation of Science, Calcutta, India

https://doi.org/10.1515/9783110625110-009

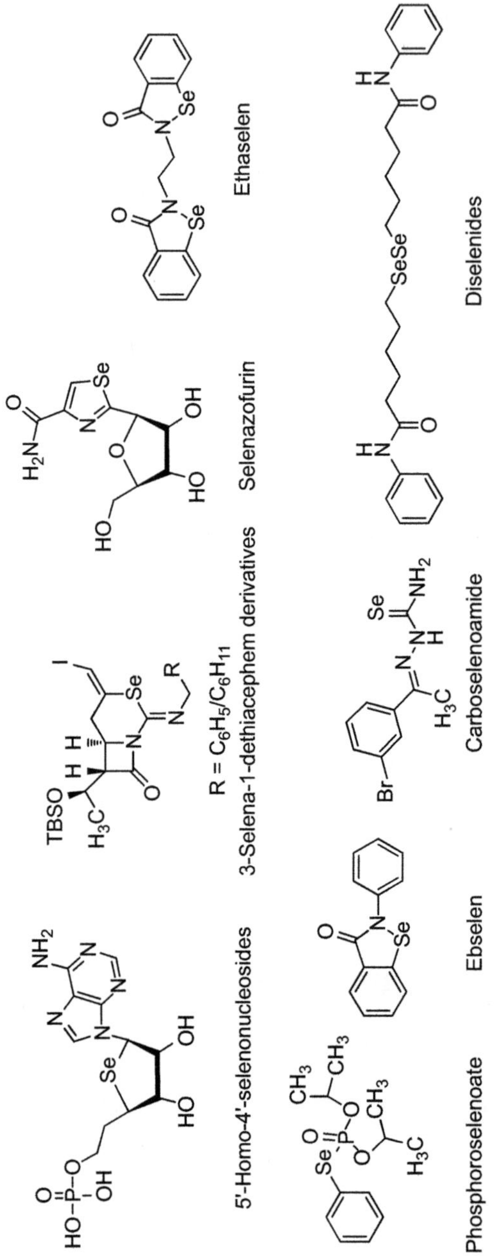

Figure 9.1: Glimpse of bioactive organoselenium compounds.

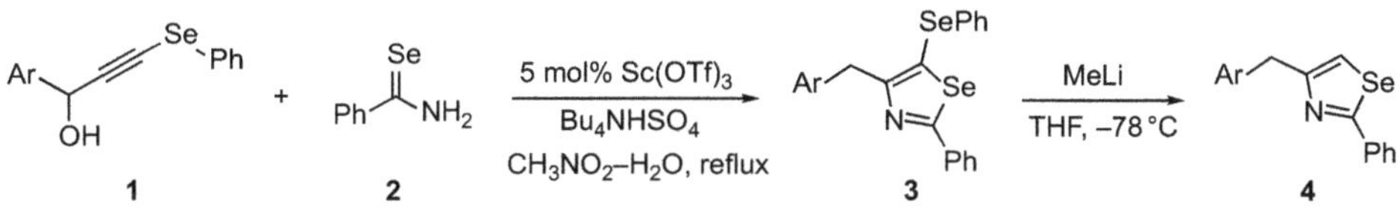

Ar	(%) yield of **4**
4-MeOC$_6$H$_4$	99
2,4,6-triMeC$_6$H$_2$	70
4-FC$_6$H$_4$	78

Figure 9.2: Sc(OTf)$_3$-catalyzed synthesis of 4-benzyl-2-phenyl-1,3-selenazole derivatives in aqueous nitromethane.

presence of tetrabutylammonium hydrogensulfate in aqueous nitromethane under reflux conditions (Figure 9.2) [25]. It was proposed that the reaction underwent through the formation of α-selanyl propadienyl cation (**5**). Bu$_4$NHSO$_4$ may act as a scavenger of the eliminated hydroxyl group. The plausible mechanism of this reaction is shown in Figure 9.3. Deselanylation of the cycloadducts (**3**) using methyl lithium (MeLi) gave the desired products with excellent yields.

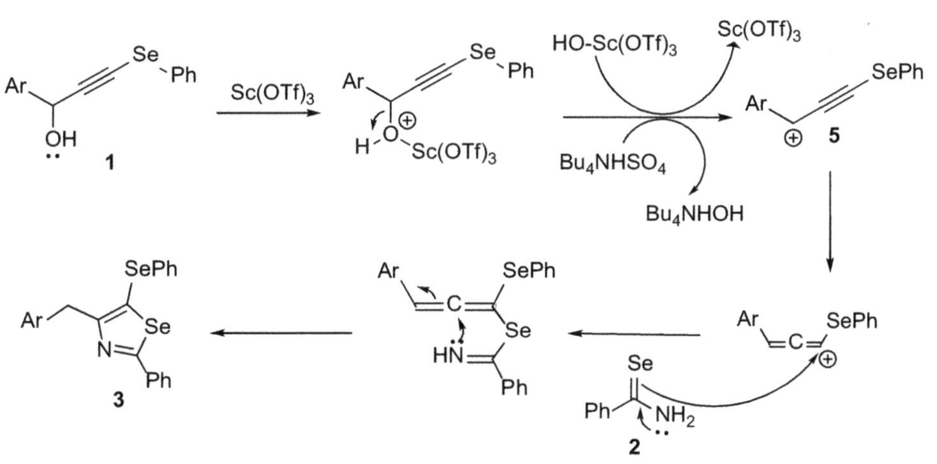

Figure 9.3: Plausible mechanism for the synthesis of 4-benzyl-2-phenyl-1,3-selenazole derivatives.

9.2.2 Synthesis of α/β-hydroxy selenoamides

Deprotonation of selenoformamides (**6**) using lithium diisopropylamide afforded the corresponding selenocarbamoyllithiums (**7**) as umpolung reagents. Murai et al. [26] synthesized a series of α-hydroxyselenoamides (**8**) via the reactions of structurally diverse ketones (**8,9**)/aldehydes (**10**) with in situ generated selenocarbamoyllithiums

(7) at –40 °C (Figure 9.4). The same group also synthesized β-hydroxyselenoamides (8) from the reactions of cyclohexanone (8) and in situ generated lithium eneselenolates (15) starting from selenoamide **14** (Figure 9.5) [27].

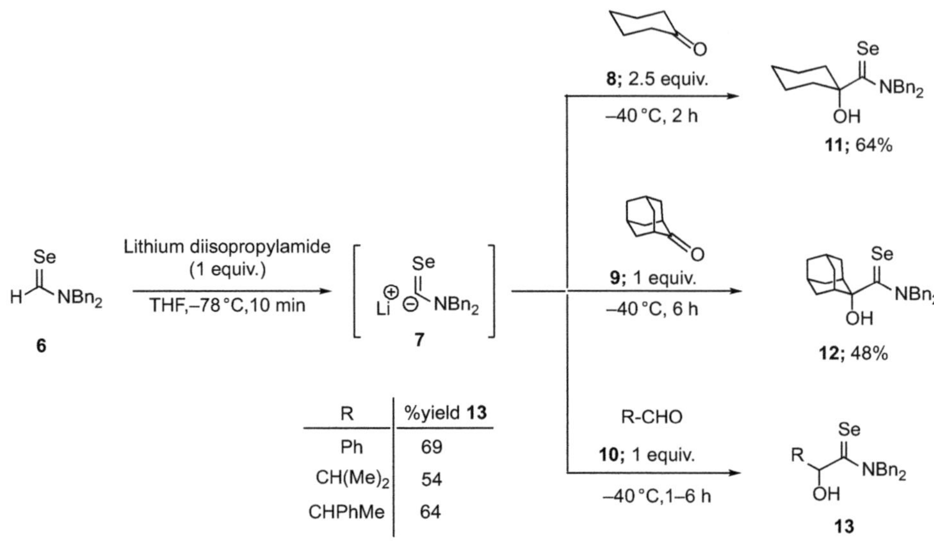

R	%yield **13**
Ph	69
CH(Me)$_2$	54
CHPhMe	64

Figure 9.4: Synthesis of α-hydroxy selenoamides in tetrahydrofuran.

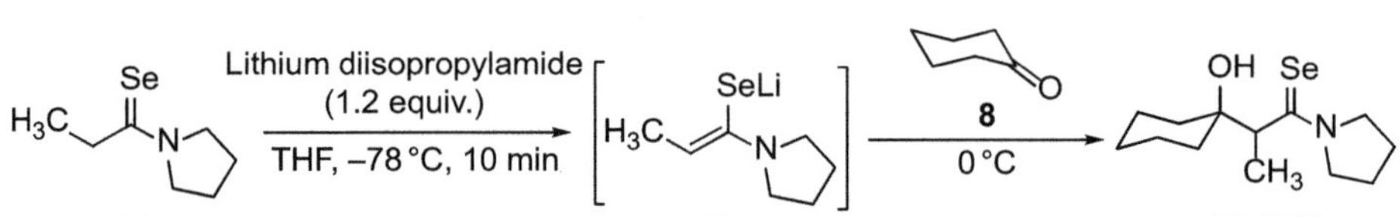

Figure 9.5: Synthesis of β-hydroxy selenoamides in tetrahydrofuran.

9.2.3 Synthesis γ,δ-unsaturated selenoamides

In situ generated lithium eneselenolate (15) was also employed for the synthesis of γ,δ-unsaturated selenoamides (18) in good yields via seleno-Claisen rearrangement reactions with 3-bromoprop-1-ene (17) at 0 °C (Figure 9.6) [28]. This reaction is diastereoselective as the diastereomeric ratio of isomers in reactant **17a** was found to be retained in the product **18a**. Cyclohex-2-enyl bromide (**17b**) also afforded the corresponding product **18b** with excellent yields though it required a higher temperature and longer reaction time.

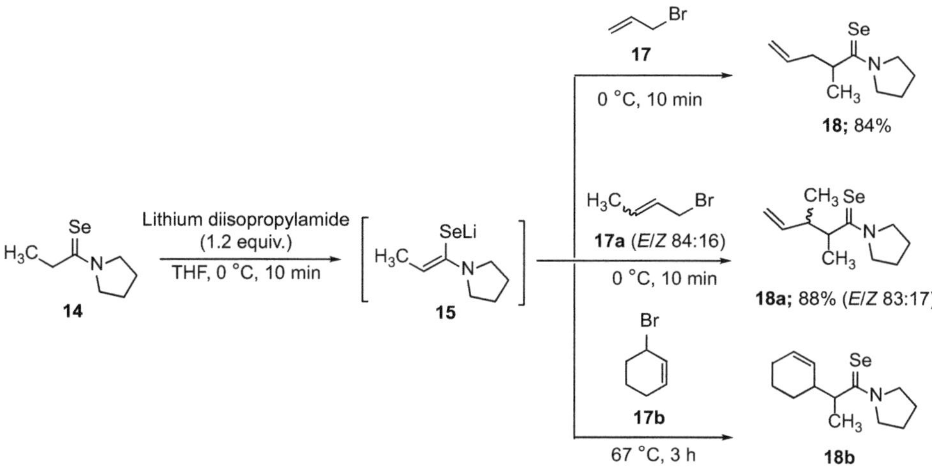

Figure 9.6: Synthesis γ,δ-unsaturated selenoamides in tetrahydrofuran.

9.2.4 Synthesis of 1,3-selenazoles

1,3-Selenazoles are generally reported as antibiotics and anticancer agents [29–31]. Selenamides are considered as the most important starting material to synthesize functionalized 1,3-selenazoles under various reaction conditions. A simple, efficient, and novel method was developed for the synthesis of 2,4-disubstituted 1,3-selenazoles (**20**) by the cyclocondensation of primary selenoamides (**2**) and alkynyl (phenyl)iodomium salts (**19**) in the presence of triethyl amine as base in methanol under reflux conditions (Figure 9.7) [32]. Geisler et al. [33] synthesized another series of 2,4-disubstituted selenazoles (**20**) by the reaction of primary selenoamides (**2**) and α-bromoacetaldehydes (**21**) without using any catalyst in ethanol under reflux (Figure 9.8). 1,4-Bis(4-phenyl-2-selenazolyl)benzene (**23**) was synthesized via the reactions 1,4-benzenedicarboselenoamide (**22**) and excess α-bromoacetaldehydes (**21**) in the absence of any catalyst using dimethylformamide and ethanol mixture as solvent under reflux conditions (Figure 9.9) [34].

4-Aryl-1,3-selenazoles (**20a**) were synthesized by the cyclization reactions of selenoformamide (**24**) and α-bromoacetaldehydes (**21**) using pyridine as catalyst in methanol at 35 °C (Figure 9.10) [35]. Geisler et al. [36] carried out reactions of α-bromoacetophenones (**21**) and phenylselenoacetic amide (**25**) without using any catalyst in ethanol under reflux conditions, which afforded the corresponding 2-benzyl-1,3-selenazoles (**20b**) with good yields (Figure 9.11). By refluxing the synthesized compounds (**20b**) with selenium dioxide in dioxane, the same group further prepared 2-benzoyl-1,3-selenazoles (**26**) with good yields (Figure 9.12) [37].

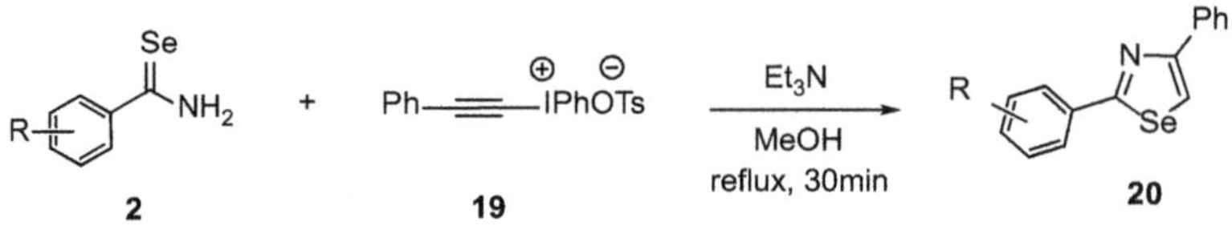

R	(%) yield of **20**
H	82
2,4-diCl	74
4-NMe$_2$	77

R	(%) yield of **20**
4-Cl	56
4-Me	83
4-OMe	84

Figure 9.7: Synthesis of 2,4-disubstituted selenazoles using alkynyl(phenyl)iodomium salts in methanol.

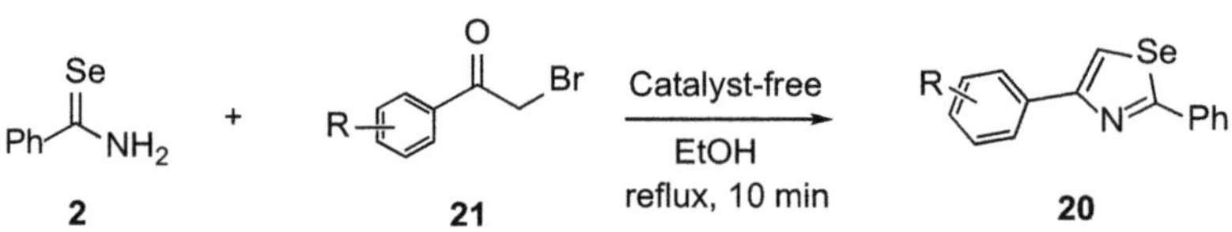

R	(%) yield of **20**
H	87
4-Me	86
4-Br	98
4-NO$_2$	98

Figure 9.8: Catalyst-free synthesis of 2,4-disubstituted selenazoles using phenacyl bromides in ethanol.

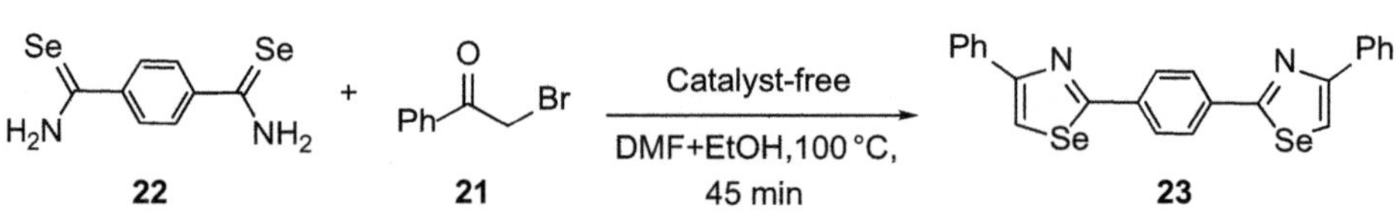

Figure 9.9: Catalyst-free synthesis of 1,4-bis(4-phenyl-2-selenazolyl)benzene.

Zhang et al. [38] reported synthesis of 2,4-disubstituted selenazoles (**20**) by the reaction of primary selenoamide (**2**) and in situ generated α-tosylated ketones (**29**) in methanol under reflux. Compound **29** was prepared by the reaction of acetophenones (**27**) and [hydroxy(tosyloxy)iodo] benzene (**28**) in acetonitrile under reflux (Figure 9.13). Acetophenones (**27**) with both electron-donating and electron-withdrawing substituent are well tolerated under this reaction conditions, and excellent yields of products are obtained.

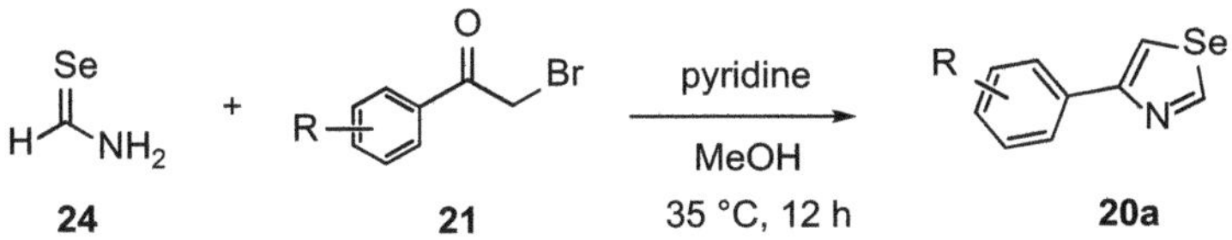

R	(%) yield of **20a**
4-Cl	22
4-F	20
4-MeO	06

Figure 9.10: Synthesis of 4-aryl-1,3-selenazoles using pyridine as catalyst.

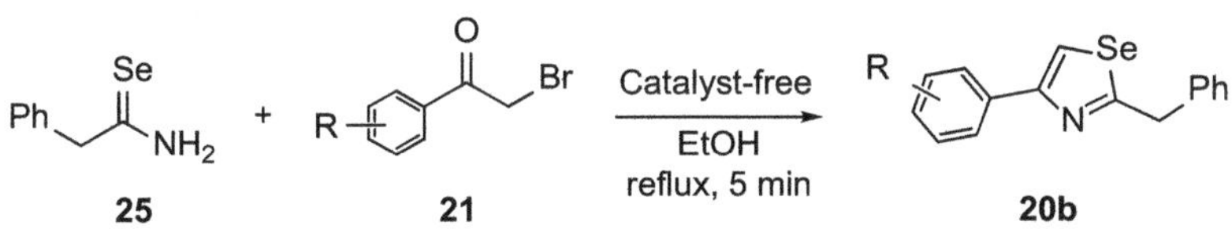

R	(%) yield of **20b**
H	91
4-Me	77
4-Br	100
4-NO_2	88

Figure 9.11: Catalyst-free synthesis of 2-benzyl-1,3-selenazoles in ethanol.

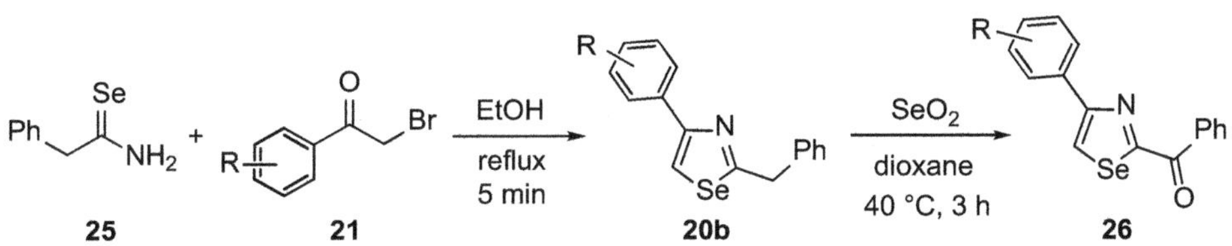

R	% yield of **26**
H	70
4-Me	74
4-Br	80
4-NO_2	70

Figure 9.12: Synthesis of 2-benzoyl-1,3-selenazole derivatives.

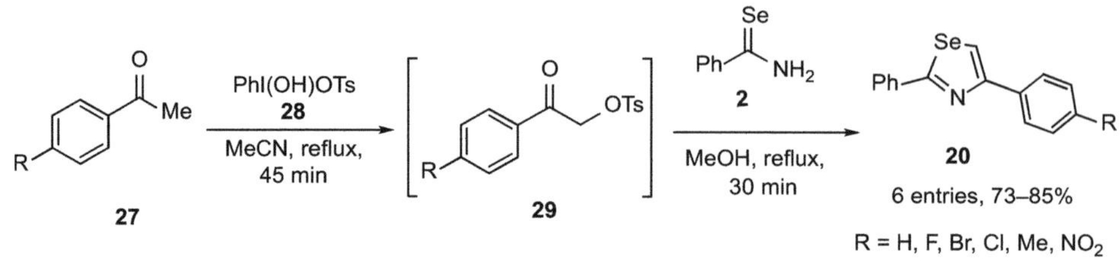

Figure 9.13: Synthesis of 2,4-disubstituted selenazoles starting from acetophenones.

Starting from compound **30**, Pizzo et al. [39] prepared propargyl selenoamides (**31**) by using the Ishihara reagent (LiAlHSeH). Cycloisomerization of this in situ generated propargyl selenoamides (**31**) produced the corresponding product **32**, which on further treatment with piperidine in acetic acid under reflux conditions yielded the desired 2,5-disubstituted selenazoles (**33**) with excellent yields (Figure 9.14).

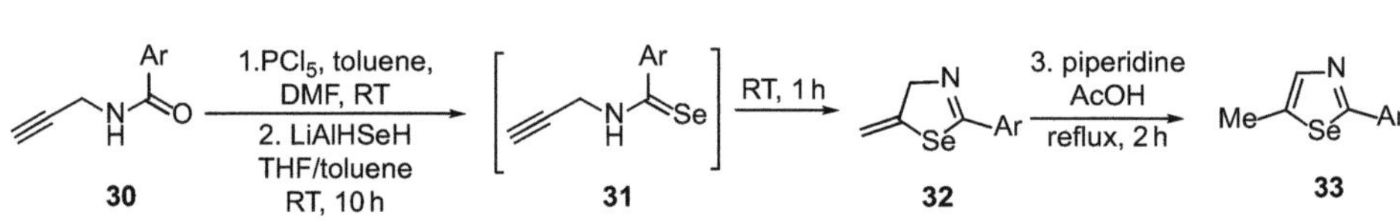

Ar	% yield of **33**
C_6H_5	74
4-BrC_6H_4	70
4-$CF_3C_6H_4$	55

Ar	% yield of **33**
4-MeOC_6H_4	88
4-CH$(Me)_2C_6H_4$	61
3-ClC_6H_4	42

Figure 9.14: Synthesis of 2,5-disubstituted selenazoles via the cycloisomerization of propargyl selenoamides.

5-Amino-2-selenazolines (**36**) were prepared by the reactions of *N*,*N*-dimethylselenoformamide (**6a**) and selenoamide dianions (**35**), which were generated in situ from the secondary selenoamides (**34**) using BuLi (Figure 9.15) [40]. On further treatment with molecular iodine, 5-amino-2-selenazolines (**36**) were oxidized to 5-aminoselenazoles (**37**) with moderate to good yields (Figure 9.15).

Attanasi et al. [41] studied the reactions of 1,2-diaza-1,3-butadienes (**38**) and selenobenzamide (**2**) in dichloromethane at 0 °C, producing two isomers of 2-selenazoline derivatives (**39** and **39a**). By following *anti*-elimination pathway, specifically the stereoisomer **39a** underwent further aromatization by the removal of one molecule of substituted hydrazine under basic conditions to generate fully functionalized selenazoles (**37**) (Figure 9.16).

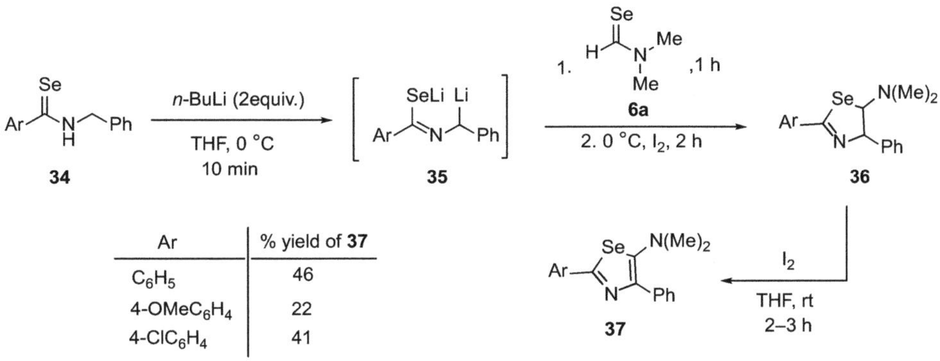

Ar	% yield of **37**
C_6H_5	46
4-OMeC_6H_4	22
4-ClC_6H_4	41

Figure 9.15: Synthesis of 5-aminoselenazoles from secondary selenoamides.

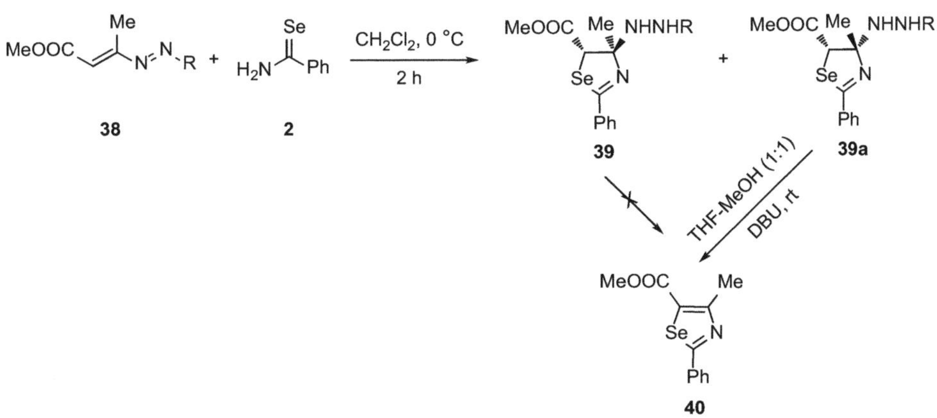

Figure 9.16: Synthesis of fully functionalized selenazoles in dichloromethane.

9.2.5 Synthesis of 2-aryl-4,5-dihydroselenazolo[4,5-*f*]quinoline

Shafiee et al. [42] reported the reactions of primary selenamides (**2**) and neutralized 6-bromo-7,8-dihydroquinolin-5(6*H*)-one (**41**) in acetone as solvent under reflux conditions, which afforded the corresponding 2-aryl-4,5-dihydroselenazolo[4,5-*f*]quinoline (**42**) (Figure 9.17).

9.2.6 Synthesis of 1,3-selenazol-4-ones

A simple and facile method was developed by Koketsu et al. [43] for the synthesis of 2-aryl-5-methoxycarbonylmethylene-4,5-dihydro-1,3-selenazol-4-ones (**45**) by the reaction of primary selenoamides (**2**) and dimethyl acetylenedicarboxylate (**43**) in

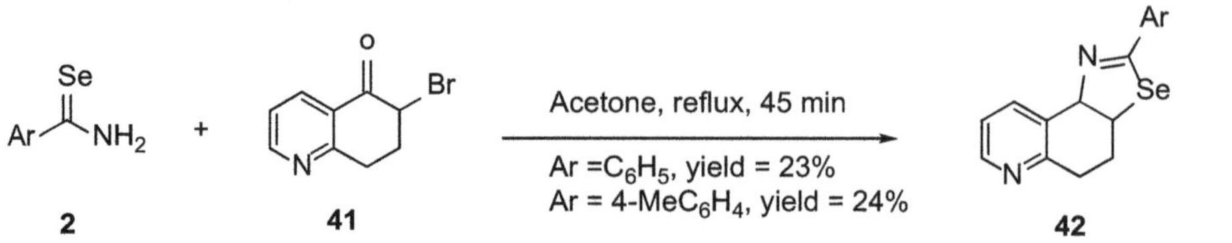

Figure 9.17: Synthesis of 2-aryl-4,5-dihydroselenazolo [4,5-*f*] quinoline using primary selenamides.

ethanol at room temperature. Under the same reaction conditions when acetylenedicarboxylic acid (**44**) was employed instead of dimethyl acetylenedicarboxylate (**43**), 2-aryl-5-carboxymethylene-4-ethoxy-4,5-dihydro-1,3-selenazol-4-ols (**46**) was obtained with moderate yields (Figure 9.18). 2-Aryl-1,3-selenazol-4-ones (**49**) were also synthesized by the same group [44] by the reaction of primary selenoamides (**2**) and haloacyl halides (**47**) through the formation of **48** and followed by **48a** in the presence of pyridine as base at room temperature (Figure 9.19).

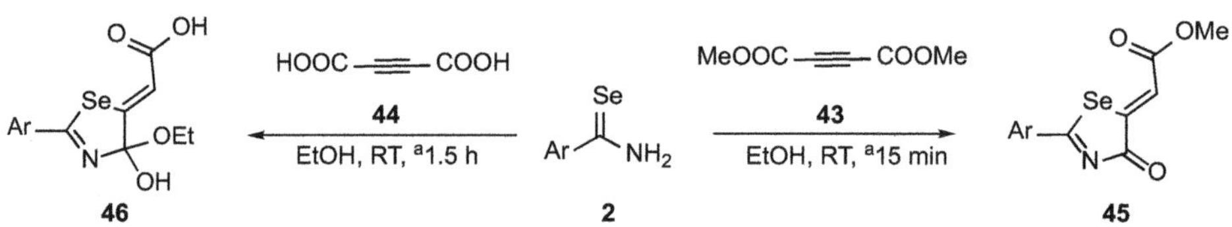

Ar	% yield of **46**
C_6H_5	52
4-MeC_6H_4	53
4-ClC_6H_4	34

Ar	% yield of **45**
C_6H_5	61
4-MeC_6H_4	57
4-OMeC_6H_4	91

[a]Exact temperature was not mentioned

Figure 9.18: Synthesis of 2-aryl-5-methoxycarbonylmethylene-4,5-dihydro-1,3-selenazol-4-ones using selenamides.

9.2.7 Synthesis of 8-methyl-4-selena-6-aza-spiro[2.5]oct-5-en-7-ols

A simple, efficient catalyst-free protocol was developed for the synthesis of 4-selena -6-aza-spiro[2.5]oct-5-en-7-ol (**51**) by the nucleophilic tandem Michael addition of 2-cyclopropylidenepropionaldehyde (**50**) and benzoselenoamides (**2**) in chloroform at 60 °C (Figure 9.20) [45].

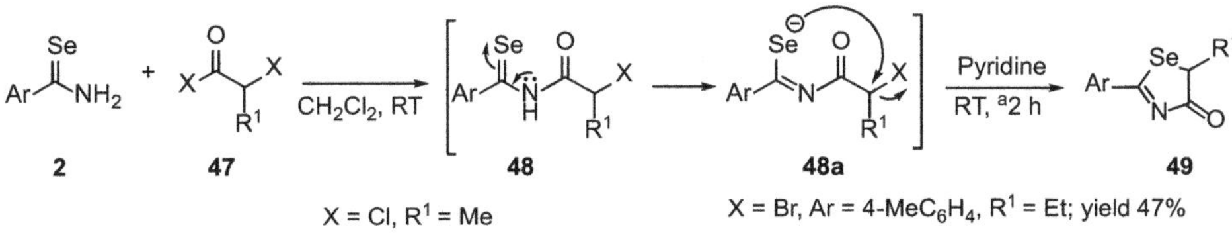

Ar	% yield of **49**
4-MeC_6H_4	74
C_6H_5	66
4-ClC_6H_4	61

[a]Exact temperature was not mentioned

Figure 9.19: Synthesis of 1,3-selenazol-4-one at room temperature.

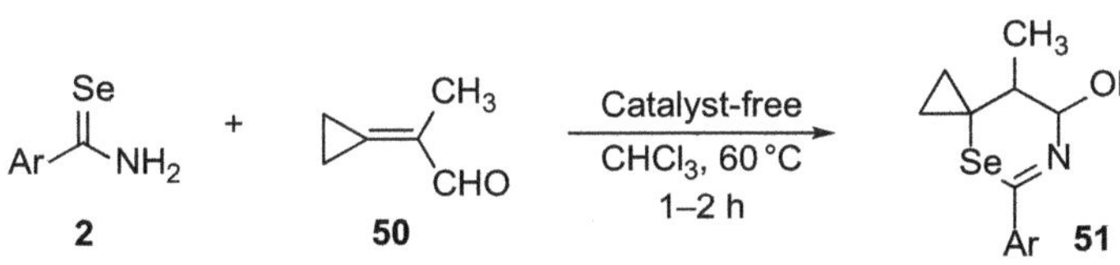

Ar	% yield of **51**
C_6H_5	92
4-BrC_6H_4	84
3-MeC_6H_4	82

Figure 9.20: Catalyst-free synthesis of 8-methyl-4-selena-6-aza-spiro[2.5]oct-5-en-7-ols.

9.2.8 Synthesis of 5-spirocyclopropane-annulated selenazoline-4-carboxylates

A series of 5-spirocyclopropane-annulated selenazoline-4-carboxylates (**53**) was prepared via the Michael addition followed by intramolecular substitution reactions of primary selenoamides (**2**) and 2-bromo-2-cyclopropylideneacetate (**52**) using sodium bicarbonate as a base in acetonitrile at 80 °C (Figure 9.21) [46].

9.2.9 Synthesis of 1,3-selenazine derivatives

1,3-Selenazine scaffolds are found to possess significant antibacterial, antitumor activities [47]. Koketsu et al. [48] developed another efficient protocol for the facile synthesis of 4-alkyl-4-hydroxy-1,3-selenazine derivatives (**55**) by the reactions of primary selenoamides (**2**) and various α,β-unsaturated ketones (**54**) using $BF_3.Et_2O$ as

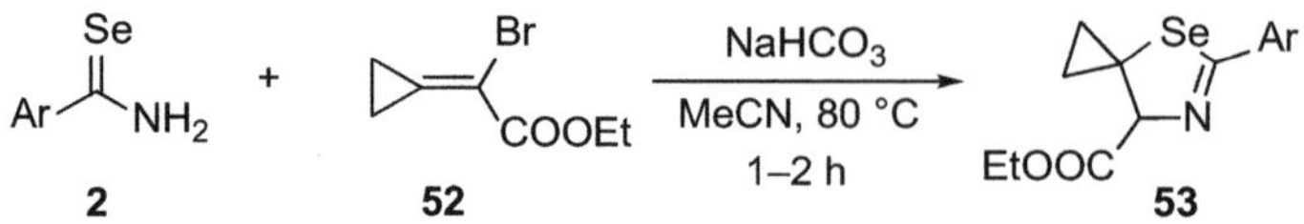

Ar	% yield of **53**	Ar	% yield of **53**
C_6H_5	85	4-MeC_6H_4	79
4-ClC_6H_4	83	3-MeC_6H_4	88
4-BrC_6H_4	75	4-FC_6H_4	84
3-BrC_6H_4	82	3,4-(OCH_2O)C_6H_3	82

Figure 9.21: Synthesis of 5-spirocyclopropane-annulated selenazoline-4-carboxylates under basic medium.

catalyst in dichloromethane at 0 °C (Figure 9.22). Later on, in 2001, the same group also synthesized another series of 4-hydroxy-1,3-selenazine derivatives (**57**) by employing α,β-unsaturated aldehydes (**56**) instead of α,β-unsaturated ketones (**54**) and using the same catalyst in chloroform under reflux (Figure 9.23) [49].

Representative entries

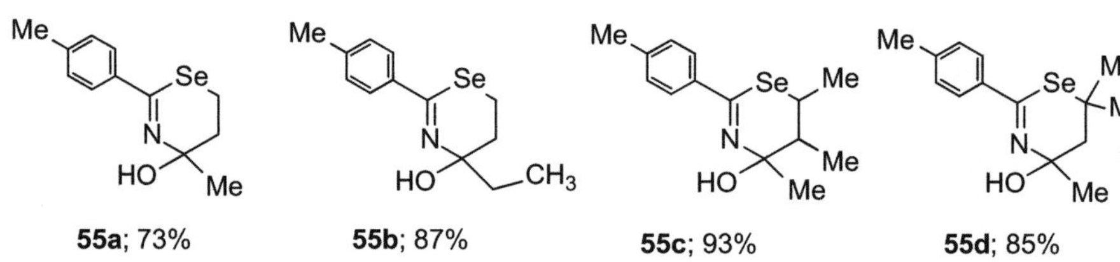

Figure 9.22: $BF_3.Et_2O$-catalyzed synthesis of 4-alkyl-4-hydroxy-1,3-selenazines in dichloromethane.

9.2.10 Synthesis of 3,5-diaryl-1,2,4-selenadiazoles

In 1999, Shafiee et al. [50] demonstrated the synthesis of 3,5-diphenyl-1,2,4-selenadiazole (**59**) by the reaction of selenobenzamide (**2**) and α-bromo-α-phenylsulfonylacetophenone (**58**) in the absence of any catalyst using dry acetone at room

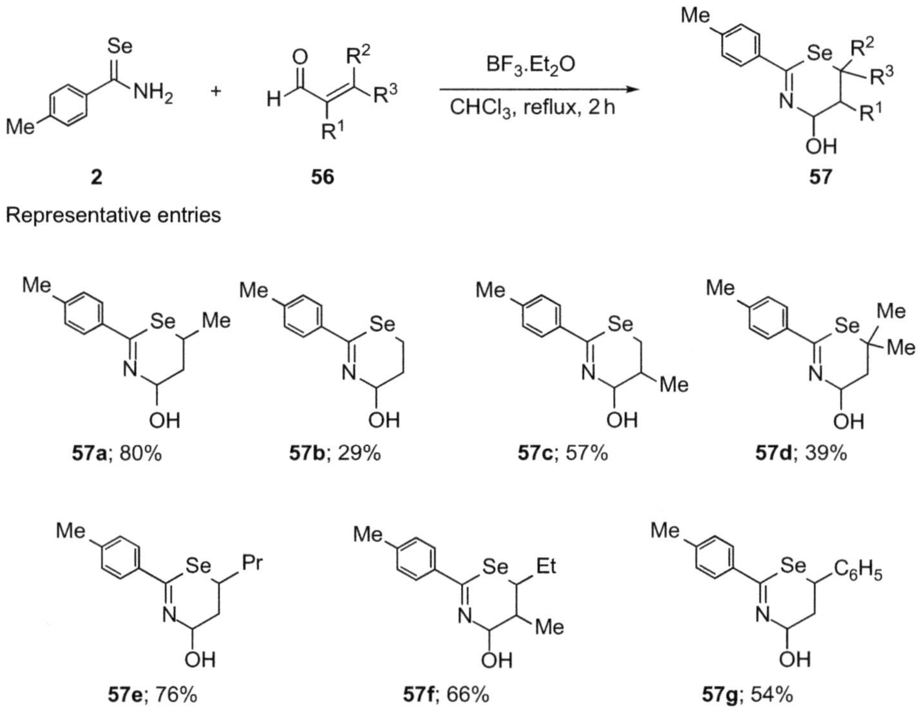

Figure 9.23: $BF_3.Et_2O$-catalyzed synthesis of 4-hydroxy-1,3-selenazines in chloroform.

temperature (Figure 9.24). Later on, in 2003, Huang et al. [51] synthesized a series of 3,5-diaryl-1,2,4-selenadiazole derivatives (**59**) in excellent yields via the dimerization reaction of primary selenoamides (**2**) using poly[-styrene(iodosodiacetate)] as an oxidant in dichloromethane at room temperature (Figure 9.25). After completion of the reaction, the polymer-supported oxidant was recovered by filtration and reused for several runs. Dimerization of benzoselenoamides (**2**) was also achieved by using tert-butyl nitrite to get the desired 1,2,4-selenadiazoles (**59**) in dichloromethane at room temperature (Figure 9.26) [52].

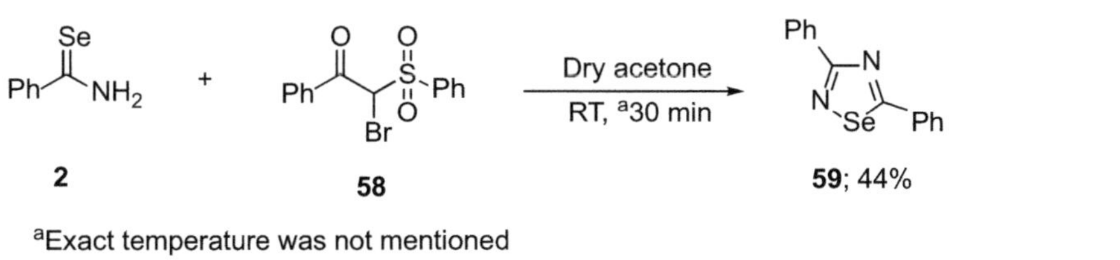

Figure 9.24: Synthesis of 3,5-diphenyl-1,2,4-selenadiazoles at room temperature.

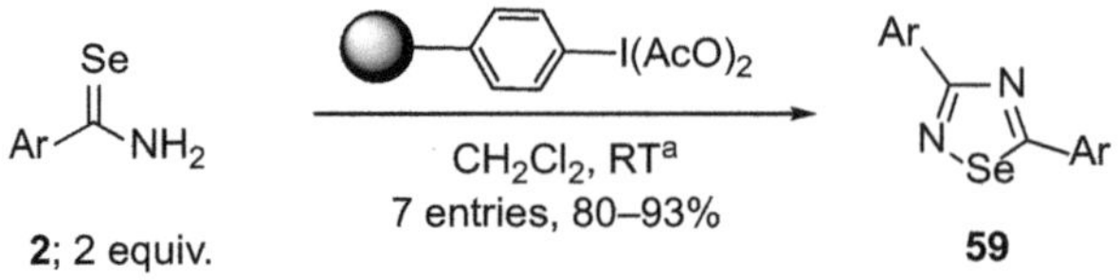

2; 2 equiv. **59**

Ar = C_6H_5, 4-ClC_6H_4, 4-MeC_6H_4, 4-BrC_6H_4, 4-FC_6H_4, 4-$MeOC_6H_4$, 4-$N(Me)_2C_6H_4$

[a]Exact temperature was not mentioned

Figure 9.25: Synthesis of 3,5-diaryl-1,2,4-selenadiazoles via dimerization of primary selenoamides.

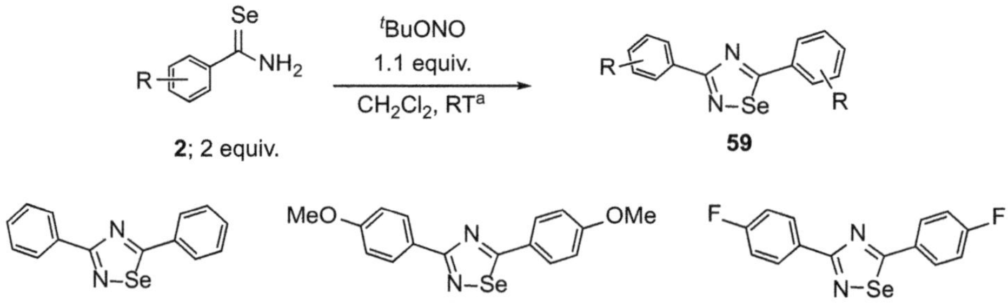

59a; 15 min, 89% **59b**; 15 min, 87% **59c**; 20 min, 80%

[a]Exact temperature was not mentioned

Figure 9.26: Synthesis of 1,2,4-selenadiazoles via the dimerization of benzoselenoamides at room temperature.

9.2.11 Synthesis of 2,5-diaryl-1,3,4-selenodiazoles

A simple, convenient, and catalyst-free protocol was developed for the efficient synthesis of a series of 2,5-diaryl-1,3,4-selenodiazoles (**61**) through the reactions of primary selenoamides (**2**) and hydrazine hydrate (**60**) in methanol at room temperature (Figure 9.27) [53]. The plausible mechanism of this transformation is shown in Figure 9.28.

9.2.12 Synthesis of 6-methyl-2-phenylselenolo[3,4-*d*]selenazole

Shafiee et al. [54] also synthesized 1-(4-methyl-2-phenyl-1,3-selenazol-5-yl)ethanone (**63**) by refluxing selenobenzamide (**2**) and 3-chloropentane-2,4-dione (**62**) mixture in dry acetone. The product **63** was then treated with *N*-bromosuccinamide (NBS) under reflux conditions in carbon tetrachloride, which afforded the corresponding

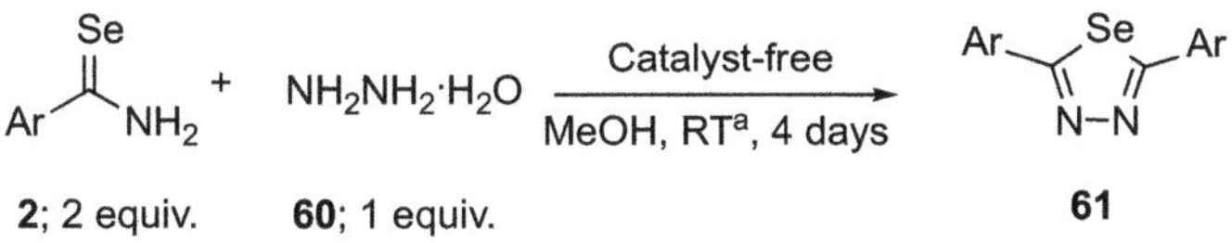

Ar	% yield of **61**	Ar	% yield of **61**
C_6H_5	17	4-MeC_6H_4	23
4-BrC_6H_4	24	4-MeOC_6H_4	35
4-ClC_6H_4	18	3,5-$(MeO)_2C_6H_3$	45
		2-Thienyl	15

[a]Exact temperature was not mentioned

Figure 9.27: Catalyst-free synthesis of 2,5-diaryl-1,3,4-selenodiazoles at room temperature.

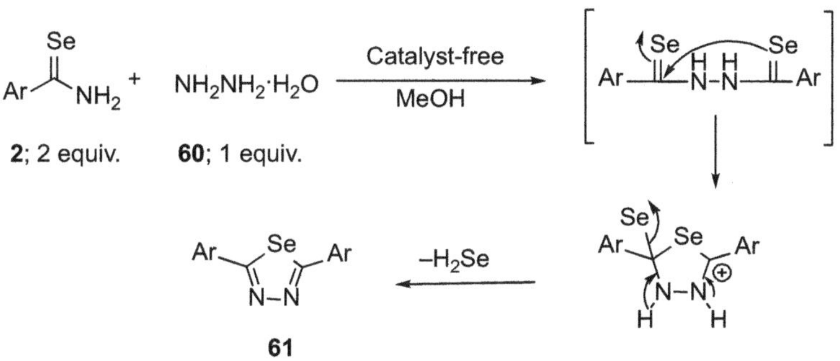

Figure 9.28: Plausible mechanism for the synthesis of 2,5-diaryl-1,3,4-selenodiazoles.

1-(4-(bromomethyl)-2-phenyl-1,3-selenazol-5-yl)ethanone (**64**). Reaction with **64** and *N,N*-diethylselenopropionamide (**65**) yielded the noble product 6-methyl-2-phenylselenolo[3,4-*d*]selenazole (**66**) in ethanol under reflux conditions (Figure 9.29).

9.2.13 Synthesis of diacyl selenides

Diacyl selenides are being used as good acylating agents in organic synthesis [55]. A simple and facile method was reported for the efficient synthesis of diacyl selenides (**68**) with excellent yields by the reactions of primary selenoamides (**2**) and benzoyl chlorides (**67**) in chloroform under nitrogen atmosphere at 60 °C (Figure 9.30) [56].

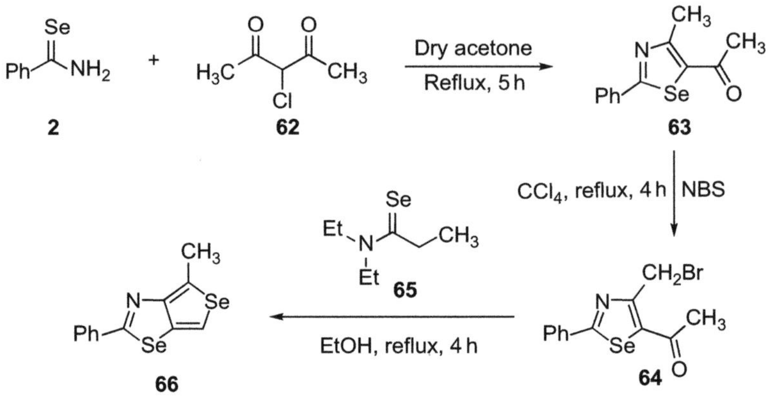

Figure 9.29: Synthesis of 6-methyl-2-phenylselenolo[3,4-*d*]selenazole from selenazoles.

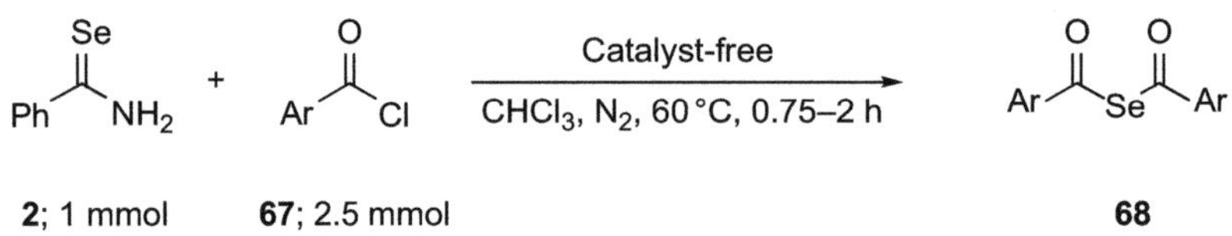

Ar	% yield of **68**
C_6H_5	80
4-$MeOC_6H_4$	85
4-MeC_6H_4	81
4-ClC_6H_4	75
3-ClC_6H_4	69

Figure 9.30: Catalyst-free synthesis of diacyl selenides.

9.2.14 Synthesis of 6*H*-1,3,5-oxaselenazines

Sekiguchi et al. [57] reported an efficient stereoselective method for the synthesis of a series of noble 6*H*-1,3,5-oxaselenazine derivatives (**69**) by the reactions of benzoselenoamides (**2**) and various aldehydes (**10**) in the presence of boron trifluoride etherate as promoter in chloroform at 20 °C (Figure 9.31). The reaction was completed within 1 h using acetaldehyde, whereas other bulky aldehydes required 20 h to complete.

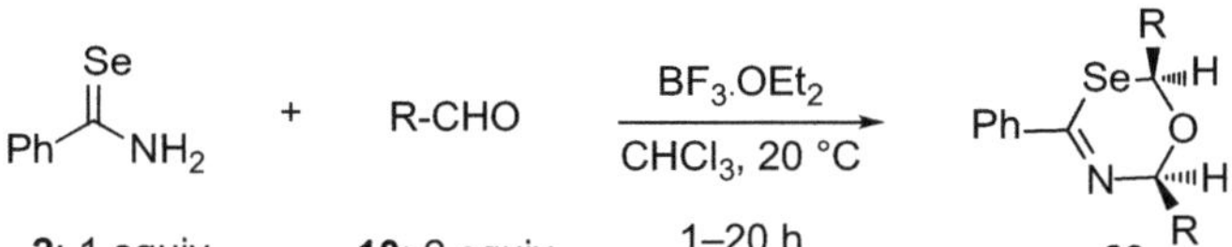

2; 1 equiv. **10**; 2 equiv. **69**

R	% yield of **69**
Me	82
$CH_2(CH_2)_5CH_3$	69
$C(CH_3)_3$	86
Cyclohexyl	97

Figure 9.31: Boron trifluoride etherate promoted synthesis of 6*H*-1,3,5-oxaselenazines.

9.2.15 Synthesis of 6-hydroxy-1,3-selenazin-4-ones

A number of 6-hydroxy-1,3-selenazin-4-ones (**71**) were synthesized starting by the reactions of primary selenoamides (**2**) and bisacyl chloride (**70**) using triethyl amine as base in dichloromethane at 0 °C (Figure 9.32) [58].

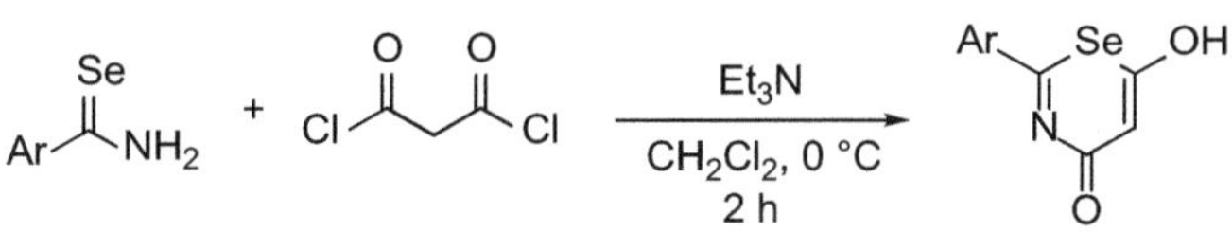

2 **70** **71**

Ar	% yield of **71**
4-MeC_6H_4	63
C_6H_5	63
4-$MeOC_6H_4$	72
4-ClC_6H_4	55

Figure 9.32: Synthesis of 6-hydroxy-1,3-selenazin-4-ones in dichloromethane.

9.2.16 Synthesis of symmetrical diaryldiselenides

Organic diselenides were found to possess various pharmacological efficacies, including anticancer [59], antiulcer [60], anti-inflammatory [61], and antioxidant [62] activities. A facile and efficient method was developed for the synthesis of symmetrical diaryldiselenides (**73**) from the reactions of various aryl halides (**72**) and *N*,*N*-diethyl selenoformamide (**6b**) as a Se source in the presence of catalytic amount of nano-copper(II) oxide in DMF–water mixture as solvent at 110 °C (Figure 9.33) [63].

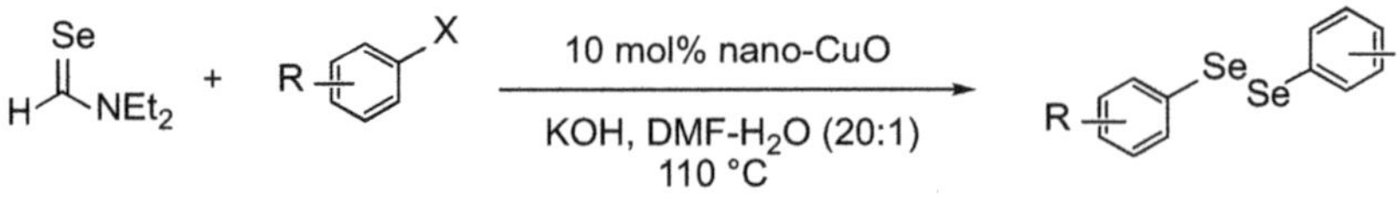

6b **72** **73**

X = I

R	Time (h)	% yield of **73**
2-Me	12	83
4-OMe	24	75
2-OMe	24	67
4-Br	4.5	81

R = H

X	Time (h)	% yield of **73**
I	3	91
Br	5	79
OTs	5	71
Cl	7	54

X = Br

R	Time (h)	% yield of **73**
4-Me	12	79
4-Cl	6.5	72

Figure 9.33: Nano-copper(II) oxide catalyzed synthesis of diaryldiselenides.

9.2.17 Synthesis of 5,6-dihydro-4*H*-1,3-selenazine

Reactions of selenobenzamide (**2**) and *N*,*N*-dimethylformamide dimethyl acetal (**74**) in dichloromethane generated *N*-selenoacylamidine (**75**) in situ, which behaves as a 4π heterodienic systems in [4 + 2] cycloaddition reactions with excess of methyl acrylate (**76**) to afford the corresponding 5,6-dihydro-4*H*-1,3-selenazine (**77**) with excellent yield at 40 °C (Figure 9.34) [64].

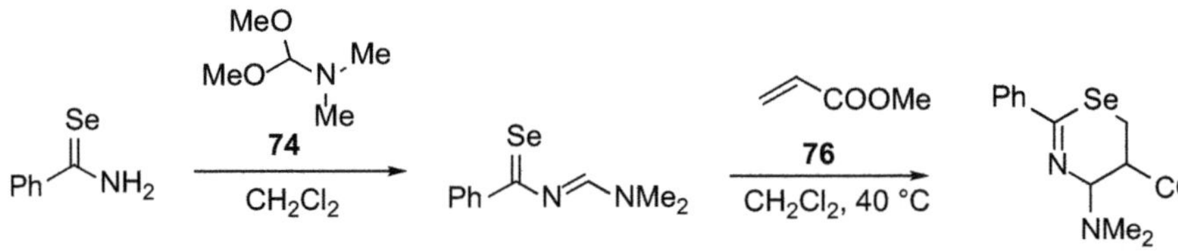

2 **75** **77**, 91%

Figure 9.34: Synthesis of 5,6-dihydro-4*H*-1,3-selenazine in dichloromethane.

9.2.18 Synthesis of selenophenes

N-Selenoacylamidine (**75**), generated from selenoamide (**2**), undergoes Diels–Alder reaction with 4.,4-diethoxy-2-butyn-1-a1 (**78**) affording 4*H*-1,3-selenazine (**79**), which under reflux conditions in tetrahydrofuran produced the corresponding selenabutadiene (**80**) with 65% yield. In this step, benzonitrile produced as a byproduct. Later on, oxidative cyclization of **80** under reflux conditions in ethanol generated the desired selenophene (**81**) with 50% yield (Figure 9.35) [65].

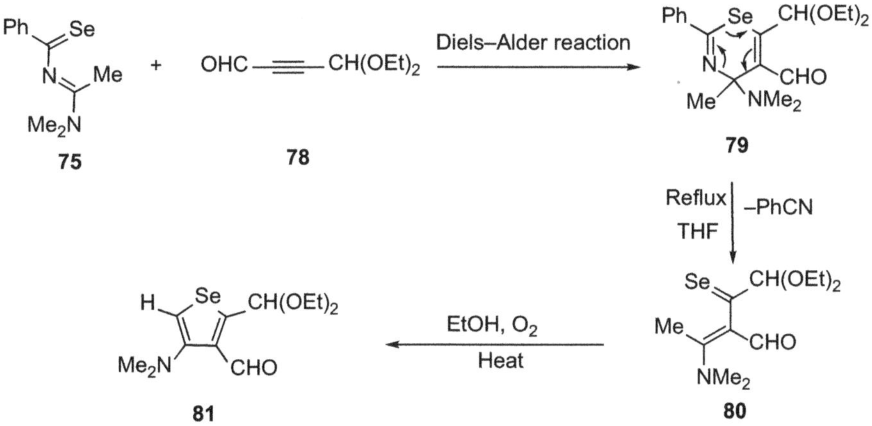

Figure 9.35: Synthesis of selenophenes starting from *N*-selenoacylamidine derivatives.

9.3 Synthesis of organoselenium scaffolds involving selenazadienes

9.3.1 Synthesis of 2-amino-1,3-selenazoles

Koketsu and his group successfully employed selenazadienes (**82**) for the synthesis of various selenium containing scaffolds. In 2005, they demonstrated the base-catalyzed reactions between selenazadienes (**82**) and chloroacetonitrile (**83**), which yielded the corresponding 2-amino-5-cyano-1,3-selenazoles (**86**) in moderate to high yields through the formation of intermediate compounds **84** and **85** in acetonitrile under reflux conditions (Figure 9.36) [66]. They further carried out the reactions of selenazadienes (**82**) and chloroacetyl chloride (**87**) either in tetrahydrofuran or in acetonitrile as solvent under reflux conditions, which produced the corresponding acyl chloride intermediates (**90**) through the formation of **88** followed by **89** (Figure 9.36). Washing of the resulting acyl chloride intermediates (**90**) with water afforded the corresponding 2-amino-1,3-selenazole-5-carboxylic acid (**91**) (Figure 9.37) [67]. When the same intermediate **90** was treated with various alcohols (**92**), it produced a series of 2-amino-1,3-selenazole-5-carboxylate derivatives (**93**) with good yields (Figure 9.37) [67]. Reactions of various amines (**94**) with the intermediate **90** yielded the

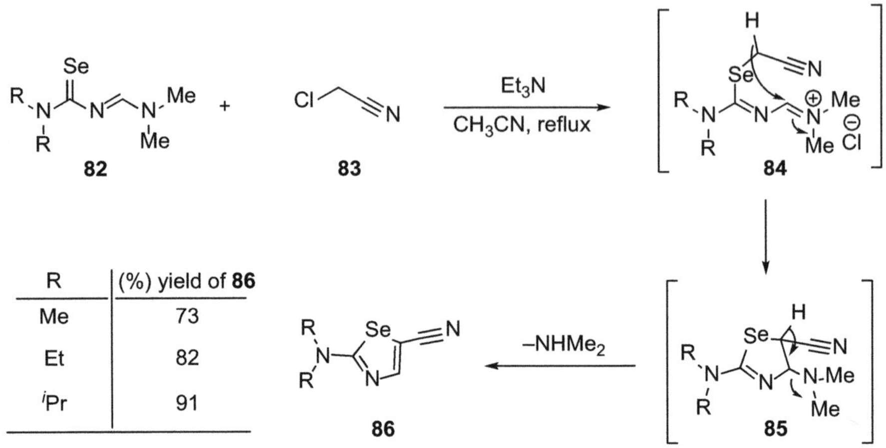

R	(%) yield of **86**
Me	73
Et	82
^{i}Pr	91

Figure 9.36: Synthesis of 2-amino-5-cyano-1,3-selenazoles in acetonitrile under reflux conditions.

corresponding 2-amino-1,3-selenazole-5-carboxamides (**95**) (Figure 9.37) [66]. In 2006, by using selenazadienes (**82**), the same group also synthesized 5-acyl-2-amino -1,3-selenazoles (**96**) in high yields by the reactions of α-haloketones (**67**) in methanol under reflux conditions (Figure 9.38) [68].

9.3.2 Synthesis of 1,3-selenazol-4-one

Koketsu et al. [69] also reported a simple and efficient method for the synthesis of 1,3-selenazol-4-one derivatives (**97**) via hetero-Diels–Alder reactions of selenazadiene (**82**) and dimethyl acetylenedicarboxylate (**43**) in methanol at 0 °C (Figure 9.39). The mechanism of this conversation is shown in Figure 9.40.

9.4 Synthesis of organoselenium scaffolds involving selenocarbonyls

9.4.1 Synthesis of Diels–Alder adducts

Selenocarbonyls are generally ignored in organic synthesis probably due to their less stability. Segi et al. [70] successfully synthesized benzoselenaldehyde (**100**) starting from acetal derivatives (**98**) and bis(dimethylaluminum)selenide [$(Me_2Al)_2Se$] (**99**) as a selenylating agent. The synthesized benzoselenaldehyde (**100**) was then employed for the Diels–Alder reaction with various dienes such as cyclopentadiene

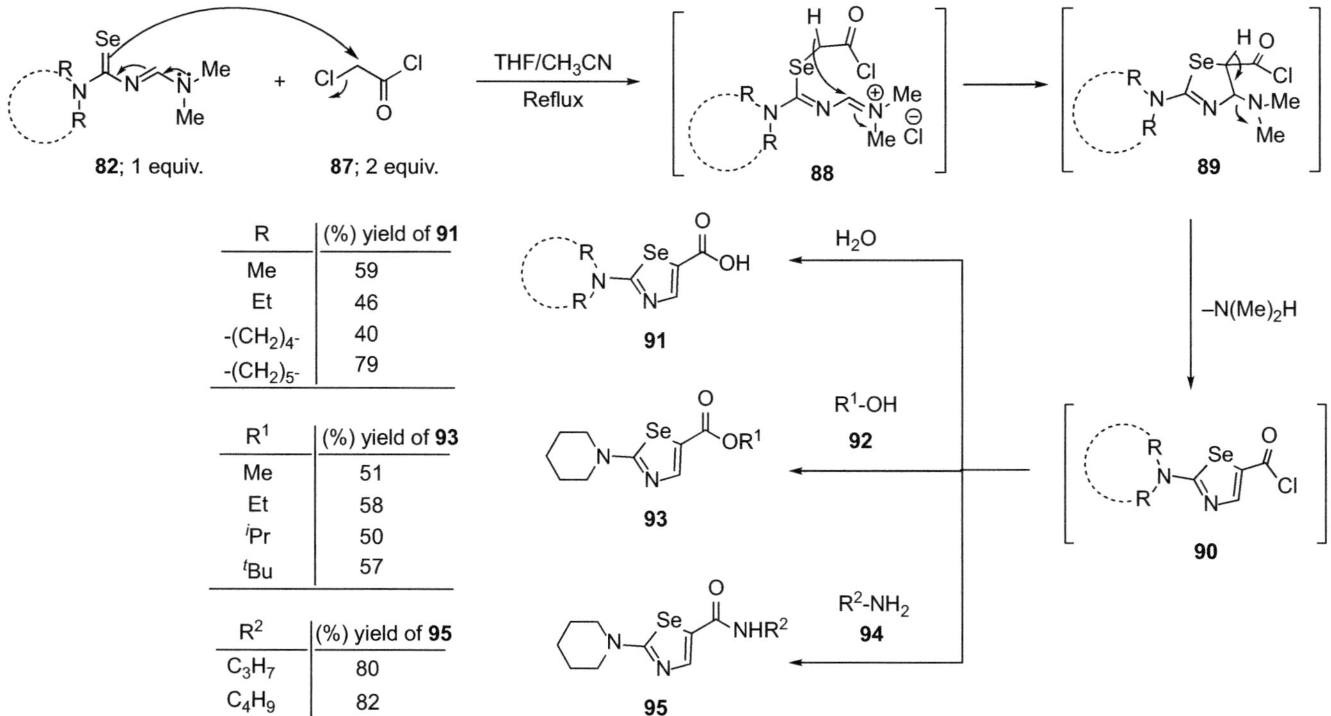

R	(%) yield of **91**
Me	59
Et	46
-$(CH_2)_4$-	40
-$(CH_2)_5$-	79

R^1	(%) yield of **93**
Me	51
Et	58
^{i}Pr	50
^{t}Bu	57

R^2	(%) yield of **95**
C_3H_7	80
C_4H_9	82
$C_6H_5CH_2$	75

Figure 9.37: Synthesis of 2-amino-1,3-selenazole-5-carboxy/carboxylate/carboxamide derivatives.

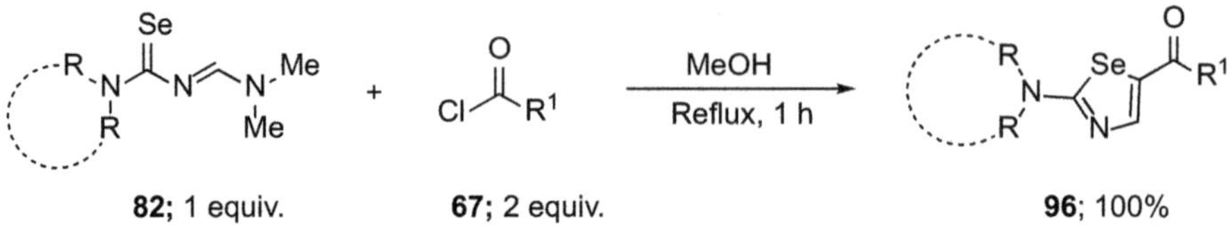

R	R¹	% yield of **96**
Me	Me	98
-(CH$_2$)$_5$-	Me	93
Me	C$_6$H$_5$	100
-(CH$_2$)$_5$-	C$_6$H$_5$	100

Figure 9.38: Synthesis of 5-acyl-2-amino-1,3-selenazoles in methanol under reflux conditions.

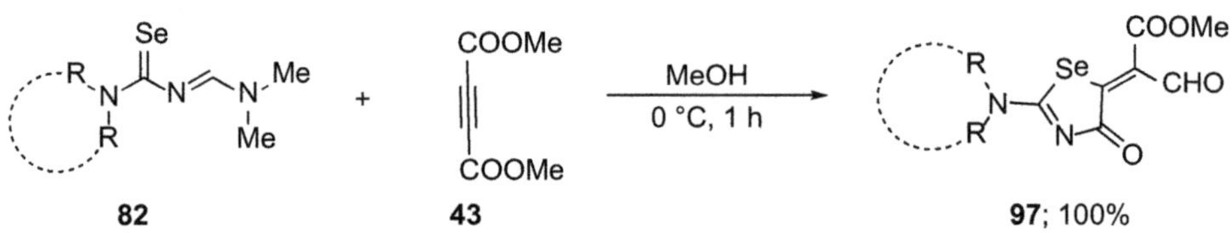

R	% yield of **97**
Me	70
Et	80
-C$_4$H$_8$-	81
-C$_5$H$_{10}$-	84

Figure 9.39: Synthesis of 1,3-selenazol-4-one derivatives in methanol.

(**101**) or 2,3-dimethyl-1,3-butadiene (**102**) at 100 °C, which yielded the corresponding adducts **103** or **104**, respectively, with excellent yields (Figure 9.41). By following the retro-Diels–Alder reaction of the corresponding cycloadducts (**103**), a series of selenoaldehydes (**100**) was formed in situ, which on further reactions with *trans*-1-acetoxy-1,3-butadiene (**105**) at 110 °C in toluene gave another series of 3,6-dihydro-2*H*-selenopyran derivatives (**104a**) (Figure 9.42) [71]. In other reports, the same group showed that the in situ generated selenoaldehydes (**100**) can be trapped by anthracene (**106**) to produce the corresponding [4 + 2] cycloaddition adducts (**107**) at 100 °C (Figure 9.43) [72–74]. This time selenoaldehydes (**100**) were generated from the reactions of aromatic aldehydes (**10**) and bis(dimethylaluminum)selenide [(Me$_2$Al)$_2$Se] (**99**).

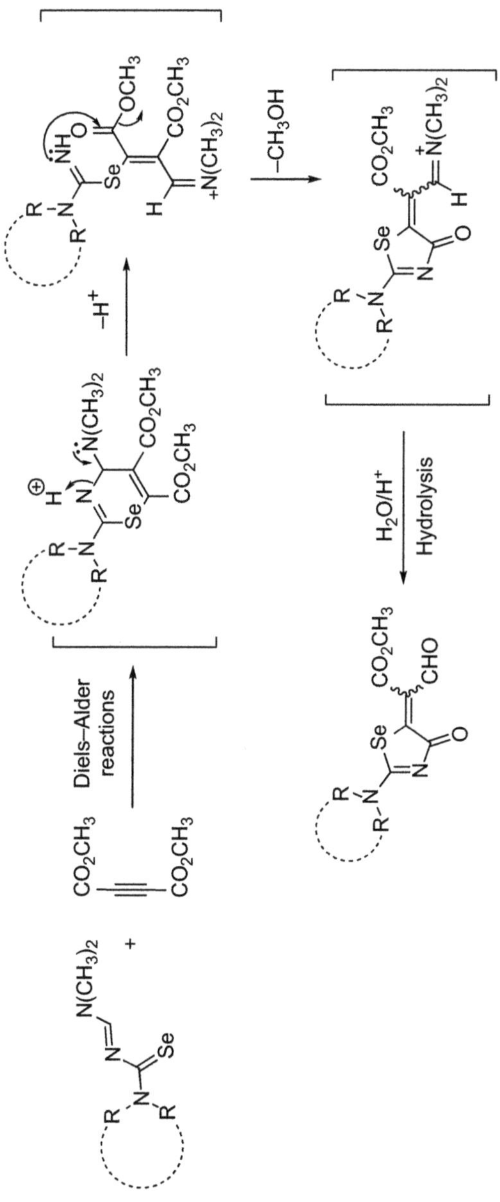

Figure 9.40: Plausible mechanism for the synthesis of 1,3-selenazol-4-ones.

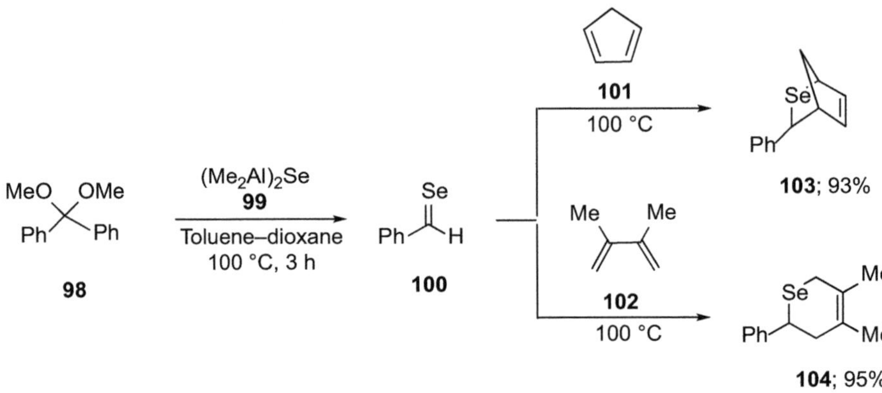

Figure 9.41: Synthesis of Diels–Alder reaction adducts involving selenoaldehydes.

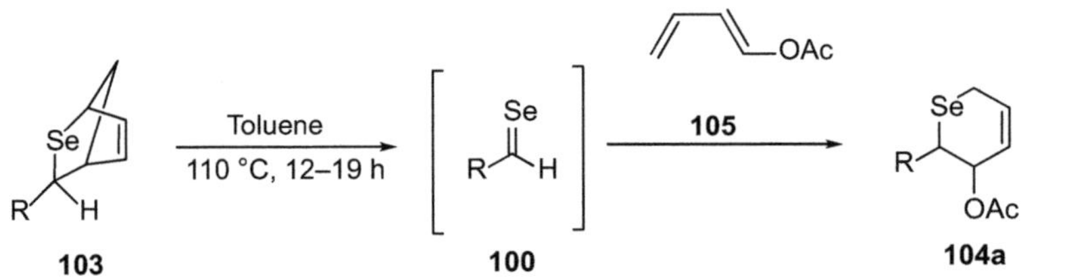

R	% yield of **104a**
C_6H_5	90
4-$OMeC_6H_4$	97
4-$CF_3C_6H_4$	80
COOMe	35

Figure 9.42: Synthesis of 3,6-dihydro-2*H*-selenopyran derivatives via Diels–Alder reaction.

Segi et al. [74] regenerated selenoaldehydes (**100**) by the thermolysis of anthracene cycloadducts (**107**) following retro-Diels–Alder reaction pathway. [3 + 2] Cycloaddition reactions of 2,4,6-trimethylbenzonitrile *N*-oxide (**108**) with this in situ generated benzoselenaldehyde (**100**) produced the desired 1,4,2-oxaselenazole (**109**) with good yield (Figure 9.44). The same group [75] also synthesized a series of 3-aryl-2-selena-7-phosphabicyclo[2.2.1]hept-5-ene-7-selenide derivatives (**111**) via the [4 + 2] cycloaddition reactions of 3,4-dimethylphosphole selenides (**110**) and selenoaldehydes (**100**), generated in situ by retro Diels–Alder reaction of anthracene cycloadducts (**107**) under thermal conditions (Figure 9.45). Considering the steric effect, it was proposed that the *endo* addition of selenoaldehyde was predominant over the *exo* addition, and as a result, **111a** was not formed, only one diastereomer (**111**) was obtained solely

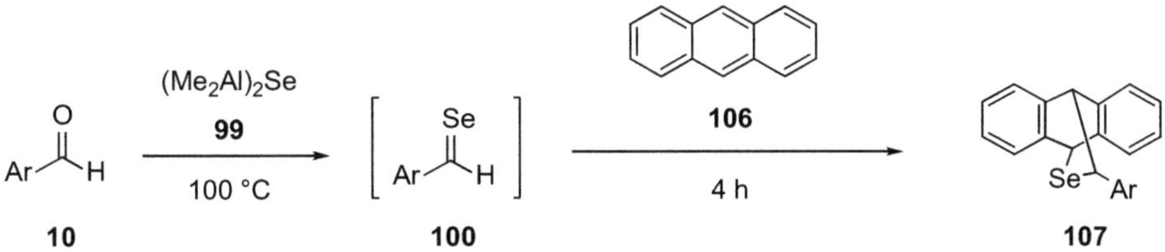

Ar	% yield of **107**
C_6H_5	85
4-F-C_6H_4	91
4-$CF_3C_6H_4$	96
4-OMeC_6H_4	66
4-SMeC_6H_4	73
4-CNC_6H_4	90

Figure 9.43: Synthesis of Diels–Alder reaction adducts involving selenoaldehydes and anthracene.

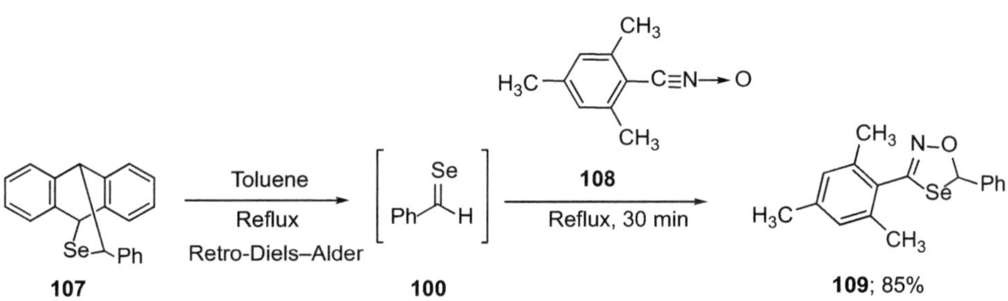

Figure 9.44: Formation of selenoaldehydes via retro-Diels–Alder reaction of anthracene cycloadducts.

(Figure 9.46). The same group also reported the reactions of selenoaldehydes (**100**), generated in situ by retro-Diels–Alder reaction of anthracene cycloadducts (**107**) and 2-methoxyfuran (**112**), which surprisingly produced penta-2,4-dienoate derivatives (**113**) with good yields by the removal of elementary selenium in toluene at 160 °C (Figure 9.47) [73]. The same in situ generated selenoaldehydes (**100**) when reacted with 5-ethoxyoxazoles (**114**) gave the corresponding 3-selenazoline derivatives (**115**) with good yields via the [4 + 2] cycloaddition reaction involving carbon–selenium bond breakage followed by successive recyclization (Figure 9.48) [72].

Selenoaldehydes (**100**) were also generated using bis(trimethylsilyl)selenide as a selenylating agent from various aromatic aldehydes (**10**) in the presence of a catalytic amount of cobalt chloride. The in situ generated so formed selenoaldehydes (**100**) were then trapped by 2,3-dimethyl-1,3-butadiene (**102**), which produced the corresponding Diels–Alder adducts (**104**) in acetonitrile at room temperature (Figure 9.49) [76].

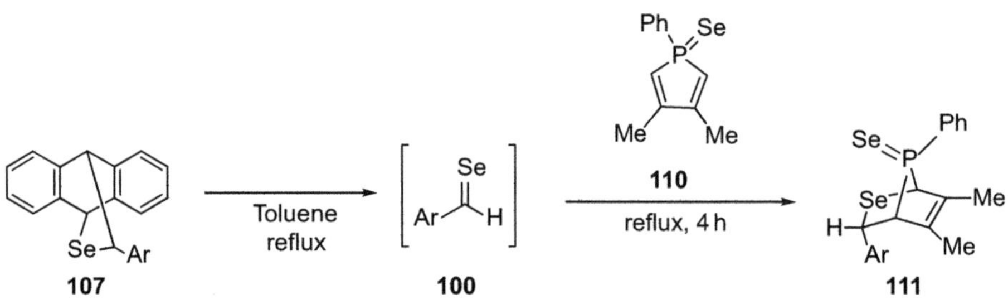

Figure 9.45: Synthesis of 3-aryl-2-selena-7-phosphabicyclo[2.2.1]hept-5-ene-7-selenide.

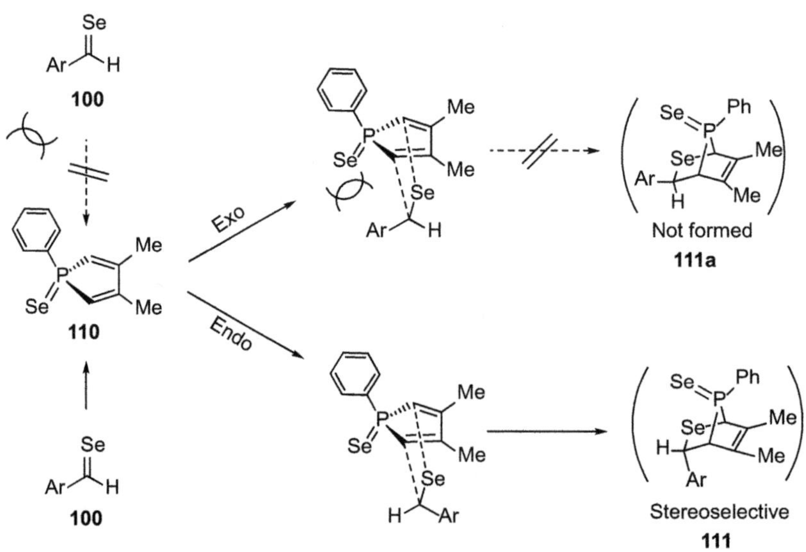

Figure 9.46: Plausible mechanism for the synthesis of 111.

In 1991, Abelman [77] prepared diethyl selenoxomalonates (**100a**) from reactions of diethyl chloromalonate (**116**) and elemental selenium in the presence of cesium carbonate as base. The selenocarbonyls generated in situ were then trapped with 2,3-dimethyl-1,3-butadiene (**102**) or anthracene (**106**) to produce the corresponding Diels–Alder adducts **104b** or **107a**, respectively, with excellent yields (Figure 9.50).

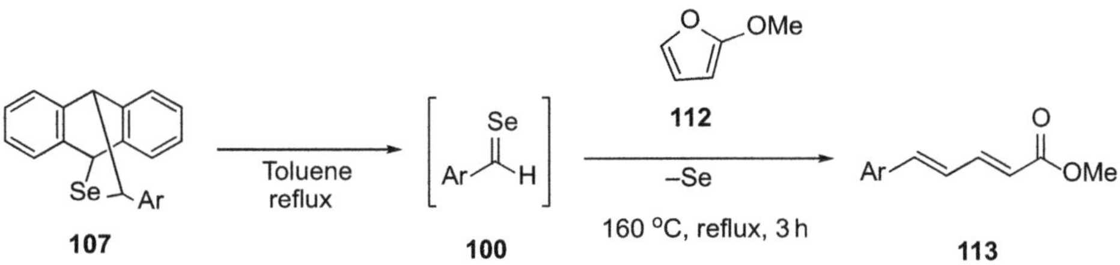

Ar	% yield of **113**
C_6H_5	87
4-OMeC_6H_4	98
4-$CF_3C_6H_4$	91

Figure 9.47: Synthesis of penta-2,4-dienoate derivatives in toluene.

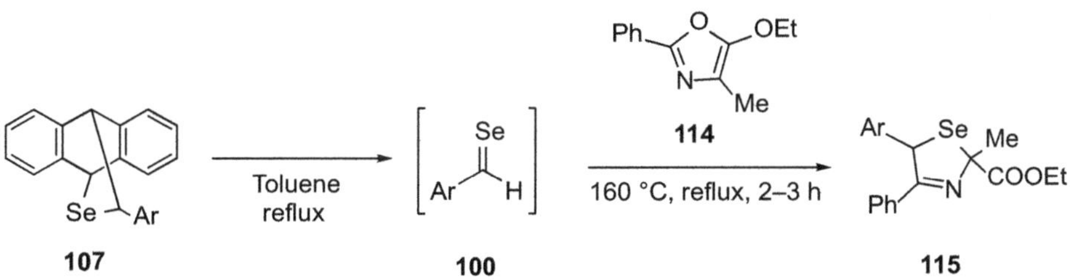

Ar	% yield of **115**
C_6H_5	95
4-OMeC_6H_4	91
4-CNC_6H_4	88

Figure 9.48: Synthesis of 3-selenazoline from in situ generated selenoaldehydes.

9.4.2 Synthesis of 2,4,6-trisubstituted 1,3,5-triselenanes

Takikawa et al. [78, 79] synthesized a new series of 2,4,6-trisubstituted 1,3,5-triselenane derivatives (**117**) through the formation of selenoaldehydes (**100**), generated in situ from the reactions of aromatic aldehydes (**10**) with bis(trimethylsilyl)selenide using $BF_3.OEt_2$ as catalyst in dichloromethane at 0 °C (Figure 9.51).

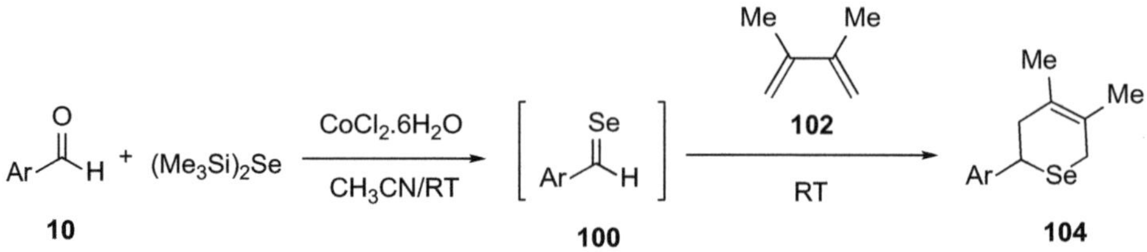

Ar	% yield of **104**
C_6H_5	45
4-F-C_6H_4	51
4-$CF_3C_6H_4$	53
4-$OMeC_6H_4$	41

Figure 9.49: Formation selenoaldehydes using bis(trimethylsilyl)selenide as a selenylating agent.

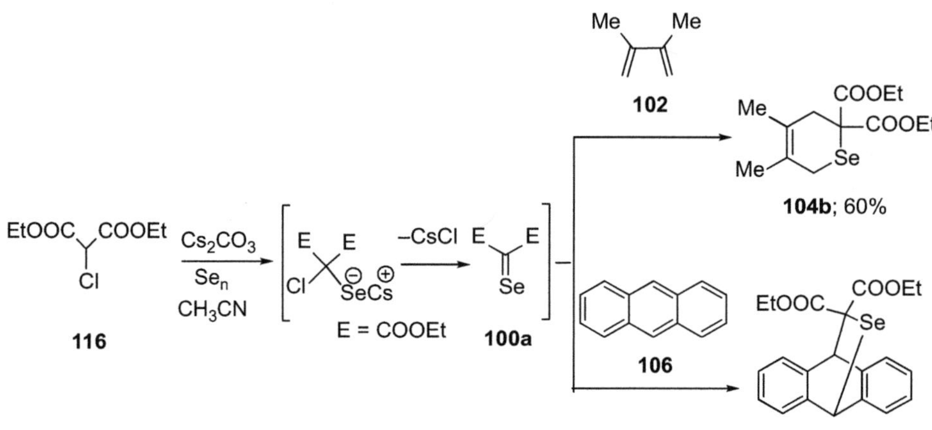

107a; 70%

Figure 9.50: Synthesis of Diels–Alder adducts involving selenocarbonyls generated from diethyl chloromalonate.

9.4.3 Synthesis of *trans*-olefins using in situ generated selenocarbonyls

Okuma et al. [79] reported the synthesis of selenoaldehydes (**100**) from the reactions of elemental selenium and phosphonium ylides (**118**) through the formation of intermediate **119** (Figure 9.52). Selenoaldehydes (**100**) so formed were then further reacted with the phosphonium ylides (**118**) to produce the corresponding *trans* olefins (**121**) via the formation of Wittig intermediate (**120**) in toluene under reflux conditions (Figure 9.53).

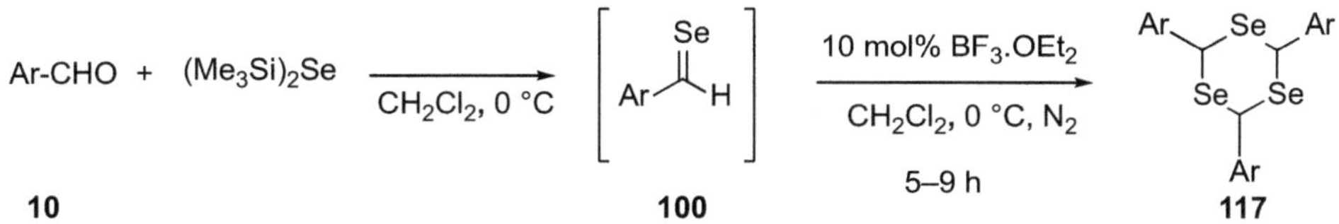

Ar	% yield of **117**
C_6H_5	79
$4\text{-}MeC_6H_4$	48
$4\text{-}OMeC_6H_4$	91

Figure 9.51: $BF_3.OEt_2$-catalyzed synthesis of 2,4,6-trisubstituted 1,3,5-triselenanes.

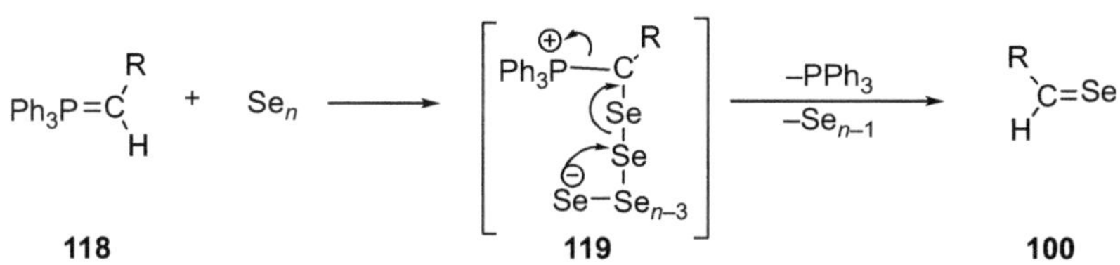

Figure 9.52: Synthesis of selenocarbonyls from elemental selenium and phosphonium ylides.

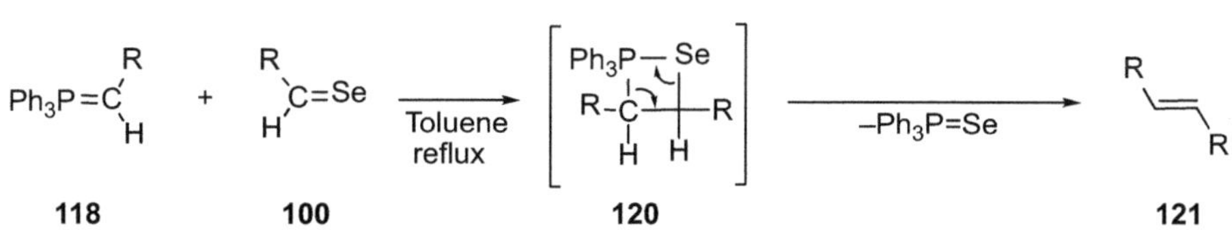

R	% yield of **121**
COOMe	74
COOEt	97
CN	52
COMe	56

Figure 9.53: Synthesis of *trans*-olefins using in situ generated selenocarbonyls via Wittig reaction.

9.5 Conclusions

Organoselenium compounds are found to possess a wide range of biological activities such as anticancer, antimelanogenesis, anti-inflammatory, and neuroprotective activities. As a result, the last decade has seen tremendous outburst to design new protocols for the synthesis of various biologically promising organoselenium scaffolds under various reaction conditions. Under this direction, among other key precursors, recently selenoamides, selenazadienes, and selenocarbonyls have gained considerable attention. This chapter summarizes the applications of selenoamide, selenazadiene, and selenocarbonyl scaffolds as the key precursors for the synthesis of various organoselenium scaffolds such as diaryldiselenides, selenazoles, selenazolo [4,5-*f*]quinolines, selenazolones, selenazines, selenadiazoles, phenylselenolo[3,4-*d*] selenazole, oxaselenazines, selenazinones, selenophenes, and 1,3,5-triselenanes.

Acknowledgments: The first author thanks the Indus International University as well as Kartha Education Society, Mumbai, India, for their kind support. B C Ranu thanks INSA, New Delhi for the award of INSA Honorary Scientist.

References

[1] Garud, DR., Koketsu, M., and Ishihara, H. Isoselenocyanates: A powerful tool for the synthesis of selenium-containing heterocycles. Molecules 2007, 12, 504–535.
[2] Quin, CF., Galeas, ML., Freeman, JL., and Pilon-Smits, EAH. Selenium: Deterrence, toxicity, and adaptation. Integr Environ Assess Manag 2007, 3, 460–462.
[3] Naithani, R. Organoselenium compounds in cancer chemoprevention. Mini-Rev Med Chem 2008, 8, 657–668.
[4] Wessjohann, LA., Schneider, A., Abbas, M., and Brandt, W. Selenium in chemistry and biochemistry in comparison to sulfur. Biol Chem 2007, 388, 997–1006.
[5] Ninomiya, M., Garud, DR., and Koketsu, M. Biologically significant selenium-containing heterocyclic. Coord Chem Rev 2011, 255, 2968–2990.
[6] Wu, W., Murakami, K., Koketsu, M., Yamada, Y., and Sakai, I. Induction of apoptosis in human gastric adenocarcinoma cells by two novel organoselenium compounds, TS-2 and TS-6. Anticancer Res 1999, 19, 5375–5381.
[7] El-Bayoumy, K., and Sinha, R. Mechanisms of mammary cancer chemoprevention by organoselenium compounds. Mutat Res 2004, 551, 181–197.
[8] Patrick, L. Selenium biochemistry and cancer: A review of the literature. Altern Med Chem 2004, 9, 239–259.
[9] Ahn, HJ., Koketsu, M., Yang, EM., Kim, YM., Ishihara, H., and Yang, HO. 2-(4-Methylphenyl) -1,3-selenazol-4-one induces apoptosis by different mechanisms in SKOV3 and HL 60 cells. J Cell Biochem 2006, 99, 807–815.
[10] Nam, KN., Koketsu, M., and Lee, EH. 5-Chloroacetyl-2-amino-1,3-selenazoles attenuate microglial inflammatory responses through NF-kappaB inhibition. Eur J Pharmacol 2008, 589, 53–57.
[11] Choi, SY., Jo, YO., Koketsu, M., Ishihara, H., Kim, SH., and Kim, SY. Inhibitory effects of 2-(4-chlorophenyl)-1,3-selenazol-4-one on lipopolysaccharide-induced nitric oxide production in RAW 264.7 cells. J Korean Soc Appl Biol Chem 2009, 52, 371–374.
[12] Koketsu, M., Choi, SY., Ishihara, H., Lim, BO., Kim, H., and Kim, SY. Inhibitory effects of 1,3-selenazol-4-one derivatives on mushroom tyrosinase. Chem Pharm Bull 2002, 50, 1594–1596.
[13] Lee, EH., Lim, Y-J., Ha, SK., Kang, TH., Koketsu, M., Kang, CH., Kim, SY., and Park, J-H. Inhibitory effects of 5-chloroacetyl-2-piperidino-1,3-selenazole, a novel selenium-containing compound, on skin melanin biosynthesis. J Pharm Pharmacol 2010, 62, 352–359.

[14] Nishina, A., Sekiguchi, A., Fukumoto, R., Koketsu, M., and Furukawa, S. Selenazoles (selenium compounds) facilitate survival of cultured rat pheochromocytoma PC12 cells after serum-deprivation and stimulate their neuronal differentiation *via* activation of Akt and mitogen-activated protein kinase, respectively. Biochem Biophys Res Commun 2007, 352, 360–365.
[15] Engman, L., Cotgreave, I., Angulo, M., Taylor, CW., Paine-Murrieta, GD., and Powis, G. Diaryl chalcogenides as selective inhibitors of thioredoxin reductase and potential antitumor agents. Anticancer Res 1997, 17, 4599–4605.
[16] Goudgaon, NM., Naguib, FN., El Kouni, M.H., and Schinazi, R.F. Phenylselenenyl- and phenylthio-substituted pyrimidines as inhibitors of dihydrouracil dehydrogenase and uridine phosphorylase. J Med Chem 1993, 36, 4250–4254.
[17] Woods, JA., Hadfield, JA., McGown, AT., and Fox, BW. Bioactivity and molecular modelling of diphenylsulfides and diphenylselenides. Bioorg Med Chem 1993, 1, 333–340.
[18] Engman, L., Stern, D., Frisell, H., Vessman, K., Berglund, M., Ek, B., and Andersson, CM. Synthesis, antioxidant properties, biological activity and molecular modelling of a series of chalcogen analogues of the 5-lipoxygenase inhibitor DuP 654. Org Biomol Chem 1995, 3, 1255–1262.
[19] Millois, C., and Diaz, P. Solution-phase synthesis of diaryl selenides using polymer-supported borohydride. Org Lett 2000, 2, 1705–1708.
[20] Rafique, J., Saba, S., Canto, RFS., Frizon, TEA., Hassan, W., Waczuk, EP., Jan, M., Back, DF., Rocha, JBTD., and Braga, AL. Synthesis and biological evaluation of 2-picolylamide-based diselenides with non-bonded interactions. Molecules 2015, 20, 10095–10109.
[21] Souza, D., Mariano, DOC., Nedel, F., Schultze, E., Campos, VF., Seixas, F., da Silva, RS., Munchen, T.S., Ilha, V., Dornelles, L., Braga, AL., Rocha, JBT., Collares, T., and Rodrigues, OED. New organochalcogen multitarget drug: Synthesis and antioxidant and antitumoral activities of chalcogenozidovudine derivatives. J Med Chem 2015, 58, 3329–3339.
[22] Terazawa, R., Garud, DR., Hamada, N., Fujita, Y., Itoh, T., Nozawa, Y., Nakane, K., Deguchi, T., Koketsu, M., and Ito, M. Identification of organoselenium compounds that possess chemopreventive properties in human prostate cancer LNCaP cells. Bioorg Med Chem 2010, 18, 7001–7008.
[23] Banerjee, B., and Koketsu, M. Recent developments in the synthesis of biologically relevant selenium-containing scaffolds. Coord Chem Rev 2017, 339, 104–127.
[24] Ninomiya, M., Garud, DR., and Koketsu, M. Selenium-containing heterocycles using selenoamides, selenoureas, selenazadienes, and isoselenocyanates. Heterocycles 2010, 81, 2027–2055.
[25] Yoshimatsu, M., Yamamoto, T., Sawa, A., Kato, T., Tanabe, G., and Muraoka, O. α-Sulfanyl and α-selanyl propadienyl cations: Regioselective generations and cycloadditions with thioamides and selemides controlled by $MeNO_2$-H_2O system. Org Lett 2009, 11, 2952–2955.
[26] Murai, T., Mizutani, T., Ebihara, M., and Maruyama, T. α-Hydroxy and α-oxo selenoamides: Synthesis *via* nucleophilic selenocarbamoylation of carbonyl compounds and characterization. J Org Chem 2015, 80, 6903–6907.
[27] Murai, T., Ishizuka, M., Suzuki, A., and Kato, S. Reaction of lithium eneselenolates derived from selenoamides with ketones: A highly diastereoselective synthetic route to β, β-disubstituted β-hydroxy selenoamides. Tetrahedron Lett 2003, 44, 1343–1346.
[28] Murai, T., Ezaka, T., and Kato, S. Synthesis of γ,δ-unsaturated selenoamides *via* the seleno-Claisen rearrangement of *in situ* generated allylic vinyl selenides from selenoamides and allylic bromides. Synthesis 2012, 44, 3197–3201.
[29] Goldstein, BM., Kennedy, SD., and Hennen, WJ. Selenium-77 NMR and crystallographic studies of selenazofurin and its 5-amino derivative. J Am Chem Soc 1990, 112, 8265–8268.
[30] Shafiee, A., Khashayarmanesh, Z., and Kamal, F. Selenium heterocycles XXXVIII. Synthesis and antimicrobial activity of 2-arylselenazoles. J. Sci. Islam. Repub. Iran 1990, 1, 111.

[31] Angeli, A., Trallori, E., Ferraroni, M., Di Cesare Mannelli, L., Ghelardini, C., and Supuran, CT. Discovery of new 2, 5-disubstituted 1,3-selenazoles as selective human carbonic anhydrase IX inhibitors with potent anti-tumor activity. Eur J Med Chem 2018, 157, 1214–1222.
[32] Zhang, P.-F., and Chen, Z.-C. Hypervalent iodine in synthesis 50: A novel method of synthesis of selenazoles by cyclocondensation of selenoamides and alkynyl(phenyl)iodonium salts. J Heterocyclic Chem 2001, 38, 503–505.
[33] Geisler, K., Pfeiffer, W.-D., Künzler, A., Below, H., Bulka, E., and Langer, P. Synthesis of 1,3-selenazoles and bis(selenazoles) from primary selenocarboxylic amides and selenourea. Synthesis 2004, 0875–84.
[34] Lai, LL., and Reid, DH. Synthesis of primary selenocarboxamides and conversion of alkyl selenocarboxamides into selenazoles. Synthesis 1993, 870–872.
[35] Below, H., Pfeiffer, W-D., Geisler, K., Lalk, M., and Langer, P. 1,3-Selenazole. Eur J Org Chem 2005, 3637–3639.
[36] Geisler, K., Künzler, A., Below, H., Bulka, E., Pfeiffer, W.-D., and Langer, P. Synthesis and reactivity of 2-acyl-1,3-selenazoles. Synthesis 2004, 97–05.
[37] Geisler, K., Künzler, A., Below, H., Bulka, E., Pfeiffer, W-D., and Langer, P. Efficient Synthesis of 2-unsubstituted 1,3-selenazoles. Synlett 2003, 1195–1197.
[38] Zhang, P-F., and Chen, Z-C. Hypervalent iodine in synthesis 53: Synthesis of 2,4-disubstituted and 2,4,5-trisubstituted 1,3-selenazoles. Synthesis 2000, 1219–1222.
[39] Pizzo, C., and Mahler, SG. Synthesis of selenazoles by *in situ* cycloisomerization of propargyl selenoamides using oxygen-selenium exchange reaction. J Org Chem 2014, 79, 1856–1860.
[40] Murai, T., Yamaguchi, K., Hori, F., and Maruyama, T. Reaction of selenoamide dianions with thio- and selenoformamides leading to the formation of 5-aminoselenazoles: Photophysical and electrochemical properties. J Org Chem 2014, 79, 4930–4939.
[41] Attanasi, OA., Filippone, P., Perrulli, F.R., and Santeusanio, S. 1,2-Diaza-1,3-butadienes: A new approach to the synthesis of selenoheterocycles. Eur J Org Chem 2002, 2323–2330.
[42] Shafiee, J., and Rezayazdi, M. Selenium heterocycles. XLI. Syntheses of substituted thiazolo, selenazolo, 1,2,3-thiadiazolo and 1,2,3-selenadiazoloquinolines. J Heterocycl Chem 1995, 32, 177–178.
[43] Koketsu, M., Sasaki, T., Ando, H., and Ishihara, H. Preparation of 4,5-dihydro-1,3-selenazoles by reaction of aromatic primary selenoamides with acetylenedicarboxylate. J Heterocycl Chem 2007, 44, 231–232.
[44] Koketsu, M., Takenaka, Y., and Ishihara, H. Preparation of 1,3-selenazol-4-one derivatives from primary selenoamides and haloacyl halides. Synthesis 2001, 731–734.
[45] Huang, X., Yu, L., and Chen, Z-H. Tandem Michael-nucleophilic addition of 2-cyclopropylidene propionaldehyde – A novel method for the synthesis of spirocyclopropane-annulated heterocycles. Synth Commun 2005, 35, 1253–1261.
[46] Huang, X., Chen, W.-L., and Zhou, H.-W. An efficient access to selenazoline-4-carboxylate derivatives incorporating cyclopropyl groups. Synlett 2004, 329–331.
[47] Koketsu, M., and Ishihara, H. Synthesis of 1,3-selenazine and 1,3-selenazole and their biological activities. Current Org Chem 2003, 7, 175–185.
[48] Koketsu, M., Senda, T., Yoshimura, K., and Ishihara, H. Synthesis and characterization of novel 1,3-selenazine derivatives. $BF_3.Et_2O$-assisted reaction of primary selenoamides with α, β-unsaturated ketones. J Chem Soc Perkin Trans 1999, 1, 453–456.
[49] Koketsu, M., Takenaka, Y., Hiramatsu, S., and Ishihara, H. Facile preparation of 1,3-selenazine derivatives by reaction of primary selenoamides with α,β-unsaturated aldehydes of in the presence of $BF_3.Et_2O$. Heterocycles 2001, 55, 1181–1188.

[50] Shafiee, A., Ebrahimzadeh, MA., and Maleki, A. Selenium heterocycles. XLIII. Syntheses of 3,5-diaryl-1,2,4-thiadiazoles and 3,5-diaryl-1,2,4-selenadiazoles. J Heterocycl Chem 1999, 36, 901–903.
[51] Huang, X., and Chen, J. A convenient synthesis of 3,5-diaryl-1,2,4-selenadiazoles. Synth Commun 2003, 33, 2823–2827.
[52] Chauhan, S., Chaudhary, P., Singh, AK., Verma, P., Srivastava, V., and Kandasamy, J. *tert*-Butyl nitrite induced radical dimerization of primary thioamides and selenoamides at room temperature. Tetrahedron Lett 2018, 59, 272–276.
[53] Cohen, VI. Reaction of hydrazine with aryl and heterocyclic selenoamides and alkylselenone esters. Synthesis of 2,5-diaryl and diheterocyclic-1,3,4-selenodiazoles and dialkylselenone ester-hydrazones. J Heterocycl Chem 1979, 16, 365–368.
[54] Shafiee, A., Shafaati, A., and Habibi-Khameneh, B. Selenium heterocycles. XXXIX. Synthesis of thieno[3,4-d]thiazole, thieno[3,4-*d*]selenazole, selenolo[3,4-d]thiazole and selenolo[3,4-*d*] selenazole. J Heterocycl Chem 1989, 26, 709–711.
[55] Zhang, Y-M., Jia, X-S., and Zhou, X-J. A new method for the synthesis of diacyl selenides. Org Prep Proc Int 1993, 25, 681–683.
[56] Zhao, H-R., Zhao, X.-J., and Huang, X. Selenium transfer reaction of primary selenoamides: A novel method for the synthesis of diacylselenides from acyl chlorides. Synth Commun 2002, 32, 3383–3387.
[57] Sekiguchi, M., Ogawa, A., Fujiwara, S-I., Ryu, I., Kambe, N., and Sonoda, N. A novel selenium-containing heterocycle. Lewis acid-assisted reaction of selenoamides with aldehydes leading to 6*H*-1,3,5-oxaselenazines. Chem Lett 1990, 913–916.
[58] Koketsu, M., Hiramatsu, S., and Ishihara, H. Reaction of primary selenoamides with bisacyl chlorides: Synthesis of 6-hydroxy-1,3-selenazin-4-ones and selenoanhydrides. Chem Lett 1999, 485–486.
[59] Kim, C., Lee, J., and Park, MS. Synthesis of new diorganodiselenides from organic halides: Their antiproliferative effects against human breast cancer MCF-7 cells. Arch Pharm Res 2015, 38, 659–665.
[60] Savegnago, L., Trevisan, M., Alves, D., Batista, J., Rocha, T., Nogueira, CW., and Zeni, G. Antisecretory and antiulcer effects of diphenyl diselenide. Environ Toxicol Pharmacol 2006, 21, 86–92.
[61] Savegnago, L., Pinto, LG., Jesse, CR., Alves, D., Rocha, JBT., Nogueira, CW., and Zeni, G. Antinociceptive properties of diphenyl diselenide: Evidences for the mechanism of action. Eur J Pharmacol 2007, 555, 129–138.
[62] Ibrahim, M., Hassan, W., Meinerz, DF., Santos, M., Klimaczewski, C., Deobald, A., Costa, MS., Nogueira, CW., Barbosa, NB., and Rocha, JB. Antioxidant properties of diorganoyl diselenides and ditellurides: Modulation by organic aryl or naphthyl moiety. Mol Cell Biochem 2012, 371, 97–04.
[63] Soleiman-Beigi, M., Yavari, I., and Sadeghizadeh, F. CuO nanoparticles-catalyzed a novel method to the synthesis of symmetrical diselenides from aryl halides: Selenoamide as an organic Se-donor reagent. Chem Papers 2018, 72, 2239–2246.
[64] Purseigle, F., Dubreuil, D., Marchand, A., Pradere, JP., Goli, M., and Toupet, L. Synthesis and reactivity of *N*-selenoacylamidines precursors of selenoheterocycles. Tetrahedron 1998, 54, 2545–2562.
[65] Heuzk, K., Purseigle, F., Dubreuil, D., and Pradkre, JP. Easy synthesis of substituted 1,3-dithiole-2-ylidene selenophenes from selenoheterodienes. Synth Commun 1998, 28, 301–310.
[66] Koketsu, M., Imagawa, M., Mio, T., and Ishihara, H. Preparation of 1,3-selenazoles using selenazadienes. J Heterocyclic Chem 2005, 42, 831–834.

[67] Koketsu, M., Mio, T., and Ishihara, H. Facile preparation of 1,3-selenazole-5-carboxylic acids and the carboxylates by reaction of selenazadienes with chloroacetyl chloride. Synthesis 2004, 233–236.

[68] Koketsu, M., Kogami, M., Ando, H., and Ishihara, H. Preparation of 5-acyl-2-amino-1,3-selenazoles by the reaction of selenazadienes with α-haloketone. Synthesis 2006, 31–36.

[69] Koketsu, M., Nada, F., Mio, T., and Ishihara, H. Syntheses of 4-selenazolones. Hetero Diels-Alder reaction of the selenazadienes with DMAD. Heterocycles 2003, 60, 1211–1218.

[70] Segi, M., Takahashi, T., Ichinose, H., Li, GM., and Nakajima, T. An efficient construction of a selenocarbonyl unit by the reaction of acetal derivatives with bis(dimethylaluminum) selenide. Tetrahedron Lett 1992, 33, 7865–7868.

[71] Segi, M., Yamamoto, H., Hori, T., and Nakajma, T. Reaction of selenocarbonyl compounds with oxygen-functionalized conjugated dienes. Phosphorus Sulfur Silicon Relat Elem 1998, 136, 599–602.

[72] Segi, M., Zhou, A., and Honda, M. Generation of selenoaldehydes *via* retro Diels-Alder reaction and their behavior in the reaction with some enophiles. Phosphorus Sulfur Silicon Relat Elem 2005, 180, 1045–1049.

[73] Zhou, A., Segi, M., and Nakajima, T. The reaction of selenoaldehydes with 2-methoxyfuran using their generation by retro Diels-Alder reaction. Tetrahedron Lett 2003, 44, 1179–1182.

[74] Segi, M., Tanno, K., Kojima, M., Honda, M., and Nakajima, T. An efficient 1,3-dipolar cycloaddition between aromatic selenoaldehydes and nitrile oxides or nitrile imines: An easy access to selenium-containing five-membered heterocyclic ring system. Tetrahedron Lett 2007, 48, 2303–2306.

[75] Segi, M., Kawaai, K., Honda, M., and Fujinam, S. Stereoselective hetero Diels-Alder reaction of selenoaldehydes with pentavalent phosphole chalcogenides. Tetrahedron Lett 2007, 48, 3349–3354.

[76] Degl'Innocenti, A., Capperucci, A., Acciai, M., and Tiberi, C. Silicon-mediated synthesis of selenoaldehydes and selenoacylsilanes and their hetero Diels-Alder reactions. Phosphorus Sulfur Silicon Relat Elem 2009, 184, 1621–1626.

[77] Abelman, M.M. Preparation and cycloaddition chemistry of thio- and selenocarbonyls derived from reaction of elemental sulfur and selenium with stabilized α-halo anions. Tetrahedron Lett 1991, 32, 7389–7392.

[78] Takikawa, Y., Uwano, A., Watanabe, H., Asanuma, M., and Shimada, K. Novel synthesis of 1,3,5-triselenanes from aldehydes, and novel generation of selenoaldehydes by fragmentation of 1,3,5-triselenanes. Tetrahedron Lett 1989, 30, 6047–6050.

[79] Okuma, K., Sakata, J.-I., Tachibana, Y., Honda, T., and Ohta, H. Stereoselective synthesis of *trans* olefins by the reaction of Wittig reagents with selenium. Formation of selenocarbonyl compounds. Tetrahedron Lett 1987, 28, 6649–6652.

Kishorkumar M. Reddy and Govindasamy Mugesh

10 Understanding the Chemistry of Selenoenzymes by Synthetic Organoselenium Compounds

10.1 Introduction

Selenium, which mainly exists as selenocysteine (Sec, U), plays a crucial role in many biological functions. Selenocysteine, which is coded by a stop codon UGA, is considered as the 21st amino acid. The insertion of selenocysteine onto growing polypeptide chain is a complex process involving specific factors. Till date, around 25 selenoproteins have been identified that contain selenocysteine. However, the structure and function of many of these proteins are unknown. The mutation of Sec to Cys results in a significant decrease in the activity of that specific protein, highlighting the importance of Sec in the active site of many of these identified proteins. Glutathione peroxidases (GPx), iodothyronine deiodinases (IDs), and thioredoxin reductases (TrxR) are some of the well-known proteins that contain selenocysteine in their active site. As the recombinant DNA technology failed to produce active selenoproteins, it has become a challenge to find the role and mechanism of selenocysteine at the active site. Therefore, synthetic organoselenium compounds were prepared as mimetic or as inhibitors to understand its role and mechanism of action. Earlier scientific reviews and book chapters highlighted the GPx mimetic activity of selenium compounds. In this chapter, we provide an overview of the applications of organoselenium compounds in understanding the chemistry of selenoenzymes.

Selenium plays an important role in many physiological functions in mammalians [1–4]. Selenium presents as selenocysteine, a cysteine analogue. It is coded by a dual function stop codon UGA [5, 6]. As it is cotranslationally coded into proteins and fulfill all the requirements for a proteinogenic amino acid, it is considered as the 21st amino acid [7]. Selenocysteine is present mainly in redox enzymes and because of its low pK_a and low negative redox potential it maintains the redox state of a cell [8, 9]. Till now, around 25 enzymes were identified to have selenocysteine, which include the major mammalian enzymes GPx [10, 11], IDs [12], and TrxR [13] that contain selenocysteine in their active site (Table 10.1).

Kishorkumar M. Reddy, Govindasamy Mugesh, Department of Inorganic and Physical Chemistry, Indian Institute of Science, Bangalore, India

https://doi.org/10.1515/9783110625110-010

Table 10.1: Human and bacterial selenoproteins and their physiological roles [14].

Selenoprotein	Physiological role
Glutathione peroxidase (GPx-1–4 and GPx6)	First line of defense against oxidative stress. Catalyzes the reduction of H_2O_2 and/or lipid peroxides by using glutathione as cofactor.
Thioredoxin reductases (TrxR)	Catalyze the reduction of oxidized thioredoxin to the corresponding dithiol and play an important role in the regulation of various signaling cascades.
Iodothyronine deiodinases (ID 1–3)	Membrane-bound proteins that activate/inactivate the thyroid hormones such as thyroxine T4 and hence play key roles in the thyroid function.
Integral membrane protein (SelN)	Functions in calcium mobilization by direct modulation of the ryanodine receptor.
Methionine sulfoxide reductase (SelR)	Catalyzes the reduction of R-form of methionine sulfoxides, oxidized methionines.
Selenophosphate synthetase (SPS 2)	Converts hydrogen selenide into selenophosphate during the selenocysteine biosynthesis.
Integral membrane protein (Sel K)	Not known.
Thioredoxin fold-like proteins (Sel V and Sel T)	Not known.
Formate dehydrogenase*	Conversion of formic acid to carbon dioxide.
Glycine reductases*	Reduction of glycine to acetate and ammonium salt.

* Bacterial enzymes.

10.2 Glutathione peroxidase (GPx)

GPx is an antioxidant selenoenzyme that protects the cells from harmful reactive oxygen species (ROS) such as hydroperoxides. There are four different isoforms of GPx: cytosolic GPx (cGPx), phospholipid hydroperoxides GPx (PHGPx), plasma GPx (pGPx), and gastrointestinal GPx (giGPx). All of these isoforms require selenocysteine in the form of selenol in the active site for their activity [15–19]. The catalytic activity of these enzymes depends on the hydroperoxide and thiol cofactor it uses. The classical GPx mainly utilizes glutathione (GSH) as the thiol cofactor in reducing hydrogen peroxide (H_2O_2) and some organic peroxides such as cumene hydroperoxides (Cum-OOH) and *tert*-butyl hydroperoxides (*t*BuOOH). PHGPx has very broad substrate specificity. It can reduce hydroperoxides, fatty acid hydroperoxides, tBuOOH, Cum-OOH, cholesterol hydroperoxides, and H_2O_2 with good catalytic rate using GSH as the cofactor [20]. In contrast to cGPx and PHGPx, pGPx exhibits very narrow hydroperoxide substrate specificity. Although it can reduce H_2O_2 and organic peroxides, its catalytic rate is less than cGPx. In addition, GSH is a very poor cofactor for this enzyme as the reducing GSH form concentration is very less in plasma [21].

The catalytic cycle for this enzyme mainly involves three steps. In the first step, the reduced selenoate residues reduce hydroperoxides to water (or alcohol) to form oxidized selenenic acid (E-SeOH) [22, 23], which upon reaction with one equivalent of GSH generates selenenyl sulfide (E-SeSG) intermediate. A second equivalent of GSH attacks at –Se–S– bond to regenerate the active selenol species with elimination of oxidized GSH (GSSG) and thus completing the catalytic cycle (Scheme 10.1A). The attack of second equivalent of GSH at the Se–S bond is the rate-determining step in the catalytic cycle [16, 23]. The GSH concentration in the cellular level is maintained by the enzyme glutathione reductase (GR), which reduces oxidized GSSG to reduced GSH by using NADPH as the cofactor [24]. In this overall catalytic process, 2 equivalents of GSH are used to reduce one equivalent of hydroperoxide to the corresponding

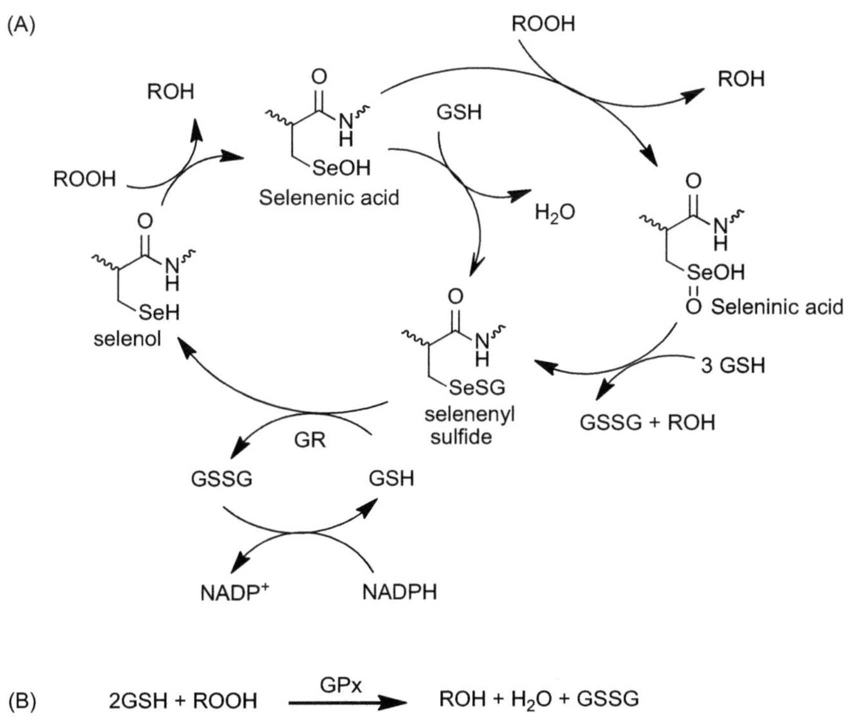

Scheme 10.1: (A) Proposed mechanism for GPx-catalyzed reduction of hydroperoxides involving the formation of intermediates selenenic acid, seleninic acid, selenenyl sulfide, and active selenol; **(B)** Overall reaction catalyzed by GPx enzyme.

water or alcohol (Scheme 10.1B) [22, 23]. At high hydroperoxide concentration and at depleting levels of GSH, the selenium center may undergo overoxidation to produce the corresponding seleninic acid (E-SeO_2H) and selenonic acid (E-SeO_3H). Seleninic acid may be converted rapidly to selenenyl sulfide by reaction with GSH. However, the overoxidized selenonic acid decreases the catalytic activity. Nevertheless, there are no reports to demonstrate the formation of these overoxidized products in vivo.

The crystal structure of GPx from bovine erythrocyte suggests that the selenium center exits in seleninic acid form in the active site, which indicates that seleninic acid form is more stable in air (Figure 10.1A) [25]. From crystal structure it is also observed that Sec is very close to two amino acids, Glutamine (Gln80) and Tryptophan (Trp158) (Figure 10.1B), by noncovalent interactions forming a catalytic triad [26]. These two amino acid residues are very important for the catalytic activity of GPx, as the hydrogen bonding stabilizes the selenol moiety in the active site.

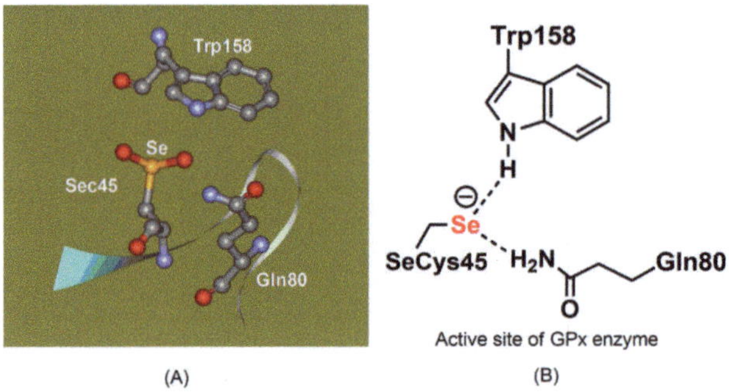

Figure 10.1: (A) Active site of GPx in seleninic acid form (PDB code: 1GP1) determined by X-ray crystallography [25]. (B) Structural representation of catalytic traid at active site of GPx [26].

Due to many therapeutic applications, several research groups worked on to synthesis GPx mimetics. These mimetic studies also helped with understanding the chemistry at the active site of GPx. Based on the structure, these GPx mimetics were classified as two categories. First category of compounds contains heteroatoms such as nitrogen, oxygen, or sulfur directly bonded to selenium center. The second category of compounds do not have direct selenium–heteroatom bond, but a heteroatom is placed near selenium leading to selenium–heteroatom nonbonding interactions [27–29].

10.2.1 Ebselen and cyclic amide-based GPx mimetics

2-Phenyl-1,2-benzisoselenazole-3-(2*H*)-one (**1**, ebselen) is the first selenium compound prepared and studied as GPx mimetic. It is one of the most therapeutically important compounds, which exhibits many biological activities both in vitro and in vivo [29–33]. It reduces hydrogen peroxide and lipid peroxides, effectively scavenges highly reactive peroxynitrite, and it also inhibits a variety of free radical-generating enzymes such as nitric oxide synthetase, NADPH oxidase, lipoxygenase, and cyclooxygenase [34–36]. It is found that it is less toxic to cells because of its stable selenazole moiety. However, some evidence showed that it can be toxic to cells as it inhibits some cell growth enzymes and induces apoptosis.

Ebselen was first synthesized by Lesser and Weiss [37]. Muller and coworkers [30] and Wendel and coworkers [31] first demonstrated its in vitro GPx activity. After discovery of its therapeutic applications, extensive research was carried out to synthesize ebselen and its structural analogues. The simple method was reported by Engman and Hallberg [38], which involves ortho-lithiation of benzanilide, followed by selenium insertion. Cyclization was carried out in the presence of CuBr as shown in the Scheme 10.2. A homologue of ebselen, compound **2**, was synthesized to increase the solubility by increasing the ring size from five membered to six membered [39]. Compound **3**, which does not have any aromatic ring, was synthesized and was extensively used to understand the redox chemistry of Selenocysteine at GPx active site [40]. Compound **4**, which does not have carbonyl group, was also able to reduce hydrogen peroxide [41]. Introduction of NO_2 group in the aromatic ring (compound **5**) was found to increase GPx-like activity due to the electron-withdrawing group at *ortho*-position [42]. On the other hand, introduction of NO_2 group at the para position for compound **2** generates compound **b** where its activity is decreased significantly [43]. The ring size was found to play a key role as the ring size when increased from five membered (compound **4**) to six membered (compound **7**) exhibits much high activity [44]. However, introduction of methoxy group to compound **7** at para position (compound **8**) does not seem to alter the activity [45]. Satheeshkumar and Mugesh reported the synthesis and antioxidant activity of a number of peptides containing ebselen analogues (**9–15**) [46]. The GPx activity of these analogues was found to be highly dependent on the nature of peptide chain attached to nitrogen atom and also the peroxide used. Compound **9** having Tyr–Val residue was found to be highly active than ebselen itself in the presence of Cum-OOH, whereas the activity decreased considerably when other peroxides such as H_2O_2 and *t*-BuOOH were used. Compound **10–12** with Phe–Val, Phe–Ile, or Phe–Ala residues, respectively, were found to be very poor catalysts in the presence of Cu-OOH and *t*-BuOOH. However, in the presence of H_2O_2, compound **12** shows better activity than ebselen. The Trp-based ebselen analogues **13**

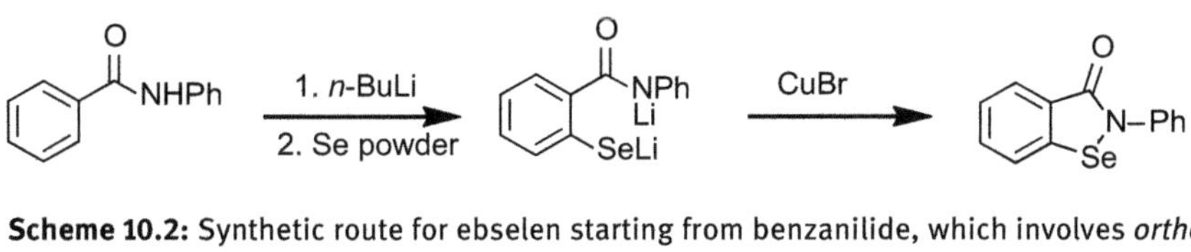

Scheme 10.2: Synthetic route for ebselen starting from benzanilide, which involves *ortho* lithiation, selenium insertion followed by cyclization [38].

and **14** were much more active than ebselen with both H_2O_2 and Cum-OOH, but the activity of these compounds was slightly less than ebselen when *t*-BuOOH was used. Compound 15 having Val–Ala residue was found to be very active in the presence of all three peroxide systems. It was reported that the difference in the activity of these compounds was mainly because of their reactivity toward GSH (Figure 10.2).

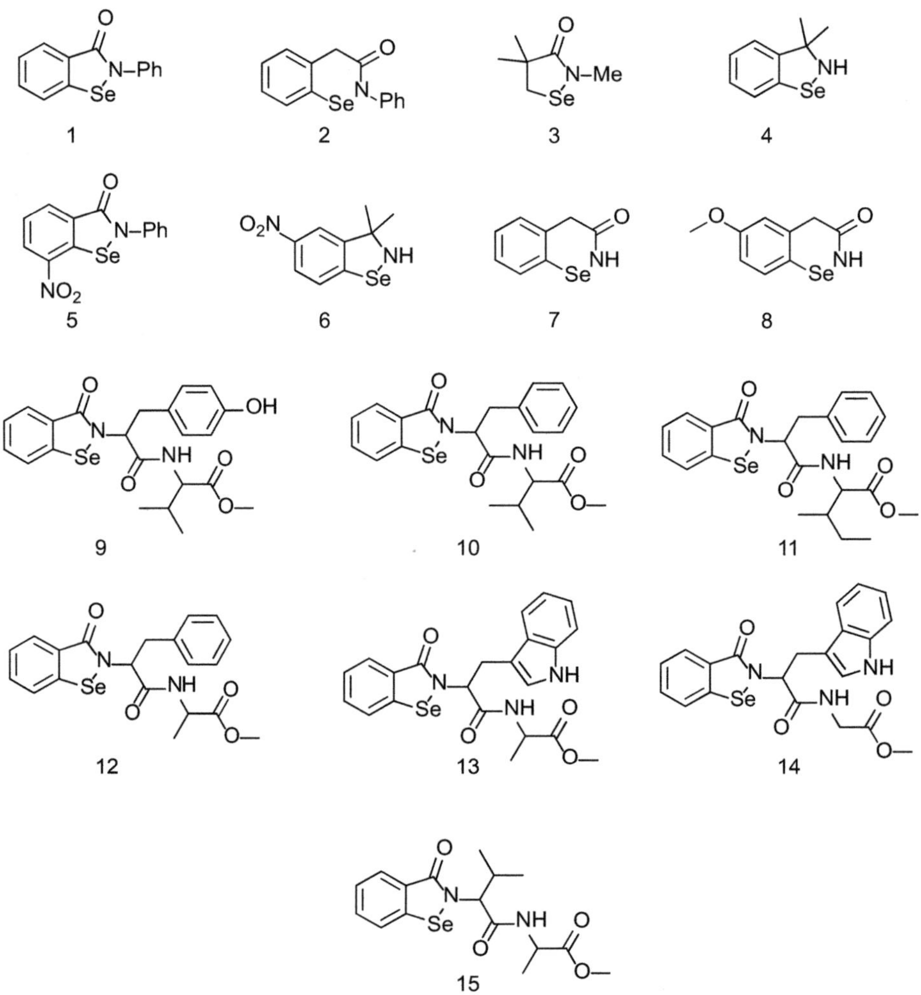

Figure 10.2: Chemical structure of Ebselen and its analogues (**1–8**) and peptide coupled ebselen derivatives (**9–15**).

Although extensive work has been done on the reduction of peroxides by ebselen, its catalytic mechanism is controversial [32, 47–49]. This is probably because of the usage of different thiols and different peroxides in different assays. Initially, it was postulated that ebselen first reacts with one equivalent of thiol to produce selenenyl sulfide **16**, which is then converted to selenol species at the expense of another equivalent of thiol cofactor. The seleol species is believed to be active species and reacts with H_2O_2 to produce selenenic acid **18** as shown in Scheme 10.3. The catalytic cycle is completed with the release of a water molecule to regenerate ebselen. In the presence of excess thiol, the selenenic acid **18** reacts with the thiol to produce the corresponding selenenyl sulfide **16**.

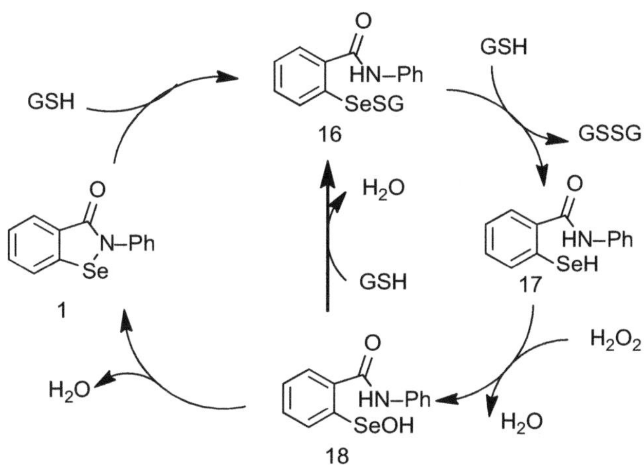

Scheme 10.3: Initially postulated catalytic mechanism of ebselen involving intermediates selenenyl sulfide (**16**), selenol (**17**), and selenenic acid (**18**).

Recent studies have shown that ebselen is a poor catalyst in reducing peroxides when aryl thiols such as PhSH or BnSH are used as thiol cofactors [50, 51]. Mugesh and coworkers have synthesized and studied GPx-like activity of ebselen analogues in the presence of aryl thiols as cofactors in assays [51–53]. It was found that the reaction of thiols with ebselen does not produce selenol. This is due to extensive thiol-exchange reaction. Once it forms selenenyl sulfide **16** species, the attack of another thiol molecule can take place either at selenium center or at the sulfur center as shown in Scheme 10.4. If thiol attacks at the sulfur center, the selenol is produced with the elimination of a disulfide. However, when the incoming thiol attacks at selenium center, it leads to the generation of another selenenyl sulfide intermediate. In the proposed catalytic cycle of ebselen, the selenium center is more electrophilic than the sulfur due to strong Se···O interaction, which favors the incoming thiol to attack at the selenium center [51]. As a result, the selenol required for the reduction of peroxides is not formed in sufficient quantities, which accounts for poor GPx-like activity of ebselen analogues in aromatic thiol assays.

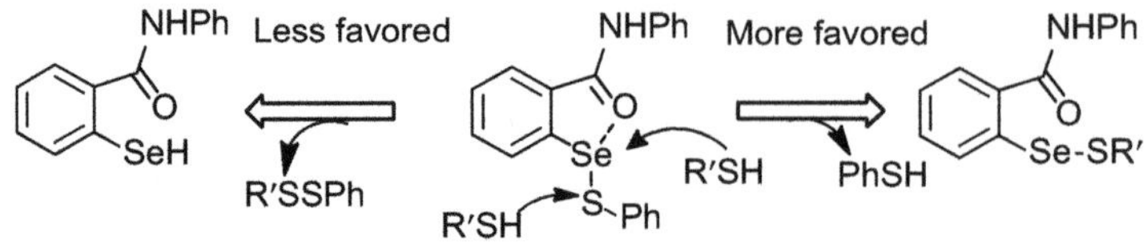

Scheme 10.4: Thiol-exchange reaction taking place at the selenenyl sulfide intermediate [51].

To understand the different substitution at the N-atom of ebselen, compounds **19–29** (Figure 10.3) were synthesized and their GPx-like activity was studied using aromatic thiols as cosubstrate and different peroxides as substrate [52]. The synthetic scheme for these compounds is shown in Scheme 10.5. The diselenide is treated with thionyl chloride ($SOCl_2$) to obtain 2-(chloroseleno)benzoyl chloride, which is then treated with the corresponding amines in acetonitrile to obtain desired ebselen analogues. All the N-substituted compounds were found to be more active than **24,** indicating the importance of substitution at nitrogen atom. While the strength of Se–N bond has little effect on the GPx-like activity, the nonbonded Se–O interaction in the selenenyl sulfide intermediate influences the activity significantly.

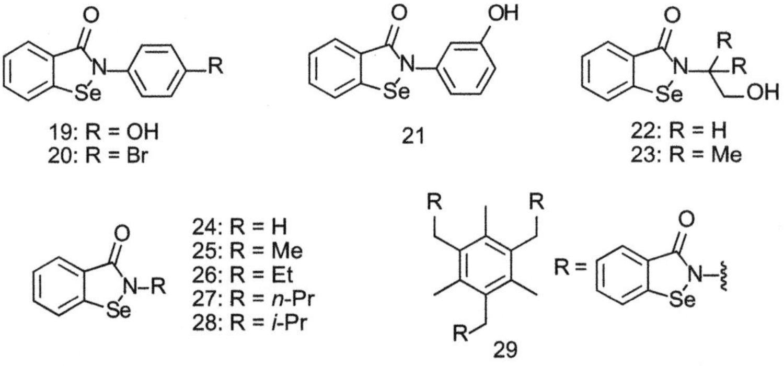

Figure 10.3: Chemical structure of GPx mimetics where the benzene ring attached to nitrogen atom has different substituents **19–29**.

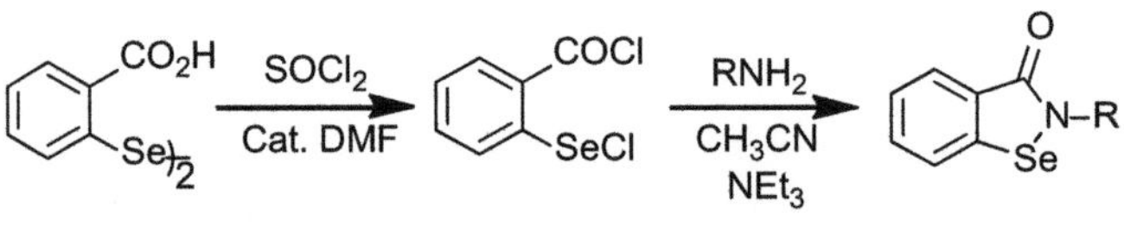

Scheme 10.5: General synthetic scheme for the preparation of ebselen derivatives **19–29** [52].

^{77}Se NMR spectroscopy and theoretical calculations confirm that irrespective of substitution, all the compounds were found to have strong Se–O nonbonded interactions. All these selenenyl sulfide intermediates exhibit significant downfield shift in ^{77}Se NMR as compared to PhSeSPh (526 ppm). This indicates that all these selenenyl sulfides undergo thiol-exchange reactions in the presence of aromatic thiols. As discussed, strong Se–O interactions in the selenenyl sulfide intermediate derived from selenenyl amide derivatives reduces GPx-like activity due to thiol-exchange reactions. Therefore, weakening such interactions is beneficial for the catalytic efficiency. This can be achieved by the following modifications.

i) Introduction of substituent that will preferentially interact with sulfur center in the selenenyl sulfide intermediate. For example, compound (Figure 10.4) has both Se–O and Se–N interactions. The Se–N interaction reduces the Se–O interaction considerably and thus decreases the electropositive character of selenium center. Thus, nucleophilic attack of incoming thiol takes place preferentially at the sulfur center [54].
ii) Use of dithiol instead of monothiol as the cosubstrate. In this case, the second free thiol group is expected to attack at the nearby sulfur center. For example, the attack of second free thiol in selenenyl sulfide intermediates **31** and **32** (Figure 10.4) takes place at sulfur center leading to the generation of free selenol. This was proved by experimentally observing higher activity of ebselen when dihydrolipoic acid was used instead of GSH as the cofactor [55].

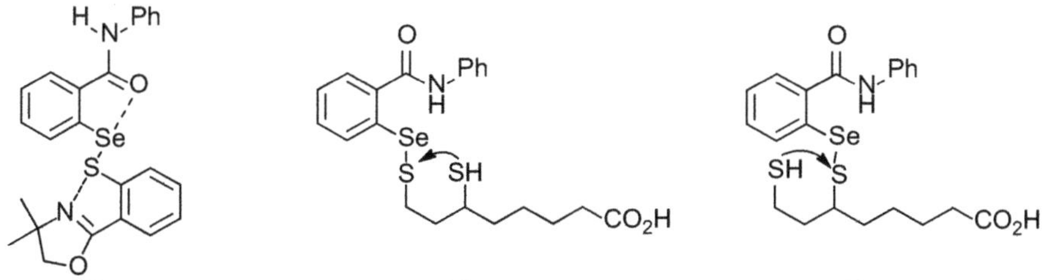

30 31 32

Figure 10.4: Chemical structures of selenenyl sulfides that reduce thiol-exchange reactions.

As catalytically active selenol was not generated in the catalytic mechanism of ebselen, Mugesh and coworkers have carried out a detailed mechanistic study on the catalytic reduction of peroxides in the presence of thiol. A revised catalytic mechanism was proposed based on the intermediates confirmed by ^{77}Se NMR spectroscopy. Some of the intermediates were isolated and characterized completely. It was observed that ebselen can readily react with peroxides even in the absence of thiol to produce seleninic acid **18a**. Treatment with an excess amount of thiol converts seleninic acid **18a** into corresponding selenenic acid **18**, which upon reaction with another equivalent of thiol generates selenenyl sulfide **33** (Scheme 10.6). In the absence of thiol, compound **18** eliminates water to regenerate ebselen **1.** The disproportionation of selenenyl sulfide **16** to diselenide **34** is the rate-determining step and the rate of disproportionation depends on the nature of thiol employed.

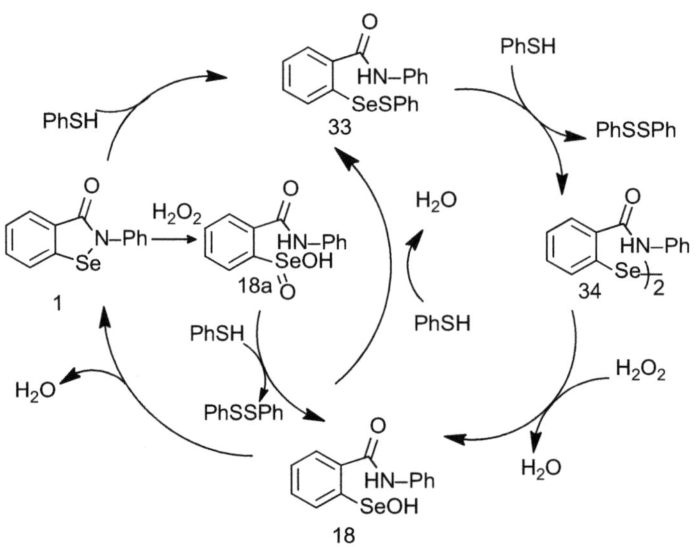

Scheme 10.6: Revised catalytic mechanism of ebselen analogues using thiophenol (PhSH) as thiol cofactor involving intermediates, seleneyl sulfide (**33**), selenenic acid (**18**), seleninic acid (**18a**), and diselenide (**34**).

As diselenide **34** was found to be the key intermediate in the catalytic mechanism of ebselen, Bhabak and Mugesh synthesized some *sec* and *tert*-amide-based diselenides **35–43** and reported their GPx-like activity (Figure 10.5) [56]. These diselenides were inactive toward aromatic thiol PhSH, and therefore, their reaction with H_2O_2 is important for their catalytic activity. The *tert*-amide-based diselenides exhibit much better activity than the corresponding *sec*-amide-based diselenides. Although the reactivity toward thiols cannot be altered by introduction of *tert*-amide groups, the Se–O interactions in selenenyl sulfides derived from *tert*-amide -based diselenide compounds (**40–43**) were found to be much weaker than the interactions observed for selenenyl sulfides obtained from *sec*-amide-based diselenides (**35–39**). This may account for the higher catalytic activity of tert-amide-based diselenides. The catalytic mechanism of these compounds follows similar to that of ebselen shown in Scheme 10.6.

As mentioned earlier, the crystal structure of GPx consists of Sec, Trp, and Gln, which activate the selenol moiety at the active site through hydrogen bonding. This

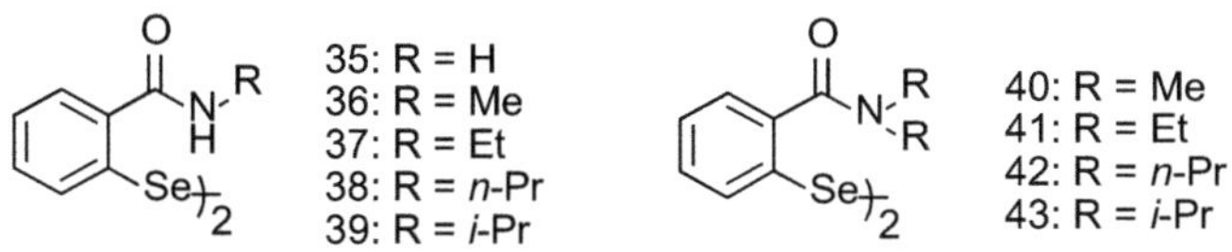

Figure 10.5: Chemical structure of *sec* and *tert*-amide-based diselenides as GPx mimetics.

concept leads to the development of *tert*-amine-based diaryl diselenides such as **44–51** (Figure 10.6) [57] and *sec*-amine-based diaryl diselenides **52–54** [58] having basic amino functionality near the selenium center. For example, reduction of **44** or **45** produces active selenol, which was stabilized by basic amino group present nearby. It was reported that the substitution of the *tert*-amino group by a *sec*-amino moiety (**52–57**) results in a significant enhancement in the catalytic activity [58]. The higher basicity of the *sec*-amino group facilitates the deprotonation of the selenol, thereby increasing the selenolate ion concentration. Thus, increasing the nucleophilicity of the selenol by deprotonation with a stronger base appears to be an efficient way to increase the antioxidant activity of the amine-based diaryl diselenides. Interestingly, the introduction of methoxy group at the *ortho* position significantly enhances the catalytic activity of parent compound. It assumes that the methoxy group blocks the attack of incoming thiol at the selenium center. Hence, it attacks at the sulfur center of selenenyl sulfide to generate the catalytically active selenol species.

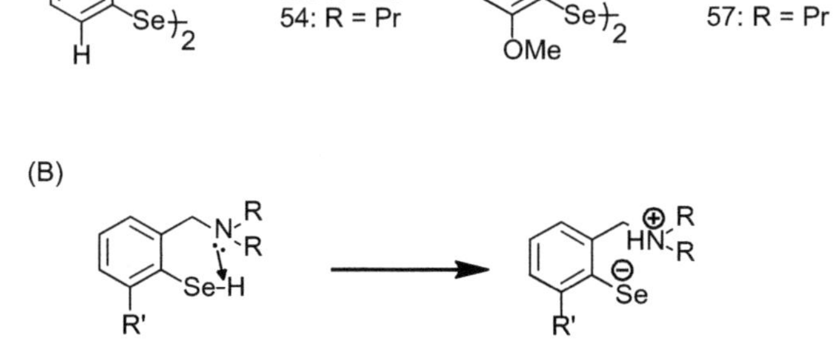

Figure 10.6: (A) Chemical structure of amine-based diselenides as GPx mimetics. (B) Activation of selenol moiety by basic amino group.

Later Bhowmick and Mugesh, inspired from these results, synthesized numerous diaryl diselenides **58–64** with different substitutions at aromatic ring and also at the nitrogen atom (Figure 10.7C) [59]. These compounds having an additional amino group on the benzylic nitrogen atom can behave as stabilizer for the selenol group. This structure can be related to the catalytic triad comprising of Sec, Trp, Gln (Figure 10.7A, B). All these compounds were tested for its GPx-like activity using all three peroxides. Compound **60** was found to be more active in reducing H_2O_2 than corresponding diselenide **44** when PhSH was used as a thiol cofactor. In addition, the *sec*-amine-based compound **59** showed higher activity than **52**. The *sec*-amine-based diselenides (**52** and **59**) showed higher catalytic activities compared to that of the corresponding *tert*-amine-based diselenides (**44** and **60**). It clearly suggests that the introduction of a *sec*-amino moiety in the benzylic position is more effective to increase the catalytic activity than the introduction of an additional amine on the nitrogen atom, although the additional amine leads to a further increase in the activity.

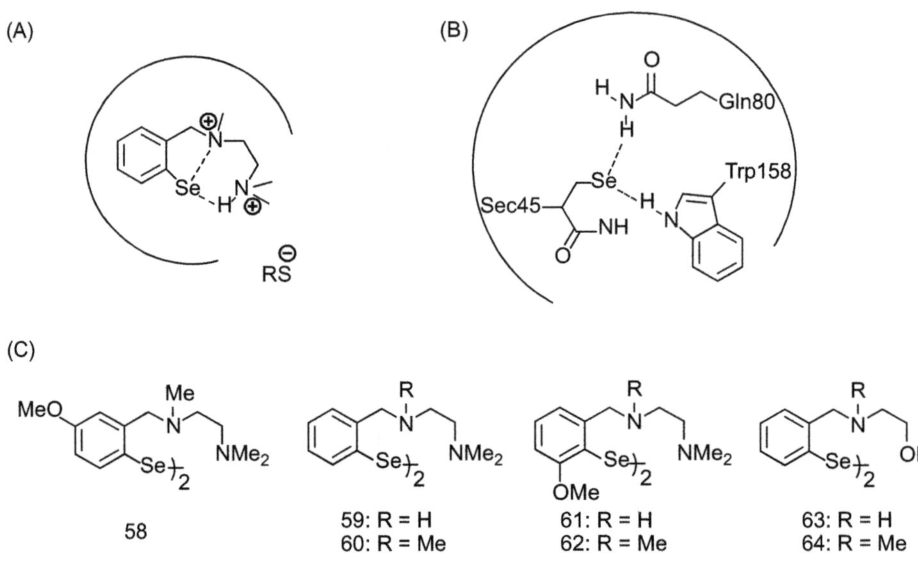

Figure 10.7: Proposed catalytic triad model (A) for synthetic mimics and (B) in native GPx enzyme. (C) Chemical structures of the diselenides containing an additional amine and an alcohol moiety on the nitrogen atom.

The *ortho*-methoxy-substituted diselenides **48** and **62** showed almost similar catalytic activity, suggesting that the additional amino group does not have any effect on the activity. Interestingly, the activity of the *sec*-amine-based diselenide **61** was higher than that of compound **48**, which exhibited exceptionally high catalytic activity in all the three peroxide systems. When GSH was used as thiol cofactor compound **59** and **60** showed much less activity than that by compounds **44** and **52**, indicating that compounds **59–60** act as poor catalysts in the presence of GSH as the cofactor in the reduction of H_2O_2. Similarly, the initial rates for compounds **48**, **62**, and **61** (534.3 μM min^{-1}, 357.2 μM min^{-1}, and 443.3 μM min^{-1}, respectively), in the presence of H_2O_2 as the substrate, indicates that compounds **61–62** display lower activities as compared to that of the diselenide **48**. The poor catalytic activities of compounds **58–62** can be ascribed to the severe steric interaction induced by the larger amino substituent on the nitrogen atom with the GSH molecule, which is also larger in size compared to PhSH. Compounds **63–64** were found to be more active than compounds **59–60** when H_2O_2 is used as the substrate. However, the activity is much lower in the other two peroxide (Cum-OOH and t-BuOOH) systems. The enhanced activities of compounds **63–64** in the presence of H_2O_2 are probably due to the less steric interaction between GSH and the relatively smaller alcohol moiety.

Singh and coworkers reported novel isoselenazolines **65–66** and isoselenazoline Se-oxides **67–68** (Figure 10.8) [60], which are stabilized by intramolecular Se⋯O interactions. These compounds exhibited higher GPx-like activities compared to that of ebselen **1** that contains a C = O group. Particularly, the selenoxides **67–68** were almost 3–4 times more active than ebselen. Based on the experimental studies, they have proposed a catalytic mechanism for compound **67**.

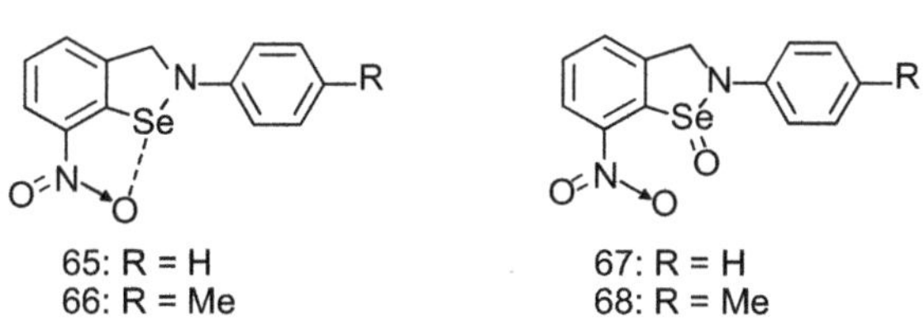

Figure 10.8: Chemical structures of isoselenazolines **65–66** and their oxides **67–68**.

The proposed mechanism followed as shown in Scheme 10.7. First, the selenenyl amide bond is cleaved by thiol (PhSH) to produce the selenenyl sulfide intermediate **69**, which acts as a true catalyst in the catalytic mechanism involving compounds selenol **70** and selenenic acid derivative **71.** Although this pathway is different from that of ebselen **1**, the formation of the diselenide **72** from the corresponding selenenyl sulfide **69** by disproportionation reaction is quite similar to that observed during the reduction of peroxides by ebselen.

Recently, Mugesh and coworkers have synthesized similar types of isoselenazole compounds **74–77** (Scheme 10.8) [58] that contain a methoxy substituent in the ortho-position and alkyl substituents on the nitrogen atom. Interestingly, it was observed that compounds **74–77** display excellent GPx activity both in vitro and inside

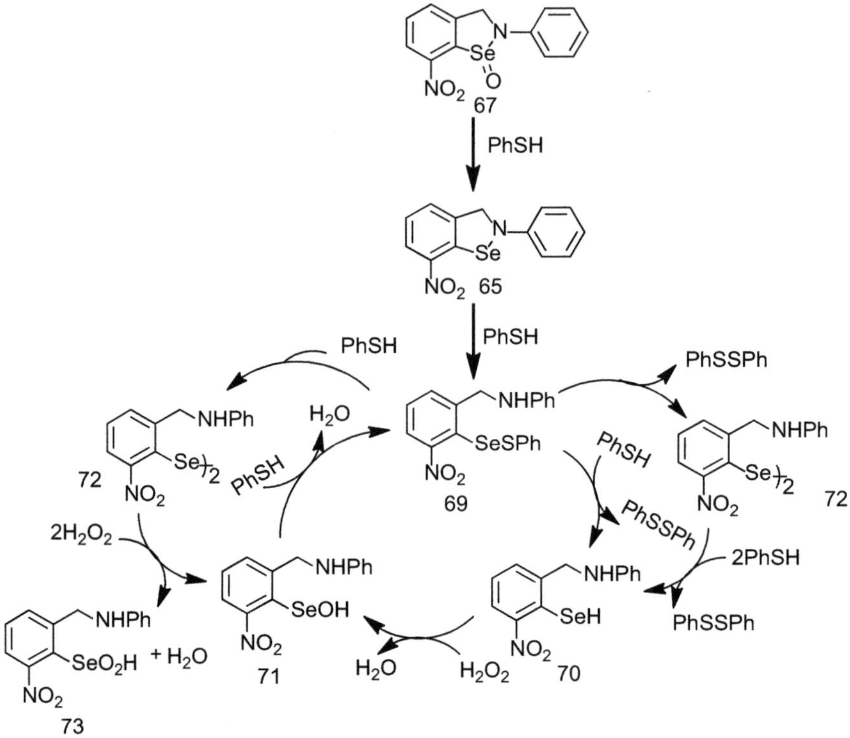

Scheme 10.7: Plausible catalytic cycle for the reduction of H_2O_2 by compound **67** that involves intermediates selenenyl sulfide (**69**), selenol (**70**), diselenide (**72**), selenenic acid (**71**), and seleninic acid (**73**) [58].

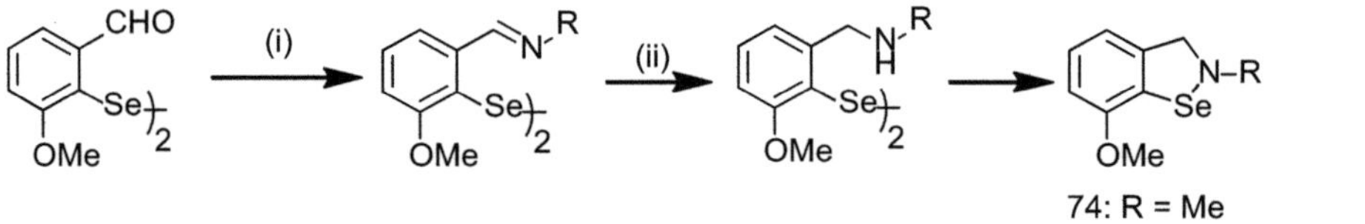

Scheme 10.8: Chemical structures of isoselenazolines **74–77**. Reagents and conditions: (i) R-NH_2, MeCN, HCl, 4 h; (ii) $NaBH_4$, MeOH, 30 min, O_2.

human cells. A comparison of the catalytic activity showed that all the isoselenazole compounds exert very high activities when compared to that shown by ebselen. These compounds also mimic the peroxiredoxins in human cells by using cellular thioredoxin (Trx) as reducing agents. Although these isoselenazoles are structurally similar to compounds **65–68**, the reactivity toward thiol is significantly different. The GPx-like activity of compounds **74–77** was studied using GSH as the thiol cofactor and three different peroxides as the substrates and found to be two to three times more active than ebselen in all three peroxide systems. It has been shown that the presence of intramolecular secondary Se O interactions with the nitro group in compounds **65–68** enhances the reactivity of the Se–N bond toward cleavage by thiol [58]. In contrast, isoselenazoles **74–77** reacted very slowly with thiol to cleave the Se–N bond as the selenium center is not activated.

The catalytic mechanism of this compound **74** follows oxidation of selenium center with peroxides and forming selenoxide intermediate **78**, which subsequently undergoes rapid reaction with an excess amount of PhSH to produce the corresponding selenenyl sulfide **79**. Compound **79** then follows a catalytic cycle similar to the native enzyme involving the selenol **80** and the selenenic acid **81**. It is noticed that, unlike compound **67**, diselenide **82** is produced from the selenol intermediate **80** after auto-oxidation (Scheme 10.9). Although isoselenazoles **67** and **74** maintain a similar catalytic cycle that involves selenenyl sulfide, selenol, and selenenic acids as the intermediates, the reactivity toward thiol and peroxides is significantly different due to the different electronic environments around the selenium atom.

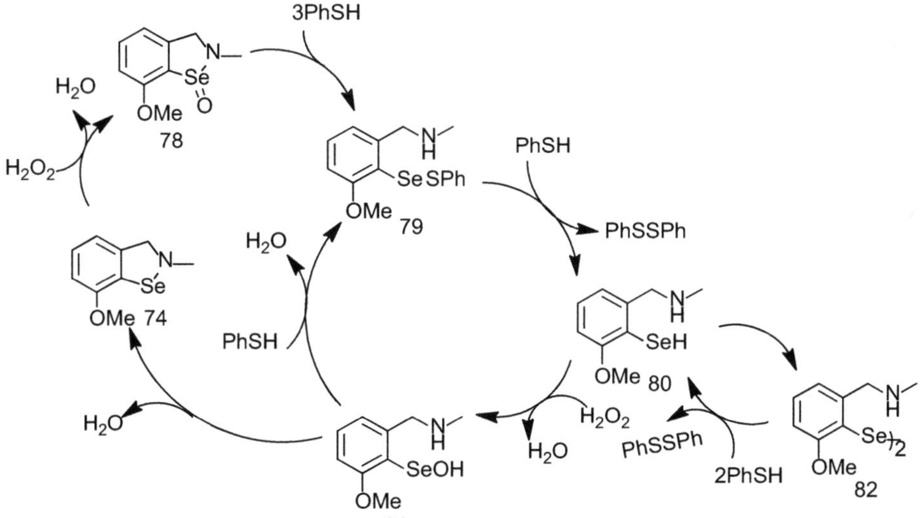

Scheme 10.9: Catalytic mechanism for the GPx-like activity of compound **74** that involves intermediates selenoxide (**78**), selenenyl sulfide (**79**), selenol (**80**), selenenic acid (**81**), and diselenide (**82**) [58].

10.3 Iodothyronine deiodinases

Thyroid hormones such as thyroxine (T4) and 3,5,5′-triiodothyronine (T3), produced by the thyroid gland, are iodine-containing compounds that regulate gene expression in every vertebrate tissue and control the metabolism in the body. T4 is a prohormone, which is produced in vivo by thyroid peroxidase that catalyzes the conversion of L-tyrosine to T4 by iodination followed by phenolic coupling. The triiodo derivative T3 is the active thyroid hormone, which is produced from T4 by an outer ring deiodination by IDs (ID-1 or ID-2, Scheme 10.10) [61–67]. The phenolic ring is referred as the outer ring and the tyrosyl ring as the inner ring. IDs catalyze the regioselective deiodination of various iodothyronines.

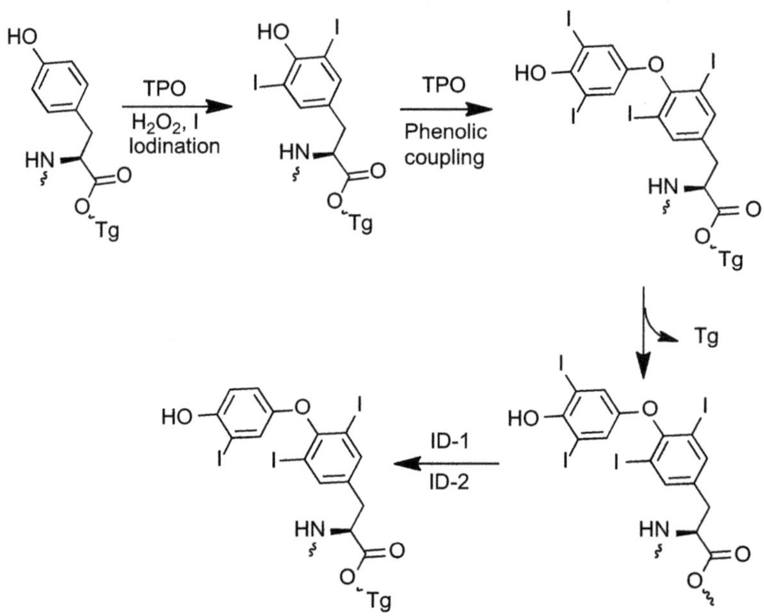

Scheme 10.10: Biosynthesis of thyroid hormone in vivo involves iodination of Tyr residues of thyroglobin by TPO, followed by a phenolic coupling reaction resulting in the formation of prohormone T4. Outer ring deiodination of T4 by ID1 or ID2 leads to the formation of active hormone T3.

These enzymes can be classified as ID-1, ID-2, and ID-3 depending on the position of deiodination and their selectivity for inner or outer ring deiodination. ID-1 removes iodine from both inner as well as outer ring iodine effectively and ID-2 removes outer ring iodine specifically, whereas ID-3 removes iodine from inner ring specifically [68–73]. Outer ring deiodination of T4 by ID-1 produces active hormone

T3 and inner ring deiodination produces reverse T3 (rT3). ID-1 also removes iodine from inner ring of rT3 to generate T2. It is known that rT3 is a better substrate for ID-1 than T4 or T3. In contrast, ID-2 removes iodine from outer ring of T4 and rT3 to generate T3 and T2, respectively (Scheme 10.11). As T4 is a better substrate for ID-2 as compared to rT3, ID-2 generally controls the production of active hormone T3. However, ID-3 removes iodine from inner ring of T4 and T3 to produce rT3 and T2, respectively. For ID-3, T3 has been shown to be a better substrate than T4. So, ID-3 is mainly responsible for inactivation of thyroid hormone.

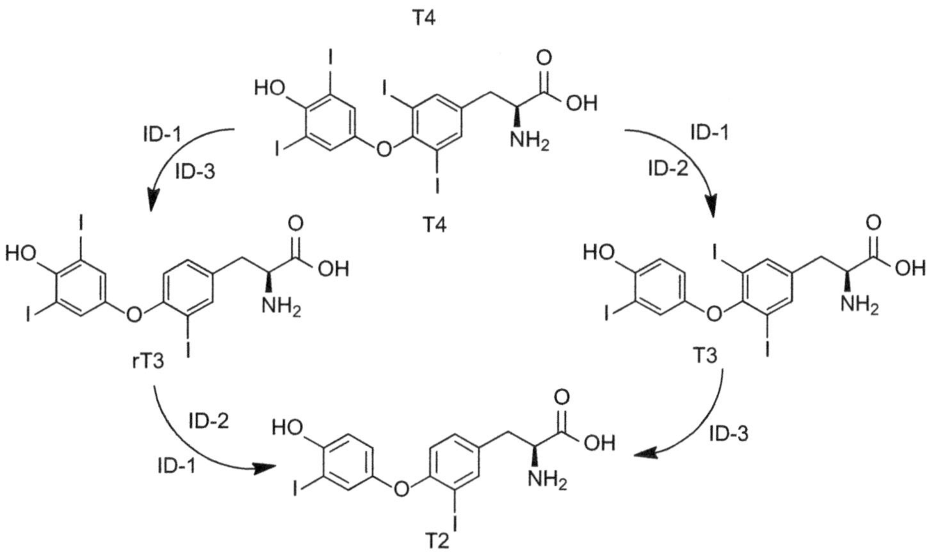

Scheme 10.11: Regioselective deiodination reactions catalyzed by different iodothyronine deiodinases.

All these three enzymes are integral membrane-bound enzymes. Because of difficulties in purification of membrane-bound proteins and expression of selenocysteine containing proteins, the crystal structures of IDs are not available. Cysteine (Cys) mutants of corresponding enzymes are functionally active, although the catalytic activity is found to be many folds lower than that of the wild-type enzyme [12, 74, 75]. Recently, the X-ray structure of the catalytic domain of mouse Dio3 (mDio3cat) has been solved [76]. The structure lacks the N-terminal membrane-associated domain as well as the linker connecting it with the catalytic domain. Furthermore, Sec170 was replaced with Cys to facilitate expression in a prokaryotic host [77]. The crystal structure reveals that the enzyme adopts a Trx fold containing a five-stranded mixed β-sheet flanked by four α-helices. A short N-terminal β-sheet followed by a 310-helix betrays an evolutionary relation to peroxiredoxins (Figure 10.9).

(A)

Human	:	RPLVLNFGSCTUPPF
Rat	:	RPLVLNFGSCTUPPF
Chicken	:	RPLILNFGSCTUPPF
Xenopus	:	RPLVVNFGSCTUPPF
Rana	:	RPLVLNFGSCTUPPF
Tilapia	:	RPLILNFGSCSUPPF
Consensus	**:**	**RPLVLNFGSCTUPPF**

(B)

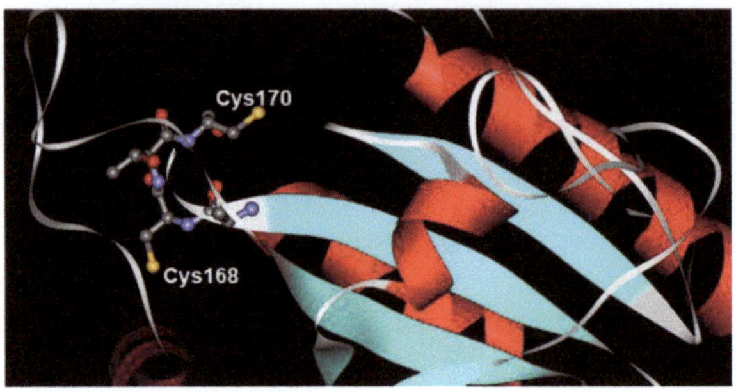

Figure 10.9: (A) Amino acid sequences at the active sites of ID-3 from different species indicating the conserved Cys (C) and Sec (U) residues; (B) crystal structure of mouse Dio3 where Sec170 was replaced by Cys (PDB: 4TR4) [76].

10.3.1 Synthetic mimetics of IDs

As deiodination mechanism of thyroxine by enzyme is very complicated, much effort was given in the direction of simple molecules as synthetic mimetic of IDs. In 1994, Reglinski and coworkers used benzeneselenol ($PhSe^-$) as deiodinating agent to remove iodine from some diiodotyrosine derivatives by refluxing in ethanol [78]. It was observed that only the activated iodo compound **85** undergoes deiodination to produce compound **88**. Other compounds **83** and **84** did not undergo deiodination to produce **86** and **87**, respectively, in the mentioned conditions (Scheme 10.12).

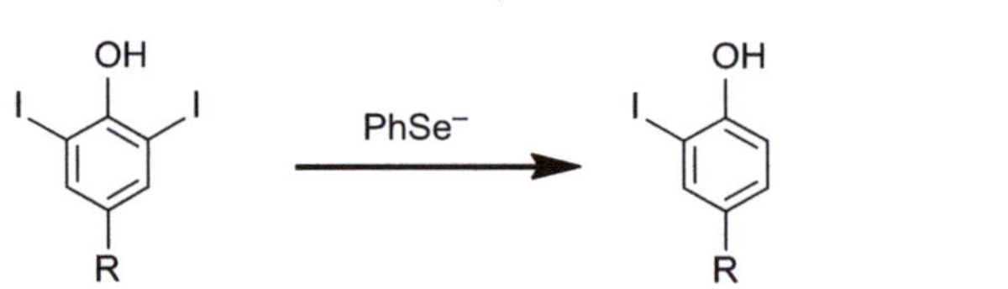

83 : R = $CH_2CH(NH_2)CO_2H$
84 : R = NH_2
85 : R = NO_2

86 : R = $CH_2CH(NH_2)CO_2H$
87 : R = NH_2
88 : R = NO_2

Scheme 10.12: Deiodination of different diiodo phenols by $PhSe^-$.

Engman and coworkers reported that phenyltellurolate ($PhTe^-$) can deiodinate phenolic ring of various model compounds. They treated the iodinating compound **89** with different sulfur, selenium, and tellurium containing compounds such as Na_2Te, NaHTe, $PhTe^-$, Na_2Se, and Na_2S [79]. Only $PhTe^-$ selectively removes iodine from compound **89** to give **90**, whereas Na_2S treatment produces compound **91** with removal of both the iodines (Scheme 10.13).

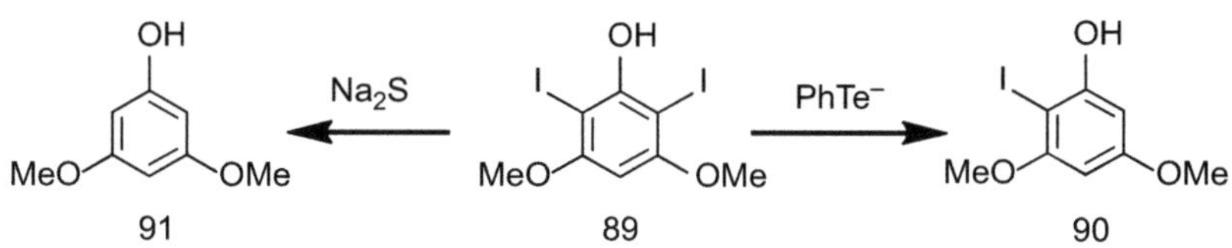

Scheme 10.13: Deiodination of model compounds by sulfur and tellurium compounds.

Recently, Goto and coworkers have demonstrated that *N*-butyrylthyroxine methyl ester (**92**), a thyroxine derivative, can be converted to the corresponding triiodo derivatives **93** by a sterically hindered selenol **94** (Scheme 10.14) [80]. In this study, the nucleophilic attack of selenol at one of the outer ring iodine produces the corresponding selenenyl iodide **95**, which was characterized by X-ray crystallography. A mechanism of this transformation was demonstrated in which the T4 first undergoes tautomerization to produce the corresponding keto derivative (Scheme 10.15). The nucleophilic attack of selenol on the positively charged iodine atom leads to the formation of selenenyl iodide and T3. The reaction was carried out in organic solvent $CDCl_3$, at higher temperature (50 °C) and longer reaction time (7 days).

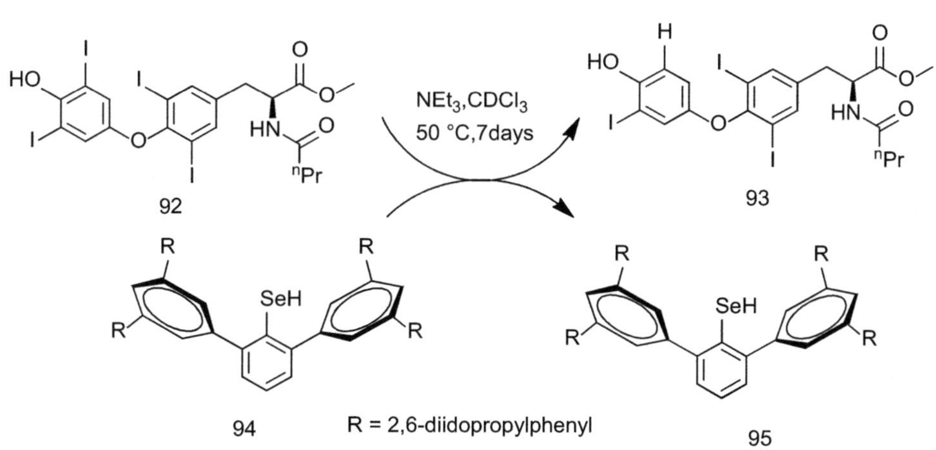

Scheme 10.14: Deiodination of thyroxine derivatives **92** by Bpq-selenol **94** to produce the corresponding selenenyl iodide **95** and T3 derivative **93**.

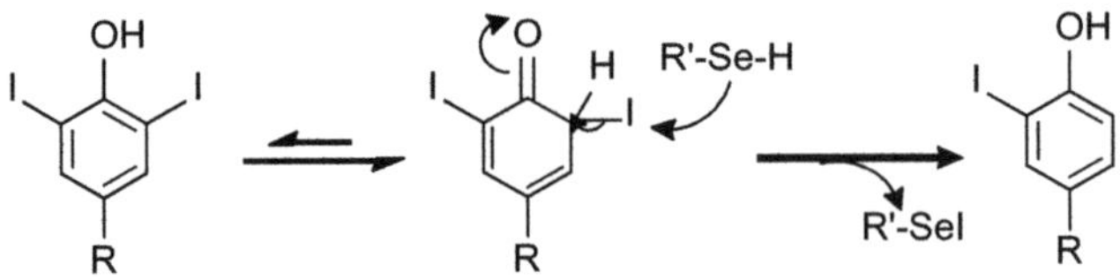

Scheme 10.15: Mechanism of deiodination via enol-keto tautomerism as proposed by Goto et al. [80].

Manna and Mugesh reported the first chemical model for the inner ring deiodination of T4 and T3 to produce rT3 and T2, respectively [81]. It was reported that the deiodination takes place at physiologically relevant conditions (pH 7.5, 37 °C). The naphthyl-based thiols and selenols **96–102** (Scheme 10.16A) were studied as ID-3 mimetics. It was observed that selenol **97** removes iodine from the inner ring of T4 more effectively than dithiol **96** (Scheme 10.16B). The lower activity of dithiol **96** is also in agreement with the mutational studies by Kuiper and coworkers, which demonstrated that the replacement of Sec with Cys in the active site reduces the substrate turnover number of T4 and T3 by six- and two-fold, respectively [75]. The comparison studies on compound **97** and **98** reveal that the presence of free selenol is important for the deiodination, as the compound **98** having a phenyl group attached to selenium was found inactive even at higher concentration. The presence of a thiol group at the 8-position of naphthyl ring is also important as the activity of compound **100** and **111** having no thiol group was found to be inactive in deiodination of T4.

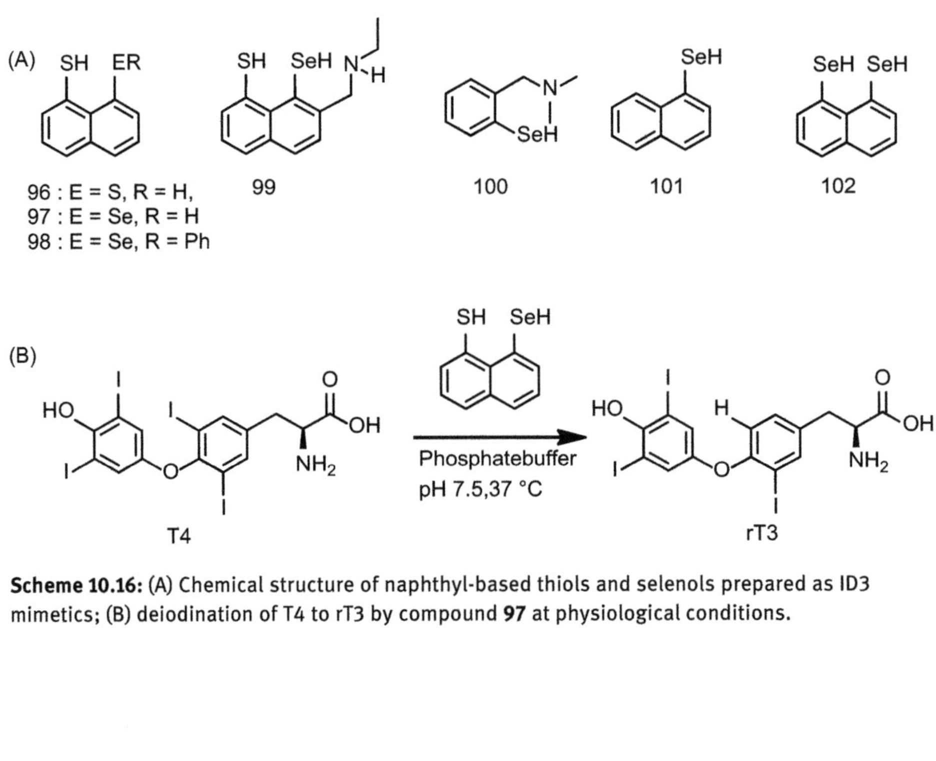

Scheme 10.16: (A) Chemical structure of naphthyl-based thiols and selenols prepared as ID3 mimetics; (B) deiodination of T4 to rT3 by compound **97** at physiological conditions.

It is reported by Köhrle and others that the presence of His residue near to active site selenol is important for the deiodination activity of ID-1, as this His residue activates the selenol by forming an imidazolium–selenolate pair [82, 83]. Goto and coworkers also reported that the deiodination of **92** by selenol **94** takes place only in the presence of triethylamine. Mugesh and coworkers compared the deiodination of compound **99**, having an amine group near to selenium, with compound **97**. It was observed that compound **99** was less active in deiodination of T4 and the addition of triethylamine also did not alter its activity [81, 84]. This observation reveals that the presence of thiol is more important than the additional amine group near to the active site of IDs. It is also reported that the presence of two selenols **102** in the peri position of naphthyl ring increases the deiodination by 91-fold [84].

The mechanism proposed by Goto and coworkers involving tautomerization may not be the actual mechanism in deiodination [80]. As the naphthyl compounds deiodinase inner ring iodine, there is no free phenolic group to tautomerize. Manna and Mugesh proposed the mechanism of deiodination of T4 by naphthyl-based compounds based on experimental and theoretical data (Scheme 10.17) [85]. It involves the initial interaction of one of the selenol moieties with iodine leading to the formation of a halogen bond. The transfer of electron density from selenium to the σ* orbital of the C–I bond generates a σ hole or partial positive charge on selenium atom, which facilitates an interaction between the halogen-bonded selenium atom and the free selenol moiety (intermediate **103**). The selenium–selenium interaction (chalcogen bond) strengthens the halogen bond, leading to a heterolytic cleavage of C–I bond. The protonation of the resulting carbanion leads to the

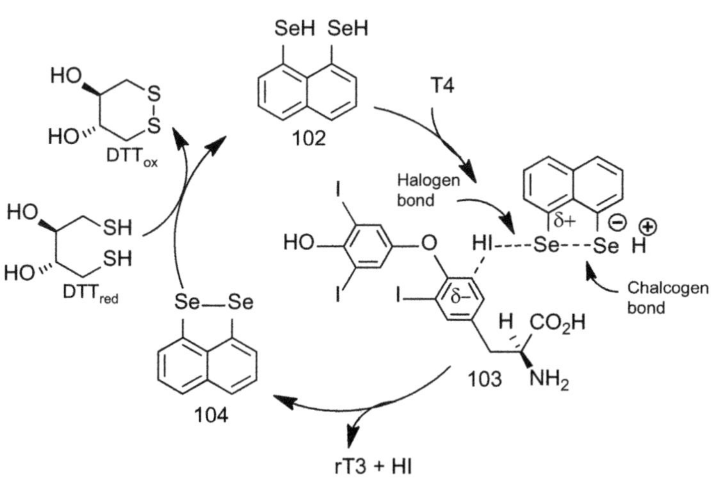

Scheme 10.17: Proposed mechanism of deiodination of T4 by naphthyl-based compound **102** by Manna et al., which involves synergic effect of halogen and chalcogen bonds in the intermediate **103** [85].

formation of rT3. On the other hand, the formation of an Se–Se bond produces the diselenide **104** with the elimination of iodide as HI. The reductive cleavage of the Se–Se bond of compound **102** regenerates the diselenol **102.**

10.4 Thioredoxin reductase

Thioredoxin reductases (TR or TrxR) (EC **1.8.1.9**) are homodimeric flavoproteins that are known to reduce Trx [86–88]. Two classes of TrxR have been identified: one, with high molecular weight (55 kDa) having selenocysteine in the active site present in higher eukaryotes and humans, and another with low molecular weight (35 kDa) present in bacteria and some eukaryotes [89]. Both classes are flavoproteins that function as homodimers. Each monomer contains a FAD prosthetic group and a NADPH-binding domain. Human genome expresses three different isoforms of TrxR: thioredoxin reductase 1 (TrxR1, cytosolic), thioredoxin reductase 2 (TrxR2, mitochondrial) [90], and thioredoxin reductase 3 (TrxR3, testis specific). All these three isoforms contain *N*-terminal conserved ^{59}Cys–Val–Asn–Val–Gly–Cys^{64} active site and a C-terminal Gly–Cys^{497}–Sec^{498}–Gly motif, which accounts for the broad substrate specificity [13]. The main function of TrxR is to reduce the oxidized Trx disulfides to dithiols [91, 92] but, because of highly nucleophilic selenol at C-terminal, it reduces hydroperoxides and regenerates some antioxidants such as lipoate, various selenium compounds, and ubiquinone [93, 94].

In *E. coli*, the NADPH-binding domain and FAD domain were separated by a rotation of 66°. NADPH domain rotates by 66° when the FAD domain remains rigid and the bound NADPH comes close to FAD domain and electron transport from NADPH to FAD occurs to active site disulfides [87]. In Mammalian TrxR, these two domains are present in proximity so that electrons pass from NADPH to FAD easily without any conformational changes. To understand the role of selenium in the active site of TrxR, several mutational studies were carried out by substituting Sec498 residue by Cys and reported their activity (Cys mutant TrxR is 100-fold less active compared with native TrxR) and also some crystal structures [95–103]. The homodimeric nature of mammalian TrxR was further confirmed by the crystal structure of native TrxR1 from rat (Figure 10.10) [104] by Arnér and coworkers. The crystal structure indicates that the N-terminal Cys59/Cys64 redox pair is located near to C-terminal redox pair Cys497/Sec498 of another domain.

10.4.1 Mechanism of action of TrxR in reducing Trx

The proposed mechanism of action in mammalians as depicted in Figure 10.11 [95, 105] first involves the electron transfers from NADPH to active N-terminal conserved

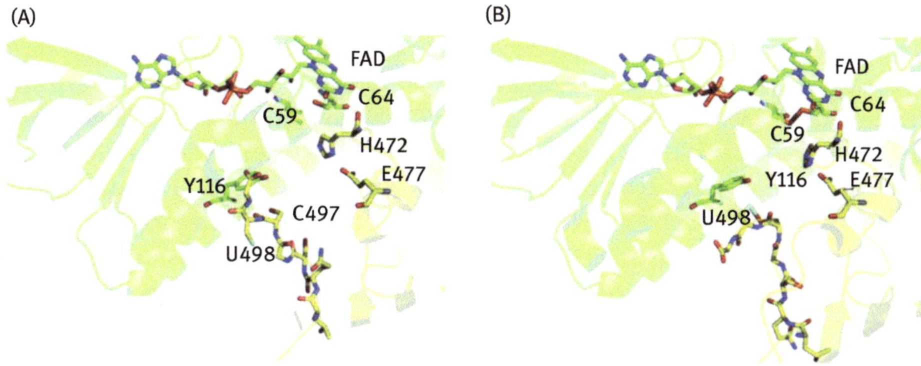

Figure 10.10: Stereo view of the TrxR1 active site of reduced (A) and oxidized (B) forms is shown. Selenocysteine is shown as U498. Image reproduced from Reference 104.

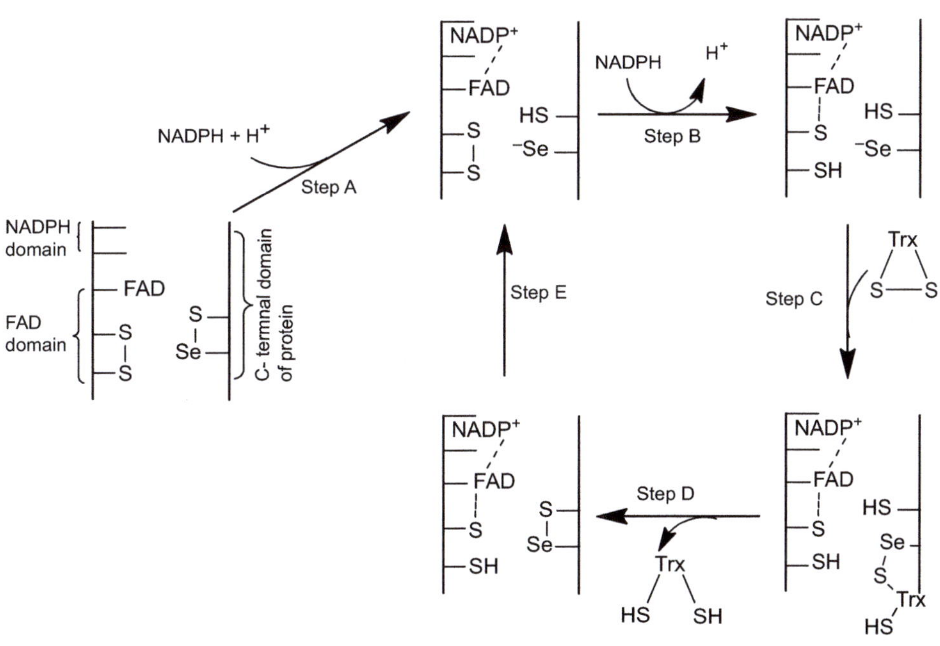

Figure 10.11: Proposed mechanism of action of TrxR in mammalians in the reduction of Trx [95, 105].

disulfide active site through FAD and reduces the disulfides to active thiols. The FAD is oriented near to N-terminal selenocysteine and cysteine redox pair and a charge transfer from Cys64 to FAD produces a charge transfer complex stabilizing the dithiol intermediate. The nucleophilic attack of Cys59 to the selenium center of Se–S bond of C-terminal domain forms an intermolecular selenenyl sulfide intermediate. Reformation of Cys59–Cys64 disulfide bond produces free selenolate. A second equivalent of NADPH produces the fully reduced form of TrxR. The formed selenolate moiety attacks at the disulfide bond of Trx to generate TrxR–Se–S–Trx (Se–S, formed by TrxR Sec498 and Cys32 of Trx) complex and the complex was cleaved by free thiol (SH of Cys497) of TrxR to regenerate oxidized TrxR at C-terminal domain for the next catalytic cycle releasing reduced Trx. The catalytic mechanism involves many steps and thiol-exchange reactions. Brandt and Wessjohann proposed that a catalytic triad consisting of His472, Glu477, and Sec498 exists as that of GPx catalytic triad. The selenocysteine at the C-terminal domain is very important for the catalytic activity, as the mutant Sec498Cys reduces k_{cat} by 100-fold [95, 106] and the truncated TrxR is completely inactive [107].

10.4.2 Mechanism of action of TrxR in reducing H_2O_2

As selenocysteine in TrxR at physiological conditions is released as selenol, which was stabilized by His and Glu, it can not only reduce Trx but also other disulfides, inorganic selenides, and hydrogen peroxide. Holmgren and Zhong proposed a mechanism for the reduction of H_2O_2 by TrxR [95]. As shown in Figure 10.12, it involves three intermediates: selenol, selenenic acid, and selenenyl sulfide. The first step involves the transfer of electrons form NADPH to active disulfide bond at N-terminal

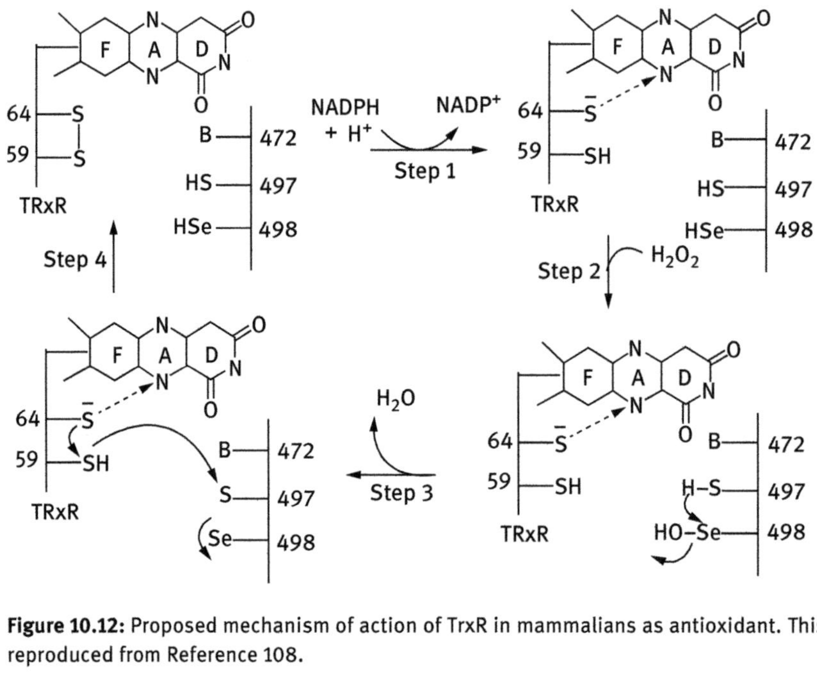

Figure 10.12: Proposed mechanism of action of TrxR in mammalians as antioxidant. This figure is reproduced from Reference 108.

domain through FAD. During this step, the reactive dithiols at N-terminal domain of first subunit generates a thiol–selenol pair in the second subunit. As selenol is more susceptible to oxidation by H_2O_2, it forms selenenic acid (–SeOH). At this step, thiol of Cys497 attacks at the selenenic acid intermediate and regenerates selenenyl–sulfide bond, releasing H_2O. Subsequently, the thiol from the N-terminal end reacts with Cys497, instead of Sec498, to regenerate the active selenol. The attack of thiol from N-terminal at Cys497 instead of Sec498 is surprising as the selenium center in the selenenyl sulfide is more electrophilic than the sulfur atom. This unusual feature can be attributed to the basic amino acid groups present nearby active site redox pair Sec498/Cys497. It is evident from Figure 10.13 that His 108 and His 472 residues are located very close to the sulfur atoms of Cys497 and Cys59, respectively [96, 108]. The distance between the nitrogen of His472 and sulfur atom of –S–S bridge is in the proximity of 3.69 Å. Similarly, nitrogen atom of H_{IS}108 is at a distance of 7.59 Å from sulfur atom of Cys497. The interaction of these His residues with sulfur or selenium can play a crucial role with regard to deprotonating thiol to enhance the nucleophilic attack of the thiolate in the selenenic acid intermediate and interacting with sulfur in the selenenyl sulfide to facilitate a thiol attack at sulfur rather than at selenium. Mugesh and coworkers have done some computational and experimental studies with model compounds having basic group near to selenium or

Figure 10.13: Active site of the SeCys498Cys mutant of rat TrxR showing the proximity of His472 and His108 to Cys59 and Cys497, respectively (PDB code: 1H6V) [96, 108].

sulfur in naphthalene-based molecules and postulated that the involvement of two His residues in the catalysis [His108–Cys497–SeCys498] and [His472–Cys59–Cys64] may be responsible for the broader substrate specificity of the mammalian systems [108].

10.4.3 Inhibitors for thioredoxin reductases

TrxR plays an important role by reducing the oxidized Trx to its reduced form, which in turn reduces many disulfide containing proteins. As mentioned, it plays a key role in maintaining the redox homeostasis in the cells by modulation of the concentration of hydrogen peroxides. Owing to their high proliferation and metabolism, cancer cells generate large amounts of ROS when compared to that of normal cells. The increase in the ROS levels in the cells leads to an imbalance between the ROS-generating and -scavenging systems, resulting in a condition called *oxidative stress* [109]. To combat the deleterious effects of ROS and oxidative stress, the levels of Trx system are elevated [110–112]. Therefore, it has been proposed that targeting TrxR/Trx is a promising strategy for cancer treatment [113–120].

TrxR has a unique selenocysteine at the C-terminal domain, which is catalytically active. As it is present as penultimate aminoamide, it is accessible to small molecules to interact more. Holmgren and coworkers reported that high electrophilic compounds **105–110** (Figure 10.14) can act as specific irreversible inhibitors

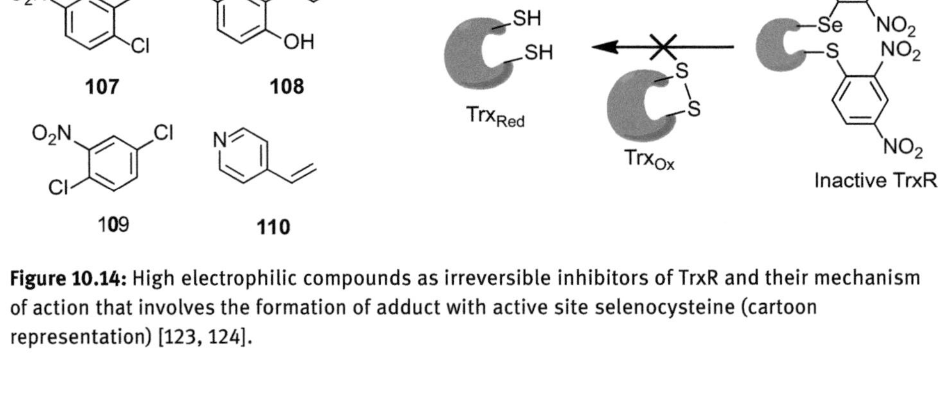

Figure 10.14: High electrophilic compounds as irreversible inhibitors of TrxR and their mechanism of action that involves the formation of adduct with active site selenocysteine (cartoon representation) [123, 124].

of the mammalian TrxR in the presence of NADPH, which in turn increases the NADPH oxidase activity [121, 122]. In the absence of NADPH, it was observed that these compounds do not have any effect on the TrxR activity, suggesting that it inhibits the enzyme only after reduction of the active site selenyl–sulfide bond by NADPH to the corresponding selenol and thiol. Further, a detailed mass spectral analysis indicated that the cysteine thiol groups of the enzyme were modified covalently by compound **105** to produce the corresponding S–Ar derivatives. The covalent bond formed between the enzyme and inhibitor can irreversibly block the enzyme activity (Figure 10.14) [123, 124]. There are many other classes of inhibitors of TrxR; as those are irrelevant to this chapter, we are excluding those (see References [125–130] for more information).

Arnér and coworkers reported that these small molecules can induce apoptosis by inhibiting the activity of TrxR, which is an integral part of the antioxidant system in the cells. The active site of TrxR comprises a C-terminal motif Gly–Cys–Sec–Gly. The formation of a selenenyl sulfide bond is critical for the redox activity. TrxR reduces Trx disulfides to dithiols, which in turn reduces the disulfide bonds in several proteins, particularly the ones involved in redox regulation. Although the main function of TrxR is to catalyze the reduction of Trx, this enzyme can also reduce hydroperoxides and regenerates some of the antioxidants such as lipoate, various selenium compounds, and ubiquinone. All these actions of TrxR are disturbed in mammalian cells upon inhibition of the enzyme by these highly electrophilic compounds (**105–110**) [131, 132]. Arnér and coworkers proposed that the mechanism by which these electrophiles inactivate the enzyme involves the reactions of both the selenolate and thiolate in the active site with the electrophiles to form S–C and Se–C bonds with the inhibitor, as shown in Figure 10.14 [133].

Mugesh and coworkers proved the mechanism of action of these small molecules under oxidative stress condition, by treating aryl selenocysteine derivatives with hydrogen peroxide. It was shown that under oxidative stress conditions the adduct of these small molecules with selenocysteine reacts with H_2O_2 and removes selenium from the enzyme by converting the selenocysteine moiety to dehydroalanine **121** (Scheme 10.18B) [124].

The aryl selenocysteine derivatives were synthesized by the following general method that involves the attack of highly nucleophilic selenol/selenolate of selenocysteine at aromatic halides. However, the yield of the reaction depends on the electrophilicity of the aryl halides. When the highly electrophilic 1-chloro-2,4-dinitrobenzene (CDNB, **105**) was used as aryl halide, the reaction afforded compound **120** in 80% yield. In other cases, particularly with the aryl halides that are less electrophilic, they observed lower yields [124]. Typically, the procedure involves the initial reduction of the protected selenocystine derivative **111** to generate the free selenol by NaBH4 in situ in methanol. The addition of the selenol or selenolate to aryl halides

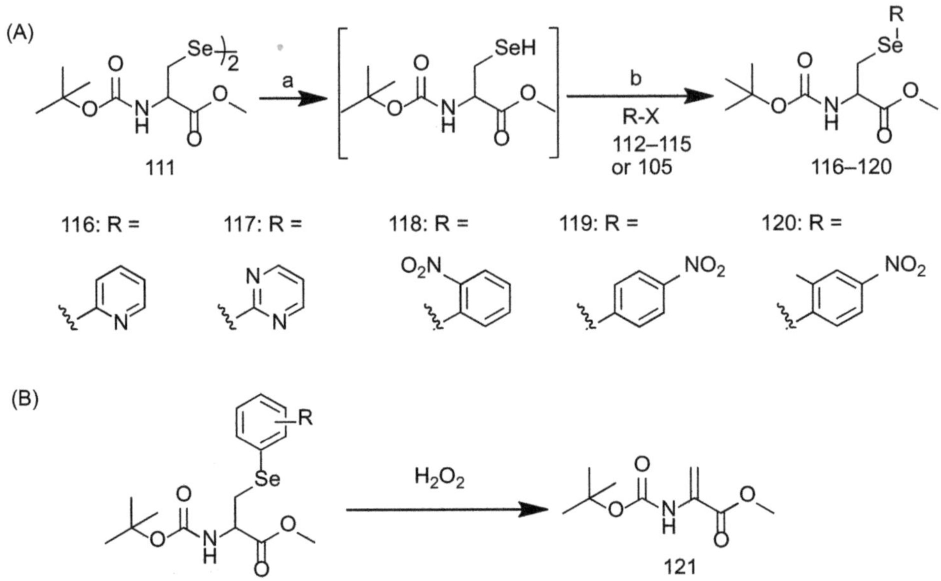

Scheme 10.18: (A) Synthesis of the selenocysteine derivatives **116–120** from the protected selenocysteine **111**, by reduction followed by treatment with appropriate aryl halides. (a) NaBH4, MeOH, 5 min 0 °C; (b) **112–115** or **105** in THF, 27 °C, 6 h. (B) Schematic representation of the formation of dehydroalanine **121** under oxidative stress conditions [124].

2-bromopyridine (**112**), 2-bromopyrimidine (**113**), 1-chloro-2-nitrobenzene (**114**), or 1-chloro-4-nitrobenzene (**115**) in THF afforded the aryl-substituted selenocysteine derivatives **116–119**, respectively. (Scheme 10.18A)

The elimination of the selenium moiety from the selenocysteine derivatives when treated with hydrogen peroxide was consistent with the proposal that the selenium moiety may be eliminated from the TrxR–inhibitor complex. Although the inhibition of TrxR by dinitrohalobenzenes such as CDNB is characterized by the arylation of both the –SeH moiety of the active site selenocysteine and the –SH group of the adjacent cysteine, only the dinitrobenzeneselenol group may be eliminated from the peptide, leading to the formation of a dehydroalanine (Figure 10.15).

This agrees with Anestål and Arnér, who reported that the enzymatically fully active selenocysteine-containing TrxR1 does not have any cell-death-promoting effects, whereas the truncated or selenium-compromised forms of TrxR1 may directly promote apoptosis. The Dha derivative of TrxR1 may be considered as the selenium-compromised form, and such a derivative cannot maintain the Trx-mediated redox balance in cells [133].

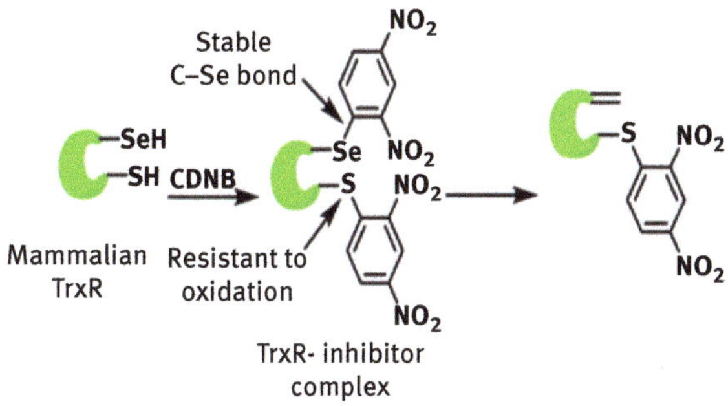

Figure 10.15: The formation of a relatively stable C–Se bond with the inhibitor and the facile oxidation of selenium lead to the elimination of the selenium moiety from TrxR, possibly under oxidative stress conditions. The arylated cysteine residue appears to be stable under these conditions [124].

10.4.4 Organoselenium compounds as inhibitors of thioredoxin reductase (TrxR)

Many organoselenium compounds were reported that can mimic GPx and ID isoforms. In contrast, there are no examples of organoselenium compounds that are mimics of TrxR or any other seleno-proteins. But there are very few reports of organoselenium compounds as inhibitors for the TrxR. Zeng and coworkers reported first selenium compound 1,2-(bis-1,2-benzisoselenazol-3(2H)-one)ethane (ethaselen **122**) as an analogue of ebselen to inhibit the activity of TrxR effectively [134, 135]. The compounds where the selenium was replaced by oxygen (**123**) or sulfur (**124**) show drastic decrease in their inhibitory activity (Figure 10.16A) [136], which shows the importance of Se–N bond in the parental compound **112**. The mechanism was postulated to involve the formation of adduct with active Sec/Cys pair at the C-terminal domain.

The interaction of ethaselen with C-terminal domain of TrxR was proved by showing that hTrxR mutants lacking selenocysteine, as well as human GR, *E. coli* TrxR, and human GPx are comparatively insensitive to ethaselen treatment. This proves that the unique TrxR Cys–Sec redox motif is crucial for the effective enzyme inhibition. The inhibition action as shown in Figure 10.16B involves ethaselen approaching Sec/Cys pair at C-terminal end initially and benzisoselenozol ring initially attacked by nucleophilic Sec498 forming an intermolecular diselenide bridge. As the benzisoselenozol ring is first opened, the structure becomes more flexible and finds a position to interact with Cys497 with second benzisoselenozol ring. This allows the nucleophilic attack by Cys497 and leads to the formation of intermolecular selenyl–sulfide intermediate [135].

The same group reported that the derivatives of ethaselen were found to be more efficient in inhibiting TrxR. To study the effects of substituents on the inhibitory

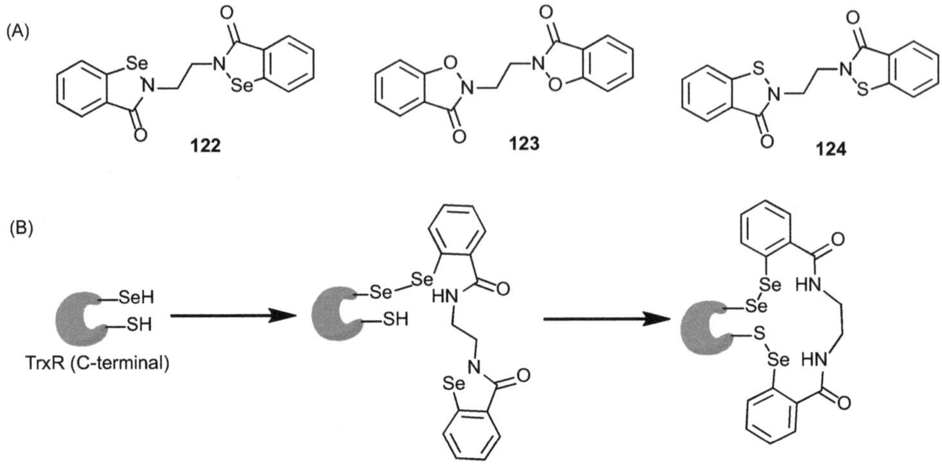

Figure 10.16: (A) Ethaselen and its analogues, where selenium is replaced with O and S; (B) its mechanism of inhibitory action that involves sequential attack of two selenium centers in the ethaselen with C-terminal active site Sec498 and Cys497 (cartoon representation).

activity of ethaselen, the core bis-1,2-benzisoselenazol-3(2H)-one structure was retained and the aromatic ring was modified (compounds **122–124, 128–137**) [136]. The synthesis of ethaselen involves that benzinediazonium salts were first generated from 2-amino benzoic acid and subsequently were treated with disodium diselenide to give 2,2′-diselenobisbenzoic acids (**126**). These compounds were refluxed with thionyl chloride to generate 2-chloroselenobezoyl chlorides (**127**), which was added to the diamines to give ethaselen (Scheme 10.19).

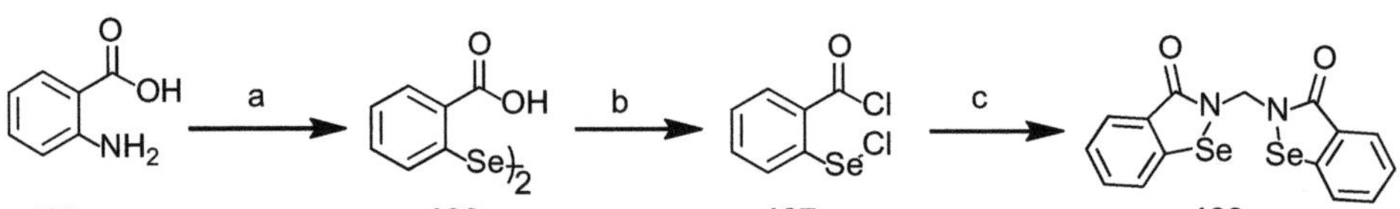

125 126 127 122

Scheme 10.19: Synthetic scheme for ethaselen (**122**): (a) H_2O, $NaNO_2$, HCl, Na_2Se_2; (b) $SOCl_2$, DMF_{cat}; (c) $NH_2{-}CH_2{-}CH_2{-}NH_2$, triethylamine, CH_3CN or CH_2Cl_2 [136].

The synthesis of all these derivatives (Figure 10.17) also involves similar steps with derivatized starting materials and was achieved in a three-step method. All these compounds were tested for the inhibitory activity of TrxR using DTNB-coupled assay. Their IC_{50} values ranged from 0.1 to more than 50 μM. These studies also showed that the R^3 group should be an alkyl chain as the replacement of R^3 with cyclohexyl or aryl

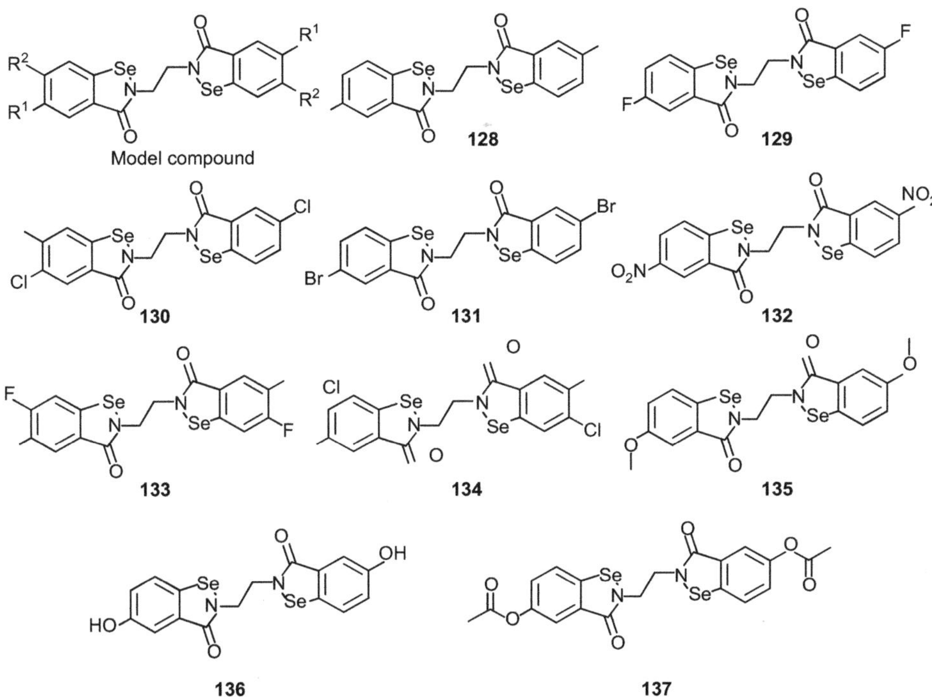

Figure 10.17: The structure of bis-1,2-benzisoselenazol-3(2H)-one and its derivatives. Substitution on the aromatic ring such as methyl, bromo, chloro, fluoro, nitro, and methoxy groups.

groups leads to a significant decrease in the inhibitory activity. In addition, for the compounds with alkyl chains, a slight decrease in the activity was observed when the number of carbon increases from 2 to 5 (Structures not included). Substituents at R^1 and R^2 also have effects on the inhibitory activity of these compounds. Compound **135** (methoxyl at R^1) showed best activity (IC_{50} = 0.13 ± 0.02), whereas compounds **129** (fluoro at R^1) and **136** (hydroxyl at R^1) were less active (IC_{50} = 0.28 ± 0.04 and 0.32 ± 0.02, respectively) but still better than parent compound **122**. Among the compounds with halogen atom, fluoro ones were more active than chloro and bromo analogues [136]. IC_{50} values of the represented compounds are mentioned in Table 10.2.

Chen and coworkers have reported some of the selenadiazoles (**141–143**) as radiosensitizers and as effective TrxR inhibitors. It was proved that these selenadiazoles sensitize cancer cells to X-ray via ROS-mediated signaling [137]. In addition, it was reported later by the same group that compounds **138–140** (Figure 10.18) were promising theranostic agents to achieve synergistic chemo-/radiotherapy in cancer cells. The results showed that the higher lipophilicity endowed compound **140** with higher cellular internalization in HeLa cells, thus

Table 10.2: In vitro inhibitory activities of title compounds [136].

Compound	IC_{50} (μM)
122	0.35
123	>100
124	48.09 ± 1.91
128	1.26 ± 0.31
129	**0.28 ± 0.04**
130	1.31 ± 0.11
131	1.58 ± 0.52
132	0.57 ± 0.01
133	0.82 ± 0.11
134	2.87 ± 0.51
135	**0.13 ± 0.02**
136	**0.32 ± 0.02**
137	22.60 ± 4.05

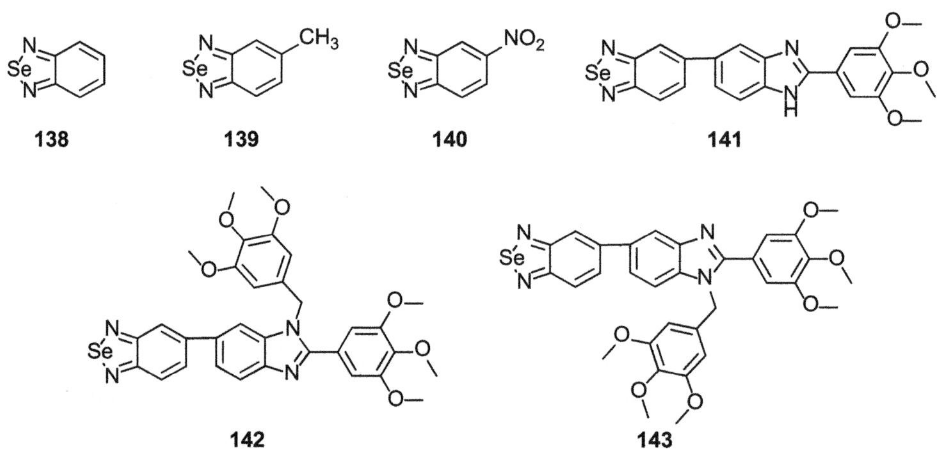

Figure 10.18: Selenadiazoles derivatives with bulky methoxy groups in the aromatic ring of the side chain and simple selenadiazoles with methyl- and nitro-substituents on the main chain aromatic ring.

resulting in a much higher anticancer activity than **138** and **139** and also it significantly enhanced the sensitivity of HeLa cervical cells to X-rays through inhibition of TrxR and triggering intracellular ROS overproduction, which activated the downstream ROS-mediated signaling pathways to regulate HeLa cell apoptosis [138].

10.4.5 Cyclic selenenyl sulfides as mimetics of TrxR

TrxR is an enzyme with active site having Sec at the C-terminal end. Thus, attention was focused to find mimetics of GPx based on this triad, keeping basic group near to selenium in organoselenium compounds. However, there are only two reports of TrxR mimetics. One is by Dawson and coworkers who have demonstrated that seleno-glutaredoxin 3, which has a sequence of C–X–X–U or U–X–X–C instead of C–X–X–C at the redox active site of glutaredoxin 3, behaves like TrxR in the reduction of oxidized Trx [139]. Another one was by Iwaoka and coworkers who have shown that series of cyclic selenenyl sulfides (**139–146**, Figure 10.19) behaved like TrxR in reducing the oxidized Trx. They have prepared many small cyclic selenenyl sulfides having an amino substituent nearby such that it can mimic the catalytic triad of TrxR [140].

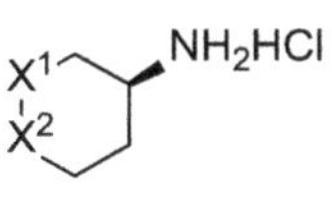

139–142

139, 143: $X^1 = X^2 = Se$
140, 144: $X^1 = Se, X^2 = S$
141, 145: $X^1 = S, X^2 = Se$
142, 146: $X^1 = X^2 = S$

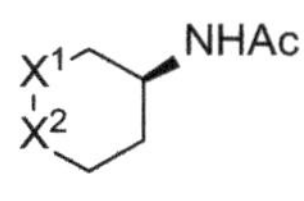

143–146

Figure 10.19: Cyclic selenenyl sulfide, disulfide, and diselenide compounds prepared as mimetics for TrxR [140].

All these compounds were tested for the reduction of bovine pancreatic insulin (BPIns) as a substrate and DTT^{red} as an activator. The reductive cleavage of disulfide bonds of native insulin was monitored by RP HPLC. The activity of compounds **139** and **143** was higher as they are having diselenide and no selenenyl sulfide bond. But the selenenyl sulfides **140, 144**, **141**, and **145** show higher activity compared to disulfides **142** and **146**. Similar trend was observed when another dithiol activator DHLA (dihydrolipoic acid) was used. However, when monothiol such as GSH was used, the activity of **139** and **143** was almost the same, but the activity of compounds **140–142 and 144–146** was drastically decreased. TrxR also possesses antioxidant-like activity by reducing H_2O_2, and so, the title compounds (**139–146**) were tested for their antioxidant activity by using DTT^{red} as activator. The rates of reduction of H_2O_2 were evaluated by monitoring UV absorbance at 310 nm owing to the formation of oxidized DTT (DTT^{ox}). It was reported that the selenenyl sulfides (**140, 141** and **144, 145**) exhibited more catalytic activity even compared to diselenide (**139, 143**) compounds. The disulfide compounds (**142, 146**) did not promote the reaction.

Based on these observations, the postulated mechanism involves reduction of cyclic selenenyl sulfides by dithiol activators to active selenolate species. This is done in two steps. The first step involves attack of thiol of activator with more favorable selenium rather than sulfur, which generates a mixed selenenyl–sulfide intermediate

I′. This species is inactive in reducing the substrate dithiol. In this step also attack of thiol of activator at the thiol of selenenyl–sulfide is less favorable until and unless it is coordinated to hetero atom such as N or O. In the second step, the other thiol of reacted dithiol activator attacks at S atom of Se–S intermediate to produce active selenolate species **I** releasing disulfide. The formed selenol reacts with oxidized insulin to be converted to reduced insulin (Figure 10.20) [140].

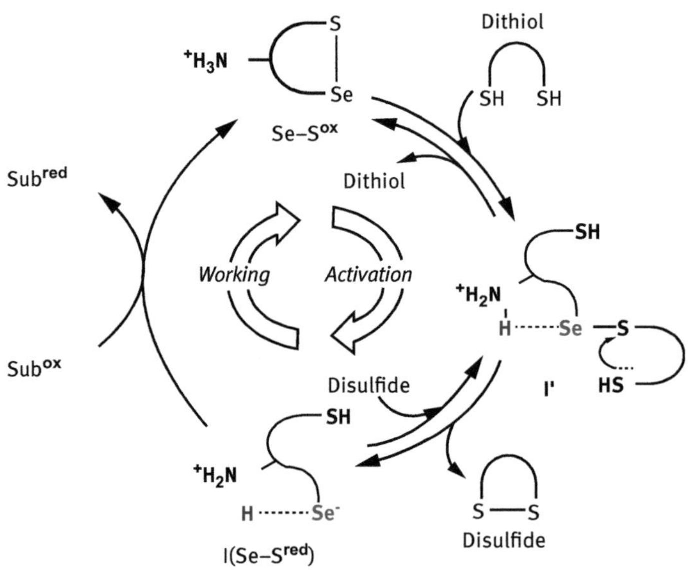

Figure 10.20: Catalytic cycle proposed for the activity of cyclic selenenyl sulfides by using DTT as the activator [140].

10.5 Conclusions

Selenium, which exists mostly in the form of selenocysteine, plays several important physiological roles. Although around 25 selenoproteins have been identified in mammals, the functions of many of these proteins are still unknown. The lower activity observed for the cysteine mutants of some of the selenoenzymes suggested that Sec is important for the biological function. Particularly, Sec present in many oxidoreductase enzymes plays crucial roles in regulating the redox balance in the cell. GPx functions as an antioxidant by reducing the harmful hydroperoxides and maintains cellular concentration of ROS in vivo. Extensive work has been carried out in the design and synthesis of mimetics for two selenoenzymes called *GPx* and *IDs*. However, successful functional mimetics for TrxR have not been reported so far. Although the main function of TrxR is to reduce the oxidized Trx, its broad substrate specificity allows this enzyme to reduce lipid peroxides, selenites, organic molecules such as DTNP, and biologically important hydrogen peroxide. It is speculated that the broad specificity of this enzyme arises because of the presence of two histidine residues, which can stabilize the active selenol group. TrxR also plays an important role in cancer therapy, as it is overexpressed in cancer cells to control the excess amount of ROS. Therefore, this enzyme is considered as an active target for the treatment of cancer. Many small molecules and metal complexes are known to inhibit the activity of TrxR. In this chapter, the development of organoselenium compounds to understand the chemistry of selenoenzymes is discussed. We also described several organoselenium compounds that have been used as inhibitors and mimetics of TrxR. Further studies are required to fully understand the biochemical mechanism of selenoenzymes and the development of functional mimetics for TrxR will be challenging.

References

[1] Muth, OH., Oldfield, JE., Remmert, LF., Schubert, JR. Effects of selenium and vitamin E on white muscle disease. Science 1958, 128, 1090.
[2] Hatfield, DL., Berry, MJ., Gladyshev, VN. Selenium: Its Molecular Biology and Role in Human Health, Springer Science+Business Media, LLC. Springer-Verlag New York, 2012.
[3] Brown, DG., Burk, RF., Selenium retention in tissues and sperm of rats fed a Torula yeast diet. J Nutr 1973, 103, 102–8.
[4] Van Vleet, JF., Ferrans, VJ., Etiologic factors and pathologic alterations in selenium-vitamin E deficiency and excess in animals and humans. Biological Trace Element Research 1992, 33, 1–21.
[5] Zinoni, F., Birkmann, A., Leinfelder, W., Böck, A. Cotranslational insertion of selenocysteine into formate dehydrogenase from Escherichia coli directed by a UGA codon. Proc Natl Acad Sci USA 1987, 84, 3156–60.
[6] Zinoni, F., Birkmann, A., Stadtman, TC., Böck, A. Nucleotide sequence and expression of the selenocysteine-containing polypeptide of formate dehydrogenase (formate-hydrogen-lyase-linked) from Escherichia coli. Proc Natl Acad Sci USA 1986, 83, 4650–4.
[7] Böck, A., Forchhammer, K., Heider, J., Leinfelder, W., Sawers, G., Veprek, B., Zinoni, F. Selenocysteine: the 21st amino acid. Mol Microbiol 1991, 5, 515–20.
[8] Steinmann, D., Nauser, T., Koppenol, WH. Selenium and Sulfur in Exchange Reactions: A Comparative Study. J Org Chem 2010, 75, 6696–9.
[9] Byun, BJ., Kang, YK. Conformational preferences and pK(a) value of selenocysteine residue. Biopolymers 2011, 95, 345–53.
[10] Flohe, L., Günzler, WA., Schock, HH. Glutathione peroxidase: a selenoenzyme. FEBS let 1973, 32, 132–4
[11] Rotruck, JT., Pope, AL., Ganther, HE., Swanson, AB., Hafeman, DG., Hoekstra, WG. Selenium: biochemical role as a component of glutathione peroxidase. Science 1973, 179, 588–90.
[12] Berry, MJ., Banu, L., Larsen, PR. Type I iodothyronine deiodinase is a selenocysteine-containing enzyme. Nature 1991, 349, 438–40.
[13] Zhong, L., Arnér, ES., Ljung, J., Aslund, F., Holmgren, A. Rat and calf thioredoxin reductase are homologous to glutathione reductase with a carboxyl-terminal elongation containing a

conserved catalytically active penultimate selenocysteine residue. J Biol Chem 1998, 273, 8581–91.
[14] Schmidt, RL., Simonović, M. Synthesis and decoding of selenocysteine and human health. Croat Med J 2012, 53, 535–50.
[15] Chu, FF., Doroshow, JH., Esworthy, RS. Expression, characterization, and tissue distribution of a new cellular selenium-dependent glutathione peroxidase, GSHPx-GI. J Biol Chem 1993, 268, 2571–6.
[16] Maiorino, M., Aumann, KD., Brigelius-Flohé, R., Doria, D., van den Heuvel, J., McCarthy, J., Roveri, A., Ursini, F., Flohé, L. Probing the presumed catalytic triad of selenium-containing peroxidases by mutational analysis of phospholipid hydroperoxide glutathione peroxidase (PHGPx). Biol Chem Hoppe Seyler 1995, 376, 651–60.
[17] Rocher, C., Lalanne, JL., Chaudière, J. Purification and properties of a recombinant sulfur analog of murine selenium-glutathione peroxidase. Eur J Biochem 1992, 205, 955–60.
[18] Maddipati, KR., Marnett, LJ. Characterization of the major hydroperoxide-reducing activity of human plasma. Purification and properties of a selenium-dependent glutathione peroxidase. J Biol Chem 1987, 262, 17398–403.
[19] Brigelius-Flohé, R. Tissue-specific functions of individual glutathione peroxidases. Free Radic Biol Med 1999, 27, 951–65.
[20] Maiorino, M., Gregolin, C., Ursini, F. Phospholipid hydroperoxide glutathione peroxidase. Methods Enzymol 1990, 186, 448–57.
[21] Björnstedt, M., Xue, J., Huang, W., Akesson, B., Holmgren, A. The thioredoxin and glutaredoxin systems are efficient electron donors to human plasma glutathione peroxidase. J Biol Chem 1994, 269, 29382–4.
[22] Flohe, L., Günzler, WA., Schock, HH. Glutathione peroxidase: a selenoenzyme. FEBS Lett 1973, 32, 132–4.
[23] Flohé, L., Glutathione Peroxidase Brought into Focus: in Free radicals in biology (Ed. W. A. Pryor), Vol. 5, Academic press, New York, 1982, 223–54.
[24] Maiorino, M., Roveri, A., Coassin, M., Ursini, F. Kinetic mechanism and substrate specificity of glutathione peroxidase activity of ebselen (PZ51). Biochem Pharmacol 1988, 37, 2267–71.
[25] Epp, O., Ladenstein, R., Wendel, A. The refined structure of the selenoenzyme glutathione peroxidase at 0.2 nm resolution. Eur J Biochem 1983, 133, 51–69.
[26] Ursini, F., Maiorino, M., Brigelius-Flohé, R., Aumann, KD., Roveri, A., Schomburg, D., Flohé, L. Diversity of glutathione peroxidases. Methods Enzymol 1995, 252, 38–53.
[27] Bhuyan, BJ., Mugesh, G. Biological and biochemical aspects of Selenium compounds: in Organoselenium chemistry. (Ed. Writh, T.), Willy-VCH Verlag GmbH & Co. KGaA. Chapter 9, 361–396, 2011.
[28] Bhowmick, D., Mugesh, G. Enzyme mimetic chemistry of organoselenium compounds: in the chemistry of organic Selenium and Tellurim compounds. (ed. Rappoport), John Willey & sons Ltd, /The Atrium, Southern Gate, Chichester, West Sussex PO19 8SQ, England, Vol. 4, Part 2, Chapter 16, 1175–1229, 2013.
[29] Mugesh, G., Singh, HB. Synthetic organoselenium compounds as antioxidants: glutathione peroxidase activity. Chem Soc Rev 2000, 29, 347–357.
[30] Müller, A., Cadenas, E., Graf, P., Sies, H. A novel biologically active seleno-organic compound–I. Glutathione peroxidase-like activity in vitro and antioxidant capacity of PZ 51 (Ebselen) 1984, 33, 3235–9.
[31] Wendel, A., Fausel, M., Safayhi, H., Tiegs, G., Otter, R. A novel biologically active seleno-organic compound–II. Activity of PZ 51 in relation to glutathione peroxidase. Biochem Pharmacol 1984, 33, 3241–5.
[32] Sies, H., Biochemistry of Oxidative Stress. Angew Chem Int Ed Engl 1986, 25, 1058–1071.

[33] Mugesh, G., du Mont, W-W., Sies, H. Chemistry of Biologically Important Synthetic Organoselenium Compounds. Chem Rev 2001, 101, 2125–80.
[34] Zembowicz, A., Hatchett, RJ., Radziszewski, W., Gryglewski, RJ. Inhibition of endothelial nitric oxide synthase by ebselen. Prevention by thiols suggests the inactivation by ebselen of a critical thiol essential for the catalytic activity of nitric oxide synthase. J Pharmacol Exp Ther 1993, 267, 1112–8.
[35] Hattori, R., Inoue, R., Sase, K., Eizawa, H., Kosuga, K., Aoyama, T., Masayasu, H., Kawai, C., Sasayama, S., Yui, Y. Preferential inhibition of inducible nitric oxide synthase by ebselen. Eur J Pharmacol 1994, 267, R1–2.
[36] Nikawa, T., Schuch, G., Wagner, G., Sies, H. Interaction of ebselen with glutathione S-transferase and papain in vitro. Biochem Pharmacol 1994, 47, 1007–12.
[37] Lesser, R., Weiss, R. About selenium containing aromatic compounds (VI). Ber Dtsch Chem Ges 1924, 57, 1077–82.
[38] Engman, L., Hallberg, A. Expedient synthesis of ebselen and related compounds. J Org Chem 1989, 54, 2964–6.
[39] Jacquemin, PV., Christiaens, LE., Renson, MJ., Evers, MJ., Dereu, N. Synthesis of 2H,3-4-Dihydro-1,2-benzoselenazin-3-one and derivatives: A new heterocyclic ring system. Tetrahedron Lett 1992, 33, 3863–6.
[40] Reich, HJ., Jasperse, CP. Organoselenium chemistry. Redox chemistry of selenocysteine model systems. J Am Chem Soc 1987, 109, 5549–51.
[41] Chaudiere, J., Yadan, JC., Erdelmeier, I., Tailhan-Lomont, C., Moutet, M. in Oxidative processes and Antioxidants (Ed. Paoletti R), Ravan press, New York, 1994.
[42] Parnham, MJ., Biedermann, J., Bittner, C., Dereu, N., Leyck, S., Wetzig, H. Structure-activity relationships of a series of anti-inflammatory benzisoselenazolones (BISAs). Agents and Actions 1989, 27, 306–8.
[43] Chaudiere, J., Erdelmeier, I., Moutet, M., Yadan, JC. One- Versus Two-Electron Transfers: Cytotoxic and Cytoprotective Effects of Seleno-Organic Catalysts. Phosphorus Sulfur Silicon Relat Elem 1998, 136, 467–470.
[44] Erdelmeier, I., Chaudiere, J., Moutet, M., Yadan, JC. Patent N FR 2718441, 1994.
[45] Moutet, M., d'Alessio, P., Malette, P., Devaux, V., Chaudière, J. Glutathione peroxidase mimics prevent TNFalpha- and neutrophil-induced endothelial alterations. Free Radic Biol Med 1998, 25, 270–81.
[46] Satheeshkumar, K., Mugesh, G. Synthesis and Antioxidant Activity of Peptide-based Ebselen Analogues. Chem Eur J 2011, 17, 4849–57.
[47] Bell, IM., Hilvert, D. Peroxide dependence of the semisynthetic enzyme selenosubtilisin. Biochemistry 1993, 32, 13969–73.
[48] Back, TG., Dyck, BP. A Novel Camphor-Derived Selenenamide That Acts as a Glutathione Peroxidase Mimetic. J Am Chem Soc 1997, 119, 2079–83.
[49] Fischer, H., Dereu, N. Mechanism of the Catalytic Reduction of Hydroperoxides by Ebselen: A Selenium – 77 NMR Study. Bull Soc Chim Belg 1987, 96, 757–68.
[50] Kice, JL., Purkiss, DW. The induced decomposition of S-tert-butyl benzenethioseleninate. J Org Chem 1987, 52, 3443–51.
[51] Sarma, BK., Mugesh, G. Glutathione Peroxidase (GPx)-like Antioxidant Activity of the Organoselenium Drug Ebselen: Unexpected Complications with Thiol Exchange Reactions. J Am Chem Soc 2005, 127, 11477–85.
[52] Bhabak, KP., Mugesh, G. Synthesis, Characterization and Antioxidant Activity of Some Ebselen Analogues. Chem Eur J 2007, 13, 4594–601.

[53] Sarma, BK., Mugesh, G. Antioxidant Activity of the Anti-inflammatory Compound Ebselen: A Reversible Cyclization Pathway via Selenenic and Seleninic Acid Intermediates. Chem Eur J 2008, 14, 10603–14.
[54] Sarma, BK., Mugesh, G. Thiol Cofactors for Selenoenzymes and their Synthetic Mimics. Org Biomol Chem 2008, 6, 965–74.
[55] Haenen, GR., De Rooij, BM., Vermeulen, NP., Bast, A. Mechanism of the reaction of ebselen with endogenous thiols: dihydrolipoate is a better cofactor than glutathione in the peroxidase activity of ebselen. Mol Pharmacol 1990, 37, 412–22.
[56] Bhabak, KP., Mugesh, G. Amide-Based Glutathione Peroxidase Mimics: Effect of Secondary and Tertiary Amide Substituents on Antioxidant Activity. Chem Asian J 2009, 4, 974–83.
[57] Bhabak, KP., Mugesh, G. A Simple and Efficient Strategy to Enhance the Antioxidant Activities of Amino-Substituted Glutathione Peroxidase Mimics. Chem Eur J 2008, 14, 8640–51.
[58] Bhowmick, D., Srivastava, S., D'Silva, P., Mugesh, G. Highly Efficient Glutathione Peroxidase and Peroxiredoxin Mimetics Protect Mammalian Cells against Oxidative Damage. Angew Chem Int Ed 2015, 54, 8449–53.
[59] Bhowmick, D., Govindasamy Mugesh, G. Introduction of a catalytic triad increases the glutathione peroxidase-like activity of diaryl diselenides. Org Biomol Chem 2015,13, 9072–82.
[60] Singh, VP., Singh, HB., Butcher, RJ. Synthesis and Glutathione Peroxidase-Like Activities of Isoselenazolines. Eur J Org Chem 2011, 5485–97.
[61] Werner, Sc., Ingbar, S. The Thyroid: A Fundamental and Clinical Text, (Eds Braverman, LE., Utigar, RD.), Lippincott, Philadelphia, 1991.
[62] Veerle, M. Darras, VM., Hume, R., Visser, TJ. Regulation of thyroid hormone metabolism during fetal development. Mol Cell Endocrinol 1999, 151, 37–47.
[63] Carayon, P., Ruf, J. Thryoperoxidase and Thyroid autoimmunity, John Libbey, London, 1990.
[64] Dunford, HB., Heme peroxidases. John Wiley & sons, Inc., New York, 1999.
[65] Visser, HJ., Leonard, JL., Kaplan, MM., Larsen, PR. Kinetic evidence suggesting two mechanisms for iodothyronine 5′-deiodination in rat cerebral cortex. Proc Natl Acad Sci USA 1982, 79, 5080–4.
[66] Mondal, S., Raja, K., Schweizer, U., Mugesh, G. Chemistry and Biology in the Biosynthesis and Action of Thyroid Hormones. Angew Chem Int Ed 2016, 55, 7606–30.
[67] Köhrle, J. Selenium and the control of thyroid hormone metabolism. Thyroid 2005, 15, 841–53.
[68] Bianco, AC., Salvatore, D., Gereben, B., Berry, MJ., Larsen, PR. Biochemistry, cellular and molecular biology, and physiological roles of the iodothyronine selenodeiodinases. Endocr Rev 2002, 23, 38–89.
[69] Köhrle, J. Local activation and inactivation of thyroid hormones: the deiodinase family. Mol Cell Endocrinol 1999, 151, 103–19.
[70] Bianco, AC., Larsen, PR. Cellular and structural biology of the deiodinases. Thyroid 2005, 15, 777–86.
[71] Leonard, JL., Visser, TJ., Leonard, DM. Characterization of the Subunit Structure of the Catalytically Active Type I Iodothyronine Deiodinase. J Biol Chem 2001, 276, 2600–7.
[72] Callebaut, I., Curcio-Morelli, C., Mornon, JP., et.al. The Iodothyronine Selenodeiodinases Are Thioredoxin-fold Family Proteins Containing a Glycoside Hydrolase Clan GH-A-like Structure. J Biol Chem 2001, 276, 36887–96.
[73] Berry, MJ., Maia, AL., Kieffer, JD., Harney, JW., Larsen, PR. Substitution of cysteine for selenocysteine in type I iodothyronine deiodinase reduces the catalytic efficiency of the protein but enhances its translation. Endocrinology 1992, 131, 1848–52.

[74] Buetner, C., Harney, JW., Larsen, PR. The role of selenocysteine 133 in catalysis by the human type 2 iodothyronine deiodinase. Endocrinology 2000, 141, 4606–12.
[75] Kuiper, GG., Klootwijk W., Visser, TJ. Substitution of cysteine for selenocysteine in the catalytic center of type III iodothyronine deiodinase reduces catalytic efficiency and alters substrate preference. Endocrinology 2003, 144, 2505–13.
[76] Schweizer, U., Schlicker, C., Braun, D., Köhrle, J., Steegborn, C. Crystal structure of mammalian selenocysteine-dependent iodothyronine deiodinase suggests a peroxiredoxin-like catalytic mechanism. Proc Natl Acad Sci USA 2014, 111, 10526–31.
[77] Kuiper, GG., Klootwijk, W., Visser, TJ. Expression of recombinant membrane-bound type I iodothyronine deiodinase in yeast. J Mol Endocrinol 2005, 34, 865–78.
[78] Beck, C., Jensen, SB., Reglinski, J. The selenium mediated de-iodination of iodophenols: a model for the mechanism of 5′ thyronine de-iodinase. Bioorg. Med. Chem, Lett 1994, 4, 1353–6.
[79] Vasil'ev, AA., Engman, L. Iodothyronine Deiodinase Mimics. Deiodination of o,o'-Diiodophenols by Selenium and Tellurium Reagents. J Org Chem 1998, 63, 3911–7.
[80] Goto, K., Sonoda, D., Shimada, K., Sase, S., Kawashima, T. Modeling of the 5′-deiodination of thyroxine by iodothyronine deiodinase: chemical corroboration of a selenenyl iodide intermediate. Angew Chem Int Ed 2010, 49, 545–7.
[81] Manna, D., Mugesh, G. A Chemical Model for the Inner-Ring Deiodination of Thyroxine by Iodothyronine Deiodinase. Angew Chem Int Ed 2010, 49, 9246–9.
[82] Köhrle, J., Hesch, RD. Biochemical characteristics of iodothyronine monodeiodination by rat liver microsomes: the interaction between iodothyronine substrate analogs and the ligand binding site of the iodothyronine deiodinase resembles that of the TBPA-iodothyronine ligand binding. Horm Metab Res Suppl 1984, 14, 42–55.
[83] Berry, MJ. Identification of essential histidine residues in rat type I iodothyronine deiodinase. J Biol Chem 1992, 267, 18055–9.
[84] Manna, D., Mugesh, G., Deiodination of Thyroid Hormones by Iodothyronine Deiodinase Mimics: Does an Increase in the Reactivity Alter the Regioselectivity?. J Am Chem Soc 2011, 133, 9980–3.
[85] Manna, D., Mugesh, G., Regioselective Deiodination of Thyroxine by Iodothyronine Deiodinase Mimics: An Unusual Mechanistic Pathway Involving Cooperative Chalcogen and Halogen Bonding. J Am Chem Soc 2012, 134, 4269–79.
[86] Luthman, M., Holmgren, A. Rat liver thioredoxin and thioredoxin reductase: purification and characterization. Biochemistry 1982, 21, 6628–33.
[87] Williams, CH. Jr. Mechanism and structure of thioredoxin reductase from Escherichia coli. FASEB J 1995, 9, 1267–76.
[88] Mustacich, D., Powis, G. Thioredoxin reductase. Biochem J 2000, 346, 1–8.
[89] Hirta, RP., Müller, S., Embley, TM., Coombs, GH. The diversity and evolution of thioredoxin reductase: new perspectives. Trends in Parasitology 2002, 18, 302–8.
[90] Conrad, M., Jakupoglu, C., Moreno, SG., et.al. Essential role for mitochondrial thioredoxin reductase in hematopoiesis, heart development, and heart function. Mol Cell Biol 24, 9414–23.
[91] Holmgren, A. Thioredoxin. Annu Rev Biochem 1985, 54, 237–71.
[92] Arnér, ES., Holmgren, A. Physiological functions of thioredoxin and thioredoxin reductase. Eur J Biochem 2000, 267, 6102–9.
[93] Zhong, L., Holmgren, A. Essential role of selenium in the catalytic activities of mammalian thioredoxin reductase revealed by characterization of recombinant enzymes with selenocysteine mutations. J Biol Chem 2000, 275, 18121–8.

[94] Björnstedt, M., Hamberg, M., Kumar, S., Xue, J., Holmgren, A. Human thioredoxin reductase directly reduces lipid hydroperoxides by NADPH and selenocystine strongly stimulates the reaction via catalytically generated selenols. J Biol Chem 1995, 270, 11761–4.
[95] Zhong, L., Arnér, ESJ., Holmgren. A. Structure and mechanism of mammalian thioredoxin reductase: The active site is a redox-active selenolthiol/selenenylsulfide formed from the conserved cysteine-selenocysteine sequence. Proc Natl Acad Sci USA 2000, 97, 5854–9.
[96] Sandalova, T., Zhong, L., Lindqvist, Y., Holmgren, A., Schneider, G. Three-dimensional structure of a mammalian thioredoxin reductase: Implications for mechanism and evolution of a selenocysteine-dependent enzyme. Proc Natl Acad Sci USA 2001, 98, 9533–8.
[97] Biterova, El., Turanov, AA., Gladyshev, VN., Barycki, JJ. Crystal structures of oxidized and reduced mitochondrial thioredoxin reductase provide molecular details of the reaction mechanism. Proc Natl Acad Sci USA 2005, 102, 15018–23.
[98] Gladyshev, VN., Jeang, KT., Stadtman, TC. Selenocysteine, identified as the penultimate C-terminal residue in human T-cell thioredoxin reductase, corresponds to TGA in the human placental gene. Proc Natl Acad Sci USA 1996, 93, 6146–6151.
[99] Zhong, L., Arnér, ESJ., Ljung, J., Åslund, F., Holmgren, A. Rat and Calf, Thioredoxin. Reductase Are Homologous to Glutathione Reductase with a Carboxyl-terminal Elongation Containing a Conserved Catalytically Active Penultimate Selenocysteine Residue. J Biol Chem 1998, 273, 8581–91.
[100] Gasdaska, PY., Berggren, MM., Berry, MJ., Powis, G. Cloning, sequencing and functional expression of a novel human thioredoxin reductase. FEBS Letters 1999, 442, 105–111.
[101] Fritz-Wolf, K., Urig, S., Becker, J. The Structure of Human Thioredoxin Reductase 1 Provides Insights into C-terminal Rearrangements During Catalysis. J Mol Biol 2007, 370, 116–27.
[102] Eckenroth, BE., Lacey, BM., Lothrop, AP., Harris, KM., Hondal, RJ. Investigation of the C-Terminal Redox Center of High-Mr Thioredoxin Reductase by Protein Engineering and Semisynthesis. Biochemistry 2007, 46, 9472–83.
[103] Eckenroth, BE., Harris, K., Turanov, AA., Gladyshev, VN., Raines, RT., Hondal, RJ. Semisynthesis and Characterization of Mammalian Thioredoxin Reductase. Biochemistry 2006, 45, 5158–70.
[104] Cheng, Q., Sandalova, T., Lindqvist, Y., Arnér, ESJ. Crystal Structure and Catalysis of the Selenoprotein Thioredoxin Reductase 1. J Biol Chem 2009, 284, 3998–4008.
[105] https://en.wikipedia.org/wiki/Thioredoxin_reductase.
[106] Zhong, L., Holmgren, A. Essential Role of Selenium in the Catalytic Activities of Mammalian Thioredoxin Reductase Revealed by Characterization of Recombinant Enzymes with Selenocysteine Mutations. The J Biol Chem 2000, 275, 18121–8.
[107] Urig, S., Lieske, J., Fritz-Wolf, K., Irmler, A., Becker, K. Truncated mutants of human thioredoxin reductase 1 do not exhibit glutathione reductase activity. FEBS Lett 2006, 580, 3595–3600.
[108] Sarma, BK., Mugesh, G. Biomimetic Studies on Selenoenzymes: Modeling the Role of Proximal Histidines in Thioredoxin Reductases. Inorg Chem 2006, 45, 5307–14.
[109] Uttara, B., Singh, AV., Zamboni, P., Mahajan, RT. Oxidative Stress and Neurodegenerative Diseases: A Review of Upstream and Downstream Antioxidant Therapeutic Options. Curr Neuropharmacol 2009, 7, 65–74.
[110] Lincoln, DT., Ali Emadi, EM., Tonissen, KF., Clarke, FM. The thioredoxin-thioredoxin reductase system: over-expression in human cancer. Anticancer Res 2003, 23, 2425–33.
[111] Lincoln, DT., Al-Yatama, F., Mohammed, FMA., Al-Banaw, AG., Al-Bader, M., Burge, M., Sinowatz, F., Singal, PK. Thioredoxin and Thioredoxin Reductase Expression in Thyroid Cancer Depends on Tumour Aggressiveness. Anticancer Res 2010, 30, 767–75.

[112] Soini, Y., Kahlos, K., Näpänkangas, U., et.al. Widespread Expression of Thioredoxin and Thioredoxin Reductase in Non-Small Cell Lung Carcinoma. Clin Cancer Res 2001, 7, 1750–7.
[113] Karlenius, TC., Tonissen, KF. Thioredoxin and Cancer: A Role for Thioredoxin in all States of Tumor Oxygenation. Cancers 2010, 2, 209–32.
[114] Lu, J., Chew, EH., Holmgren, A. Targeting thioredoxin reductase is a basis for cancer therapy by arsenic trioxide. Proc Natl Acad Sci USA 2007, 104, 12288–93.
[115] Zhang, J., Li, X., Han, X., Liu, R., Fang, J. Targeting the Thioredoxin System for Cancer Therapy. Trends Pharmacol Sci 2017, 38, 794–808.
[116] Yoo, MH., Xu, XM., Carlson, BA., Patterson, AD., Gladyshev, VN., Hatfield, DL. Targeting thioredoxin reductase 1 reduction in cancer cells inhibits self-sufficient growth and DNA replication. PLoS One 2007, 2, e1112.
[117] Powis, G., Kirkpatrick, DL. Thioredoxin signaling as a target for cancer therapy. Curr Opin Pharmacol 2007, 7, 392–7.
[118] Nguyen, P., Awwad, RT., Smart, DD., Spitz, DR., Gius, D. Thioredoxin reductase as a novel molecular target for cancer therapy. Cancer Lett 2006, 236, 164–74.
[119] Tonissen, KF., Di Trapani, G. Thioredoxin system inhibitors as mediators of apoptosis for cancer therapy. Mol Nutr Food Res 2009, 53, 87–103.
[120] Arnér, ES., Holmgren, A. The thioredoxin system in cancer. Semin Cancer Biol 2006, 16, 420–6.
[121] Arnér, ESJ., Björnstedt, M., Holmgren, A. 1-Chloro-2,4-dinitrobenzene Is an Irreversible Inhibitor of Human Thioredoxin Reductase: Loss of thioredoxin disulfide reductase activity is accompanied by a large increase in nadph oxidase activity. J Biol Chem 1995, 270, 3479–82.
[122] Cenas, N., Prast, S., Nivinskas, H., Sarlauskas, J., Arnér, ESJ. Interactions of Nitroaromatic Compounds with the Mammalian Selenoprotein Thioredoxin Reductase and the Relation to Induction of Apoptosis in Human Cancer Cells. J Biol Chem 2006, 281, 5593–603.
[123] Nordberg, J., Zhong, L., Holmgren, A., Arnér, ESJ. Mammalian Thioredoxin Reductase Is Irreversibly Inhibited by Dinitrohalobenzenes by Alkylation of Both the Redox Active Selenocysteine and Its Neighboring Cysteine Residue. J Biol Chem 1998, 273, 10835–42.
[124] Reddy, KM. Mugesh, G. Modelling the Inhibition of Selenoproteins by Small Molecules Using Cysteine and Selenocysteine Derivatives. Chem Eur J 2019, 25, 8875–83.
[125] Saccoccia, F., Angelucci, F., Boumis, G., Carotti, D., Desiato, G., Miele, AE, Bellelli A. Thioredoxin reductase and its inhibitors. Curr Protein Pept Sci 2014, 15, 621–46.
[126] Zhang, B., Zhang, J., Peng, S., et.al. Thioredoxin reductase inhibitors: a patent review. Expert Opin Ther Pat 2017, 27, 547–56.
[127] Bindoli, A., Rigobello, MP., Scutari, G., Gabbiani, C., Casini, A., Messoric, L. Thioredoxin reductase: A target for gold compounds acting as potential anticancer drugs. Thioredoxin reductase: A target for gold compounds acting as potential anticancer drugs. Coord Chem Rev 2009, 253, 1692–1707.
[128] Cai, W., Zhang, L., Song, Y., et.al. Small molecule inhibitors of mammalian thioredoxin reductase. Free Radic Biol Med 2012, 52, 257–65.
[129] Zhang, B., Liu, Y., Li, X., Xu, J., Fang, J. Small Molecules to Target the Selenoprotein Thioredoxin Reductase. Chem Asian J 2018, 13, 3593–3600.
[130] Zhang, J., Zhang, B., Li, X., Han, X., Liu, R., Fang, J. Small molecule inhibitors of mammalian thioredoxin reductase as potential anticancer agents: An update. Med Res Rev, 2019, 39, 5–39.
[131] Lillig, CH., Holmgren, A. Thioredoxin and related molecules–from biology to health and disease. Antioxid Redox Signal 2007, 9, 25–47.
[132] Holmgren, A., Lu, J. Thioredoxin and thioredoxin reductase: Current research with special reference to human disease. Biochem Biophys Res Commun 2010, 396, 120–4.

[133] Anestål, K., Arnér, ESJ. Rapid Induction of Cell Death by Selenium-compromised Thioredoxin Reductase 1 but Not by the Fully Active Enzyme Containing Selenocysteine. J Biol Chem 2003, 278, 15966–72.

[134] Shia, C., Yu, L., Yanga, F., Yan, J., Zeng, H. A novel organoselenium compound induces cell cycle arrest and apoptosis in prostate cancer cell lines. Biochem Biophys Res Commun 2003, 309, 578–83.

[135] Wang, L., Yang, Z., Fu, J., et.al. Ethaselen: a potent mammalian thioredoxin reductase 1 inhibitor and novel organoselenium anticancer agent. Free Radic Biol Med 2012, 52, 898–908.

[136] He, J., Li, D., Xiong, K., et.al. Inhibition of thioredoxin reductase by a novel series of bis-1,2-benzisoselenazol-3(2H)-ones: Organoselenium compounds for cancer therapy. Bioorg Med Chem 2012, 20, 3816–27.

[137] Liang, YW., Zheng, J., Li, X., Zheng, W., Chen, T. Selenadiazole derivatives as potent thioredoxin reductase inhibitors that enhance the radiosensitivity of cancer cells. Eur J Med Chem 2014, 84, 335–42.

[138] He, L., Ji, S., Lai, H., Chen, T. Selenadiazole derivatives as theranostic agents for simultaneous cancer chemo-/radiotherapy by targeting thioredoxin reductase. J Mater Chem B 2015,3, 8383–93

[139] Metanis, N., Keinan, E., Dawson, PE. Synthetic Seleno-Glutaredoxin 3 Analogues Are Highly Reducing Oxidoreductases with Enhanced Catalytic Efficiency. J Am Chem Soc 2006, 128, 16684–91.

[140] Arai, K., Matsunaga, T., Ueno, H., Akahoshi, N., Sato, Y., Chakrabarty, G., Mugesh, G., Iwaoka, M. Modeling Thioredoxin Reductase-Like Activity with Cyclic Selenenyl Sulfides: Participation of an NH···Se Hydrogen Bond through Stabilization of the Mixed Se–S Intermediate. Chem Eur J 2019, 25, 12751–60.

Index

https://doi.org/10.1515/9783110625110-011

Bei Fragen zur Produktsicherheit wenden Sie sich bitte an:
If you have any questions regarding product safety,
please contact:

Walter de Gruyter GmbH
Genthiner Straße 13
10785 Berlin
productsafety@degruyterbrill.com